Hugh L. Applewhite

Prentice Hall Advanced Reference Series

Engineering

ANTOGNETTI AND MILUTINOVIC, EDS. *Neural Networks: Concepts, Applications, and Implementations, Vol. I*

DENNO *Power System Design and Applications for Alternative Energy Sources*

ESOGBUE, ED. *Dynamic Programming for Optimal Water Resources Systems Analysis*

FERRY, AKERS, AND GREENEICH *Ultra Large Scale Integrated Microelectronics*

GNADT AND LAWLER, EDS. *Automating Electric Utility Distribution Systems: The Athens Automation and Control Experiment*

HAYKIN, ED. *Advances in Spectrum Analysis and Array Processing, Vol. I*

HAYKIN, ED. *Advances in Spectrum Analysis and Array Processing, Vol. II*

HAYKIN, ED. *Selected Topics in Signal Processing*

HENSEL *Inverse Theory and Applications for Engineers*

JOHNSON *Lectures on Adaptive Parameter Estimation*

MILUTINOVIC, ED. *Microprocessor Design for GaAs Technology*

QUACKENBUSH, BARNWELL III, AND CLEMENTS *Objective Measures of Speech Quality*

ROFFEL, VERMEER, AND CHIN *Simulation and Implementation of Self-Tuning Controllers*

SASTRY AND BODSON *Adaptive Control: Stability, Convergence, and Robustness*

PRENTICE HALL SIGNAL PROCESSING SERIES

Alan V. Oppenheim, Series Editor

ANDREWS AND HUNT *Digital Image Restoration*
BRIGHAM *The Fast Fourier Transform*
BRIGHAM *The Fast Fourier Transform and Its Applications*
BURDIC *Underwater Acoustic System Analysis, 2/E*
CASTLEMAN *Digital Image Processing*
COWAN AND GRANT *Adaptive Filters*
CROCHIERE AND RABINER *Multirate Digital Signal Processing*
DUDGEON AND MERSEREAU *Multidimensional Digital Signal Processing*
HAMMING *Digital Filters, 3/E*
HAYKIN, ED. *Advances in Spectrum Analysis and Array Processing, Vols. I & II*
HAYKIN, ED. *Array Signal Processing*
JAYANT AND NOLL *Digital Coding of Waveforms*
KAY *Modern Spectral Estimation*
KINO *Acoustic Waves: Devices, Imaging, and Analog Signal Processing*
LEA, ED. *Trends in Speech Recognition*
LIM *Two-Dimensional Signal and Image Processing*
LIM, ED. *Speech Enhancement*
LIM AND OPPENHEIM, EDS. *Advanced Topics in Signal Processing*
MARPLE *Digital Spectral Analysis with Applications*
MCCLELLAN AND RADER *Number Theory in Digital Signal Processing*
MENDEL *Lessons in Digital Estimation Theory*
OPPENHEIM, ED. *Applications of Digital Signal Processing*
OPPENHEIM, WILLSKY, WITH YOUNG *Signals and Systems*
OPPENHEIM AND SCHAFER *Digital Signal Processing*
OPPENHEIM AND SCHAFER *Discrete-Time Signal Processing*
QUACKENBUSH ET AL. *Objective Measures of Speech Quality*
RABINER AND GOLD *Theory and Applications of Digital Signal Processing*
RABINER AND SCHAFER *Digital Processing of Speech Signals*
ROBINSON AND TREITEL *Geophysical Signal Analysis*
STEARNS AND DAVID *Signal Processing Algorithms*
STEARNS AND HUSH *Digital Signal Analysis, 2/E*
TRIBOLET *Seismic Applications of Homomorphic Signal Processing*
WIDROW AND STEARNS *Adaptive Signal Processing*

Advances in Spectrum Analysis and Array Processing

Volume II

Simon Haykin, editor

McMaster University
Hamilton, Ontario
Canada

PRENTICE HALL, Englewood Cliffs, New Jersey 07632

Library of Congress Cataloging-in-Publication Data

Advances in spectrum analysis and array processing.

Includes bibliographical references.
1. Signal processing. 2. Signal theory
(Telecommunication) I. Haykin, Simon S.
TK5102.5.A334 1991 621.382'23 89-26459
ISBN 0-13-007444-6 (v. 1)
ISBN 0-13-008574-X (v. 2)

Editorial/production supervision
and interior design: *Brendan M. Stewart*
Cover design: *Karen Stephens*
Manufacturing buyers: *Kelly Behr* and *Susan Brunke*

A Division of Simon & Schuster
Englewood Cliffs, New Jersey 07632

This book can be made available to businesses and organizations at a special discount when ordered in large quantities. For more information contact:

Prentice-Hall, Inc.
Special Sales and Markets
College Division
Englewood Cliffs, N.J. 07632

Printed in the United States of America
10 9 8 7 6 5 4 3 2 1

ISBN 0-13-008574-X

Prentice-Hall International (UK) Limited, *London*
Prentice-Hall of Australia Pty. Limited, *Sydney*
Prentice-Hall Canada Inc., *Toronto*
Prentice-Hall Hispanoamericana, S.A., *Mexico*
Prentice-Hall of India Private Limited, *New Delhi*
Prentice-Hall of Japan, Inc., *Tokyo*
Simon & Schuster Asia Pte. Ltd., *Singapore*
Editora Prentice-Hall do Brasil, Ltda., *Rio de Janeiro*

Contents

Contributors

Z.D. BAI, Penn State University, 123 Pond Laboratory, Department of Statistics, University Park, Pennsylvania 16802, USA.

JOHANN F. BÖHME, Lehrstuhl fur Signaltheorie, Ruhr-Universitat, 4630 Bochum, Federal Republic of Germany.

IRA CLARKE, Royal Signals & Radar Establishment, St. Andrews Road, Great Malvern, Worcs WR14 3PS, United Kingdom.

BENJAMIN FRIEDLANDER, Signal Processing Technology Ltd., 703 Coastland Drive, Palo Alto, California 94303, USA.

DANIEL R. FUHRMANN, Washington University in St. Louis, Electronic Systems and Signals Research Laboratory, Department of Electrical Engineering, 1 Brookings Drive, St. Louis, Missouri 63130, USA.

SIMON HAYKIN, McMaster University, Communications Research Laboratory, 1280 Main Street West, Hamilton, Ontario, L8S 4K1, Canada.

MOS KAVEH, University of Minnesota, 123 Church Street S.E., Minneapolis, Minnesota 55455, USA.

VYTAS KEZYS, McMaster University, Communications Research Laboratory, 1280 Main Street West, Hamilton, Ontario, L8S 4K1, Canada.

JEFFREY KROLIK, Concordia University, Department of Electrical Engineering, 1455 Maisonneuve Blvd. West, Montreal, Quebec, H3G, 1M8, Canada.

B.Q. MIAO, Penn State University, 123 Pond Laboratory, Department of Statistics, University Park, Pennsylvania 16802, USA.

MICHAEL I. MILLER, Washington University in St. Louis, Department of Electrical Engineering, 1 Brookings Drive, St. Louis, Missouri 63130, USA.

ARYE NEHORAI, Yale University, Department of Electrical Engineering, P.O. Box 2157 Yale Station, (Becton Center), New Haven, Connecticut 06520-2157, USA.

JOSEPH A. O'SULLIVAN, Washington University in St. Louis, Electronic Systems and Signals Research Laboratory, Department of Electrical Engineering, 1 Brookings Drive, St. Louis, Missouri 63130, USA.

C. RADHAKRISHNA RAO, Penn State University, 123 Pond Laboratory, Department of Statistics, University Park, Pennsylvania 16802, USA.

DONALD L. SNYDER, Washington University in St. Louis, Electronic Systems and Signals Research Laboratory, Department of Electrical Engineering, 1 Brookings Drive, St. Louis, Missouri 63130, USA.

PETRE STOICA, Polytechnic Institute of Bucharest, Department of Automatic Control, Splaiul Independentei 313, R-77 206 Bucharest, Romania.

EDWARD VERTATSCHITSCH, High Technology Center, Boeing Electronics Company, P.O. Box 24969, MS 7J-65, Seattle, Washington 98124-6269, USA.

H. WANG, Syracuse University, Department of Electrical Engineering, Syracuse, New York, USA.

ANTHONY J. WEISS, Tel-Aviv University, Department of Electrical Engineering, Tel Aviv, Israel.

Preface

The *power spectral density, power spectrum,* or simply *spectrum* of a stochastic process describes the distribution of the average power of the process as a function of frequency. By definition, the power spectrum is the Fourier transform of the *autocorrelation function* of the process for varying time lag. The Fourier transform pairs of autocorrelation function and power spectrum, and those of their higher-order counterparts, namely, *cumulants* and *polyspectra,* are basic to the study of the ever-expanding field of signal processing. In radar, sonar, geophysics, astronomy, communications, and biomedical engineering (just to mention some notable disciplines), signal processing plays a dominant role; hence, the importance of spectrum analysis.

In 1979 I edited a book entitled "Nonlinear Methods of Spectral Analysis", which provided the first treatment of parametric methods of spectral analysis in book form [1]. The subject matter of that book is currently referred to as "modern spectrum analysis". Then in 1982 I had the honor of co-editing a special issue of the *Proceedings of the Institute of Electrical and Electronic Engineers* that was devoted to a historical perspective, fundamentals, and applications of spectral analysis [2]. In 1985 I edited another book on "Array Signal Processing," which presented the first treatment of *array processing* techniques for passive systems in book form [3]. During the 1980s, *nonparametic* (classical) and *parametric* (modern) methods of spectrum analysis and array processing, and their applications, went through a period of consolidations and advances to justify another major contribution on the subject; hence, this new book entitled *Advances in Spectrum Analysis and Array Processing.* The terms "spectrum analysis" and "spectral analysis" are used interchangeably in the literature.

The book is in two volumes, with Volume I consisting of nine chapters and Volume II consisting of ten chapters. These two volumes provide:

1. Detailed discussions of fundamental issues in spectrum analysis and array processing.
2. Detailed treatments of popular and new algorithms for spectrum analysis and array processing.
3. Applications.

The book is unique in the sense that the material in both volumes, covering a very broad range of topics (see the list of contents) is being presented in book form for the first time.

The material in Volume I has an underlying slant toward the spectrum analysis of time series and time-frequency signal analysis. On the other hand, the material in Volume II has an underlying slant toward array processing techniques applied to passive systems. Naturally, the fundamentals of spectrum analysis formulated in the time domain and those of array processing in a spatial context are closely linked. Hence, despite the division of the book into two volumes, brought on by the sheer size of the book, there is a great deal of interplay between the chapters in Volume I and those in Volume II. Indeed, it is hoped that by having these two volumes side-by-side, the book will become a source of cross-fertilization, and a tool for the development of new ideas and applications.

—*Simon Haykin*
Hamilton, Ontario

REFERENCES

1. Haykin, S. (editor), "Nonlinear Methods of Spectral Analysis," First Edition, 1979, and Second Edition, (Springer-Verlag) 1983.
2. Haykin, S., and J.A. Cadzow (co-editors), Proc. IEEE Special Issue on Spectral Estimation, Vol. 70, September 1982.
3. Haykin, S. (editor), "Array Signal Processing," (Prentice Hall) 1985.

1

Array Processing

Johann F. Böhme

1.1 INTRODUCTION

Array processing deals with methods for processing the output data of an array of sensors located at different points in space in a wavefield. The purpose of this signal processing is to obtain insight into the structure of the waves carrying information and traversing the array. For example, location parameters of the sources that generate the signals, propagation velocities of the waves transmitting the signals, and spectral properties of the signals have to be measured. Array processing methods have been developed in several fields with very different wave phenomena. Typical examples with radiating sources are *radio telescopy,* in which radio sources are measured by means of antenna groups; *passive* (listening only) *sonar,* with hydrophone arrays and ship noise as signals; and *geophysical* work with seismometers for measuring earthquakes. Array processing is also applied in the so-called *active* methods such as active radar and sonar, where the received signals are pulses diffracted or reflected by objects.

Correspondingly, array processing methods have been extensively studied, and results can be found in the literature on radio astronomy, electrical engineering, acoustics, geophysics, and statistics. Probably, the first book devoted completely to array processing and a variety of applications was edited by Simon Haykin [1]. That volume contains chapters on exploration seismology by J. H. Justice, on sonar by N. L. Owsley, on radar by S. Haykin, on radio telescopy by J. L. Yen, and on tomography by A. C. Kak. Because our contribution can only present a very limited view on the topic, we ignore the historical perspective and give only some further references in connection with applications to radar and sonar. The collection of papers by Haykin [2], the monographs by Monzingo and Miller [3] and by Hudson [4] present detailed treatments on adaptive array antennas, especially with applications to radar. Extensive introductions into sonar signal processing are the papers by Baggeroer [5] and Knight and others [6]. The NATO-ASI series on underwater acoustics and signal processing, for example, edited by Urban [7] and by Chan [8], give introductions and overviews on new research directions. Important journals in the field are, for example, the *IEEE Transactions on Acoustics, Speech, and Signal Processing,* the *Journal of the Acoustical Society of America,* and in Europe the *IEE Proceedings Pt. F.*

This chapter considers only some fundamentals of array processing methods, concentrating mainly on passive systems. Relatively simple wave models and ideal point sensors are used to investigate concepts and both classical and more recent array processing methods. We describe the wavefields by suitable stochastic processes and then interpret array processing as a statistical problem. This allows us to apply well-known results to design techniques for spectrum or parameter estimation and, possibly, to predict the behavior of the resulting array processing methods. The material presented here has similarity with the lecture notes published in [9]. However, the classical results are reworked, and both high-resolution and parametric methods are presented in the light of more recent research.

Outlining this chapter, we begin in Section 1.2 with the definition of elementary plane waves and the parameters of interest. The role of time delays is discussed. We introduce homogeneous wavefields as suitable stationary stochastic processes and as a superposition of plane waves. Filtering and sampling of wavefields are defined. We then study phased array beamforming, in which time delays are compensated when estimating the signals of a source. The second moments of homogeneous wavefields are characterized by a frequency-wavenumber spectrum. Its estimation is the next point of interest. One method to do this is classical beamforming. Properties of and relations between different beamforming methods are investigated. The use of line and of plane sensor arrays are discussed if source directions or the direction of arrivals of the signal waves have to be estimated. A disadvantage in applying one of these spectrum estimation techniques to direction estimation is the limited spatial resolution. Therefore, the so-called high-resolution methods, for example, eigensystem methods and optimum beamformers, are studied in Section 1.3. The wave models used up to this point assume uncorrelated signals from different sources and plane waves. Multipath propagation and sources within finite range, for example, cannot be handled with these methods. We therefore address the problem with a different approach in the last section. We use parametric models in the frequency domain, assuming certain stationarity properties of the array output. Maximum-likelihood and least-squares techniques are applied to estimate the source locations and the spectral parameters of interest. Asymptotic properties and numerical procedures are indicated. A discussion of estimation in unknown noise fields follows. Results of numerical experiments are presented. We comment how parametric methods can be used in practice and conclude with remarks on different approaches to handle the problem.

1.2. WAVEFIELDS AND SPECTRUM ESTIMATION

1.2.1 Elementary Waves and Delayed Signals

Traveling waves can be observed if a driving force is coupled to an open medium. They travel away from the source of the disturbance and transport energy radiated by the source. In an ideal, infinite medium (or ocean) and in a large distance away from the source, the observer would measure essentially plane waves. The concept of *coherent plane waves* is the basis of the investigation of this section. We do not discuss details of wave equations and of propagation. Rather, we refer to a physics textbook, for example [10].

The most simple model of a plane wave measured at a position described by a vector $\boldsymbol{p}$ in Cartesian coordinates and at time t is a complex exponential

$$x(t,\mathbf{p}) = a\, e^{j(\omega t + \mathbf{k}^T\mathbf{p})} \tag{1.1}$$

and is called *elementary wave*. In (1.1) the parameter a is a complex amplitude, j is the imaginary unit, $\omega = 2\pi f$ is the regular frequency, $\mathbf{k}$ is the *wavenumber vector,* and $\mathbf{k}^T$ denotes the transpose of a column vector $\mathbf{k}$. We denote $\mathbf{k} = \omega\boldsymbol{\xi}$ and call $\boldsymbol{\xi}$ the *slowness vector* because $v = 1/|\boldsymbol{\xi}|$ is the *propagation velocity* of a wavefront in the direction described by the unit vector $-\boldsymbol{\xi}/|\boldsymbol{\xi}|$. A wavefront is a connected set of points in space with similar phases of the wave for a given time and is a plane for an elementary wave. The period of such wavefronts in the direction of propagation is the *wavelength* $\lambda = 2\pi v/|\omega|$. If $\mathbf{k} = (k_1, k_2, k_3)^T$, then $\xi_i = k_i/\omega$ is the slowness in the direction of the ith coordinate. In a similar manner, we can define elementary waves and their parameters propagating in a plane. Interpreting (1.1), an elementary wave carries a monochromatic signal of frequency ω radiated by a source in the direction that is pointed by the vector $\boldsymbol{\xi}$.

Let us now assume that we wish to measure an elementary wave at three positions in space described by linearly independent vectors $\mathbf{p}_1$, $\mathbf{p}_2$, and $\mathbf{p}_3$. We furthermore assume the wave to be propagating so that a *wavefront* will first traverse the point $\mathbf{p}_n$ and then the origin $\mathbf{O}$. We now look for the time difference τ_n, which can be measured between the wavefront traversed $\mathbf{p}_n$ and $\mathbf{O}$. This difference is called *time delay.* Because the projection of the position vector on the negative direction of propagation describes the way of the wavefront, we calculate the time delays by dividing the distance traversed by the velocity of propagation,

$$\tau_n = \mathbf{p}_n^{\mathrm{T}}(\boldsymbol{\xi}/|\boldsymbol{\xi}|)/v = \mathbf{p}_n^{\mathrm{T}}\boldsymbol{\xi}, \qquad n = 1, 2, 3. \tag{1.2}$$

Assuming no amplitude loss through propagation, we have, from (1.1),

$$x(t, \mathbf{p}_n) = x(t + \tau_n, \mathbf{0}). \tag{1.3}$$

If we know the positions $\mathbf{p}_n$ and can measure the time delays τ_n, we can compute the slowness vector $\boldsymbol{\xi}$ of the wave by solving the system of linear equations (1.2). This means that the direction of the source and the velocity of propagation of the waves can be determined. The frequency ω and the complex amplitude a of the signal is calculated as usual if we observe $x(t, \mathbf{p}_n)$ for some time instances at position $\mathbf{p}_n$. If we know the slowness vectors of the waves generated by three sources and can measure the time delays at a fixed position, we can calculate the coordinate vector $\mathbf{p}$ in a similar manner.

These elementary arguments indicate that the measurement of time delays seems to be a fundamental problem in array processing. In applications, however, we never find such a simple situation. We have, for example, to consider the possibility that there are several sources in different directions and radiating at the same time, that waves with different velocities of propagation carry signals generally not monochromatic and having different strengths. The sum of signals is usually additively disturbed by noise from the amplifiers of the sensors (*sensor noise*) and by many weak and not resolvable sources (*ambient noise*). In this section, we do not consider other effects such as reflections of waves or dispersivity of the medium.

In Section 1.2, we approach the problem to localize the sources and to characterize their waves and signals by developing signal processing methods for the estimation of the following:

- The distribution of sources over directions.
- The distribution of velocities per direction.
- The distribution of power over frequencies per direction and velocity.

The methods are *nonparametric*; that is, the models cannot be characterized by a finite set of parameters.

The passive sonar problem is an example of interest. A surface ship tracks a towed array by a tow cable in the depth. The array is a chain of hydrophones. The sources of interest are submarines emanating machinery noise, flow noise, and so on. The hydrophones measure these signals disturbed by a lot of other marine noises, for example, noise of the own and other surface ships, noise of the surface wave motion, noise of animals (see Wenz [11] for corresponding empirical spectral distributions). The signals of the submarines have to be detected, a topic a little outside the scope of this chapter. The submarines have to be localized, classified by careful frequency analysis of the signals, and tracked. The last is not discussed in this chapter, and we refer, for example, to Moura [12].

1.2.2 Homogeneous Wavefields

The first problem we have to solve is the definition of a simple model of wavefields that allows us to investigate methods for estimating the distributions as indicated in Section 1.2.1. We assume the wavefield to be a superposition of elementary waves, where arbitrary combinations of frequencies and wavenumbers are tolerated. We proceed similarly to [13] to [16] and do not ask whether these so defined wavefields are solutions of certain wave equations. We do not investigate either more accurate models as discussed, for example, by Middleton [17]. The idea here is to define suitable stationary stochastic processes.

A stationary process defined over time is a family of random variables $x(t)$, $t \in \mathbb{R}$ which has, under weak assumptions, a *Cramér representation*,

$$x(t) = \int e^{j\omega t} dZ_x(\omega). \tag{1.4}$$

The integral is understood as a *Stieltjes integral* in the sense of mean squares (m.s.) and is taken over all frequencies ω. $Z_x(\omega)$ is a complex random function with expectation $E\{Z_x(\omega)\} = 0$, assuming $x(t)$ has zero mean, and independent increments $dZ_x(\omega)$. The latter has the consequence that the covariances formally satisfy

$$\mathrm{Cov}\{dZ_x(\omega), dZ_x(\nu)\} = \delta(\omega - \nu)\, d\nu\, dF_{xx}(\omega), \tag{1.5}$$

where δ denotes the delta distribution and $F_{xx}(\omega)$ is the *power spectral distribution* of $x(t)$. We assume that $F_{xx}(\omega)$ has a density $C_{xx}(\omega)$, simply called *spectrum,* such that $dF_{xx}(\omega) = C_{xx}(\omega)\, d\omega/(2\pi)$. According to (1.4), we may think of $x(t)$ as a superposition of complex oscillations $\exp(j\omega t)$ with complex amplitudes $dZ_x(\omega)$ that are independent random variables for different frequencies.

In a similar manner, we define a homogeneous wavefield as a stochastic process stationary over time and space that has a Cramér representation. The random variable describing the wavefield at time t and at position $\mathbf{p}$ is represented (m.s.) by

$$x(t,\mathbf{p}) = \int e^{j(\omega t + \mathbf{k}^T\mathbf{p})}\, dZ_x(\omega, \mathbf{k}). \tag{1.6}$$

The Stieltjes integral is taken over all frequencies ω and over all wavenumber vectors $\mathbf{k}$. $Z_x(\omega, \mathbf{k})$ is a complex random function with $E\{Z_x(\omega, \mathbf{k})\} = 0$ and independent increments $dZ_x(\omega, \mathbf{k})$ satisfying

$$\mathrm{Cov}\{dZ_x(\omega, \mathbf{k}), dZ_x(\nu, \mathbf{g})\} = \delta(\omega - \nu, \mathbf{k} - \mathbf{g}) C_{xx}(\omega, \mathbf{k})\, d\omega\, d\mathbf{k}\, d\nu\, d\mathbf{g}/(2\pi)^4. \tag{1.7}$$

We assume the existence of the density $C_{xx}(\omega, \mathbf{k})$, which is called the *frequency-wavenumber spectrum* of the wavefield $x(t, \mathbf{p})$. The increments are symbolically defined by

$$\begin{aligned} dZ_x(\omega, \mathbf{k}) &= d_\omega d_{k_1} d_{k_2} d_{k_3} Z_x(\omega, \mathbf{k}) \\ &= d_\omega d_{k_1} d_{k_2}[Z_x(\omega, k_1, k_2, k_3 + dk_3) \\ &\quad - Z_x(\omega, k_1, k_2, k_3)] = \cdots. \end{aligned} \tag{1.8}$$

Because (1.6) is defined in mean-squared sense, we have $\mathrm{E}\{x(t, \mathbf{p})\} = 0$ and the *covariance function,* using (1.7),

$$\begin{aligned} c_{xx}(t, \mathbf{p}) &= \mathrm{Cov}\{x(t + u, \mathbf{p} + \mathbf{q}), x(u, \mathbf{q})\} \\ &= \int e^{j(\omega t + \mathbf{k}^{\mathrm{T}}\mathbf{p})} C_{xx}(\omega, \mathbf{k})\, d\omega\, d\mathbf{k}/(2\pi)^4. \end{aligned} \tag{1.9}$$

We thus see that the frequency-wavenumber spectrum is the four-dimensional Fourier transform of the covariance function and has properties similar to a frequency spectrum of a stationary process over time. $C_{xx}(\omega, \mathbf{k})$ is nonnegative and describes the distribution of the average power $\mathrm{E}\{|x(t, \mathbf{p})|^2\} = \mathrm{Var}\{x(t, \mathbf{p})\} = c_{xx}(0, \mathbf{0})$ over frequencies and wavenumbers.

For our first example, we characterize a wavefield generated by one source, consisting of waves with a fixed slowness vector $\boldsymbol{\xi}_0$ and carrying a stationary signal. Let this signal received at $\mathbf{p} = \mathbf{0}$ be denoted by $x(t, \mathbf{0}) = s(t)$, with a Cramér representation corresponding to (1.4) with increments $dZ_s(\omega)$. Then the wavefield at position $\mathbf{p}$ must be

$$x(t, \mathbf{p}) = \int e^{j(\omega t + \mathbf{k}^{\mathrm{T}}\mathbf{p})} dZ_x(\omega, \mathbf{k}) = \int e^{j\omega(t + \boldsymbol{\xi}_0^{\mathrm{T}}\mathbf{p})} dZ_s(\omega) = s(t + \boldsymbol{\xi}_0^{\mathrm{T}}\mathbf{p}) \tag{1.10}$$

because the possible wavenumber vectors are $\mathbf{k} = \omega\boldsymbol{\xi}_0$. The frequency-wavenumber spectrum is as follows,

$$C_{xx}(\omega, \mathbf{k}) = C_{ss}(\omega)(2\pi)^3\delta(\mathbf{k} - \omega\boldsymbol{\xi}_0), \tag{1.11}$$

which is the Fourier transform of the covariance function

$$c_{xx}(t, \mathbf{p}) = \mathrm{Cov}[x(t + u, p + \mathbf{q}), x(u, \mathbf{q})] = c_{ss}(t + \boldsymbol{\xi}_0^{\mathrm{T}}\mathbf{p}), \tag{1.12}$$

where $C_{ss}(\omega)$ is the frequency spectrum of $s(t)$.

Our second example deals with a wavefield consisting of waves propagating with a fixed velocity v_0. We are interested in the distribution of power of the wavefield over frequencies and directions. We use polar coordinates

$$\mathbf{k} = k(\cos\varphi \cos\alpha, \cos\varphi \sin\alpha, \sin\varphi)^T = k\mathbf{e}(\alpha, \varphi)^T$$

in (1.9), where the angles α and φ are called *azimuth* and *elevation,* respectively. Knowing that $k = |\omega|/v_0$, we may write

$$C_{xx}[\omega, k\mathbf{e}(\alpha, \varphi)] = \overline{C}_{xx}(\omega, \alpha, \varphi)2\pi\delta(k - |\omega|/v_0) \tag{1.13}$$

and so obtain

$$c_{xx}(t, \mathbf{p}) = (2\pi)^{-3}\int_{-\infty}^{\infty} d\omega\, e^{j\omega t}\int_{0}^{2\pi} d\alpha \int_{-\pi/2}^{\pi/2} d\varphi \cos\varphi$$
$$(\omega/v_0)^2\bar{C}_{xx}(\omega, \alpha, \varphi)\, e^{j(|\omega|/v_0)\mathbf{e}(\alpha,\varphi)^T\mathbf{p}}, \tag{1.14}$$

as the desired spectral representation. If $\bar{C}_{xx}(\omega, \alpha, \varphi)$ does not contain any δ-contributions with respect to α and φ, we have a suitable model for ambient noise. If $\bar{C}_{xx}(\omega, \alpha, \varphi)$ does not depend on α and φ, that is, the direction spectrum is white, the wavefield is isotropic. The integrals over α and φ in (1.14) can be solved, we thus obtain

$$c_{xx}(t, \mathbf{p}) = \int_{-\infty}^{\infty} e^{j\omega t}\frac{\sin(|\mathbf{p}|\,|\omega|/v_0)}{\pi|\mathbf{p}|\,|\omega|/v_0}\left(\frac{\omega}{v_0}\right)^2\bar{C}_{xx}(\omega)\frac{d\omega}{2\pi}, \tag{1.15}$$

which only depends on t and $|\mathbf{p}|$. For more results on isotropic wavefields, we refer to Brillinger [15] and Adler [16].

Wavefields can be *spatially filtered.* If $h(t, \mathbf{p})$ is the *impulse response* of a *linear time- and space-invariant system*, the filtered wavefield can be expressed (in the mean squares sense) by

$$y(t, \mathbf{p}) = \int h(u, \mathbf{q})x(t-u, \mathbf{p}-\mathbf{q})\, du\, d\mathbf{q}. \tag{1.16}$$

The frequency-wavenumber response of the system is

$$H(\omega, \mathbf{k}) = \int h(u, \mathbf{q})e^{-j(\omega u + \mathbf{k}^T\mathbf{q})}\, du\, d\mathbf{q}. \tag{1.17}$$

As with stationary processes, we find the frequency-wavenumber spectrum of the system output $y(t, \mathbf{p})$,

$$C_{yy}(\omega, \mathbf{k}) = |H(\omega, \mathbf{k})|^2 C_{xx}(\omega, \mathbf{k}). \tag{1.18}$$

We may now describe the outputs of N sensors of an array. The sensors are ideal point receivers and do not influence the wavefield. The position of the nth sensor is $\mathbf{p}_n$ $(n = 1, \ldots, N)$. The sensor outputs are then given by sampling the wavefield, $x_n(t) = x(t, \mathbf{p}_n)$ $(n = 1, \ldots, N)$, combined into the array output,

$$\mathbf{x}(t) = [x_1(t), \ldots, x_N(t)]^T. \tag{1.19}$$

The random function $\mathbf{x}(t)$ is a stationary vector process with zero mean. The components have the Cramér representation

$$x_n(t) = \int_{\omega} e^{j\omega t}\int_{\mathbf{k}} e^{j\mathbf{k}^T\mathbf{p}_n}\, dZ_x(\omega, \mathbf{k}) \tag{1.20}$$

such that the inner integral can be identified as the increment $dZ_{x_n}(\omega)$. The *spectral (density) matrix* $\mathbf{C}_x(\omega)$ of $\mathbf{x}(t)$ has the typical element:

$$C_{x_n x_i}(\omega) = \int e^{-j\omega t}c_{xx}(t, \mathbf{p}_n - \mathbf{p}_i)\, dt = \int e^{j\mathbf{k}^T(\mathbf{p}_n - \mathbf{p}_i)}C_{xx}(\omega, \mathbf{k})\, d\mathbf{k}/(2\pi)^3. \tag{1.21}$$

In the following example, we discuss the structure of the spectral matrix $\mathbf{C}_x(\omega)$. We assume a wavefield consisting of signals from M sources and of noise $u(t, \mathbf{p})$. The resulting frequency-wavenumber spectrum is

$$C_{xx}(\omega, \mathbf{k}) = \sum_{m=1}^{M} C_{s_m s_m}(\omega)\,(2\pi)^3\delta(\omega\xi_m - \mathbf{k}) + C_{uu}(\omega, \mathbf{k}) \tag{1.22}$$

for which the typical element (1.21) takes the form

$$C_{x_n x_i}(\omega) = \sum_{m=1}^{M} C_{s_m s_m}(\omega)\, e^{j\omega\xi_m^{\mathrm{T}}(\mathbf{p}_n - \mathbf{p}_i)} + C_{u_n u_i}(\omega). \tag{1.23}$$

Define the so-called *steering vector,*

$$\mathbf{d}(\mathbf{k}) = (e^{j\mathbf{k}^{\mathrm{T}}\mathbf{p}_1}, \ldots, e^{j\mathbf{k}^{\mathrm{T}}\mathbf{p}_N})^{\mathrm{T}} \tag{1.24}$$

The spectral matrix may then be written as

$$\mathbf{C}_x(\omega) = \sum_{m=1}^{M} C_{s_m s_m}(\omega)\,\mathbf{d}(\omega\xi_m)\mathbf{d}(\omega\xi_m)^{\dagger} + \mathbf{C}_u(\omega). \tag{1.25}$$

where the superscript † denotes the conjugate (Hermitian) transpose. The additive noise may, for example, be ambient noise with a frequency-wavenumber spectrum described by (1.13), or sensor noise, for which we assume uncorrelated noise from sensor to sensor and identical frequency spectra $C_{uu}(\omega)$. If $\mathbf{I}$ denotes the unit matrix, sensor noise has the spectral matrix

$$\mathbf{C}_u(\omega) = C_{uu}(\omega)\mathbf{I}. \tag{1.26}$$

The construction of a homogeneous wavefield with this spectral matrix is not difficult. We only have to find a covariance function satisfying $c_{uu}(t, \mathbf{p}_n - \mathbf{p}_i) = c_{uu}(t)\delta_{ni}$, where δ_{ni} is Kronecker's delta and $\mathbf{p}_n$ is the position of the nth sensor. For example, sensor positions generated by a periodic sampling scheme, as discussed in the next paragraphs, a suitable $\mathbf{k}$-domain limitation and a white wavenumber spectrum are sufficient. We remark that model (1.25), especially in connection with (1.26) is basic for certain high-resolution methods to be discussed in Section 1.3.

Periodic sampling of a wavefield can be described as follows. Using the notation, $\mathbf{t} = (t, \mathbf{p}^{\mathrm{T}})^{\mathrm{T}}$ and $\boldsymbol{\omega} = (\omega, \mathbf{k}^{\mathrm{T}})^{\mathrm{T}}$, integer vectors $\mathbf{n}$ and $\mathbf{m}$ with each one having four components, and a (4×4) nonsingular sampling matrix $\boldsymbol{\Delta}$, then the *sampled wavefield* is

$$x(\boldsymbol{\Delta}\mathbf{n}) = \int e^{j\boldsymbol{\omega}^{\mathrm{T}}\boldsymbol{\Delta}\mathbf{n}}\, dZ_x(\boldsymbol{\omega})$$

$$= \int_{\boldsymbol{\Delta}^{\mathrm{T}}\mathbf{v}\in[-\pi,\pi)^4} e^{j\boldsymbol{\nu}^{\mathrm{T}}\boldsymbol{\Delta}\mathbf{n}} \sum_{\mathbf{m}} dZ_x(\boldsymbol{\nu} + 2\pi\boldsymbol{\Delta}^{\mathrm{T}}\mathbf{m}) = \int_{\boldsymbol{\lambda}\in[-\pi,\pi)^4} e^{j\boldsymbol{\lambda}^{\mathrm{T}}\mathbf{n}}\, dZ_x^D(\boldsymbol{\lambda}), \tag{1.27}$$

where $\boldsymbol{\lambda} = \boldsymbol{\Delta}^{\mathrm{T}}\boldsymbol{\nu}$ is the normalized frequency-wavenumber vector and $dZ_x^D(\boldsymbol{\lambda})$ corresponds to $\Sigma_{\mathbf{m}} dZ_x[\boldsymbol{\Delta}^{-\mathrm{T}}(\boldsymbol{\lambda} + 2\pi\mathbf{m})]$. The last term in (1.27) is the Cramér representation of the discrete wavefield $x^D(\mathbf{n}) = x(\mathbf{\Delta n})$. The corresponding frequency-wavenumber spectrum is

$$C_{xx}^D(\boldsymbol{\lambda}) = |\det\mathbf{\Delta}|^{-1}\sum_{\mathbf{m}} C_{xx}[\mathbf{\Delta}^{-\mathrm{T}}(\boldsymbol{\lambda} + 2\pi\mathbf{m})]. \tag{1.28}$$

A reconstruction of the wavefield from the discrete one is possible in the mean-squared sense if the spectrum $C_{xx}(\boldsymbol{\omega})$ is band limited,

$$C_{xx}(\boldsymbol{\omega}) = 0 \qquad \text{if } \mathbf{\Delta}^{\mathrm{T}}\boldsymbol{\omega} \notin [-\pi, \pi)^4. \tag{1.29}$$

If the sampling matrix is diagonal, $\mathbf{\Delta} = \mathrm{diag}(\Delta_t, \Delta_1, \Delta_2, \Delta_3)$, then the wavefield is measured at positions $(p_1, p_2, p_3)^{\mathrm{T}}$ with coordinates p_i being a multiple of Δ_i and in parallel at all positions at time instances being a multiple of Δ_t. Various choices of sampling matrices allow to implement very different sampling schemes, for example, hexagonal ones. Interesting results in this direction can be found in [18].

Let us finally remark that homogeneous wavefields in a plane can be defined similarly. Alternatively, we use a cylindrical model in space and apply the results just given.

1.2.3 Compensation of Time Delays

Let us now discuss a simple array processing method that can be interpreted as a typical filtering problem. We assume to sample the wavefield by means of sensors at positions $\mathbf{p}_n (n = 1, \ldots, N)$. We look for the signal of a possible source carried by waves with a fixed slowness vector $\boldsymbol{\xi}_0$. The waves of interest are plane waves propagating in the direction of $-\boldsymbol{\xi}_0$ and with velocity $1/|\boldsymbol{\xi}_0|$ not dependent on frequency. This means that we can use (1.2) to calculate the time delay $\tau_n = \boldsymbol{\xi}_0^{\mathrm{T}}\mathbf{p}_n$ corresponding to the travel time of the wavefront traversing $\mathbf{p}_n$ and $\mathbf{O}$. A method to filter out the signal is now obvious. If the sensor outputs $x_n(t)$ are delayed by τ_n, the signal becomes in phase in all delayed sensor outputs. Summing up the delayed outputs means to filter the signal out of the set of other signals and of noise. More generally, the signal is estimated by compensating the time delays, shading (weighting), and summing up,

$$\hat{s}(t) = \sum_{n=1}^{N} w_n x_n(t - \tau_n) = \sum_{n=1}^{N} w_n x(t - \boldsymbol{\xi}_0^{\mathrm{T}}\mathbf{p}_n, \mathbf{p}_n). \tag{1.30}$$

This method is called *phased array beamforming*, or more accurately *delay-and-sum beamforming*, and $\hat{s}(t)$ is the beamsignal.

Now, we interpret phased array beamforming as filtering of the wavefield $x(t, \mathbf{p})$. We assume the filtered wavefield $y(t, \mathbf{p})$ in the sense of (1.16) so that $y(t, \mathbf{O}) = \hat{s}(t)$. If we choose

$$h(t, \mathbf{q}) = \sum_{n=1}^{N} w_n \delta(t - \boldsymbol{\xi}_0^{\mathrm{T}}\mathbf{p}_n, \mathbf{q} + \mathbf{p}_n), \tag{1.31}$$

the use of (1.16) results in (1.30). The frequency-wavenumber response of (1.31) is

$$H(\omega, \mathbf{k}) = \sum_{n=1}^{N} w_n e^{-j(\omega\xi_0 - \mathbf{k})^T \mathbf{p}_n} = B(\omega\xi_0 - \mathbf{k}). \tag{1.32}$$

$B(\omega\xi_0 - \mathbf{k})$ is called the response pattern of the array steered for $\omega\xi_0$. The frequency-wavenumber spectrum of $y(t, \mathbf{p})$ is

$$C_{yy}(\omega, \mathbf{k}) = |B(\omega\xi_0 - \mathbf{k})|^2 C_{xx}(\omega, \mathbf{k}), \tag{1.33}$$

and the frequency spectrum of $\hat{s}(t)$,

$$\begin{aligned} C_{\hat{s}\hat{s}}(\omega) &= \int |B(\omega\xi_0 - \mathbf{k})|^2 C_{xx}(\omega, \mathbf{k})\, d\mathbf{k}/(2\pi)^3 \\ &= \sum_{n,i=1}^{N} w_n w_i^* e^{-j\omega\xi^T{}_0(\mathbf{p}_n - \mathbf{p}_i)} C_{x_n x_i}(\omega), \end{aligned} \tag{1.34}$$

where (1.21) was used. The asterisk is required if complex numbers are considered.

The spectrum (1.34) of the beamsignal (1.30) is a smoothed version of the frequency wavenumber spectrum of the wavefield. Smoothing is done over wavenumber vectors with a kernel centered at the wavenumber vectors $\omega\xi_0$ of interest. The kernel is the beampattern $|B(\omega\xi_0 - \mathbf{k})|^2$, which is a finite sum of periodic functions; in general, it has many lobes. Therefore, complicated overlapping has to be expected in (1.34). If the beampattern had properties as $(2\pi)^3\delta(\omega\xi_0 - \mathbf{k})$, we would find $C_{\hat{s}\hat{s}}(\omega) = C_{xx}(\omega, \omega\xi_0)$, that is, the estimated signal would have the same spectrum as the signal carried by the waves propagating with slowness vector ξ_0. In Section 1.2.5, we discuss similar problems when using line arrays.

Modifications and extensions of phased array beamforming are used in applications, such as *split-beam processors* and *null-steering processors* in sonar; see Baggeroer [5] and Owsley's chapter in [1]. Null steering, for example, means that the influence of a noise source is minimized when the array is focused in the direction of the source of interest. First, the phased array is focused to the noise source characterized by the time delays $\tau_{2,n}$,

$$y_2(t) = N^{-1} \sum_{n=1}^{N} x_n(t - \tau_{2,n}). \tag{1.35}$$

If $\tau \geq \max_n \tau_{2,n}$, we calculate

$$z_n(t) = x_n(t - \tau) - y_2(t - \tau + \tau_{2,n}), \qquad (n = 1, \ldots, N)$$

to remove causally the estimated noise signal from the sensor outputs with corrected phases. Then, we focus to the source of interest with time delays $\tau_{1,n}$,

$$y_1(t) = N^{-1} \sum_{n=1}^{N} z_n(t - \tau_{1,n}). \tag{1.36}$$

If, for example,

$$x_n(t) = s_1(t + \boldsymbol{\xi}_1^{\mathrm{T}}\mathbf{p}_n) + s_2(t + \boldsymbol{\xi}_2^{\mathrm{T}}\mathbf{p}_n) \tag{1.37}$$

and

$$\tau_{i,n} = \boldsymbol{\xi}_i^{\mathrm{T}}\mathbf{p}_n \ (i = 1, 2;\ n = 1, \ldots, N),$$

then

$$y_1(t) = N^{-1}(N-1)s_1(t-\tau) - N^{-2} \sum_{n \neq k} s_1[t - \tau + (\boldsymbol{\xi}_2 - \boldsymbol{\xi}_1)^{\mathrm{T}}(\mathbf{p}_k - \mathbf{p}_n)]. \tag{1.38}$$

This signal is zero for $\boldsymbol{\xi}_1 = \boldsymbol{\xi}_2$ as expected and essentially a delayed version of the desired signal otherwise.

1.2.4 Spectrum Estimation

The frequency-wavenumber spectrum contains information about the homogeneous wavefield we are interested in. $C_{xx}(\omega, \mathbf{k})$ describes the distribution of discrete sources over directions, the density of the distribution of nonresolvable sources (e.g., ambient noise), the distribution of velocities of propagation per direction, and the spectrum of the sum of signals carried by waves arriving from a given direction and with a given velocity. The objective is now to estimate this spectrum. We assume that the wavefield has been observed over a finite range of time, for example, $0 \leq t \leq T$ and in a finite spatial domain D. Practically, we can do it only by means of a finite set of sensors placed at different positions in that domain.

Because we consider homogeneous wavefields as stationary processes, we first formulate nonparametric estimation techniques for wavefields such as they are known for stationary processes, (see Capon [14], Brillinger [15], and so on). In a manner similar to [15], we start with the properties of the finite Fourier transform of the observed piece of a real wavefield, namely, $x(t, \mathbf{p}_n)$: $0 \leq t \leq T$, $n = 1, \ldots, N$. We first assume that the positions $\mathbf{p}_n \in \mathbb{R}^r$ are defined by a periodic sampling scheme with a nonsingular $(r \times r)$ sampling matrix $\boldsymbol{\Delta}$ and integer r-vectors $\mathbf{n}$, $\mathbf{p}_n = \boldsymbol{\Delta}\mathbf{n}$. We choose $r = 2$ for wavefields in a plane and $r = 3$ in space. We assume a general frequency and wavenumber band limitation and sample the time as well, $t = \Delta i$, $i = 0, \ldots, K-1$, with $T = \Delta K$. Similar to spectrum estimation of processes stationary over time, we use data windows to reduce leakage effects. Let $w_T(t) = 0$ for $t \notin [0, T]$ and $\int w_T(t)^2 dt = \Delta K$, and let $w_D(\mathbf{p}) = 0$ for $\mathbf{p} \notin D \subset \mathbb{R}^r$ and $\int w_D(\mathbf{p})^2\, d\mathbf{p} = \int_D d\mathbf{p} = V = N\det\boldsymbol{\Delta}$, where det means the determinant. The Fourier transform of the observed data is

$$X(\omega, \mathbf{k}) = \Delta|\det\boldsymbol{\Delta}| \sum_i \sum_{\mathbf{n}} w_T(\Delta i)\, w_D(\boldsymbol{\Delta}\mathbf{n})\, x(\Delta i, \boldsymbol{\Delta}\mathbf{n})\, e^{-j(\omega\Delta i + \mathbf{k}^{\mathrm{T}}\boldsymbol{\Delta}\mathbf{n})}. \tag{1.39}$$

This is the *Riemann approximation* of the corresponding continuous finite Fourier transform.

The asymptotic statistical properties of the finite Fourier transform motivate different spectrum estimation techniques. Assuming certain regularity conditions with respect to the smoothness of the spectrum, the higher-moment spectra and the windows, allowing frequencies and wavenumbers with $(\omega\Delta, \mathbf{k}^T\boldsymbol{\Delta})^T \in (0, \pi)^{r+1}$, we can show the following results for large D extended in all dimensions and large T:

1. $X(\omega, \mathbf{k})$ is complex normally distributed with zero mean and variance of $TVC_{xx}(\omega, \mathbf{k})$. (1.40)

2. If $[0, T]$ is divided into L pieces of length T', that is, $T = LT'$, and if we calculate finite Fourier transforms of successive data pieces, similar to (1.39), with time windows $w_{T'}[t - (l - 1)T']$ and call the results $X^l(\omega, \mathbf{k}), l = 1, \ldots, L$, then these random variables are mutually independent and distributed as in (1.40) with a variance of $T'VC_{xx}(\omega, \mathbf{k})$. (1.41)

3. If $(\omega_l, \mathbf{k}_l), l = 1, \ldots, L$, are different in pairs, then $X(\omega_l, \mathbf{k}_l), l = 1, \ldots, L$, are mutually independent random variables. (1.42)

Using the Cramér representation of (1.6), we obtain a different expression for the finite Fourier transform (1.39),

$$X(\omega, \mathbf{k}) = \int W_T(\omega - \nu)W_D(\mathbf{k} - \mathbf{g})\, dZ_x(\nu, \mathbf{g}), \tag{1.43}$$

where

$$W_T(\omega) = \Delta \sum_i w_T(\Delta i)\, e^{-j\omega\Delta i}, \tag{1.44}$$

$$W_D(\mathbf{k}) = |\det \boldsymbol{\Delta}| \sum_{\mathbf{n}} w_D(\boldsymbol{\Delta}\mathbf{n})\, e^{-j\mathbf{k}^T\boldsymbol{\Delta}\mathbf{n}}. \tag{1.45}$$

To construct spectrum estimates, we use the periodogram

$$I_{xx}(\omega, \mathbf{k}) = (TV)^{-1}|X(\omega, \mathbf{k})|^2 \tag{1.46}$$

The expected value of the periodogram, is, using (1.43) and (1.7),

$$E\{I_{xx}(\omega, \mathbf{k})\} = (TV)^{-1} \int |W_T(\omega - \nu)|^2 |W_D(\mathbf{k} - g)|^2 C_{xx}(\nu, \mathbf{g})\, d\nu\, d\mathbf{g}/(2\pi)^{r+1}. \tag{1.47}$$

This is a smoothed version of the spectrum around $(\omega, \mathbf{k})$. If the kernel $(TV)^{-1}|W_T(\omega - \nu)|^2|W_D(\mathbf{k} - \mathbf{g})|^2$ behaves approximately as $(2\pi)^{r+1}\delta(\omega - \nu, \mathbf{k} - \mathbf{g})$, the result is approximately $C_{xx}(\omega, \mathbf{k})$ in accordance with (1.40). The number of sensors, N, is limited in practice. By contrast, however, we may increase the observation time T so that the smoothing effect over frequencies can be ignored. We may interpret $V^{-1}|W_D(\mathbf{k} - \mathbf{g})|^2$ as the beampattern $|B(\mathbf{k} - \mathbf{g})|^2$ of the sensor array steered for $\mathbf{k}$ if we identify (1.45) with (1.32) suitably by choosing

$$w_n = |\det \boldsymbol{\Delta}| V^{-1/2} W_D(\mathbf{p}_n), \qquad n = 1, \ldots, N, \tag{1.48}$$

with $\Sigma_n |w_n|^2 = V/N$. The mean of the periodogram can then be expressed by

$$E\{I_{xx}(\omega, \mathbf{k})\} = \int |B(\mathbf{k} - \mathbf{g})|^2 C_{xx}(\omega, \mathbf{g})\, d\mathbf{g}/(2\pi)^r. \tag{1.49}$$

Using (1.40), the variance of the periodogram is approximately given by

$$\text{Var}\{I_{xx}(\omega, \mathbf{k})\} = C_{xx}(\omega, \mathbf{k})^2 \tag{1.50}$$

for $(\omega\Delta, \mathbf{k}^T\boldsymbol{\Delta}) \in (0, \pi)^{r+1}$. This means that the variance is approximately the square of the value we want to estimate.

Because property (1.41) is also asymptotically true for the periodogram, the variance of a spectrum estimate using periodograms can be reduced by taking the mean of periodograms of successive pieces of the array output, as shown by

$$\hat{C}_{xx}(\omega, \mathbf{k}) = \frac{1}{LT'V} \sum_{l=1}^{L} |X^l(\omega, \mathbf{k})|^2. \tag{1.51}$$

The expected value of $\hat{C}_{xx}(\omega, \mathbf{k})$ is given by (1.47), where T is replaced by T', or by (1.49) if T' is large enough; its variance is given by (1.50), divided by L. Correspondingly, we can apply (1.42) and smooth the periodogram over frequency-wavenumbers $(\omega_l, \mathbf{k}_l)$, $l = 1, \ldots, L$, in the neighborhood of the pair $(\omega, \mathbf{k})$ of interest,

$$\hat{C}_{xx}(\omega, \mathbf{k}) = \frac{1}{LTV} \sum_{l=1}^{L} |X(\omega_l, \mathbf{k}_l)|^2. \tag{1.52}$$

The estimate has the same variance reduction as (1.51); however, the mean is a smoothed version of (1.47) in the sense of (1.52).

We have assumed the sensor positions $\mathbf{p}_n$ to be generated by a periodic sampling scheme. In addition, we assume that a fast Fourier transform algorithm is available for this scheme to have an efficient method for computing (1.39) at discrete frequencies $2\pi i/T$ and discrete wavenumbers. These transformations are of course the most time-consuming computations to obtain the spectrum estimates (1.51) and (1.52). In practice, we are interested in the spectrum values of a finite set of selected frequencies, velocities of propagation, and directions. Accordingly, interpolation or special smoothing of the spectrum calculated for discrete frequencies and wavenumbers is required.

If there exists no fast Fourier transform, for example, the sensors are located at arbitrary positions $\mathbf{p}_n$ $(n = 1, \ldots, N)$, a different algorithm is usually applied. With $\mathbf{x}(t) = [x_1(t), \ldots, x_N(t)]^T$ as the array output, where $x_n(t) = x(t, \mathbf{p}_n)$, we first compute, via fast Fourier transform over time,

$$\mathbf{X}^l(\omega) = \Delta \sum_i w_{T'}(\Delta i - (l-1)T')\, \mathbf{x}(\Delta i)\, e^{-j\omega\Delta i} \tag{1.53}$$

of successive pieces of the array output and then compute the estimate

$$\hat{\mathbf{C}}_x(\omega) = \frac{1}{LT'} \sum_{l=1}^{L} \mathbf{X}^l(\omega)\, \mathbf{X}^l(\omega)^\dagger \tag{1.54}$$

of the spectral matrix $\mathbf{C}_x(\omega)$ of $\mathbf{x}(t)$, see (1.21). This random matrix is described by, under certain regularity conditions and asymptotically for large T',

$$\left.\begin{array}{l}\text{A complex Wishart distribution with } L \text{ degrees of}\\ \text{freedom and parameter matrix } \mathbf{C}_x(\omega)/L.\end{array}\right\} \tag{1.55}$$

An estimate of the frequency-wavenumber spectrum, sometimes called the *classical beamformer,* is found by interchanging the summations involved in (1.51) which can be written in compact form as follows:

$$\tilde{C}_{xx}(\omega, \mathbf{k}) = \mathbf{d}(\mathbf{k})^\dagger \mathbf{w}\, \hat{\mathbf{C}}_x(\omega)\mathbf{w}^\dagger \mathbf{d}(\mathbf{k}). \tag{1.56}$$

Here, $\mathbf{d}(\mathbf{k})$ is the steering vector (1.24) and $\mathbf{w}$ denotes the diagonal matrix of weights w_n normalized in the sense of (1.48). This estimate has to be calculated for all combinations of frequencies, directions, and velocities of interest. If we ignore the smoothing effect over frequencies and apply (1.55), $\tilde{C}_{xx}(\omega, \mathbf{k})$ is approximately proportional to a χ^2_{2L}-random variable with mean,

$$E\{\tilde{C}_{xx}(\omega, \mathbf{k})\} = \mathbf{d}(\mathbf{k})^\dagger \mathbf{w} \mathbf{C}_x(\omega)\mathbf{w}^\dagger \mathbf{d}(\mathbf{k}), \tag{1.57}$$

which corresponds to the right side of (1.49). The variance of $\tilde{C}_{xx}(\omega, \mathbf{k})$ is

$$\mathrm{Var}\{\tilde{C}_{xx}(\omega, \mathbf{k})\} = [\mathbf{d}(\mathbf{k})^\dagger \mathbf{w} \mathbf{C}_x(\omega)\mathbf{w}^\dagger \mathbf{d}(\mathbf{k})]^2/L. \tag{1.58}$$

There is an interesting connection between phased array beamforming and the classical beamformer. If we try to estimate the spectrum (1.34) of the beamsignal (1.30) by the mean of periodograms of successive pieces of the beamsignal $\hat{s}(t)$, we obtain

$$\hat{C}_{\hat{s}\hat{s}}(\omega) = \frac{1}{LT} \sum_{l=1}^{L} |\hat{S}^l(\omega)|^2, \tag{1.59}$$

where $\hat{S}^1(\omega)$ is calculated as $\mathbf{X}^1(\omega)$ in (1.53) if $\mathbf{x}(t)$ is replaced by $\hat{s}(t)$. The expected value of this estimate is

$$E\{\hat{C}_{\hat{s}\hat{s}}(\omega)\} = \frac{1}{T'} \int |W(\omega - \nu)|^2 |B(\nu\xi_0 - \mathbf{g})|^2 C_{xx}(\nu, \mathbf{g})\, d\nu\, d\mathbf{g}/(2\pi)^{r+1}, \tag{1.60}$$

where (1.33) was applied. The right side of (1.60) also describes the mean of the classical beamformer of (1.56) if we replace $\nu\xi_0$ by $\omega\xi_0$. The variance of (1.59) has a similar relation to that of (1.56). Note that phased array beamforming and classical beamforming can only result in similar spectrum estimates for special cases, for example, if the wavefield is narrowband with respect to frequencies and wavenumbers or if the smoothing effect over frequencies can be neglected.

We could stabilize the estimate of (1.56) by smoothing either (1.54) or (1.56) for $\mathbf{k} = \omega\boldsymbol{\xi}$ over discrete frequencies $\omega_l = 2\pi n_l/T'$ in the neighborhood of ω. However, this is suitable only if $\mathbf{C}_x(\nu)$ and $\mathbf{d}(\nu\boldsymbol{\xi})$ slowly change with ν in the neighborhood of $\nu = \omega$. The additional bias could otherwise destroy the information we are interested in. We finally mention that related results on classical beamformers and phased array beamforming can be also derived for spectra that are evolutionary in time; see Böhme [19].

1.2.5 Use of Spectrum Estimates

Some examples are discussed in this section to show how the spectrum estimation techniques of the last section can be used. We first investigate the precision of the classical beamformer applied to the wavemodel described by (1.22). Even though the regularity conditions in connection with (1.40) to (1.42) are not satisfied for this model, we still apply (1.56). For smooth frequency spectra and large time windows, we can ignore the smoothing effect over frequencies. The expected value of (1.56) becomes

$$E\{\tilde{C}_{xx}(\omega, \omega\boldsymbol{\xi})\} = \sum_{m=1}^{M} |B[\omega(\boldsymbol{\xi} - \boldsymbol{\xi}_m)]|^2 C_{s_m s_m}(\omega) + \int |B(\omega\boldsymbol{\xi} - \mathbf{g})|^2 C_{uu}(\omega, \mathbf{g})\, d\mathbf{g}/(2\pi)^r. \tag{1.61}$$

Instead of M spectral lines in a continuous-noise spectrum as in (1.22), we can at best estimate a superposition of beampatterns steered for $\omega\boldsymbol{\xi}_m$ and weighted by $C_{s_m s_m}(\omega)$ plus a smoothed version of the noise spectrum. If we like to estimate the parameters $\boldsymbol{\xi}_m$ of the sources, we can search for the local maxima of the diagrams $\boldsymbol{\xi} \rightarrow \tilde{C}_{xx}(\omega, \omega\boldsymbol{\xi})$ for frequencies ω indicating high signal-to-noise ratios. The number of sources can also be roughly estimated in this way. However, the lobes of the beampattern have a finite width such that a variety of overlapping can appear if slowness vectors $\boldsymbol{\xi}_m$ and $\boldsymbol{\xi}_i$ are not well separated. Consequently, the peaks cannot always be resolved and the spurious peaks are generated by the sidelobes without even considering the random effects of the estimates. The resolution problem of classical methods has been frequently investigated in the literature, for example, by Cox [20] and motivated the design of high-resolution methods to be discussed in the following section.

Next, we like to apply the classical beamformer to data from a wavefield consisting of waves propagating with a fixed velocity v_0. We assume the discrete frequency-wavenumber spectrum of (1.13) and a sensor array in a plane as in radar or seismic applications.

Suppose $\mathbf{p}_n - \mathbf{p}_m = \mathbf{p} = (p_1, p_2, 0)^T$ for each pair of sensor positions of the array. Wavefronts arriving with elevation φ as in Fig. 1.1 have straight lines as traces in the plane of sensors propagating with velocity $v_0/\cos\varphi$. Because the wavefield is

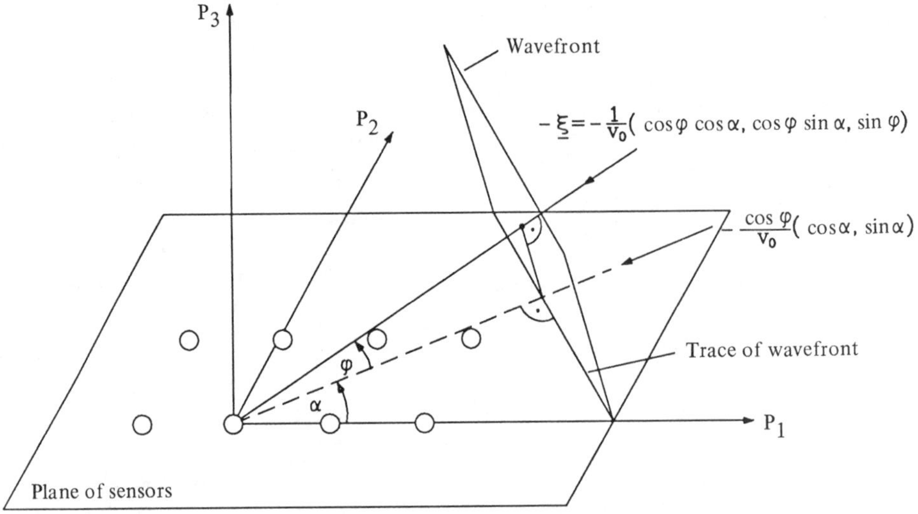

Figure 1.1 Plane sensor array and plane waves in space.

observed in the plane of sensors, we would like to have a model of the wavefield projected in this plane. Let this wavefield be y, and define $\mathbf{q} = (p_1, p_2)^T$ and wavenumber vectors $\mathbf{g} = \omega \cos\varphi/v_0(\cos\alpha, \sin\alpha)^T$. We obtain from (1.14), letting $g = |\omega|\cos\varphi/v_0$ and after some manipulations,

$$c_{xx}(t, \mathbf{p}) = \frac{1}{(2\pi)^2} \int_{-\infty}^{\infty} d\omega e^{j\omega t} \int_0^{2\pi} d\alpha \int_0^{|\omega|/v_0} dg\, g\, e^{jg(p_1 \cos\alpha + p_2 \sin\alpha)}$$

$$[\bar{C}_{xx}(\omega, \alpha, \varphi) + \bar{C}_{xx}(\omega, \alpha, -\varphi)](1 - g^2 v_0^2/\omega^2)^{-1/2} \tag{1.62}$$

$$= \frac{1}{(2\pi)^3} \int_{-\infty}^{\infty} d\omega \int\!\!\int_{|\mathbf{g}| \le |\omega|/v_0} d\mathbf{g}\, e^{j(\omega t + \mathbf{g}^T\mathbf{q})} C_{yy}(\omega, \mathbf{g}) = c_{yy}(t, \mathbf{q}),$$

that is, the spectral representation of the covariance function of the wavefield in the plane. While $\bar{C}_{xx}(\omega, \alpha, \varphi)$ is sufficiently smooth in ω, α, φ, so is $C_{yy}(\omega, \mathbf{g})$ in ω and $\mathbf{g}$ for $|\mathbf{g}| < |\omega|/v_0$, and we can apply the results of the last section for the estimation of $C_{yy}(\omega, \mathbf{g})$. That means, if we like to analyze a wavefield consisting of waves propagating with a known velocity by means of a plane sensor array, we can estimate first the distribution of power over frequencies and wavenumbers $\mathbf{g}$ in the plane of sensors. Taking polar coordinates, we obtain the distribution over azimuth α and wavenumber $g = |\omega|\cos\varphi/v_0$ in the direction of propagation and then over φ except for $\varphi = \pi/2$. To determine whether the waves arrive from the upper half-space ($\varphi > 0$)

or lower half-space ($\varphi < 0$) is not possible, which is obvious from the second row of (1.62).

The third example deals with the properties of a line array beampattern that indicates the degree of overlapping caused by smoothing the spectrum with the beampattern. Line arrays are, for example, models for towed arrays in sonar and are used frequently in numerical experiments. The sensors are equispaced on a straight line. We assume $\mathbf{p}_n = (n - 1)(a, 0, 0)^T$, $n = 1, \ldots, N$; that is, the line array on the first coordinate axis with the most left sensor lies at the origin. The response pattern (1.32) of the array steered for $\mathbf{0}$ and then the beampattern becomes

$$|B(\mathbf{k})|^2 = |\sum_{n=1}^{N} w_n e^{-jk_1(n-1)a}|^2, \tag{1.63}$$

which simplifies, for $w_n^2 = V/N^2$, n = 1, . . ., N, to

$$|B(\mathbf{k})|^2 = V|\sin(Nk_1a/2)/[N\sin(k_1a/2)]|^2. \tag{1.64}$$

The beampattern has the following properties. $|B(\mathbf{k})|^2$ is rotationally symmetric around the array axis. It is periodic in the wavenumber of the direction of the array axis with period $2\pi/a$. If $k_1 = (\omega/v)\cos\varphi\sin\beta$, where φ is the elevation and $\beta = \pi/2 - \alpha$ is the bearing measured from broadside ($p_1 = 0$ in Fig. 1.1), and if λ is the wavelength in the direction of propagation, then $|B(\mathbf{k})|^2$ is a periodic function of $\sin\beta$ with period λ/a. This means that the pattern can be interpreted uniquely in the domain $|\beta| \leq \pi/2$ and $\varphi = 0$ if $a/\lambda \leq 1/2$. Otherwise, so called *grating lobes* appear. This is illustrated in Fig. 1.2, where (1.64) in the form of $10\log(|B|^2/V)$ is depicted as a function of bearing β and a/λ for a line array with 15 sensors. Different illustrations of beampatterns and some discussions about possible misinterpretations, while estimating the source directions by means of such arrays, can be found in Monzingo and Miller [3].

We finally discuss the numerical problems encountered in calculating the spectrum estimates of the last section. For example, Dudgeon [21] investigates in detail the digital signal-processing problems in array processing. Phased array beamforming (1.30) usually has, for example, for surveillance purposes, to be computed for a large number of possible slowness vectors $\boldsymbol{\xi}$. The resulting time delays are approximated by an integer multiple of a small delay increment. Then, a hard-wired tapped-delay line multichannel processor is used, for example, in certain sonar systems. The beamsignals for the different slowness vectors are spectrum analysed as in (1.59) in parallel via *fast Fourier transforms* (*FFT*). If only narrow-band signals are expected, all computations can be approximately executed in the frequency domain. To calculate the classical beamformer, there are two possibilities, (1.51) and (1.56), as discussed in the last section. If the spectrum estimation is to be computed for many frequencies and slowness vectors, (1.51) is more efficient by using a four-dimensional FFT followed by suitable interpolations. The dimension of the FFT can be reduced

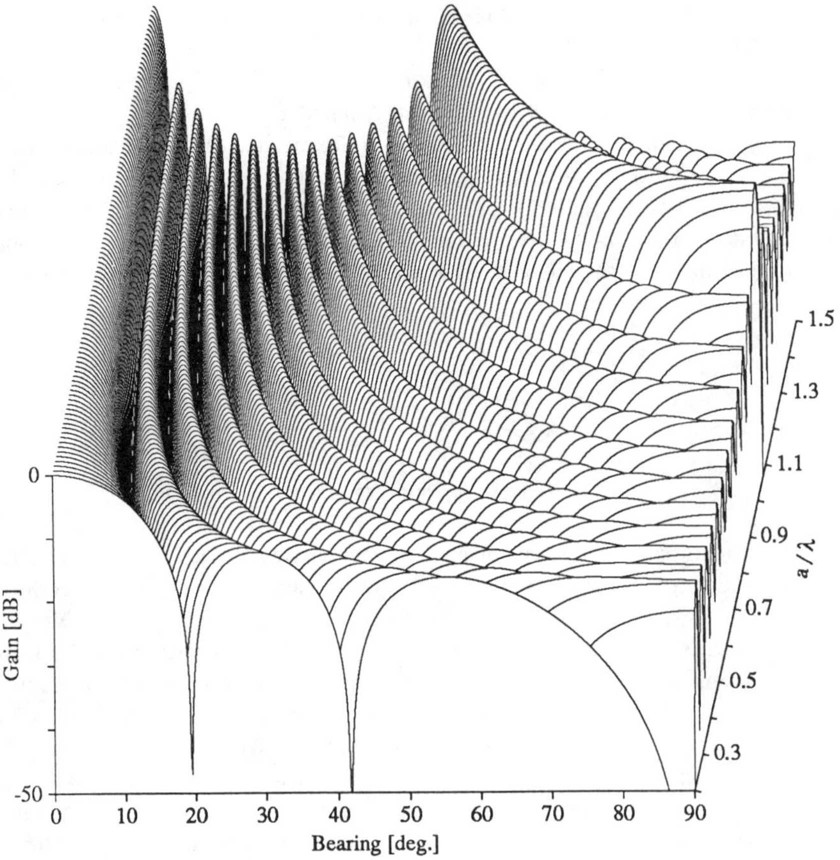

Figure 1.2 Beampattern of a line array with 15 sensors.

depending on the dimension of the array. A line array requires, for example, a two-dimensional FFT. The FFTs are usually computed by special-purpose machines.

Up to now we assumed a stationary model consisting of plane waves. For applications as in passive sonar, the sources may be located not very distant from the array such that the wavefronts are spherical. The time delays then depend on directions and ranges. Though the beamformers we have discussed were designed for direction estimation of plane wave arrivals, they can be used also for range estimation. For example, the classical beamformer (1.56) supplies a rough description of the power spectral distribution over directions and ranges if the time delays in the steering vectors are determined properly. The accuracy of such range estimation is, however, very limited. A different method is to split the array into two groups, to estimate the direction of arrival separately and to apply the parallax. When separating the array in three groups, the curvature of the wavefront can be estimated and then the range. We discuss range estimation in Section 1.3 using parameter estimation methods.

1.3. HIGH-RESOLUTION METHODS

1.3.1 High-Resolution Diagrams

Many ideas and methods are known from the literature that can be used to overcome the poor resolving power of the classical beamformer. We briefly discuss methods producing diagrams that are comparable to those of the classical beamformer, however, that should have sharper peaks at points indicating sources. Any such diagram computed from an estimate of the spectral matrix $\mathbf{C}_x(\omega)$, for example, (1.54), is called a high-resolution diagram or a peak estimate. The parameters ξ_m of the sources in model (1.22) can, for example, be estimated by high-resolution diagrams, but not (in general) the distribution of power over frequencies and wavenumbers, especially the relative power of signals and the noise distribution.

Many of the known peak estimates can be motivated by certain properties of the spectral matrix of a model similar to (1.25) and (1.26). We fix a frequency ω of interest, and accordingly write

$$\mathbf{C}_x = \mathbf{H}\mathbf{C}_S\mathbf{H}^* + \nu\mathbf{I}. \tag{1.65}$$

Here, $\mathbf{C}_s$ is the spectral matrix of the vector $\mathbf{s}(t) = [s_1(t), \ldots, s_M(t)]^T$ of signals; the matrix $\mathbf{C}_s$ is not necessarily diagonal as in (1.25); $\mathbf{H}$ is an $N \times M$ matrix with columns $\mathbf{d}_m = \mathbf{d}(\omega\xi_m)$, $m = 1, \ldots, M$, ν is the spectral power C_{uu} of noise in (1.26), and $\mathbf{I}$ is the identity matrix. We assume $M < N$ and both, $\mathbf{C}_s$ and $\mathbf{H}$ are of rank M. We note that the signals are not fully coherent, and the array is well designed for the wavenumbers of interest. Let $\lambda_1, \ldots, \lambda_N$ be the eigenvalues and $\mathbf{u}_1, \ldots, \mathbf{u}_N$ the corresponding orthogonal eigenvectors of the nonnegative definite, Hermitian matrix $\mathbf{C}_x$, that is, $\mathbf{C}_x\mathbf{u}_n = \lambda_n\mathbf{u}_n$, $n = 1, \ldots, N$, where $\lambda_1 \geq \lambda_2 \geq \ldots \geq \lambda_N$. We have $\lambda_{M+1} = \cdots = \lambda_N = \nu$ and, because of $\lambda_M > \lambda_{M+1}$, the eigenvectors $\mathbf{u}_n$ spanning the *noise space* are orthogonal to all the steering vectors $\mathbf{d}_m$. If the signals are uncorrelated and the steering vectors are orthogonal, it is easy to show that

$$N\lambda_i = \max \mathbf{d}(\omega\xi)^{\dagger}(\mathbf{C}_x - \sum_{l=1}^{i-1} \lambda_l\mathbf{u}_l\mathbf{u}_l^{\dagger})\mathbf{d}(\omega\xi), \qquad i = 1, \ldots, M \tag{1.66}$$

with maximizing parameters ξ_{m_i} and $\mathbf{u}_i = \boldsymbol{d}(\omega\xi_{m_i})$, $i = 1, \ldots, M$. If, however, the steering vectors are not orthogonal, the eigenvectors $\mathbf{u}_1, \ldots, \mathbf{u}_M$ are linear combinations of the steering vectors $\mathbf{d}_1, \ldots, \mathbf{d}_M$, and vice versa. The identity matrix in (1.65) can be replaced by any known positive definite matrix to obtain similar properties. Additional results of interest can be found in Reddi [22], Bienvenu and Kopp [23], and in a nice report by Mermoz [24].

The properties of the model matrix $\mathbf{C}_x$ can be applied to construct high-resolution diagrams if the estimate $\hat{\mathbf{C}}_x$ has approximately the same properties as $\mathbf{C}_x$. If $\hat{\lambda}_1 \geq \cdots \geq \hat{\lambda}_N$ are the eigenvalues and $\hat{\mathbf{u}}_1, \ldots, \hat{\mathbf{u}}_N$ are the corresponding eigenvectors of $\hat{\mathbf{C}}_x$ as in (1.54), and if we assume $\lambda_1 > \cdots \lambda_M > \lambda_{M+1}$ for the eigenvalues of $\mathbf{C}_x$, the following results are true asymptotically for large L; see Brillinger [15]. $\hat{\lambda}_1, \ldots, \hat{\lambda}_M$, $\hat{\mathbf{u}}_1, \ldots, \hat{\mathbf{u}}_M$ are independently and normally distributed random variables and consistent estimates of $\lambda_1, \ldots, \lambda_M, \mathbf{u}_1, \ldots, \mathbf{u}_M$ if possible scalings $\exp(j\varphi)$ of the eigenvectors are ignored. If $\min\{\lambda_i - \lambda_{i+1}: i+1, \ldots, M\}$ is sufficiently separated from 0, a stable estimate $\hat{\mathbf{C}}_x$ results in useful estimates $\hat{\lambda}_1, \ldots, \hat{\lambda}_M, \hat{\mathbf{u}}_1, \ldots, \hat{\mathbf{u}}_M$.

We try to define the high-resolution diagrams by means of the eigensystem of $\hat{\mathbf{C}}_x$ and apply

$$\hat{\mathbf{C}}_x = \sum_{i=1}^{N} \hat{\lambda}_i \hat{\mathbf{u}}_i \hat{\mathbf{u}}_i^\dagger. \tag{1.67}$$

The diagrams are denoted by the symbol f, where we omit the notation of $\mathbf{k} = \omega\boldsymbol{\xi}$. The *classical beamformer* (1.56) with equal weights $w_n = 1$ is written in this way as

$$\tilde{f} = \mathbf{d}^\dagger \hat{\mathbf{C}}_x \mathbf{d} = \sum_{i=1}^{N} \hat{\lambda}_i |\hat{\mathbf{u}}_i^\dagger \mathbf{d}|^2. \tag{1.68}$$

Capon [14] introduced the *adaptive beamformer,*

$$f_C = (\mathbf{d}^\dagger \hat{\mathbf{C}}_x^{-1} \mathbf{d})^{-1} = \left(\sum_{i=1}^{N} \hat{\lambda}_i^{-1} |\hat{\mathbf{u}}_i^\dagger \mathbf{d}|^2 \right)^{-1}, \tag{1.69}$$

where the nonsingularity of $\hat{\mathbf{C}}_x$, that is, $\hat{\lambda}_N > 0$, is assumed. For a more general beamforming, Pisarenko [25] used

$$f_h = h^{-1}\left[\sum_{i=1}^{N} h(\hat{\lambda}_i) |\hat{\mathbf{u}}_i^\dagger \mathbf{d}|^2 \right], \tag{1.70}$$

where $h(x)$ is a strictly monotonic function and positive for $x > 0$ and $h^{-1}(x)$ is its inverse. Equations (1.68) and (1.69) are special cases of (1.70) with $h(x) = x$ and $h(x) = x^{-1}$, respectively.

The property that the steering vectors of the sources are orthogonal to the eigenvectors spanning the noise space was applied by Pisarenko [26], Bienvenu [27], Cantoni and Godara [28]:

$$f_P = |\hat{\mathbf{u}}_N^\dagger \mathbf{d}|^{-2}. \tag{1.71}$$

Schmidt [29] and Ziegenbein [30] assumed to know the number M of sources and used the so called MUSIC (*multiple signal classification*) diagram:

$$f_S = \left(\sum_{i=M+1}^{N} |\hat{\mathbf{u}}_i^\dagger \mathbf{d}|^2 \right)^{-1}. \tag{1.72}$$

Instead of a projection of $\mathbf{d}$ in the noise space, we may equivalently use a projection in the *signal space* spanned by $\mathbf{u}_1, \ldots, \hat{\mathbf{u}}_M$ because of $\sum_{i=1}^{N} |\hat{\mathbf{u}}_i^\dagger \mathbf{d}|^2 = N$. Owsley [31] and Liggett [32] have investigated the beamformer,

$$f_i = |\hat{\mathbf{u}}_i^\dagger \mathbf{d}|^2, \qquad i = 1, \ldots, M \tag{1.73}$$

and maximized f_i to estimate ξ_{m_i} of the source with power rank i. These methods tend to put the estimated steering vectors to be orthogonal as the eigenvectors are. A similar behavior is exhibited by Böhme's [33] method that calculates (1.66) applied to $\hat{\mathbf{C}}_x$ to obtain initial estimates of the ξ_{m_i} that can be iteratively enhanced by methods investigated in Section 1.4. Eigenvectors are not needed in this method. The estimate ξ_{m_1} obviously maximizes the classical beamformer (1.68).

We finally mention two methods originally designed for line arrays. These are Burg's [34] so called *maximum entropy method* (*MEM*), described by

$$f_B = \sum_{i=1}^{N} \hat{\lambda}_i |\hat{u}_{i1}|^2 / |\sum_{k=1}^{N} \hat{\lambda}_k^{-1} \hat{u}_{k1} \hat{\mathbf{u}}_k^\dagger \mathbf{d}|^2 = \mathbf{e}_1^\dagger \hat{\mathbf{C}}_x^{-1} / |\mathbf{e}_1^\dagger \hat{\mathbf{C}}_x^{-1} \mathbf{d}|^2, \tag{1.74}$$

where $\hat{u}_{i1}$ is the first component of $\hat{\mathbf{u}}_i$ and $\mathbf{e}_1 = (1, 0, \ldots, 0)^T$, and the *minimum norm method* or Kumeresan and Tufts' [35] (KTM) described by

$$f_{KT} = |\mathbf{a}^\dagger \mathbf{d}|^{-2}. \tag{1.75}$$

Here, $\mathbf{a}$ is a vector from the noise space spanned by $\hat{\mathbf{u}}_{M+1}, \ldots, \hat{\mathbf{u}}_N$, having unity for the first component and having minimum length in the noise space. The KTM is related to Reddi's method in [22].

Nickel [36] has shown that these various beamformers are interrelated: Indeed, the relations between them remain true for line arrays that are not necessarily equispaced. First, we note that the adaptive beamformer f_C in (1.69) can be presented as a harmonic mean of MEM diagrams f_B^k, $k = 1, \ldots, N$, defined as in (1.74), where only the last k sensors of the line array are used for the calculation of f_B^k; see [37]. This averaging explains the greater fluctuations of f_B compared with f_C as observed in experiments. Next we note that an obvious relation between f_C and the MUSIC diagram f_S exists. If $\hat{\mathbf{C}}_x = \mathbf{C}_x$ is assumed, then f_C approaches f_S for $\nu = 1$ and $\mathbf{C}_S^{-1} \to \mathbf{0}$ in model (1.65). MUSIC may then be interpreted as a Capon-type method, but one which uses a covariance matrix corresponding to infinite SNR. This explains the superior resolution of MUSIC compared to the Capon method. A similar relation can be shown between MEM and KTM in (1.75) for $\hat{\mathbf{C}}_x = \mathbf{C}_x$. If $\mathbf{P} = \sum_{k=M+1}^{N} \mathbf{u}_k \mathbf{u}_k^\dagger$ is the projector in the noise space, then $f_B \to \mathbf{e}_1^\dagger \mathbf{P} \mathbf{e}_1 / |\mathbf{e}_1^\dagger \mathbf{P} \mathbf{d}|^2$ if $\nu = 1$ and $\mathbf{C}_S^{-1} \to \mathbf{0}$. The limit is proportional to f_{KT}. This may explain the better resolution of KTM compared with MEM. Finally we note for $\hat{\mathbf{C}}_x = \mathbf{C}_x$ that f_S can be represented as a weighted harmonic mean of diagrams f_{KT}^k ($k = 1, \ldots, N$), where (1.75) is calculated by using the last k sensors of the array. This averaging possibly explains the greater fluctuations of KTM diagrams compared with MUSIC diagrams as observed in practice.

For equispaced line arrays, other interesting connections between MEM and different linear prediction methods exist. Detailed discussions can be found for example in Haykin's chapter in [1] and in Tufts and Kumaresan [38]. McDonough [39] and Johnson [40] are further references for a detailed investigation of the relations of MEM to other methods. Extensions of MEM to more general array configurations are investigated, for example, in Lang [41]. Generalizations of Capon's adaptive beamformer and certain optimality properties are discussed by Lagunas [42]. There exists a variety of modifications and generalizations of the MUSIC method. Examples are the multifrequency case, (see Wax et al. [43] and Wang and Kaveh [44]), singular signal spectral matrices (see Shan et al. [45]), the ESPRIT method by Roy and others [46] that exploits certain invariances resulting, for example, from two temporally displaced data sets and more general noise models; see Kopp and Bienvenu [47] and Paulraj and Kailath [48]. Very early, Liggett [49] investigated possibilities to remove an unknown noise covariance structure from the estimated spectral matrix. Finally, there exist certain methods to enhance iteratively diagrams tending to give biased estimates of the source parameters ξ_m. We refer to Liggett [32], d'Assumpcao [50], and Böhme [9].

The statistical properties and, in particular, the stability of high-resolution diagrams are of special interest. Specifically, the asymptotic distribution of the diagrams can be derived from property (1.55) of $\hat{\mathbf{C}}_x$ and from the distributional properties of its eigensystem as discussed earlier in this section. The classical beamformer is proportional to a χ^2 variable with $2L$ degrees of freedom, where L is the number of snapshots for calculating $\hat{\mathbf{C}}_x$, see (1.54). Capon and Goodman [51] have described the behavior of f_C in (1.69) by means of a χ^2 variable with $2(L - N + 1)$ degrees of freedom, where N is the number of sensors. Pisarenko [25] has proved an asymptotic normal distribution of (1.70). Baggeroer [52] has found a Cauchy-type distribution for the MEM in (1.74). This means relatively high probabilities for large deviations of f_B from the corresponding limit for $\hat{\mathbf{C}}_x \rightarrow \mathbf{C}_x$. Applying the approximate normal distribution of the eigenvectors $\hat{\mathbf{u}}_i$ of $\hat{\mathbf{C}}_x$, the distribution of f_i in (1.73) is proportional to a noncentral χ^2 with two degrees of freedom. Using the moments of $\hat{u}_i$ as calculated in Brillinger [15], expected values and variances of $f_i, \bar{f}_S = N - f^{-1}_S$, where f_S is the MUSIC diagram (1.72), as well as of different diagrams using the eigenvalues in addition can be found; see Böhme [53]. Related works on MUSIC are due to Kaveh and Barabell [54], Jeffries and Farrier [55], and recently Porat and Friedlander [56], who have investigated the asymptotic relative efficiency of the direction of arrival estimates. The use of the asymptotic moments is limited for predicting, for example, stability of MUSIC in practice. The reason is that we have to calculate the moments by means of the estimated eigenvectors. The stability of these random variables, however, strongly depends on the separation in pairs of the true eigenvalues. Therefore, in [53] and in Knötsch and Böhme [57], results from perturbation theory of Hermitian operators were applied successfully to predict the maximum deviation to the limit for $\hat{\mathbf{C}}_x \rightarrow \mathbf{C}_x$. Additional information on the statistical properties of high-resolution beamformers may be found in Chapters 5–8.

The number M of sources is assumed to be known when, for example, MUSIC is applied. A wrong choice of M can result in a breakdown of the method. Therefore, the estimation of M using the eigenstructure of $\hat{\mathbf{C}}_x$ is of special interest. Liggett [32] first applied the ratio of the arithmetic mean of the m smallest eigenvalues $\hat{\lambda}_i$ to the geometric one for testing whether the m smallest eigenvalues of $\mathbf{C}_x$ are equal; in so doing he modified the corresponding procedure known from real multivariable analysis. In [23], it was shown that this test function is essentially the *generalized likelihood ratio test.* For estimating M, a multihypothesis test is used. However, the calculation of the thresholds is a difficult problem to guarantee a global false-alarm rate. This is one reason for applying information-description criteria such as *Akaike's AIC* (an information criterion) or *Rissanen's MDL* (minimum distance length) *criterion* for estimating of M; see Wax and Kailath [58]. In these criteria, the maximum of the likelihood function is required which is, apart from constant terms, the ratio of means. For example, Rissanen's criterion is, assuming m sources, described by

$$MDL(m) = -L(N-m)\log\left[\left(\prod_{i=m+1}^{N}\hat{\lambda}_i\right)^{1/(N-m)}\Bigg/\left(\sum_{i=m+1}^{N}\hat{\lambda}_i/(N-m)\right)\right]$$
$$+\frac{\log L}{2}m(2N-m), \qquad m = 0, 1, \ldots, N-1. \tag{1.76}$$

Unknown parameters in the model (1.65) with m signals are assumed to be the eigenvalues $\lambda_1, \ldots, \lambda_m$ and $\lambda_{m+1} = \cdots = \lambda_N = \nu$ and the eigenvectors $\mathbf{u}_1, \cdots, \mathbf{u}_m$ of $\mathbf{C}_x$. It can be shown [58] that $\hat{M}$ satisfying $MDL(\hat{M}) = \min\{MDL(0), \ldots, MDL(N-1)\}$ is a consistent estimate of the number M of signals. Recently, Yin and Krishnaiah [59] have found a more flexible class of methods for estimating the number of signals that are strongly consistent; these methods have a faster rate of convergence and can be applied also for coherent signals with a singular spectral matrix $\mathbf{C}_S$. For more information on this topic, see Chapter 9.

1.3.2 Optimum Beamforming

In Section 1.2.3, we investigated phased array beamforming. The beamsignal (1.30) estimates the signal transmitted by a plane wave characterized by a slowness vector ξ_0. If we calculate the finite Fourier transform in the sense of (1.53), we obtain approximately, from (1.30),

$$\hat{S}^{\mathrm{l}}(\omega) = \sum_{n=1}^{N} w_n e^{-j\omega\tau_n} X_n^{\mathrm{l}}(\omega) \tag{1.77}$$

if the observation time is large compared with the delays $\tau_n = \boldsymbol{\xi}_0^{\mathrm{T}}\mathbf{p}_n$. In matrix form, for a fixed ω, we have

$$\hat{S}^{\mathrm{l}} = \mathbf{W}^{\dagger}\mathbf{X}^{\mathrm{l}}, \tag{1.78}$$

where $\mathbf{W} = \mathbf{w}^{\dagger}\mathbf{d}(\mathbf{k}_0)$, $\mathbf{k}_0 = \omega\xi_0$, $\mathbf{d}(\mathbf{k})$ is the steering vector (1.24) and $\mathbf{w}$ is the diagonal matrix of weights w_n. With this approximation, the classical beamformer (1.56) coincides with the spectrum estimate (1.59) of the beamsignal.

$$\hat{C}_{\hat{S}\hat{S}} = \frac{1}{LT}\sum_{l=1}^{L} |\mathbf{W}^{\dagger}\mathbf{X}^{1}|^2 = \mathbf{W}^{\dagger}\hat{\mathbf{C}}_x\mathbf{W}, \tag{1.79}$$

with $\hat{\mathbf{C}}_x$ denoting the estimate of the spectral matrix of the array output as in (1.54).

A linear combination of the array output in the frequency domain such as (1.78), that is, a filter operation for estimating a signal is called beamforming. If the vector $\mathbf{W} = \mathbf{W}_0$ is chosen so that a certain criterion is optimized, (1.78) is called optimum beamforming. A power estimate as (1.79) calculated with the optimum vector $\mathbf{W}_0$ is then an optimum beamformer. If the criterion only depends on a priori knowledge about the statistics of the array output, $\mathbf{W}_0$ can be calculated beforehand. For example, if only noise is received from a known direction, $\mathbf{W}$ can be determined so that the beampattern $|B(\mathbf{k})|^2 = |\mathbf{W}^{\dagger}\mathbf{d}(\mathbf{k})|^2$ has a null in this direction. In practice, the statistics of the array output are only partly known, and we have to estimate them from the received data. A beamforming method that calculates (1.78) and that updates the vector $\mathbf{W}$, in accordance with some algorithm and depending on data $\mathbf{X}^{l}$, is called *adaptive beamforming*. For example, the influence of certain misadjustments of the sensor group and of stationary noise can be reduced, or the vector $\mathbf{W}$ has to be adapted to a slowly changing noise structure. In radar applications, the number of sensors N can be very large, and a lot of hardware would be required to apply adaptive beamforming to the sensor output directly. Instead, for a fixed set of slowness vectors or of subgroups of sensors, beamsignals via (1.30) are calculated, followed by adaptive beamforming applied to the beamsignals. This is sometimes called *postbeamforming adaption* or *adaptive beam weighting*.

There has been much activity in the development of optimum beamformers in different fields of application since the work of Bryn [60] and Widrow [61]. The reader may refer to Monzingo and Miller [3] and Hudson [4] as introductory textbooks, to Owsley's chapter in [1], Gabriel [62], and special issues on adaptive arrays, for example in the *IEE Proceedings,* **130,** part F (1983), *IEEE Transactions on Antennas and Propagation,* AP-24 (1976) and **AP-34** (1986) for further reading. In the following, we only discuss some basic principles and some relationships with high-resolution diagrams and parameter estimation methods to be discussed in the next section.

We start with a simple signal-plus-noise model of the array output in the frequency domain,

$$\mathbf{X} = S\mathbf{d} + \mathbf{U}, \tag{1.80}$$

where S and $\mathbf{U}$ are uncorrelated random variables and $\mathbf{d}$ is the known steering vector of the waves carrying the signal, S is the complex amplitude of the signal and $\mathbf{U}$ is the array output for noise alone. In the case of post beamforming adaption, we write, for example, $\mathbf{d} = \mathbf{v} = (1, 0, \ldots, 0)^T$ if the first beam is steered to the source of interest and the other beams receive only noise and **U** as the postbeamformed noise vector. If we like to estimate the unknown signal in the least *mean-squared error* (*MSE*) sense, the optimum vector $\mathbf{W}_0$ minimizes $E\{|S - \mathbf{W}^\dagger \mathbf{X}|^2\}$ over **W**. The desired solution is easily found by setting the gradient of the criterion to be zero. Because

$$\mathbf{C}_x = C_{SS}\mathbf{dd}^\dagger + \mathbf{C}_u, \tag{1.81}$$

we find

$$\mathbf{W}_0 = C_{SS}\mathbf{C}_x^{-1}\mathbf{d} = \mu\mathbf{C}_u^{-1}\mathbf{d}, \tag{1.82}$$

where we assumed $\mathbf{C}_u$ to be nonsingular. We applied the well known matrix inversion lemma to (1.81) and found the scaling factor to be $\mu = C_{SS}(1 + C_{SS}\boldsymbol{d}^\dagger\mathbf{C}_u^{-1}\boldsymbol{d})^{-1}$. The power of the beamsignal $W_0^\dagger X$, which is also the expected value of the optimum beamformer in the sense of (1.79) and assuming (1.55), is

$$E|\mathbf{W}_0^\dagger X|^2 = \mathbf{W}_0^\dagger\mathbf{C}_x\mathbf{W}_0 = C_{SS}\mathbf{d}^\dagger\mathbf{C}_x^{-1}\mathbf{d} = C_{SS}\mu\mathbf{d}^\dagger\mathbf{C}_u^{-1}\mathbf{d}. \tag{1.83}$$

This means that the beamformer gives an approximately unbiased estimate of the signal spectrum C_{SS} if $C_{SS}\boldsymbol{d}^\dagger\mathbf{C}_u^{-1}\mathbf{d} >> 1$.

We remark that the optimization of different criteria results in optimum weight vectors similar to (1.82) except for the scaling factor μ, see [3]. For example, the output signal to noise ratio $\boldsymbol{W}^\dagger C_{SS}\boldsymbol{dd}^\dagger\boldsymbol{W}/\boldsymbol{W}^\dagger\mathbf{C}_u\boldsymbol{W}$ is maximized by (1.82) with arbitrary μ and maximum value $C_{SS}d^\dagger\mathbf{C}_u^{-1}d$. A minimum distance estimate of S that minimizes $(\mathbf{X} - S\mathbf{d})^\dagger\mathbf{C}_u^{-1}(\mathbf{X} - S\mathbf{d})$, which is a maximum likelihood estimate for Gaussian noise, see Section 1.4.2(c) of this chapter, is given by (1.78) and (1.82), where $\mu = (\mathbf{d}^\dagger\mathbf{C}_u^{-1}\mathbf{d})^{-1}$. We obtain the same solution if we ask for an unbiased estimate of S that minimizes the output noise power $\mathbf{W}^\dagger\mathbf{C}_u\mathbf{W}$. That means, W_0 is the solution of the so called minimum variance distortionless response (MVDR) beamforming problem,

$$\min_{w} W^\dagger\mathbf{C}_u W \text{ subject to } \mathbf{d}^\dagger\mathbf{W} = 1. \tag{1.84}$$

Because of (1.81) and the matrix inversion lemma, we can write

$$\mathbf{W}_0 = (\mathbf{d}^\dagger\mathbf{C}_u^{-1}\mathbf{d})^{-1}\mathbf{C}_u^{-1}\mathbf{d} = (\mathbf{d}^\dagger\mathbf{C}_x^{-1}\mathbf{d})^{-1}\mathbf{C}_x^{-1}\mathbf{d}. \tag{1.85}$$

Thus, W_0 is also the solution of the problem (1.84) if $\mathbf{C}_u$ is replaced by $\mathbf{C}_x$. The power of the beamsignal in the sense of (1.83) is now

$$\mathbf{W}_0^\dagger\mathbf{C}_x\mathbf{W}_0 = (\mathbf{d}^\dagger\boldsymbol{C}_x^{-1}\mathbf{d})^{-1} = C_{SS} + (\mathbf{d}^\dagger\mathbf{C}_u^{-1}\mathbf{d})^{-1}. \tag{1.86}$$

Capon's [14] adaptive beamformer (1.69) then estimates this value. A generalization is the multiple constraint optimum beamforming problem,

$$\min_{\mathbf{W}} \mathbf{W}^\dagger\mathbf{C}_x\mathbf{W} \text{ subject to } \mathbf{D}^\dagger\mathbf{W} = \mathbf{f}, \tag{1.87}$$

where $\mathbf{D}$ is a known, nonsingular matrix and $\mathbf{f}$ is a known vector. For example, we may use additional constraints of the derivative of the steering vector of the signal with respect to directions for introducing certain robustness properties in the case of signal plus interference; see Frost [63] and Owsley's chapter in [1]. The solution is straightforward to obtain by using the method of Lagrange multipliers; we thus have

$$\mathbf{W}_0 = \mathbf{C}_x^{-1}\mathbf{D}(\mathbf{D}^\dagger \mathbf{C}_x^{-1}\mathbf{D})^{-1}\mathbf{f}. \tag{1.88}$$

The spectral matrix (1.81) of the array output is not known in most applications and has to be estimated from the successive array outputs $\mathbf{X}^1$. We can use the mean of periodograms $\hat{\mathbf{C}}_x$ in (1.54) and then solve the system of linear equations $\hat{\mathbf{C}}_x\mathbf{W} = \mathbf{d}$ to obtain the weight vector, except for scaling. The estimate of (1.85) becomes $\hat{\mathbf{W}}_0 = (\mathbf{d}^\dagger \mathbf{W})^{-1}\mathbf{W}$. This is a good estimate; however, the calculation is laborious if $\hat{\mathbf{C}}_x$ has to be updated with new data. The implicit sample matrix inversion of this method can be avoided by the well known recursive update of $\hat{\mathbf{C}}_x^{-1}$, which can be easily derived using the matrix inversion lemma. Stochastic approximation of the optimum weight vector for minimizing, for example, $E\{|S - \mathbf{W}^\dagger\mathbf{X}|^2\}$ has been used for a long time; see [61]. Good convergence properties are obtained by using Kalman filters for updating the weight vector; see Krücker [64].

Efficient algorithms with good numerical properties have been investigated for some years. These algorithms avoid the implicit update of $\hat{\mathbf{C}}_x$ or of its inverse. McWhirter and Shepherd [65], for example, have considered the problem as a linear regression if the signal is known sufficiently accurately in a post beamforming situation. In particular, they have used a QR-decomposition of the data matrix of successive array outputs to obtain a *systolic array* that updates the QR-factors and calculates the minimum sum of squares without determining the weight vector. Schreiber [66] has proposed to solve the MVDR problem by updating the Cholesky factors of $\hat{\mathbf{C}}_x$ and has investigated numerically stable and efficient update algorithms for estimating (1.85). The dominant costs of these algorithms grow only linearly with the number N of sensors, as do the stochastic approximation methods. Straightforward use of matrix inverse or triangular factorization incurs a quadratic cost. Developing Schreiber's ideas further, Yang and Böhme [67] have presented a linear systolic array consisting of CORDIC processors that update the Cholesky factor $\mathbf{L}$ of $\hat{\mathbf{C}}_x = \mathbf{L}\mathbf{L}^\dagger$, the solutions, $\mathbf{V}$ and $\mathbf{W}$, of $\mathbf{L}\mathbf{V} = \mathbf{d}$ and $\mathbf{L}^\dagger\mathbf{W} = \mathbf{V}$, respectively, and calculate directly the beamsignal $\mathbf{W}^\dagger\mathbf{X}/\mathbf{W}^\dagger\mathbf{d}$. A related solution has been developed in [68]. Systolic arrays that solve the multiple constraints problem (1.87) have been recently proposed by McWhirter and Shepherd [69] and by Yang and Böhme [70] who generalized the methods from [65] and [67], respectively.

The signal direction is unknown in certain applications. This direction is frequently estimated by maximizing an optimum beamformer

$$b(\mathbf{k}) = \mathbf{W}(\mathbf{k})^\dagger\hat{\mathbf{C}}_x\mathbf{W}(\mathbf{k}) \tag{1.89}$$

over the wavenumber vectors $\mathbf{k}$ of interest, where $\mathbf{W}(\mathbf{k})$ is an estimate of (1.82) depending on $\mathbf{d} = \mathbf{d}(\mathbf{k})$. If $\mathbf{W}(\mathbf{k})$ is known and (1.81) is used for $\mathbf{d} = \mathbf{d}(\mathbf{k}_0)$, the expected value of (1.89) is the adapted scan pattern

$$E\{b(\mathbf{k})\} = C_{ss}|\mathbf{W}(\mathbf{k})^{\dagger}\mathbf{d}(\mathbf{k}_0)|^2 + \mathbf{W}(\mathbf{k}^{\dagger})\mathbf{C}_u\mathbf{W}(\mathbf{k}). \tag{1.90}$$

If this pattern has a unique maximum for $\mathbf{k} = \mathbf{k}_0$, we obtain an unbiased estimate of the signal direction. However, this function can have a very different behavior, dependent on the scaling factor $\mu = \mu(\mathbf{k})$ of the optimum weights and on the noise spectral matrix $\mathbf{C}_u$, for example, in the case of a mainbeam interference. This problem has been investigated in Nickel [71] and Cox [20].

If $\mathbf{C}_u$ is known, but the signal S and its wavenumber $\mathbf{k}_0$ are not, we can apply minimum distance estimation to the array output $\mathbf{X}$, which corresponds to maximum-likelihood estimation of S and $\mathbf{k}_0$ for gaussian noise; see Davis and others [72] and Section 1.4.3 of this chapter. Minimizing of $[\mathbf{X} - S\mathbf{d}(\mathbf{k})]^{\dagger}\, \mathbf{C}_u^{-1}[\mathbf{X} - S\mathbf{d}(\mathbf{k})]$ over S and $\mathbf{k}$ results in

$$\hat{S} = \mathbf{W}_0(\hat{\mathbf{k}})^{\dagger}\mathbf{X}, \tag{1.91}$$

where $\mathbf{W}_0(\hat{\mathbf{k}})$ is (1.85) calculated for $\mathbf{d} = \mathbf{d}(\hat{\mathbf{k}})$, and $\hat{\mathbf{k}}$ solves

$$\max_{\mathbf{k}}|\mathbf{d}(\mathbf{k})^{\dagger}\mathbf{C}_u^{-1}\mathbf{X}|^2/\mathbf{d}(\mathbf{k})^{\dagger}\mathbf{C}_u^{-1}\mathbf{d}(\mathbf{k}). \tag{1.92}$$

This operation corresponds to maximizing the beamformer (1.89) with a weight vector different from $\mathbf{W}_0(\mathbf{k})$, as shown by

$$\mathbf{W}(\mathbf{k}) = [\mathbf{d}(\mathbf{k})^{\dagger}\mathbf{C}_u^{-1}\mathbf{d}(\mathbf{k})]^{-1/2}\mathbf{C}_u^{-1}\mathbf{d}(\mathbf{k}). \tag{1.93}$$

We now assume that noise consists of interfering signals and of sensor noise with spectral matrix,

$$\mathbf{C}_u = \mathbf{H}\mathbf{C}_i\mathbf{H}^{\dagger} + \nu\mathbf{I}, \tag{1.94}$$

in addition to (1.81). This model corresponds to (1.65) with the difference that only one discrete plane wave arrival is interpreted as a signal and the others as noise and that signal and noise are uncorrelated. The columns of $\mathbf{H}$ are the known steering vectors $\mathbf{d}_1, \ldots, \mathbf{d}_M$ of the interference signals. We assume $\mathbf{H}$ and $\mathbf{C}_i$ to have a rank M less than the number N of sensors. The inverse of (1.94) is found by the matrix inversion lemma, given $\nu > 0$,

$$\mathbf{C}_u^{-1} = \nu^{-1}[\mathbf{I} - \mathbf{H}(\mathbf{H}^{\dagger}\mathbf{H} + \nu\mathbf{C}_i^{-1})^{-1}\mathbf{H}^{\dagger}]. \tag{1.95}$$

Interferences are assumed to be strong in comparison with sensor noise, that is, $\nu\mathbf{C}_i^{-1} \approx \mathbf{0}$. Thus,

$$\nu\mathbf{C}_u^{-1} \approx \mathbf{I} - \mathbf{H}(\mathbf{H}^{\dagger}\mathbf{H})^{-1}\mathbf{H}^{\dagger} = \mathbf{P} \tag{1.96}$$

is the matrix that projects a vector in the space orthogonal to the space spanned by the steering vectors of the interference signals, the *interference space*. This projection matrix can also be expressed in terms of the eigenvectors of $\mathbf{C}_u$ as in the previous section. If $\mu_1 \geq \mu_2 \geq \cdots \mu_m > \mu_{M+1} = \cdots = \mu_N = \nu$ are the eigenvalues and $\mathbf{v}_i$ the corresponding orthonormal eigenvectors of $\mathbf{C}_u$, then

$$\mathbf{P} = \sum_{i=M+1}^{N} \mathbf{v}_i\mathbf{v}_i^\dagger = \mathbf{I} - \sum_{i=1}^{M} \mathbf{v}_i\mathbf{v}_i^\dagger. \tag{1.97}$$

If two of the interference steering vectors are approximately parallel, the projector (1.96) has to be stabilized by substituting $(\mathbf{H}^\dagger\mathbf{H} + \varepsilon\mathbf{I})^{-1}$ for $(\mathbf{H}^\dagger\mathbf{H})^{-1}$ with $\varepsilon > 0$. This means that an interference-to-noise ratio equal to ε^{-1} is introduced.

Beamforming, for example, with weights (1.85) and with (1.96), is

$$\mathbf{W}^\dagger\mathbf{X} = (\mathbf{d}^\dagger\mathbf{P}\mathbf{d})^{-1}\mathbf{d}^\dagger\mathbf{P}\mathbf{X} = |\mathbf{P}\mathbf{d}|^{-2}\mathbf{d}^\dagger\mathbf{P}\mathbf{X} \tag{1.98}$$

because of $\mathbf{P} = \mathbf{P}^\dagger = \mathbf{P}\mathbf{P}^\dagger$. This means that we first remove the possible interferences in $\mathbf{X}$ by projection in the space orthogonal to the interference space and then apply beamforming with the weighted steering vector $|\mathbf{P}\mathbf{d}|^{-2}\mathbf{d}$. This is sometimes called *deterministic nulling* because $\mathbf{H}$ is assumed to be known. Equivalently, beamforming is applied to $\mathbf{X}$ with the weighted and projected steering vector $|\mathbf{P}\mathbf{d}|^{-2}(\mathbf{P}\mathbf{d})$. Direction finding via (1.92) can be interpreted correspondingly. Because (1.92) is equivalent to minimizing

$$\mathbf{X}^\dagger(\mathbf{P} - \mathbf{P}\mathbf{d}(\mathbf{d}^\dagger\mathbf{P}\mathbf{d})^{-1}\mathbf{d}^\dagger\mathbf{P})\mathbf{X} = \mathbf{X}^\dagger(\mathbf{I} - \tilde{\mathbf{H}}(\tilde{\mathbf{H}}^\dagger\tilde{\mathbf{H}})^{-1}\tilde{\mathbf{H}}^\dagger)\mathbf{X} \tag{1.99}$$

over $\mathbf{k}$, where $\tilde{\mathbf{H}} = (\mathbf{d}, \mathbf{H})$, the estimate $\hat{\mathbf{k}}$ of the signal wavenumbers minimizes the length of the projection of $\mathbf{X}$ into the space orthogonal to the signal and to the interference steering vectors. Efficient algorithms are available for minimizing the right side of (1.99), which will be discussed in connection with parameter estimation in Section 1.4.3.

If the direction of the strong interferences in model (1.94) slowly changes with time, we have to update the projection matrix $\mathbf{P}$ to get updates of beamforming weight vectors as in (1.98). In the absence of a signal, we could update the right side of (1.97) by the eigenvectors of the M largest eigenvalues of the updated $\hat{\mathbf{C}}_x$. This is very costly. Efficient methods for the direct updating of the eigenvectors $\mathbf{v}_i$ with good numerical properties are also discussed in Schreiber [66]. If an upper bound M' of the number M of interferences is known, additional savings are possible; see Karasalo [73]. We remark that these methods can be applied also for updating high-resolution diagrams as discussed in Section 1.3.1 if we interpret the interferences again as signals; see, for example, Pisarenko's nonlinear method (1.70) and MUSIC (1.72).

Let us now assume that the signal and the interferences are not necessarily uncorrelated such that the model spectral matrix of the array output is

$$\mathbf{C}_x = \tilde{\mathbf{H}}\begin{pmatrix} C_{ss} & \mathbf{c}^\dagger \\ \mathbf{c} & \mathbf{C}_i \end{pmatrix}\tilde{\mathbf{H}}^\dagger + \nu\mathbf{I}, \tag{1.100}$$

where $\tilde{\mathbf{H}} = (\mathbf{d}, \mathbf{H})$ as in (1.99) and $\mathbf{c}$ is the vector of cross-spectra between the signal and the interferences. If the correlation is ignored in designing the beamforming methods $\hat{S} = \mathbf{W}^{\dagger}\mathbf{X}$, signal cancellation may take place as well reported in the literature, see [20], [62], and [74]. Bresler and others [75] recently investigated the design of optimum beamforming for data from an equispaced line array, taking into account a model similar to (1.100). The authors assumed the directions of the signal and of the interferences as well as all spectral parameters in (1.100) to be unknown. They calculated, among other things, the optimum weight vectors for the MSE criterion and the MVDR criterion (1.84), where $\mathbf{C}_u$ is replaced by $\mathbf{C}_x$. These weights are very different from those in the case of uncorrelated signal and interference. The direct estimation of the directions is avoided by a special parametrization of the noise space that is orthogonal to the steering vectors of the signal and the interferences. *Maximum-likelihood estimates (MLE)* of these parameters such as in Bresler and Macovski [76] and MLE of the spectral parameters given $\hat{\mathbf{C}}_x$, which is discussed in the next section, allow the explicit determination of the optimum weight vectors. Signal cancellations were not observed with these weight vectors.

The signal cancellation, for example, indicates that optimum beamforming methods are not robust in general against slight changes of the underlying model of the spectral matrix. In practice, it is difficult to choose a good model and a suitable beamforming method which have to be identified from array data. Clarke [77] recently proposed a multistages algorithm that progressively whitens the different components of the array data by adaptively adjusting the weight vector of a beamformer. It may be a solution to the problem.

1.4 PARAMETRIC METHODS

1.4.1 Data Models

Homogeneous wavefields, as discussed in Section 1.2, consist of uncorrelated plane waves. The classical beamformer (1.56), for example, tries to estimate the distribution of sources in space, the velocities of propagation of the waves and the frequency spectra of the signals. Wavefields consisting of curved waves transmitting correlated signals cannot be suitably treated with such models. Correlated signals received from different directions appear, for example, in underwater acoustics and multipath propagation. One possibility to include these phenomena in theory consists of the development of a suitable model of the array output. The model has to be perfectly specified except for some parameters. The problem is to estimate the parameters of interest given an observed piece of the array output.

Parameter estimation methods in this sense have been investigated for a long time. The estimation of time delays as in (1.2) has been of special interest. A corresponding special issue of the *IEEE Transactions on Acoustics, Speech, and Signal Processing,* ***ASSP-29***, no. 3 (1981), edited by G. C. Carter, can be used as a general reference. By means of time-delay estimation, we can determine, for example, the location parameters of sources generating coherent waves by fitting the parameters to the time delays. In this situation, it may be more suitable to estimate the parameters of interest directly from the data. This was done for range, bearing or velocity estimation in [78] to [83]. The authors usually started with the likelihood ratio detector for a signal in noise with known signal and noise spectra. They formulated maximum-likelihood estimates and predicted the asymptotic behavior of the estimates using the *Cramér-Rao lower bound.* More rigorous treatments in a statistical sense can be found in [84] to [87]. Knowledge about signal spectra, for example, is not used for the design of the estimates described in these papers. Section 1.4 of this chapter reviews similar methods. A related earlier tutorial paper is [88].

A conventional model is used. Figure 1.3 shows a simple example. The sources $m = 1, \ldots, M$ generate signals that are transmitted by the waves. The wavefield has known properties of propagation except for some parameters. In the example, the location vectors $\mathbf{r}_m$ of the sources (or ranges $|\mathbf{r}_m|$ and bearings β_m measured from broadside) may be unknown. The signals are described by a stationary vector process $\mathbf{s}(t) = [s_1(t), \ldots, S_M(t)]^T$, where $s_m(t)$ is the signal from the mth source if we could record it undisturbed in the origin $\mathbf{0}$. The output of the array of N sensors, $\mathbf{x}(t) = [x_1(t), \ldots, x_N(t)]^T$, is assumed to be a zero-mean stationary vector process consisting of a filtered version of the signals and noise:

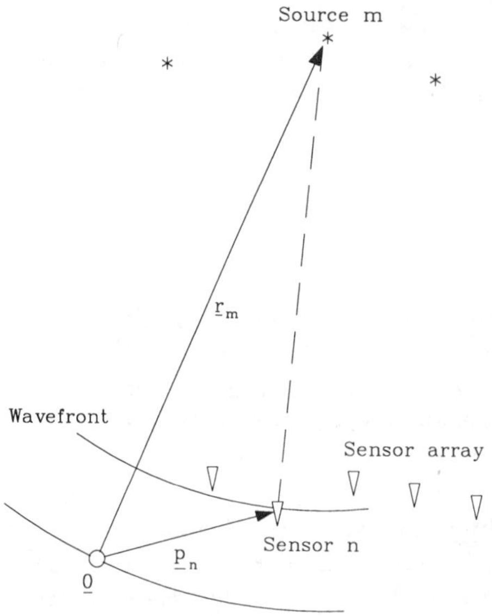

Figure 1.3 Spheric waves in the plane.

$$\mathbf{x}(t) = \int \mathbf{h}(t - t')\mathbf{s}(t')\,dt' + \mathbf{u}(t). \tag{1.101}$$

The impulse response of the filter is an $N \times M$ matrix $\mathbf{h}(t)$ and describes the propagation-reception conditions for the signals and the array. Its frequency response $\mathbf{H}(\omega)$ has columns that can be interpreted as the steering vectors $\mathbf{d}_1, \ldots, \mathbf{d}_m$ of the sources. If we assume in the example of Fig. 1.3 that the distances $|\mathbf{r}_m - \mathbf{p}_n|$ between sources and sensors are large in comparison with the $|\mathbf{p}_n|$, we may ignore possible propagation losses and write the nth component of the steering vector $\mathbf{d}_m$, $\exp(j\omega\tau_{nm})$, with the time delay $\tau_{nm} = (|\mathbf{r}_m| - |\mathbf{r}_m - \mathbf{p}_n|)/v$ of the mth signal in the

nth sensor, where v is the velocity of propagation. Assuming the signals to be independent from noise, the spectral matrix of the array output is

$$\mathbf{C}_x(\omega) = \mathbf{H}(\omega)\mathbf{C}_s\mathbf{H}(\omega)^\dagger + \mathbf{C}_u(\omega), \tag{1.102}$$

where $\mathbf{C}_s(\omega)$ is the $M \times M$ spectral matrix of the signals and $\mathbf{C}_u(\omega)$ is the $N \times N$ spectral matrix of noise. The models (1.25) and (1.65) are special cases of (1.102). A frequency-band limitation $|\omega| < \Omega$ of the signals and of noise is assumed so that the array output can be sampled with a period $\Delta \leq \pi/\Omega$. The stationarity of $\mathbf{x}(t)$ motivates an analysis of the data in the frequency domain. We take the finite Fourier transform of the array output observed in the interval $0 \leq t \leq T = \Delta K$,

$$\mathbf{X}(\omega) = \Delta \sum_i w_T(\Delta i)\, \mathbf{x}(\Delta i)\, e^{-j\omega\Delta i}, \tag{1.103}$$

calculated for discrete frequencies $\omega_i = 2\pi i/T$ satisfying $0 < \omega_i < \Omega$, where the window w_T is chosen as in (1.39). A second possibility, frequently used in practice, is to divide the interval $[0, T]$ in L pieces of length $T' = T/L$ and to calculate the Fourier transforms $\mathbf{X}^l(\omega)$ ($l = 1, \ldots, L$) as in (1.53) for frequencies $\omega_i' = 2\pi i/T' = L\omega_i$.

The asymptotic distributional properties of the Fourier-transformed data for large T and T' can be proved under certain regularity conditions; see, for example, Brillinger [15] and [89]. In particular, we may state that

1. $\mathbf{X}(\omega)$ is complex normally distributed with mean $\mathbf{O}$ and covariance matrix $T\mathbf{C}_x$. (1.104)

2. For frequencies $0 < \omega^1 < \ldots < \omega^P < \Omega$, $\mathbf{X}(\omega^1), \ldots, \mathbf{X}(\omega^P)$ are independent. The ω^l can also be discrete frequencies ω_i in a neighborhood of a fixed frequency ω. (1.105)

3. The conditional distribution of $\mathbf{X}(\omega)$, given $\mathbf{S}(\omega)$, the Fourier transform (1.103) of the signal $\mathbf{s}(t)$, is complex normal with mean $\mathbf{H}(\omega)\mathbf{S}(\omega)$ and covariance matrix $T\mathbf{C}_u(\omega)$. So we can write $\mathbf{X}(\omega) = \mathbf{H}(\omega)\mathbf{S}(\omega) + \mathbf{U}(\omega)$. (1.106)

4. $\mathbf{X}^l(\omega)$ ($l = 1, \ldots, L$) are independently and identically distributed similar to $\mathbf{X}(\omega)$ if T is replaced by T'. (1.107)

The parameters have to be described now. The parameters of the propagation-reception conditions or, briefly, the wave parameters are described by the vector $\boldsymbol{\xi}$, and we write $\mathbf{H}(\omega) = \mathbf{H}(\omega, \boldsymbol{\xi})$. For plane waves as in Section 1.2, $\boldsymbol{\xi}$ summarizes the components of all slowness vectors of the M sources. In the example of Fig. 1.3, $\boldsymbol{\xi}$ contains bearings and ranges and, possibly, some sensor positions as components. In other applications, the velocity of propagation can be frequency-dependent with certain unknown parameters to be collected in $\boldsymbol{\xi}$. We assume that $\boldsymbol{\xi}$ does not depend on ω. For a fixed frequency, the parameters of the signal spectral matrix $\mathbf{C}_S(\omega) = \mathbf{C}_S(\omega, \boldsymbol{\zeta})$ are summarized into a vector $\boldsymbol{\zeta} = \boldsymbol{\zeta}(\omega)$ that depends on ω in general. For example, the elements of the Hermitian matrix $\mathbf{C}_S(\omega)$ may be the parameters themselves. The parameters of the noise spectral matrix $\mathbf{C}_u(\omega) = \mathbf{C}_u(\omega, \boldsymbol{\nu})$ are summarized into the vector $\boldsymbol{\nu} = \boldsymbol{\nu}(\omega)$. For example, as in (1.65), $\mathbf{C}_u(\omega)$ is known to be the identity matrix $\mathbf{I}$ except for a scaling $\boldsymbol{\nu}(\omega) > 0$. $\boldsymbol{\theta} = (\boldsymbol{\zeta}^T, \boldsymbol{\nu}^T, \boldsymbol{\xi}^T)^T$ is the vector of unknown parameters of $\mathbf{C}_x(\omega) = \mathbf{C}_x(\omega, \boldsymbol{\theta})$. If we use property (1.106), the Fourier-transformed signals $\mathbf{S} = \mathbf{S}(\omega)$ can be seen as (random) parameters, and we note $\boldsymbol{\theta} = (\mathbf{S}^T, \boldsymbol{\nu}^T, \boldsymbol{\xi}^T)^T$. In addition, the number M of sources is generally unknown.

1.4.2 Estimation Concepts

The asymptotic distributional properties (1.104) to (1.105) stated in the preceding section allow us to formulate approximate likelihood functions of the data in the frequency domain. Maximizing these functions over the parameters of interest is a well-known method in time series analysis to construct parameter estimates and to predict their asymptotic properties; see Brillinger [89] and [90] and Dzhaparidze [91]. These estimates are, of course, not maximum-likelihood estimates of the original data in the time domain, $\mathbf{x}(t)$, $0 \le t \le T$, those cannot be constructed because the distribution of these data is assumed to be unknown. Nevertheless, we shall sometimes call these estimates MLE if we compare them with different estimates. In the following, we use certain combinations of properties (1.104) to (1.107) to define likelihood functions and the corresponding estimates. We first assume to know the number M of signals and indicate later how to estimate M. We furthermore state some asymptotic distributional properties of the estimates if possible in this generality. The following sections investigate special cases and provide more details.

(a) Wideband estimates

We first apply (1.104) and (1.107) in connection with a nonsingular matrix (1.102) to define the criterion,

$$L^1(\boldsymbol{\theta}) = \frac{1}{T}\sum_i \{\log \det \mathbf{C}_x(\omega_i, \boldsymbol{\theta}) + \operatorname{tr}[\mathbf{I}_x(\omega_i)\mathbf{C}_x(\omega_i, \boldsymbol{\theta})^{-1}]\}, \tag{1.108}$$

where $\mathbf{I}_x(\omega) = \mathbf{X}(\omega)\mathbf{X}(\omega)^\dagger/T$ is the periodogram (not to confuse with the identity matrix $\mathbf{I}$), det and tr denote the determinant and the trace of a matrix, respectively, and we sum up over all frequencies $0 < \omega_i < \Omega$. $L^1(\boldsymbol{\theta})$ is approximately the log-likelihood function of all data $\mathbf{X}(\omega_i)$ except for a constant and a scale factor. We use superscripts to label different likelihood functions, estimates, and parameter vectors. There exist several motivations to use $L^1(\boldsymbol{\theta})$, see [91]. The parameter vector $\boldsymbol{\theta}^1$ that uniquely maximizes (1.108) and that is an element of the parameter set of interest is the estimate of the true value $\boldsymbol{\theta}^0$ of the model.

A search for $\boldsymbol{\theta}^1$ makes sense only if the model (1.102) can be identified by the parameters; that is, there exists a one-to-one mapping between the parameters and the spectral matrices at least locally. Assuming certain additional regularity conditions, it can be shown that $\boldsymbol{\theta}^1$ asymptotically maximizes a criterion that is (1.108) if the sum over the frequencies is replaced by an integral over the frequency domain of interest; see [89] and [90].

In [89], the asymptotic behavior of $\boldsymbol{\theta}^1$ for large T is described as follows if certain regularity conditions are satisfied. $\sqrt{T}(\boldsymbol{\theta}^1 - \boldsymbol{\theta}^0)$ is asymptotically normally distributed with mean $\mathbf{0}$ and covariance matrix,

$$\mathbf{K}(\boldsymbol{\theta}^0) = \mathbf{U}^{-1}(\mathbf{U} + \mathbf{V})\mathbf{U}^{-1}, \tag{1.109}$$

where $\mathbf{U}$ has the typical element,

$$U_{ik} = \int_0^{\Omega} \mathrm{tr}\left[\frac{\partial \mathbf{C}_x}{\partial \theta_i} \mathbf{C}_x^{-1} \frac{\partial \mathbf{C}_x}{\partial \theta_k} \mathbf{C}_x^{-1}\right] \frac{d\omega}{2\pi}, \tag{1.110}$$

and we omitted the notation of the arguments ω and $\boldsymbol{\theta}^0$. The typical element V_{ik} of $\mathbf{V}$ is a similar integral over the cumulant spectra of the fourth order that is given explicitly in [89]. $\mathbf{V}$ is the zero matrix for a Gaussian process $\mathbf{x}(t)$, and $K = \mathbf{U}^{-1}$ is what we obtain from the Cramér-Rao lower bound for Gaussian data. The application of this result is difficult because the identifiability and the regularity conditions have to be verified. Therefore, the asymptotic behavior has to be derived directly in many special cases.

(b) Narrow-band estimates

In certain applications, such as passive sonar, we have to analyze the data only for one or a small number of frequencies if the data can be observed over a long time T. In particular, if the window length T' of the Fourier transform must be smaller than T, property (1.107) in connection with (1.105) can be applied. The criterion is

$$L^2(\boldsymbol{\theta}) = -\frac{1}{P} \sum_{i=1}^{P} \{\log \det \mathbf{C}_x^i + \mathrm{tr}[\hat{\mathbf{C}}_x^i (\mathbf{C}_x^i)^{-1}]\}, \tag{1.111}$$

where $\mathbf{C}_x^i = \mathbf{C}_x(\omega^i, \boldsymbol{\theta})$ and $\hat{\mathbf{C}}_x^i = \hat{\mathbf{C}}_x(\omega^i)$ is the estimate (1.54) for the data $\mathbf{X}^l(\omega^i)$ ($l = 1, \ldots, L$). The ω^i ($i = 1, \ldots, P$) are the selected frequencies or, in other applications, discrete frequencies ω_i' in the neighborhood of a special frequency. The parameter vector maximizing (1.111) is the estimate $\boldsymbol{\theta}^2$.

If the model is identifiable and certain regularity conditions are satisfied, we can show asymptotic properties of $\boldsymbol{\theta}^2$ similar to those of $\boldsymbol{\theta}^1$. If $P = 1$, then $\sqrt{L}(\boldsymbol{\theta}^2 - \boldsymbol{\theta}^0)$ is asymptotically normal with zero mean and a covariance matrix as in (1.109), if we remove the integrals in the definition (1.110) of $\mathbf{U}$ and in $\mathbf{V}$.

(c) Signal estimates

Property (1.106) leads to estimate the Fourier-transformed signals $\mathbf{S}(\omega)$ as deterministic signals together with the noise spectral parameters and the wave parameters. Such a signal estimate is one of a random variable and is only used as a vehicle for estimating $\boldsymbol{\nu}$ and $\boldsymbol{\xi}$. The criterion is, for a single frequency ω and data (1.103),

$$\tilde{L}^3(\boldsymbol{\theta}) = -\log \det \mathbf{C}_u - T^{-1}(\mathbf{X} - \mathbf{HS})^{\dagger}\mathbf{C}_u^{-1}(\mathbf{X} - \mathbf{HS}) \tag{1.112}$$

if $\boldsymbol{\theta} = (\mathbf{S}^T, \boldsymbol{\nu}^T, \boldsymbol{\xi}^T)^T$. Let the maximizing parameter be $\boldsymbol{\theta}^3$. Fixing the parameters $\boldsymbol{\nu}$ and $\boldsymbol{\xi}$, the estimate $\tilde{\mathbf{S}}^3$ follows immediately;

$$\tilde{\mathbf{S}}^3 = (\mathbf{H}^\dagger \mathbf{C}_u^{-1}\mathbf{H})^{-1}\mathbf{H}^\dagger \mathbf{C}_u^{-1}\mathbf{X} \tag{1.113}$$

if the expression in brackets exists. Otherwise, a generalized inverse has to be used. $\tilde{\mathbf{S}}^3$ is, of course, the well-known least-squares estimate of $\mathbf{S}$, with the sum of squares being the rightmost term in (1.112). The relative maximum gives us a criterion for estimating $\boldsymbol{\eta}$ that contains the components of $\boldsymbol{\nu}$ and $\boldsymbol{\xi}$,

$$\tilde{Q}^3(\boldsymbol{\eta}) = -\log \det \mathbf{C}_u - T^{-1}\mathbf{X}^\dagger \mathbf{C}_u^{-1}\mathbf{P}_{\mathbf{C}_u}\mathbf{X}. \tag{1.114}$$

Here,

$$\mathbf{P}_{\mathbf{C}_u} = \mathbf{I} - \mathbf{H}(\mathbf{H}^\dagger \mathbf{C}_u^{-1}\mathbf{H})^{-1}\mathbf{H}^\dagger \mathbf{C}_u^{-1} \tag{1.115}$$

is a projection matrix similar to $\mathbf{P} = \mathbf{P}_\mathbf{I}$ in (1.96) that projects a vector in the space orthogonal to the columns of $\mathbf{H}$ by using the inner product $\mathbf{a}^\dagger \mathbf{C}_u^{-1}\mathbf{b}$ of N-vectors $\mathbf{a}$ and $\mathbf{b}$.

To be a little more general, we can combine properties (1.106) and (1.107) for the data $\mathbf{X}^{li} = \mathbf{X}^l(\omega^i)$ ($l = 1, \ldots, L$; $i = 1, \ldots, P$) and selected frequencies as in the previous section. If we use notations as in (1.111), $\mathbf{H}^i = \mathbf{H}(\omega^i, \boldsymbol{\xi})$, and so on, and we call $\mathbf{S}^{li} = \mathbf{S}^l(\omega^i)$ the Fourier transform of the signal in the sense of (1.53) for $\omega = \omega^i$, then the criterion to be maximized is the sum from $l = 1, \ldots, L$ and $i = 1, \ldots, P$ over the right side of equation (1.112) with the indicated replacements. The estimates of the $\mathbf{S}^{li}$ are least-squares estimates as (1.113). The criterion for estimating the noise spectral parameters and the wave parameters follows from (1.114),

$$Q^3(\boldsymbol{\eta}) = -\frac{1}{P}\sum_{i=1}^{P} \{\log \det \mathbf{C}_u^i + \mathrm{tr}[\mathbf{P}_{\mathbf{C}_u}^i \hat{\mathbf{C}}_x^i (\mathbf{C}_u^i)^{-1}]\} \tag{1.116}$$

where $\hat{\mathbf{C}}_x^i = \hat{\mathbf{C}}_x(\omega^i)$ is again the estimate (1.54). The vector $\boldsymbol{\nu}$ may now contain components $\nu_k = \nu_k(\omega^i)$ that depend on the frequency ω^i. The optimum parameter is called $\boldsymbol{\eta}^3 = (\boldsymbol{\nu}^{3T}, \boldsymbol{\xi}^{3T})^T$. The optimum estimates of $\mathbf{S}^{li}$ are calculated as in (1.113), where $\boldsymbol{\xi}^3$ and $\boldsymbol{\nu}^3$ are used in $\mathbf{H}^i$ and $\mathbf{C}_u^i$, respectively.

The criterion (1.116) has certain similarities to (1.111). The differences, however, are that the spectral matrix $\mathbf{C}_x^i$ of the data in (1.111) is replaced by the noise spectral matrix and the estimate $\hat{\mathbf{C}}_x^i$ by its projection onto the space orthogonal to the columns of $\mathbf{H}^i$. Possible useful information about the signal process cannot be applied in (1.116). On the other hand, we do not assume that the signal spectral matrix has full rank. The advantage of using (1.116) is that the estimation of signals is separated from that of noise and wave parameters and that a numerical optimization of (1.116) can be less complicated than that of (1.111).

The estimation of the signal pieces $\mathbf{S}^{li}$, of course, cannot be stabilized. This would be possible only if we could increase the number N of sensors. The asymptotic behavior of the estimates of $\boldsymbol{\xi}$ and $\boldsymbol{\nu}$, for increasing number L of data pieces and, for example, for $P = 1$, can be derived similarly to the case of (1.111) by developing the gradient of Q^3 instead of the gradient of L^2 around the true parameter vector. We present a result only for a special case below. For fixed L and increasing P, we can only expect a stabilization of the estimate of $\boldsymbol{\xi}$ in general. We finally remark that the signal estimate (1.113) is a direct generalization of the optimum beamformer with weights (1.85) to the multisignal case when $\mathbf{C}_u$ is assumed to be known.

(d) Estimation of the number of signals

The number M of signals was assumed to be known in the preceding sections. In most applications this number is not known. We then have a special detection problem to decide from how many sources we do receive signals superimposed by noise. We cannot do this by maximizing a likelihood function over possible signal numbers m in addition because the maxima of the likelihood function calculated for a fixed m do not decrease with increasing m. The introduction of suitable penalty functions can help. A heuristic application of criteria as those by Akaike [92] and Rissanen [93] is such an approach. A special case was presented in connection with (1.76). The general form of Rissanen's criterion can be written, for example, for an application to (1.111) with $P = 1$, as

$$MDL(m) = -L\,L^2(\boldsymbol{\theta}^2) + \frac{\log L}{2}\kappa(m, N), \tag{1.117}$$

where $\boldsymbol{\theta}^2$ maximizes L^2 and $\kappa(m, N)$ is the number of free parameters that has to be estimated in $\boldsymbol{\theta}$. A number $m \in \{0, \ldots, M_0\}$ that minimizes $MDL(m)$ is an estimate of M.

In certain special cases, the number of signals can be determined by multiple-hypothesis testing about the structure of $\mathbf{C}_x$. Krishnaiah, Lee, and Chang [94] is a useful reference. If, for example, the spectral matrix of noise is known except for a scaling, the test whether there is noise alone or not constitutes the well-known sphericity test. Other examples are discussed in the following two sections.

1.4.3 Known Noise Structure

We first simplify the estimation problem by the use of model (1.102) with the restriction that the noise spectral matrix is known except for a scaling factor:

$$\mathbf{C}_u(\omega) = \nu(\omega)\mathbf{F}(\omega). \tag{1.118}$$

The matrix $\mathbf{F}$ is assumed to be known and positively definite, and $\nu > 0$ is the unknown factor. We can then prewhiten the array output by filtering with the inverse of the Cholesky factor $\mathbf{F}^{1/2}$ of $\mathbf{F}$, that is, $\mathbf{Y} = \mathbf{F}^{-1/2}\mathbf{X}$, and obtain an equivalent estimation problem with the noise spectral matrix $\nu\mathbf{I}$ and the spectral matrix $\bar{\mathbf{H}}\mathbf{C}_s\bar{\mathbf{H}}^\dagger$ of the filtered signals, where $\bar{\mathbf{H}} = \mathbf{F}^{-1/2}\mathbf{H}$. We therefore assume that noise is sensor noise only as in (1.26), and model (1.102) coincides with (1.65).

Let the unknown parameters in $\boldsymbol{\theta}$ be the waveparameters $\boldsymbol{\xi}$ not depending on frequency, the elements of $\mathbf{C}_s(\omega)$ (M^2 real parameters for each frequency ω) and the noise spectral level $\nu(\omega)$. To have a finite number of parameters, we assume that $\mathbf{C}_s(\omega)$ and $\nu(\omega)$ slowly change with ω so that we can describe these spectra by their values for frequencies ω^i $(i = 1, \ldots, P)$ in a good approximation.

(a) Wide-band estimates

The criterion L^1 in (1.108) is to be maximized by considering the simplified model and that $\mathbf{C}_s(\omega)$ and $\nu(\omega)$ change slowly with frequency ω. We assume the same property for $\mathbf{H}(\omega)$. In seismic and in passive sonar applications, these assumptions are sometimes satisfied. Let the frequencies be $\omega^i = \pi(2i - 1)p/T$, $i = 1, \ldots, P$, where $2pP \approx K$ and $T = K\Delta$, for the sake of simplicity. This means that the frequency domain of interest is divided into bands of width $2\pi p/T$ and the ω^i are center frequencies. The criterion (1.108) can then be written as

$$L^1(\boldsymbol{\theta}) = -\frac{p}{T}\sum_{i=1}^{P}\{\log \det\mathbf{C}_x^i + \mathrm{tr}[\hat{\mathbf{C}}_x^i(\mathbf{C}_x^i)^{-1}]\}, \tag{1.119}$$

where

$$\mathbf{C}_x^i = \mathbf{C}_x(\omega^i)$$

and

$$\hat{\mathbf{C}}_x^i = \hat{\mathbf{C}}_x(\omega^i) = \frac{1}{p}\sum_k \mathbf{I}_x(\omega_k) \tag{1.120}$$

is the smoothed periodogram estimate of $\mathbf{C}_x(\omega^i)$. The summation is over the p discrete frequencies $\omega_k = 2\pi k/T$ around ω^i. We conclude from (1.105) and from $\mathbf{C}_x(\omega_k) \approx \mathbf{C}_x(\omega^i)$ that the estimate (1.120) has similar properties as the mean of periodograms estimate (1.54) in (1.55) if we replace ω by ω^i. Parameter estimates maximizing (1.19) have therefore similar properties as those maximizing (1.111) for $\mathbf{C}_u = \nu\mathbf{I}$, which is discussed in the next section.

(b) Narrow-band estimates

We first investigate criterion (1.111) for $P = 1$ and the simplified model assuming $\mathbf{H}$ and $\mathbf{C}_s$ of rank $M < N$,

$$L^2(\boldsymbol{\theta}) = -\log \det(\mathbf{G}_s + \nu\mathbf{I}) - \mathrm{tr}[\hat{\mathbf{C}}_x(\mathbf{G}_s + \nu\mathbf{I})^{-1}], \tag{1.121}$$

where $\mathbf{G}_s = \mathbf{H}\mathbf{C}_s\mathbf{H}^\dagger$. The parameters are the elements of $\mathbf{C}_s$, ν, and ξ.

Signal Subspace Methods. Traditionally, instead of directly estimating these parameters, the spectral matrix $\mathbf{G}_s$ of the filtered signals has been estimated together with ν. The parameters ξ and $\mathbf{C}_s$ have to be fitted to the estimate $\hat{\mathbf{G}}_s$ of $\mathbf{G}_s$ afterwards. If we ignore the special structure of the steering vectors $\mathbf{d}_m$ that are the columns of $\mathbf{H}$, the matrix $\mathbf{G}_s$ has no unique representation because $\mathbf{G}_s = \mathbf{H}\mathbf{T}(\mathbf{T}^\dagger\mathbf{C}_s\mathbf{T})\mathbf{T}^\dagger\mathbf{H}^\dagger$ for any unitary $M \times M$ matrix $\mathbf{T}$. Therefore, we can also use its representation in terms of the eigenvalues $\lambda_1 \geq \cdots \geq \lambda_M$ and orthonormal eigenvectors $\mathbf{u}_i$ $(i = 1, \ldots, M)$ of $\mathbf{C}_x$ that span the signal subspace. As we know from the discussion following (1.65), the eigenvalues have the property $\lambda_M > \lambda_{M+1} = \cdots = \lambda_N = \nu$. Thus,

$$\mathbf{G}_s = \sum_{i=1}^{M} (\lambda_i - \nu)\mathbf{u}_i \cdot \mathbf{u}_i^\dagger, \tag{1.122}$$

and the eigenvalues $\lambda_1, \ldots, \lambda_M, \lambda_{M+1} = \nu$ and the eigenvectors $\mathbf{u}_1, \ldots, \mathbf{u}_M$ can be estimated. Criterion (1.121) is given by

$$L^2 = -\sum_{i=1}^{M} \log \lambda_i - (N - M) \log \nu - \sum_{i=1}^{M} \lambda_i^{-1}\mathbf{u}_i^\dagger\hat{\mathbf{C}}_x\mathbf{u}_i - \nu^{-1}(\mathrm{tr}\hat{\mathbf{C}}_x - \sum_{i=1}^{M} \mathbf{u}_i^\dagger\mathbf{C}_x\mathbf{u}_i). \tag{1.123}$$

Optimum numbers λ_i and orthonormal vectors $\mathbf{u}_i$ are determined by the eigenvalues $\hat{\lambda}_1 \geq \ldots \geq \hat{\lambda}_M$ and eigenvectors $\hat{\mathbf{u}}_1, \ldots, \hat{\mathbf{u}}_M$ of $\hat{\mathbf{C}}_x$, and the estimate of ν becomes

$$\hat{\nu} = (N - M)^{-1}\left[\mathrm{tr}\,\hat{\mathbf{C}}_x - \sum_{i=1}^{M} \hat{\lambda}_i\right]. \tag{1.124}$$

Equation (1.124) can be interpreted as the arithmetic mean of the $N - M$ smallest eigenvalues of $\hat{\mathbf{C}}_x$. The estimate $\hat{\mathbf{G}}_s$ is given by (1.122) if we insert the $\hat{\lambda}_i$, $\hat{\mathbf{u}}_i$, and $\hat{\nu}$. These estimates of $\mathbf{G}_s$ were used by Liggett [32] and discussed in detail by Bienvenu and Kopp [23].

For fitting the wave parameters to the structured estimate of $\mathbf{C}_x$ given by the $\hat{\lambda}_i$, $\hat{\mathbf{u}}_i$, and $\hat{\nu}$, we can use high-resolution diagrams as discussed in Section 1.3.1. The MUSIC diagram (1.72) calculated with the eigenvectors $\hat{\mathbf{u}}_1, \ldots, \hat{\mathbf{u}}_M$ is an example. Having determined the waveparameters $\hat{\boldsymbol{\xi}}$ of the M sources via inspection of the maxima of the diagram, the matrix $\hat{\mathbf{H}}$ is determined. Following Schmidt [29], we solve $\mathbf{G}_s = \mathbf{H}\mathbf{C}_s\mathbf{H}^\dagger$ for $\mathbf{C}_s$ to obtain

$$\hat{\mathbf{C}}_s = (\hat{\mathbf{H}}^\dagger\hat{\mathbf{H}})^{-1}\hat{\mathbf{H}}^\dagger(\hat{\mathbf{C}}_x - \hat{\nu}\mathbf{I})\hat{\mathbf{H}}(\hat{\mathbf{H}}^\dagger\hat{\mathbf{H}})^{-1}. \tag{1.125}$$

This estimate of $\mathbf{C}_s$ is an indefinite matrix in general, because the estimated columns of $\hat{\mathbf{H}}$ need not span exactly the space spanned by $\hat{\mathbf{u}}_1, \ldots, \hat{\mathbf{u}}_M$. However, the estimates $\hat{\nu}$, $\hat{\boldsymbol{\xi}}$, and $\hat{\mathbf{C}}_s$ are consistent under regularity conditions as indicated in Section 1.3.1 and are frequently used in applications.

Liggett [32] assumed uncorrelated signals (i.e., $\mathbf{C}_s$ is diagonal) and fitted the wave parameters in a different way. He started with the vectors $\hat{\mathbf{w}}_i = (\hat{\lambda}_i - \hat{\nu})^{1/2}\hat{\mathbf{u}}_i$, $i = 1, \ldots, M$, and tried to determine iteratively a unitary transformation $\hat{\mathbf{v}}_l = \sum_{i=1}^{M} a_{il}\hat{\mathbf{w}}_i$ such that

$$b = \sum_{i=1}^{M} \max_{\xi_i} |\mathbf{d}(\xi_i)^\dagger\hat{\mathbf{v}}_i|^2/|\hat{\mathbf{v}}_i|^2 \tag{1.126}$$

is maximized. Herein, $\mathbf{d}(\xi_i)$ is the model of the steering vector for the ith source described by a parameter ξ_i, for example, bearing. The maximizing parameters give the estimates $\hat{\xi}_i$ and $\hat{C}_{s_i s_i} = |\hat{v}_i|/N$. The estimates compared favorably with MUSIC in numerical experiments, especially if resolution problems are expected and the number of snapshots is small; see Sandkühler [95]. However, the numerical burden of the method is high such that its application to range and bearing estimation can be prohibitive in practice.

A generalization of the MUSIC method was investigated by Vezzosi [96]. He proposed an algorithm that fits a phase vector $\mathbf{d}(\boldsymbol{\theta})$ with elements $\exp(j\theta_n)$ and parameters θ_n $(n = 1, \ldots, N)$ to the estimated eigenvectors $\hat{\mathbf{u}}_1, \ldots, \hat{\mathbf{u}}_M$ by maximizing $\sum_{i=1}^{M} |\mathbf{d}(\boldsymbol{\theta})^\dagger \hat{\mathbf{u}}_i|^2$ over $\boldsymbol{\theta}$. Instead of maximizing this criterion directly, he investigated an algorithm for solving the system of equations $\sum_{i=1}^{M} \mu_i \hat{\mathbf{u}}_i = \mathbf{d}(\boldsymbol{\theta})$ for the θ_n and the complex numbers μ_i. The wave parameters of a source and, possibly, locations of sensors can be fitted to the estimate $\hat{\boldsymbol{\theta}}$, see Nicolas [97]. Vezzosi also calculated the asymptotic covariance matrix of $\hat{\boldsymbol{\theta}}$.

The number M of signals can be estimated by using (1.117), that is, (1.76) in the present approach, and by methods discussed in [59]. Liggett [32] and later on other authors used successive tests whether $\lambda_{m+1} = \cdots = \lambda_N$ $(m = N-1, \ldots, 0)$ as described in Section 1.3.1. If the matrix $\mathbf{C}_s$ has not the rank M, (e.g., two signals are fully correlated), the method cannot be applied because of $\lambda_M = \lambda_{M+1}$ in this case. We reported in Section 1.3.1 on some work to overcome this problem and also how the method can be adapted to multifrequency data.

Direct Parameter Estimation. The model (1.102) can be written in a different form if we reduce $\mathbf{HC}_s\mathbf{H}^\dagger$ by means of the columns $\mathbf{d}_m$ of $\mathbf{H}$ as follows,

$$\mathbf{C}_x = \nu\mathbf{I} + \sum_m C_{s_m s_m}\mathbf{d}_m\mathbf{d}_m^\dagger + \sum_{i<m} [2\mathrm{Re}C_{s_i s_m}(\mathbf{d}_i\mathbf{d}_m^\dagger + \mathbf{d}_m\mathbf{d}_i^\dagger) + 2\mathrm{Im}C_{s_i s_m}j(\mathbf{d}_i\mathbf{d}_m^\dagger - \mathbf{d}_m\mathbf{d}_i^\dagger)] = \sum_{i=0}^{\mathbf{M}^2} \zeta_i\mathbf{V}_{i'} \tag{1.127}$$

where $\zeta_0 = \nu$, $\mathbf{V}_0 = \mathbf{I}$, and with suitable definitions of the other ζ_i and Hermitian matrices $\mathbf{V}_i$ that are known except for the wave parameters $\boldsymbol{\xi}$. This means that our problem is related to the estimation of variance-covariance components; See Rao and Kleffe [98], Böhme [99], and Burg and others. [100]. These papers do not assume matrices $\mathbf{V}_i$ dependent on nonlinear parameters in general. Similarly, we first assume to know the wave parameters.

The first problem is whether the model is identifiable. In [99], it is shown that representation (1.127) is unique if and only if the rank of $\mathbf{H}$ is M and $M < N$. D'Assumpcao [50], therefore, could not do much better to estimate $\mathbf{C}_s$ by a diagonal matrix as he assumed a large number of sources. We now turn to the maximization of the criterion (1.121). Necessary conditions for a stationarity point follow by taking the gradient of L^2 to be $\mathbf{0}$. Using (1.127), it may be shown that [50]

$$\mathrm{tr}[\mathbf{C}_x^{-1}\mathbf{V}_m] + \mathrm{tr}[\mathbf{C}_x^{-1}\hat{\mathbf{C}}_x\mathbf{C}_x^{-1}\mathbf{V}_m] = 0, \qquad m = 0, \ldots, M^2. \tag{1.128}$$

This system of equations is highly nonlinear. By some manipulations we find another system of equations that can be easily solved, yielding

$$\mathrm{tr}[\mathbf{C}_x^{-1}] = \mathrm{tr}[\mathbf{C}_x^{-1}\hat{\mathbf{C}}_x\mathbf{C}_x^{-1}], \tag{1.129}$$

$$\mathbf{H}^\dagger\mathbf{C}_x^{-1}(\mathbf{H}\mathbf{C}_s\mathbf{H}^\dagger + \nu\mathbf{I})\mathbf{C}_x^{-1}\mathbf{H} = \mathbf{H}^\dagger\mathbf{C}_x^{-1}\hat{\mathbf{C}}_x\mathbf{C}_x^{-1}\mathbf{H}. \tag{1.130}$$

The solution is

$$\tilde{\nu} = (N - M)^{-1}\mathrm{tr}[\mathbf{P}\hat{\mathbf{C}}_x], \tag{1.131}$$

$$\tilde{\mathbf{C}}_s = (\mathbf{H}^\dagger\mathbf{H})^{-1}\mathbf{H}^\dagger(\hat{\mathbf{C}}_x - \tilde{\nu}\mathbf{I})\mathbf{H}(\mathbf{H}^\dagger\mathbf{H})^{-1}, \tag{1.132}$$

where $\mathbf{P}$ is again the projection matrix (1.96). This solution is slightly different from (1.124) and (1.125). Depending on $\hat{\mathbf{C}}_x$, $\tilde{\mathbf{C}}_s$ can be indefinite. The probability of this event tends to go to zero for $\mathbf{C}_s > 0$ and increasing number L of degrees of freedom. The estimates $\tilde{\nu}$ and $\tilde{\mathbf{C}}_s$ have nice properties that are discussed in [99]. They also minimize the sum of squares $\mathrm{tr}[(\hat{\mathbf{C}}_x - \mathbf{C}_x)^2]$ and they are minimum-variance unbiased estimates for Gaussian data, if we ignore the condition $\mathbf{C}_s \geq 0$. For Gaussian data, $\mathbf{W} = L[\tilde{\mathbf{C}}_s + \tilde{\nu}(\mathbf{H}^\dagger\mathbf{H})^{-1}]$ is complex Wishart-distributed with L degrees of freedom and parameter matrix $\mathbf{C}_s + \nu(\mathbf{H}^\dagger\mathbf{H})^{-1}$, $w = \tilde{\nu}L(N - M)/\nu$ is chi-squared distributed with $2L(N - M)$ degrees of freedom, as well as $\mathbf{W}$ and w are independent random variables.

Bresler [101] recently investigated the problem to calculate the MLE of ν and $\mathbf{C}_s$ subject to the constraint that $\hat{\mathbf{C}}_s$ is nonnegative. The MLE satisfies the following conditions. Let $\varphi_1 \geq \cdots \geq \varphi_M > 0$ be the nonzero eigenvalues of $(\mathbf{I} - \mathbf{P})\hat{\mathbf{C}}_x(\mathbf{I} - \mathbf{P})$ and $\mathbf{e}_1, \ldots, \mathbf{e}_M$ be the corresponding eigenvectors. These vectors form an orthonormal basis of the signal space spanned by the columns of $\mathbf{H}$. The matrix $\mathbf{E}$ with columns $\mathbf{e}_1, \ldots, \mathbf{e}_M$ satisfies

$$\mathbf{E}\mathbf{E}^\dagger = (\mathbf{I} - \mathbf{P}) = \mathbf{H}(\mathbf{H}^\dagger\mathbf{H})^{-1}\mathbf{H}^\dagger. \tag{1.133}$$

The conditions are

$$\hat{\nu} = M^{-1}\mathrm{tr}[\mathbf{P}\hat{\mathbf{C}}_x] + M^{-1}\sum_{i=1}^{M} \min\{\varphi_i, \hat{\nu}\}, \tag{1.134}$$

$$\mathbf{H}\hat{\mathbf{C}}_s\mathbf{H} = \mathbf{E}\boldsymbol{\Psi}\mathbf{E}^\dagger, \tag{1.135}$$

where $\boldsymbol{\Psi}$ is diagonal with diagonal elements $\Psi_i = \max\{\varphi_i - \hat{\nu}, 0\}$ $(i = 1, \ldots, M)$. An explicit solution of (1.134) for $\hat{\nu}$ is not possible. However, after at most $M + 1$ calculations of

$$\nu_k = \frac{1}{N - k}\left\{\mathrm{tr}\,(\mathbf{P}\hat{\mathbf{C}}_x) + \sum_{i=k+1}^{M} \varphi_i\right\}, \tag{1.136}$$

we know which ν_k satisfies $\varphi_{k+1} \leq \nu_k \leq \varphi_k$ $(k = M, \ldots, 0)$ and then $\hat{\nu} = \nu_k$. If $\hat{\nu} \leq \varphi_M$, the estimates $\hat{\nu}$ and $\hat{\mathbf{C}}_s$ are the estimates $\tilde{\nu}$ and $\tilde{\mathbf{C}}_s$ in (1.131) and (1.132), respectively.

We now return to the problem that the waveparameters $\boldsymbol{\xi}$ are unknown. A global optimization of $L^2(\boldsymbol{\theta})$ in (1.121) remains laborious. If, however, we use the explicit expressions (1.131) and (1.132) that maximize L^2 relatively for a fixed $\boldsymbol{\xi}$, we can proceed as with (1.114). The relative maximum yields a new criterion

$$\begin{aligned}\tilde{Q}^2(\boldsymbol{\xi}) &= -\log\det(\mathbf{H}\tilde{\mathbf{C}}_s\mathbf{H}^\dagger + \tilde{\nu}\mathbf{I}) \\ &= -\log\det\{(\mathbf{I} - \mathbf{P})\hat{\mathbf{C}}_x(\mathbf{I} - \mathbf{P}) + P\,\mathrm{tr}[\mathbf{P}\hat{\mathbf{C}}_x]/(N - M)\}\end{aligned} \tag{1.137}$$

that has to be maximized over the wave parameters $\boldsymbol{\xi}$ alone. This can be a simpler problem and was investigated by Böhme and Sandkühler [9], [95], [102], and [103]. The optimum parameter $\tilde{\boldsymbol{\xi}}$ is used in (1.131) and (1.132) or in (1.134) and (1.135) for calculating the corresponding spectral parameter estimates.

A generalization to the multifrequency case is easily done as in Section 1.4.2.(c). The criterion (1.111) is first maximized over the spectral parameters $\mathbf{C}_s^i = \mathbf{C}_s(\omega^i)$ and $\nu^i = \nu(\omega^i)$ for each frequency ω^i separately, where $\boldsymbol{\xi}$ is fixed. The resulting mean of relative maxima provides

$$\mathbf{Q}^2(\boldsymbol{\xi}) = -\frac{1}{P}\sum_{i=1}^{P} \log\det\{\mathbf{I} - \mathbf{P}^i\hat{\mathbf{C}}_x^i(\mathbf{I} - \mathbf{P}^i) + \mathbf{P}^i\mathrm{tr}[(\mathbf{P}^i\hat{\mathbf{C}}_x^i)]/N - M)\}, \tag{1.138}$$

which is maximized over $\boldsymbol{\xi}$.

The use of (1.138) or (1.137) means that the estimation of the wave parameters can be separated algorithmically from that of the other parameters. We can interpret the criterion (1.137) so that the generalized variance of the estimated model $\mathbf{H}\tilde{\mathbf{C}}_s\mathbf{H}^\dagger + \tilde{\nu}\mathbf{I}$ is minimized by the optimum wave parameter $\tilde{\boldsymbol{\xi}}$. Another nice property appears if there is only one source. Thus, $\mathbf{I} - \mathbf{P} = \mathbf{d}\mathbf{d}^\dagger/\mathrm{N}$, and $Q^2(\boldsymbol{\xi})$ is a monotonically increasing function of $\mathbf{d}^\dagger\hat{\mathbf{C}}_x\mathbf{d}$. Consequently, the optimum $\tilde{\boldsymbol{\xi}}$ maximizes the classical beamformer output which has been recognized as a well-known result for a long time. The asymptotic behavior of $\tilde{\boldsymbol{\xi}}$ can be principally determined by developing the gradient, for example, of $\tilde{Q}^2$ around the true parameter vector, but this is not easy, and closed expressions are not known to the author. Another way is to argue that, for $\mathbf{C}_s > 0$, the estimates $\tilde{\mathbf{C}}_s$ nd $\tilde{\nu}$ have the same asymptotic behavior as $\hat{\mathbf{C}}_s$ and $\hat{\nu}$, respectively, and to use the result in 1.4.2(b) and the Cramér-Rao lower bound. This was done in [95] and [102]. The estimation of the number M of signals can be done, for example, by use of (1.117) if $L^2(\boldsymbol{\theta}^2)$ is replaced by $\tilde{Q}^2(\tilde{\boldsymbol{\xi}})$.

A numerical procedure for optimizing, for example, of $\tilde{Q}^2(\boldsymbol{\xi})$ requires a global search and local optimizations. Many methods, can be applied; see Bard [104] and Fletcher [105]. We do not discuss these possibilities. If we have a good initial estimate $\boldsymbol{\xi}_0$, for example, from the peaks of a high-resolution diagram, we can locally optimize by a variant of a Newton method. For $\tilde{Q}^2$ we have

$$\boldsymbol{\xi}_{k+1} = \boldsymbol{\xi}_k - \mu_k\mathbf{D}^{-1}\nabla\tilde{Q}^2|_{\boldsymbol{\xi}_k}, \qquad (k = 0, 1, \ldots), \tag{1.139}$$

where $\nabla\tilde{Q}^2$ denotes the gradient, $\mathbf{D}$ is a nonsingular random matrix that has the same limit for $\hat{\mathbf{C}}_x \rightarrow \mathbf{C}_x$ as the Hessian matrix of $\tilde{Q}^2$, and the μ_k are convergence parameters. A few iterations may be sufficient to get a good estimate of $\boldsymbol{\xi}$. This is a consequence of a theorem that can be shown by a technique investigated by Dzhaparidze [91]. This theorem states that only one correction, in the sense of (1.139), of a consistent estimate yields an estimate having the same asymptotic behavior as the optimum $\tilde{\boldsymbol{\xi}}$, if certain regularity conditions are satisfied. In [95] and [103], the elements of the gradient $\nabla\tilde{Q}^2$ are determined,

$$\tilde{Q}_i^2 = -2\mathrm{Re}\{\mathrm{tr}[(\mathbf{H}^\dagger\hat{\mathbf{C}}_x\mathbf{H})^{-1}\mathbf{H}_i\mathbf{P}\hat{\mathbf{C}}_x\mathbf{H}] + \tilde{\nu}^{-1}\mathrm{tr}[\mathbf{P}\mathbf{H}_i(\mathbf{H}^\dagger\mathbf{H})^{-1}\mathbf{H}^\dagger\hat{\mathbf{C}}_x]\}, \tag{1.140}$$

where the subscript i denotes the derivative with respect to the ith component of $\boldsymbol{\xi}$, and a matrix $\mathbf{D}$ with the following typical element is used,

$$D_{ik} = 2\mathrm{Re}\ \mathrm{tr}\{[(\mathbf{H}^\dagger\mathbf{H}\tilde{\mathbf{C}}_s + \tilde{\nu}\mathbf{I})^{-1} - \tilde{\nu}^{-1}\mathbf{I}]\mathbf{H}_i^\dagger\mathbf{P}\mathbf{H}_k\tilde{\mathbf{C}}_s\}. \tag{1.141}$$

These values are much easier to calculate than the elements of the Hessian $\nabla\nabla^{\mathrm{T}}\tilde{Q}^2$. Usually, a proper stabilization of $\mathbf{D}$ is required. The numerical effort for one iteration, however, remains considerable.

(c) Signal estimates

Because we assume $\mathbf{C}_u = \nu\mathbf{I}$, we may rewrite (1.112) as

$$\tilde{L}^3((\mathbf{S}^{\mathrm{T}}, \nu, \boldsymbol{\xi}^{\mathrm{T}})^{\mathrm{T}}) = -N\log\nu - (\nu T)^{-1}|\mathbf{X} - \mathbf{H}\mathbf{S}|^2. \tag{1.142}$$

The signal estimate minimizes (1.142) over $\mathbf{S}$,

$$\tilde{\mathbf{S}}^3 = (\mathbf{H}^\dagger\mathbf{H})^{-1}\mathbf{H}^\dagger\mathbf{X}. \tag{1.143}$$

The relative maximum of (1.142),

$$\tilde{Q}^3[(\nu,\boldsymbol{\xi}^{\mathrm{T}})^{\mathrm{T}}] = -N\log\nu - (\nu T)^{-1}\mathbf{X}^\dagger\mathbf{P}\mathbf{X}, \tag{1.144}$$

can be maximized over ν if $\boldsymbol{\xi}$ is fixed. The maximum provides

$$\tilde{q}^3(\boldsymbol{\xi}) = -\log\bar{\nu} \tag{1.145}$$

that has to be maximized over $\boldsymbol{\xi}$. The generalization to the case of multiple snapshots and multiple frequencies follows from (1.116), as shown by

$$Q^3[(\nu^T, \xi^T)^T] = -\frac{1}{P}\sum_{i=1}^{P}\left\{N\log\nu^i + \frac{1}{\nu^i}\mathrm{tr}[\mathbf{P}^i\hat{\mathbf{C}}^i]\right\}, \tag{1.146}$$

where $\nu^i = \nu(\omega^i)$, and so on. The maximization over all ν^i is solved by

$$\bar{\nu}^i = N^{-1}\mathrm{tr}[\mathbf{P}^i\hat{\mathbf{C}}^i], \tag{1.147}$$

and the relative maximum of Q^3 results in the criterion

$$q^3(\boldsymbol{\xi}) = -\frac{1}{P}\sum_{i=1}^{P}\log\bar{\nu}^i. \tag{1.148}$$

The optimum wave parameter $\bar{\xi}$ maximizes $q^3(\xi)$. The corresponding estimates of ν^i are given by (1.147) if the projection matrix $\mathbf{P}^i$ is (1.96) and $\bar{\xi}$ determines $\mathbf{H}^i$. The signal pieces are estimated via (1.143) using the same $\mathbf{H}^i$.

The maximization of q^3 over ξ means that the estimation of the wave parameters can be separated algorithmically from that of the other parameters as in the previous section. Because ν^i is the estimated noise level for frequency ω^i, the parameter $\bar{\xi}$ maximizing (1.148) also minimizes the geometric mean of the estimated noise levels. If there is only one source and $P = 1$, the optimum $\bar{\xi}$ maximizes the classical beamformer output as in the last section. The optimum $\bar{\xi}$ and the corresponding signal estimates are also the solution of the regression problem with the sum of squares $\sum_{i,1}|\mathbf{X}^{li} - \mathbf{H}^i\mathbf{S}^{li}|^2$. The method has been used for a long time in array processing; see Hinich and Shaman [84]; Hinich [106] in connection with mode propagation; Nickel [107], Böhme [33], and Wax [108].

The optimization of q^3 can be solved completely for $P = 1$, if we ignore the structure of the full rank matrix $\mathbf{H}$ as we did previously. We then search for a projector $\mathbf{P}$ in an $(N - M)$-dimensional space that minimizes $\mathrm{tr}[\mathbf{P}\hat{\mathbf{C}}_x]$. The solution is given by the eigenvectors $\hat{\mathbf{u}}_{M+1}, \ldots, \hat{\mathbf{u}}_N$ of the $N - M$ smallest eigenvalues of $\hat{\mathbf{C}}_x$. This motivates once more the use of high-resolution diagrams such as MUSIC for estimating the wave parameters.

Supposing again a structure of $\mathbf{H}$, numerical procedures have to be applied for optimization performing a global search and local optimizations. Before, we discuss the asymptotic behavior of $\bar{\xi}$. Results can be found, for example, in [84] for regularly spaced sensor arrays. We follow [9] and assume $P = 1$, properties (1.107), (1.104), and (1.65). Presuming certain regularity conditions, $\sqrt{L}(\bar{\xi} - \xi^0)$ is asymptotically normally distributed with the covariance matrix $\mathbf{M}^{-1}\mathbf{N}\mathbf{M}^{-1}$. The typical elements of $\mathbf{N}$ and $\mathbf{M}$ are

$$N_{in} = \mathrm{tr}[\mathbf{P}_i\mathbf{C}_x\mathbf{P}_n\mathbf{C}_x] \tag{1.149}$$

and

$$M_{in} = 2\mathrm{Re}\ \mathrm{tr}[\mathbf{C}_S\mathbf{H}_n^{\dagger}\mathbf{P}\mathbf{H}_i] \tag{1.150}$$

calculated for the true parameter ξ^0, respectively, where nonsingularity of $\mathbf{M}$ is assumed and $\mathbf{P}_i$ and $\mathbf{H}_i$ denote the derivatives with respect to the ith component of ξ. If we like to use these expressions in practice, we can use the actual estimate of ξ^0, $\hat{\mathbf{C}}_x$ and an estimate of $\mathbf{C}_S$, for example, (1.132), in (1.149), and (1.150) for estimating the covariance matrix.

If we have $P = 1$ and a good initial estimate of ξ_0 of ξ^0, for example, from the maxima of a high-resolution diagram, it seems to be more suitable to optimize locally in the sense of (1.139). Let us take as the gradient, the vector with elements $\mathrm{tr}[\mathbf{P}_i\hat{\mathbf{C}}_x]$ and as the matrix $\mathbf{D}$ the matrix $\mathbf{M}$ both calculated with the actual estimates. Sometimes, we have to stabilize the matrix $\mathbf{M}$. Such a correction of a consistent estimate results in an estimate that has the same asymptotic behavior as the optimum $\tilde{\xi}$. Therefore, a few iterations of a good initial estimate can provide a useful estimate that must not be worse than $\tilde{\xi}$. A generalization of the method to the multifrequency case is obvious and not discussed subsequently. Sandkühler [95] applied these iterations extensively and observed that the numerical effort for range and bearing estimation is smaller than that of Liggetts methods and comparable with that of MUSIC if the maxima are determined numerically. Different methods have been published, for example, a stochastic approximation method that allows applications to radar by Nickel [107] and simplified iterations by Ziskind and Wax [109].

The number M of sources can be estimated by use of the MDL criterion. There are many possibilities to do this. We indicate, for $P = 1$, one of them developed by Wax and Ziskind [110]. They show that (1.76) can be applied with the $N - m$ largest eigenvalues of $\mathbf{P}\hat{\mathbf{C}}_x\mathbf{P}$ instead of $\hat{\lambda}_{m+1}, \ldots, \hat{\lambda}_N$, where $\mathbf{P}$ is calculated with the optimum $\tilde{\xi}$ for m sources. Different possibilities for determining the number of signals and for deciding whether certain signals are absent or not are investigated by Shumway [111] in detail. Analysis of variance methods are consequently applied to the data in the frequency domain. Nickel [107] proposed a multihypotheses test for determining M using a chi-squared approximation of min tr[$\mathbf{P}\hat{\mathbf{C}}_x$].

We now turn to the case of data obtained from a line array with equispaced sensors. The direction of arrival estimation problem, for example, can also be interpreted as the estimation problem for superimposed exponential signals in white noise. Apart from the possibilities to apply prediction techniques as in Tufts and Kumaresan [38], we can parametrize the model suitably and calculate the MLE of the parameters. As an important example, we follow Bresler and Macovski [76] who also discuss related investigations in the literature, for example, Kumaresan and others [112]. The essential property applied to the estimation method is that $\mathbf{H}$ is a *Vandermonde matrix* with elements z_m^{n-1} ($n = 1, \ldots, N$; $m = 1, \ldots, M$) with pairwise different z_m. A polynom

$$a(z) = a_0 z^M + \cdots + a_{M-1} z + a_M, \tag{1.151}$$

is defined with the roots z_m, $a(z_m) = 0$ ($m = 1, \ldots, M$). If $a_0 = 1$, the polynom is similar to the well-known prediction error polynom. If the elements of the steering vectors are pure phase factors, then $|z_m| = 1$ and $a_i = a^*_{M-i}$. We define the $(N - M) \times N$ matrix $\mathbf{A}^\dagger$ with rows $(a_M, \ldots, a_0, 0, \ldots, 0)$, $(0, a_M, \ldots, a_0, 0, \ldots, 0)$, $\ldots$, $(0, \ldots, 0, a_M, \ldots, a_0)$ and require that $\mathbf{A}^\dagger\mathbf{H} = \mathbf{0}$. Therefore, the projection matrix (1.96) has an alternative representation,

$$\mathbf{P} = \mathbf{A}(\mathbf{A}^\dagger\mathbf{A})^{-1}\mathbf{A}. \tag{1.152}$$

With restrictions on the coefficients a_i in (1.151), there is a one-to-one mapping between $(z_1, \ldots, z_M)$ and $(a_0, \ldots, a_M)$. Instead of estimating the z_m, we can estimate the a_i by maximizing the criterion (1.147) for $P = 1$. In [76], a corresponding iterative quadratic algorithm termed IQML has been investigated that includes implementations of different constraints on a_i.

1.4.4 Unknown Noise Structure

The noise spectral matrix $\mathbf{C}_u(\boldsymbol{\omega}) = \mathbf{C}_u[\omega, \boldsymbol{\nu}(\omega)]$ may have several unknown parameters that are collected in the vector $\boldsymbol{\nu}(\omega)$. The estimation problem of the wave parameters and of the spectral parameters of both signals and noise is a more complicated problem. The first complication lies in the identifiability of the model (1.102). If, for example, the elements of $\mathbf{C}_u(\omega)$ are unknown and we do not know anything about how $\mathbf{C}_u(\omega)$ changes with ω, we cannot estimate parameters. As in the first part of Section 1.4.3, we assume that the spectral matrices $\mathbf{C}_S(\omega)$ and $\mathbf{C}_u(\omega)$, and then the spectral parameters, do not change rapidly with ω and that their values at frequencies ω^i $(i = 1, \ldots, P)$ are sufficient.

(a) Wideband estimates

As in 1.4.3(a), we assume $\mathbf{H}(\omega)$ to change slowly with ω. This allows smoothing over discrete frequencies and to apply the criterion (1.119) with the smoothed periodogram estimate (1.120) and the more general noise model. We discuss one important special case in the sequel.

Hamon and Hannan [85] investigated the problem for one signal and an arbitrary diagonal spectral matrix of noise. They assumed the elements of the signal matrix $C_{SS}\mathbf{dd}^{\dagger}$ to be $C_{SS}(\omega)\exp[j\varphi_{nk}(\omega)]$, where $\varphi_{nk}(\omega) = \varphi_{nk}(\omega, \boldsymbol{\xi})$. The vector $\boldsymbol{\xi}$ may contain parameters that characterize a dispersive medium. The criterion (1.119) can be approximated as follows, except for a constant and a scaling factor,

$$Q^H(\boldsymbol{\xi}) = -\frac{1}{P}\sum_{i=1}^{P}\sum_{n<k}\hat{\sigma}^{nk}(\omega^i)\,\hat{\sigma}_{nk}(\omega^i)\cos[\hat{\varphi}_{nk}(\omega^i) - \varphi_{nk}(\omega^i)] \tag{1.153}$$

that has to be maximized over $\boldsymbol{\xi}$. Herein, the typical element of $\hat{\mathbf{C}}_x$ reduces to $\hat{C}_{x_nx_k} = |\hat{C}_{x_nx_k}|exp(j\hat{\varphi}_{nk})$, we write $\hat{\sigma}_{nk} = |\hat{C}_{x_nx_k}|(\hat{C}_{x_nx_n}\hat{C}_{x_kx_k})^{-1/2}$ and denote the elements of the inverse of the matrix $(\hat{\sigma}_{nk})$ by $\hat{\sigma}^{nk}$. As shown in [85], the optimum $\boldsymbol{\xi}$ is asymptotically unbiased and normally distributed with covariance matrix $L^{-1}\mathbf{A}^{-1}$. The typical element of $\mathbf{A}$ can be consistently estimated by

$$\hat{A}_{bm} = -\frac{1}{P}\sum_{i=1}^{P}\hat{\sigma}^{nk}(\omega^i)\hat{\sigma}_{nk}(\omega^i)\partial\varphi_{nk}(\omega^i, \hat{\boldsymbol{\xi}})/\partial\xi_b\;\partial\varphi_{nk}(\omega^i, \hat{\boldsymbol{\xi}})/\partial\xi_m. \tag{1.154}$$

This is a very useful approximation. As an example with plane coherent wave arrivals and an unknown slowness vector $\boldsymbol{\xi}^0$, we have

$$\partial\varphi_{nk}(\omega^i, \hat{\boldsymbol{\xi}})/\partial\xi_m = \omega(p_{nm} - p_{km}), \tag{1.155}$$

where p_{nm} is the mth coordinate of the sensor position $\mathbf{p}_n$. Thus, $\hat{\mathbf{A}}$ does not depend on $\hat{\boldsymbol{\xi}}$. The numerical maximization of (1.154) can be difficult because the phases $\hat{\varphi}_{nk}$ are determined only up to a multiple of 2π. There is a possibility to generalize (1.153) to the multiple signal case, see [85] and [87], but the computations are much more complicated.

(b) Narrow-band estimates

Signal Subspace Methods. Let us assume a representation of $\mathbf{C}_u$ similar to (1.118) with a positive definite matrix $\mathbf{F}$ that depends on the additional parameters $\boldsymbol{\mu}$. If we fix these parameters first, we can prewhiten the data as described at the beginning of Section 1.4.3, or we can simply transform the estimate of the spectral matrix, $\hat{\mathbf{C}}_y = \mathbf{F}^{-1/2}\hat{\mathbf{C}}_x(\mathbf{F}^{-1/2})^{\dagger}$, that is, the mean of periodograms estimate of the transformed model matrix. If we now maximize L^2 in (1.123) applied to $\hat{\mathbf{C}}_y$, the maximum is given up to a constant by

$$Q^2(\boldsymbol{\mu}) = \sum_{i=M+1}^{N} \log \hat{\lambda}_i - (N - M) \log \hat{\nu}, \tag{1.156}$$

where the $\hat{\lambda}_i$ are the $N - M$ smallest eigenvalues of $\hat{\mathbf{C}}_x$ and $\hat{\nu}$ is their mean. The function $Q^2(\boldsymbol{\mu})$ is never positive and equals zero if and only if the $\hat{\lambda}_i$ are equal. Numerical maximization of (1.156) for estimating the parameter $\boldsymbol{\mu}$ is discussed, for example, by Kopp and Bienvenu [113]. The special case, where $\mathbf{F}$ is diagonal with positive diagonal elements that are unknown except for one, is a suitable model if the spacing between the sensors is relatively large. Also in this typical factor analysis problem, we have to optimize (1.156) numerically. Newton iterations were applied by Foka and others [114].

Kopp and Bienvenu [47] discuss some heuristic methods if no simple noise model is available. For example, let ambient noise be as in model (1.13) with a slowly varying spectrum. The array may then be split into several subarrays steered for a given direction. The spectral matrix of the beamsignals has a structure similar to that of the array output, but with the difference that the noise spectral matrix is approximately known except for a scaling factor. Liggett [49] discusses a possibility to remove the noise parts from the spectral matrix $\hat{\mathbf{C}}_x$ that cannot be described by means of plane wave arrivals. If we use an equispaced line array and assume a homogeneous noise wavefield, the noise spectral matrix is a Toeplitz matrix. This special case was investigated in detail by Le Cadre [115]. One of his approaches is to assume an autoregressive noise model in the optimization problem (1.156). He provides some tools that allow an easy application of a gradient method for computing the *AR* coefficients.

Direct Parameter Estimation. A rigorous investigation was presented by Thomson [116] for one special case. He assumed a planar array, one signal transmitted by a dispersive medium, unknown sensor gains and phases and a nonsingular diagonal spectral matrix of noise. In the neighborhood of the frequency ω_0 of interest, the signal and noise spectra and the sensor gains are assumed to be constant. The phases of the elements of the corresponding signal steering vector are approximated by a linear function of frequency, $\varphi_n(\omega) \approx \varphi_n(\omega_0) + (\omega - \omega_0)\tau_n(\omega_0)$, where $\tau_n(\omega_0)$ corresponds to the time delay of the signal in sensor n. Unknown wave parameters are the bearing β and the velocity v at frequency ω_0. The spectral parameters are the spectral noise levels, the signal-to-noise ratios and the phase shifts at the output of the sensors. Stating the likelihood function (1.111), where the sum is taken over P discrete frequencies ω^i in the neighborhood of ω_0, Thomson derives a two-stage procedure for an approximate maximization of (1.111). If $\Lambda(\omega)$ is a diagonal matrix with elements $\exp[j\tau_n(\omega_0)(\omega - \omega_0)]$, he defines the matrix

$$\mathbf{R}(\beta, v) = \frac{1}{P} \sum_{i=1}^{P} \Lambda(\omega^i)^\dagger \mathbf{I}_x(\omega^i)\Lambda(\omega^i), \tag{1.157}$$

where $\mathbf{I}_x(\omega)$ is the periodogram of the array output. The determinant of $\mathbf{R}(\beta, v)$ is minimized over β and v to obtain $\hat{\beta}$ and $\hat{v}$ in the first stage. A factor analysis on $\mathbf{R}(\hat{\beta}, \hat{v})$ for the spectral parameters is carried out in the second stage. The computational advantage of this method is obvious compared with the direct optimization of (1.111) over all parameters. Thomson proved the consistency, the asymptotic normality and calculated the asymptotic covariance matrix of the estimates.

We now use model (1.102) in a more general case, where $\mathbf{C}_u$ is nonsingular and known except for the parameter vector $\boldsymbol{\nu}$. A numerical optimization of (1.111) for $P = 1$ over the parameters in $\boldsymbol{\xi}$, $\mathbf{C}_S$, and $\boldsymbol{\nu}$ simultaneously seems to be too complicated if algorithms should be applied as in Burg and others [100]. A possibility for reducing the number of parameters is to proceed as with (1.137). We first parametrize $\mathbf{C}_u = \nu\mathbf{F}(\boldsymbol{\mu})$ as earlier and maximize L^2 for $P = 1$ over $\mathbf{C}_S$ and ν. If we ignore the constraint $\mathbf{C}_S \geq 0$, there exists an explicit solution that generalizes (1.131) and (1.132) in an obvious way. The relative maximum of L^2 results in the criterion

$$\tilde{Q}^2(\boldsymbol{\xi}, \boldsymbol{\mu}) = -\log\det\{(\mathbf{I} - \mathbf{P_F})\hat{\mathbf{C}}_x(\mathbf{I} - \mathbf{P_F}) + \mathbf{P_F}\mathbf{F}\,\mathrm{tr}[\mathbf{F}^{-1}\mathbf{P_F}\hat{\mathbf{C}}_x]/(N - M)\}, \tag{1.158}$$

where $\mathbf{P_F}$ is defined by (1.115) if we replace $\mathbf{C}_u$ by $\mathbf{F}$. Details and related approaches can be found in Kraus [117]. A further algorithmic separation of the estimation of $\boldsymbol{\xi}$ and $\boldsymbol{\mu}$ seems not to be possible. The numerical optimization of (1.158) remains difficult even for a small number of parameters.

Heuristically, we can introduce iterative two-stage methods. If we first assume to know the parameter $\boldsymbol{\mu}$, that is, $\mathbf{F}$, we can maximize (1.158) over $\boldsymbol{\xi}$ numerically, which is equivalent to maximizing (1.137) after prewhitening. The preliminary estimate of $\boldsymbol{\xi}$ is used as a fixed value while optimizing the criterion over $\boldsymbol{\mu}$ in the second stage. We then repeat the first stage and so on. It is not yet clear under which conditions these iterations converge to the true parameters. Because the computation of (1.158) and its gradient is time consuming in all cases, we can search for different criteria that are easier to optimize. After prewhitening or prebeamforming, we can maximize, as did Forster and Vezzosi [118], the criterion

$$Q^F(\boldsymbol{\theta}) = -\,\mathrm{tr}[\hat{\mathbf{C}}_y^{-1/2}\mathbf{C}_y(\hat{\mathbf{C}}_y^{-1/2})^\dagger - \mathbf{I})^2] \tag{1.159}$$

over $\boldsymbol{\theta}$, where the subscripts y indicate the prewhitened model and its estimate. Nonsingularity of $\hat{\mathbf{C}}_y$ is supposed to hold. The nice property of the optimum parameters is that they have the same asymptotic behavior as the MLE for Gaussian data. Moreover, an algorithmic separation of the estimation of the wave and spectral parameters is possible such that only Newton iterations for the wave parameters have to be executed. A disadvantage of the method is that the estimated spectral matrix needs to be nonsingular. The estimates will be poor if the number of snapshots is not much greater than the number of sensors.

(c) Signal estimates

Let us first return to the maximization of (1.112). If we knew the wave parameters $\boldsymbol{\xi}$, we could apply the result of Mardia and Marshall [119] for estimating the signals $\mathbf{S}$ and the noise spectral parameters $\boldsymbol{\nu}$. They investigated an iterative two-stage procedure using (1.113) and scoring for updating $\boldsymbol{\nu}$. They proved sufficient conditions for the asymptotic efficiency for increasing number of sensors and normal data, especially, if the sensor positions are given by a regular lattice with fixed spacings. However, the wave parameters $\boldsymbol{\xi}$ need also to be estimated, and (1.114) has to be maximized over $\boldsymbol{\xi}$ and $\boldsymbol{\nu}$.

We are now interested in the optimization of (1.116) for $P = 1$ if we again assume $\mathbf{C}_u = \nu\mathbf{F}(\boldsymbol{\mu})$ with a nonsingular matrix $\mathbf{F}$. Fixing the parameter $\boldsymbol{\xi}$ and $\boldsymbol{\mu}$, we can explicitely calculate the signals and ν that maximize (1.116). The relative maximum provides the criterion

$$q^3 = -\log \det \mathbf{F} - N \log \operatorname{tr}[\mathbf{F}^{-1}\mathbf{P}_\mathbf{F}\hat{\mathbf{C}}_x] \tag{1.160}$$

that has to be maximized over $\boldsymbol{\xi}$ and $\boldsymbol{\mu}$. The same criterion is obtained if we first prewhiten the model and data, proceed as in the beginning of Section 1.4.3(c), and transform back the likelihood function in the right way.

The numerical maximization of (1.160) could be tried by an iterative two-stage method. For a given $\hat{\boldsymbol{\mu}}$, the model and data are prewhitened, and an update of $\hat{\boldsymbol{\xi}}$ is computed by a Newton iteration step as described in Section 1.4.3(c). The updated $\hat{\boldsymbol{\xi}}$ is used to calculate an update of $\hat{\boldsymbol{\mu}}$ by using a similar step, and so on. Numerical experiments described in [117] show that this method need not result in useful estimates, especially for $\boldsymbol{\mu}$. Because simultaneous optimization is too expensive, different heuristic methods for estimating $\boldsymbol{\mu}$ have been proposed, [117] and [120]. One possibility is, after a preliminary estimation of $\boldsymbol{\xi}$ via maximizing (1.148), to fit all spectral parameters of the model matrix $\mathbf{C}_x$ in (1.102) to the estimated one, $\hat{\mathbf{C}}_x$, in a least-squares sense and then to use the updated $\hat{\boldsymbol{\mu}}$ for prewhitening and updating $\hat{\boldsymbol{\xi}}$ as earlier. For simplicity, we minimize

$$Q = \operatorname{tr}[(\hat{\mathbf{C}}_x - \mathbf{C}_x)^2] \tag{1.161}$$

over $\mathbf{C}_S$ and $\boldsymbol{\nu}$. If we ignore that the solution has to result in nonnegative matrices, the explicit solution for $\mathbf{C}_S$ is given by (1.96), where $\tilde{\nu}\mathbf{I}$ is replaced by $\mathbf{C}_u$, with $\mathbf{C}_u$ satisfying

$$\operatorname{tr}[\mathbf{C}_{ui}(\mathbf{I} - \mathbf{P})\mathbf{C}_u] = \operatorname{tr}[\mathbf{C}_{ui}(\mathbf{I} - \mathbf{P})\hat{\mathbf{C}}_x]. \tag{1.162}$$

The subscript i indicates the derivative with respect to the ith component ν_i of $\boldsymbol{\nu}$, and $\mathbf{P}$ is the projection matrix (1.132). Using a linear model of $\mathbf{C}_u$,

$$\mathbf{C}_u = \sum_i \nu_i \mathbf{J}_\mathbf{i}, \tag{1.163}$$

where the $\mathbf{J}_i$ are known Hermitian matrices, the system of equations (1.162) becomes linear in $\boldsymbol{\nu}$. Such noise models can be useful in applications as sonar; see Cron and Sherman [121]. This simple estimate of $\boldsymbol{\nu}$ was used in numerical experiments and provided satisfactory results. The numerical effort of the method is essentially determined by the iterations for the wave parameters $\boldsymbol{\xi}$. Other algorithms, as described in [107] or [109], or those in [122] and [123], possibly provide useful estimates of $\boldsymbol{\xi}$ more efficiently.

We finally indicate some important results from the literature that can be useful in special situations. Let us first assume an equispaced line array with a large number N of sensors. Because we are interested essentially in the estimation of the wave parameters, we can estimate them by nonlinear regression instead of maximizing (1.112) when noise is described by a homogeneous wavefield. Hannan's [124] results on time series can be applied if we transform the model and data in the wavenumber domain. In a first step, the sum of squares is defined by the squared residues with weights 1 and is taken over the wavenumber bins. The resulting estimates of $\boldsymbol{\xi}$ and $\mathbf{S}$ (assumed deterministic) are consistent for increasing N under regularity conditions. If we calculate the sum of squares with these estimates and use suitable weights, we obtain directly consistent estimates of the elements of the Toeplitz matrix $\mathbf{C}_u$. Transforming these estimates in the wavenumber domain provides weights for the sum of squares, which results in asymptotic minimum-variance estimates. Brillinger [125] is concerned with the likelihood ratio detection and maximum-likelihood estimation of plane wave arrivals that carry deterministic sinusoids of known frequencies. Both problems are solved by means of the so-called maximum-likelihood statistic. His development clarifies the interrelationships between the classical beamformer, Capon's method and least-squares methods. Cameron and Hannan [87] have investigated transient signals aside from stationary signals. Transient signals are nonstationary random signals and are assumed to have the property that the finite Fourier transform does not change rapidly in the neighborhood of a frequency ω^i of interest. If the steering vector of a source and the noise spectral matrix have the same property, we can smooth the data (1.53) over frequencies and obtain, with (1.106),

$$\bar{\mathbf{X}}(\omega^i) = \frac{1}{p}\sum_k \mathbf{X}(\omega_k) \approx \mathbf{H}(\omega^i)\mathbf{S}(\omega^i) + \bar{\mathbf{U}}(\omega^i), \tag{1.164}$$

where the ω_k are p discrete frequencies in the neighborhood of ω^i. This allows the estimation of $\mathbf{C}_u(\omega^i)$ by

$$\hat{\mathbf{C}}_u(\omega^i) = \frac{1}{(p-1)\mathrm{T}} \sum_k [\mathbf{X}(\omega_k) - \bar{\mathbf{X}}(\omega^i)][\mathbf{X}(\omega_k) - \bar{\mathbf{X}}(\omega^i)]^\dagger. \tag{1.165}$$

If we further assume property (1.105) to be valid for sufficiently spaced ω^i, we can estimate the wave parameters by optimization of (1.148) after prewhitening with $\hat{\mathbf{C}}_u(\omega^i)^{-1/2}$. In [87], limitations of this method and confidence regions for the parameters of one signal have been investigated.

1.4.5 Numerical Experiments

The asymptotic behavior of certain estimates discussed in the previous sections is known and can be used to predict the behavior for large degrees of freedom. In practice, the behavior for moderate degrees of freedom is of interest. Precision, stabilitity, and also the common behavior of the estimates have to be investigated by means of numerical experiments. The purpose of this section is to illustrate the behavior of the parameter estimates derived from Capon's adaptive beamformer (1.69), Schmidt's MUSIC (1.72), Kumaresan and Tufts' KTM (1.75), as well as the behavior of the estimates that maximize (1.148) numerically (called AML for approximate maximum likelihood). Ranges, and bearings, as well as spectral powers of signals and noise are to be determined.

Model (1.65) is used in a Monte Carlo simulation for a line array with $N = 15$ sensors spaced by half a wavelength. Three sources, located approximately broadside, generate uncorrelated signals, that is $\mathbf{C}_S = \text{diag}(C_{S_1S_1}, C_{S_2S_2}, C_{S_3S_3})$. The unknown wave parameters are bearings β_m and ranges $\rho_m = |\mathbf{r}_m|$ as depicted in Fig. 1.3. Two equally strong sources, separated by only 0.4 of the beamwidth, are located at $\beta_1 = -3$ and $\beta_2 = 0$ degrees, both at a distance of 20 antenna lengths and having a $\text{SNR}_1 = 10\log(NC_{S_1S_1}/\nu) = 3$ dB. The third source is well separated from the others at $\beta_3 = 10$ degrees and a distance of 35 antenna lengths and has $\text{SNR}_3 = -3$ dB. The estimates $\hat{\mathbf{C}}_x$ are simulated by complex Wishart-distributed matrices with L degrees of freedom and parameter matrix $\mathbf{C}_x/L$.

In the first experiment, we calculated $\hat{\mathbf{C}}_x$ with $L = 2000$, which is, of course, the model matrix with small distortions to demonstrate the limitations of the resolution power of the methods. Figures 1.4(a) to (c) show the corresponding diagrams of the adaptive beamformer, MUSIC and KTM over bearings and ranges, where a linear scale of the vertical axis is chosen. Figure 1.4(d) tries a similar representation for AML. For that, we introduced a fourth, hypothetical source with location parameters β and ρ and depicted the graph of the mapping $(\beta,\rho) \rightarrow -q^3(\beta_1, \rho_1, \ldots, \beta_3, \rho_3, \beta, \rho)^{-1}$, where the parameters of the other three source are as in the previous figures. Obviously, the adaptive beamformer allows neither a separation of sources nor range estimates. The other diagrams show the peaks at the right positions, however, the maximum in range direction of source 3 is very flat with the MUSIC method.

In the second experiment, we used the same model and 2048 independent matrices $\hat{\mathbf{C}}_x$ with $L = 20$ degrees of freedom and computed, for each $\hat{\mathbf{C}}_x$, the estimates of all parameters. Because the adaptive beamformer cannot resolve the sources, we computed only MUSIC, KTM, and AML. The numerical procedure was as follows. We first determined initial estimates of bearings and ranges by means of the simple method of signal extraction in [33] that calculates the recursions in (1.66) applied to $\hat{\mathbf{C}}_x$ for $M = 3$. The raster search was coarse, for example, half a beamwidth in bearing direction. This method separates the sources but estimates with a bias as indicated in Section 1.3.1. The initial estimates were corrected by numerical hill climbing in the MUSIC and the KTM diagrams. If the estimates for sources 1 and 2 converged to the same value, the sources were merged. Spectral powers were deter-

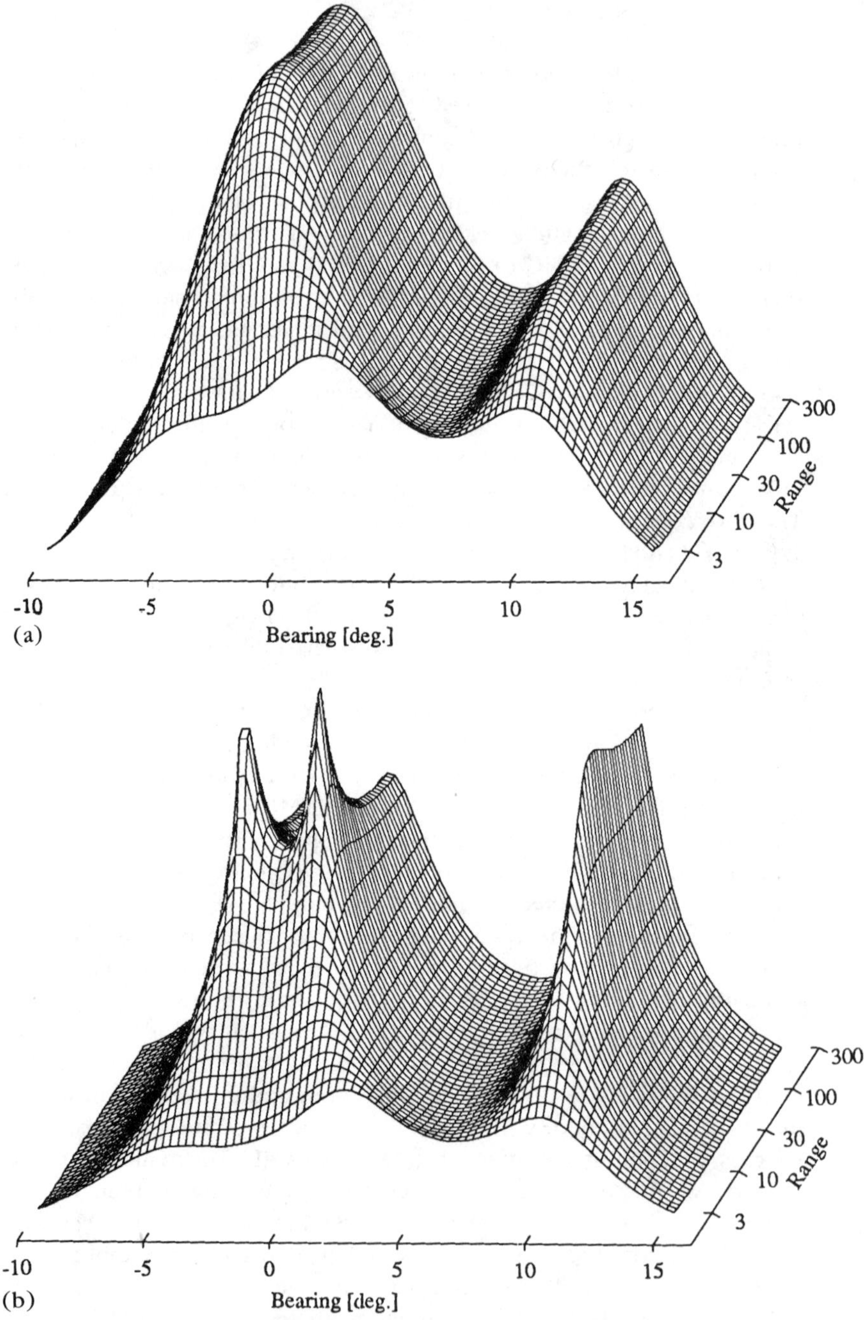

Figure 1.4 High-resolution diagrams for a large number of snapshots over bearing in degrees and range in multiples of the antenna length: (a) adoptive beamformer, (b) MUSIC.

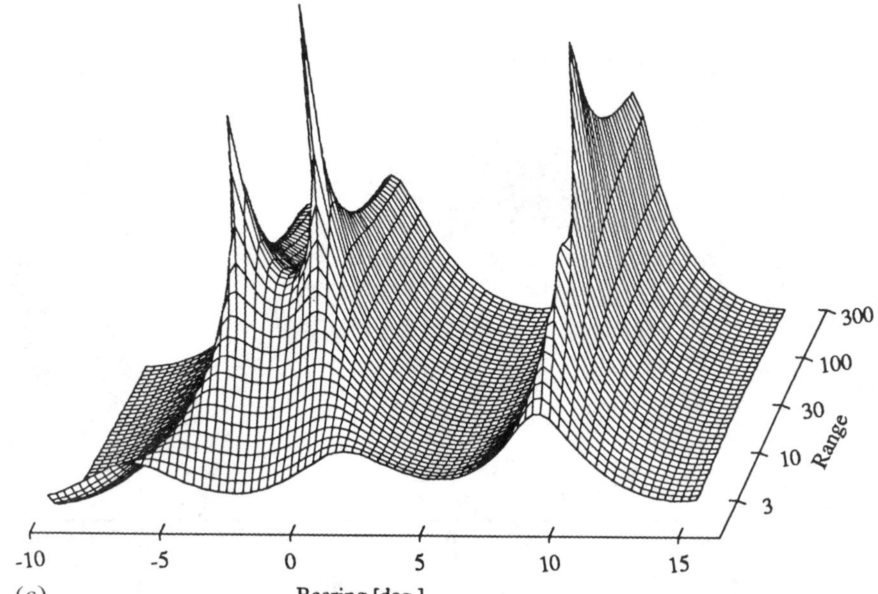

(c)

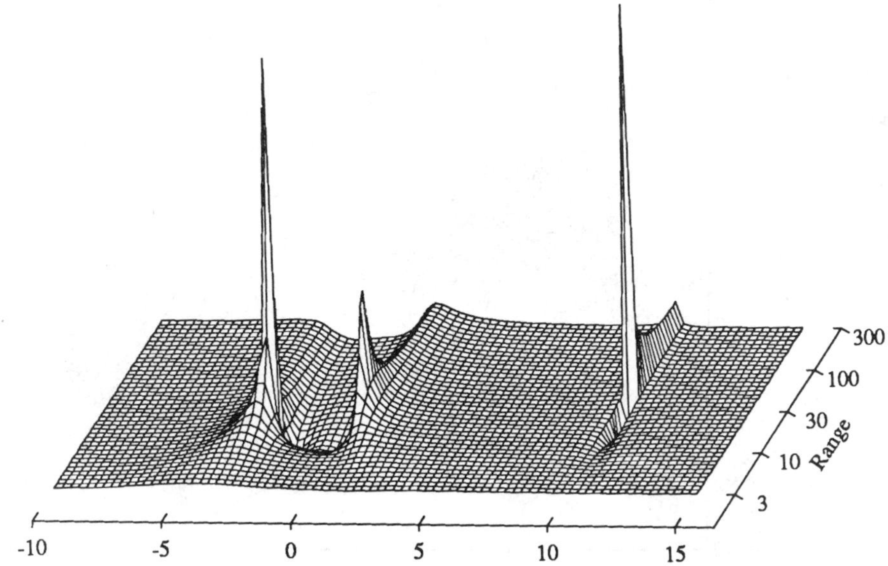

(d)

Figure 1.4 cont., (c) KTM, (d) AML.

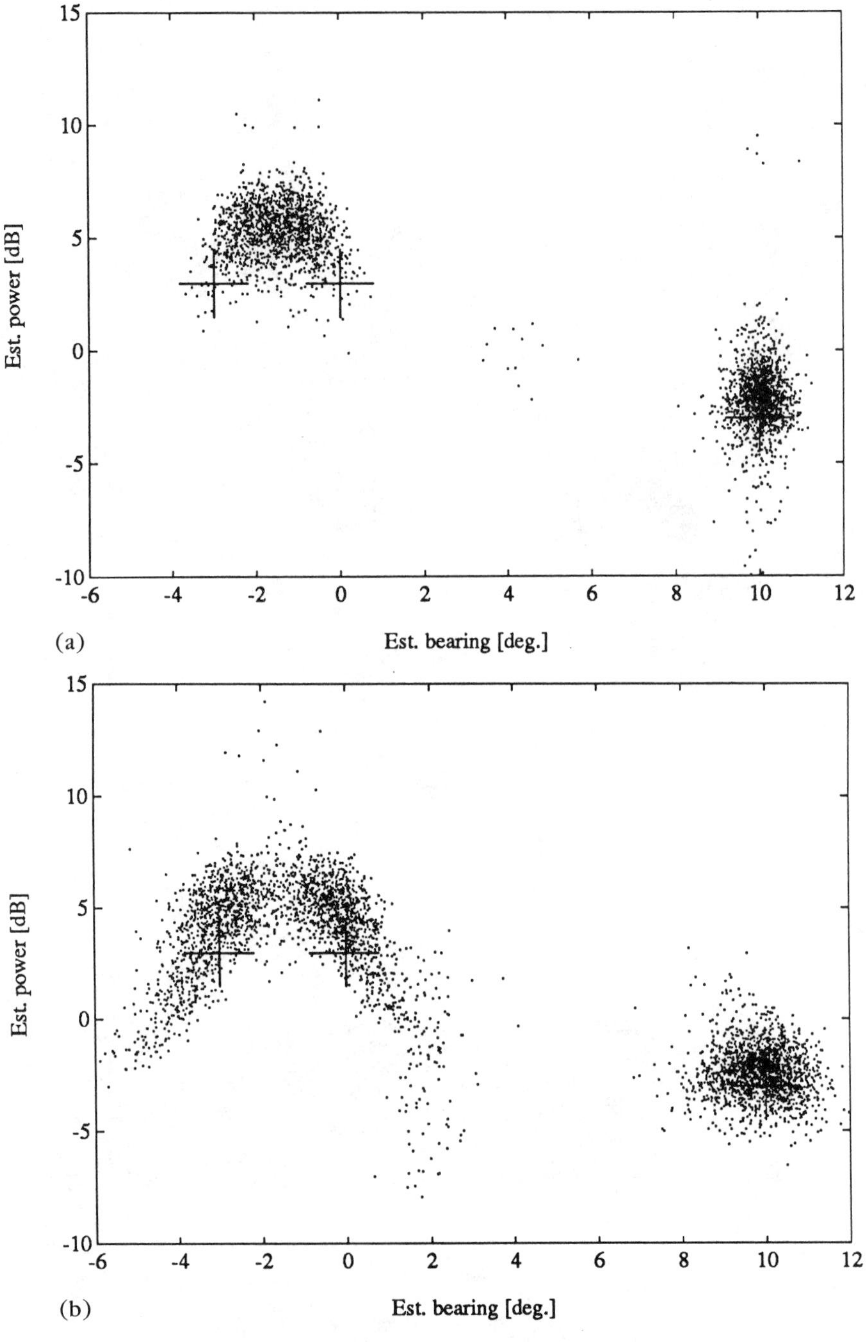

Figure 1.5 Scatter diagrams of parameter estimates: 2048 simulations, 20 snapshots, estimated signal power over estimated bearing, (a) MUSIC, (b) KTM.

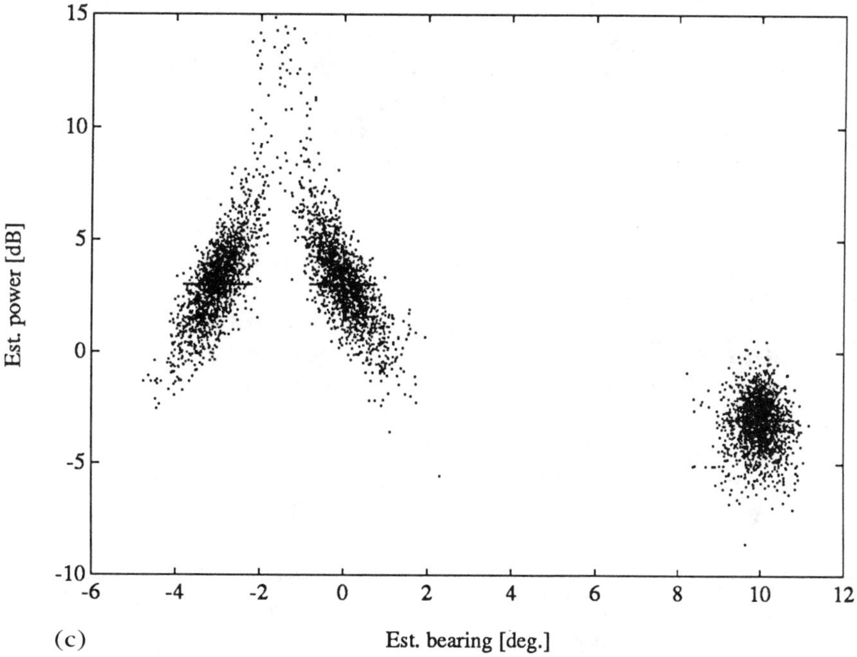

(c)

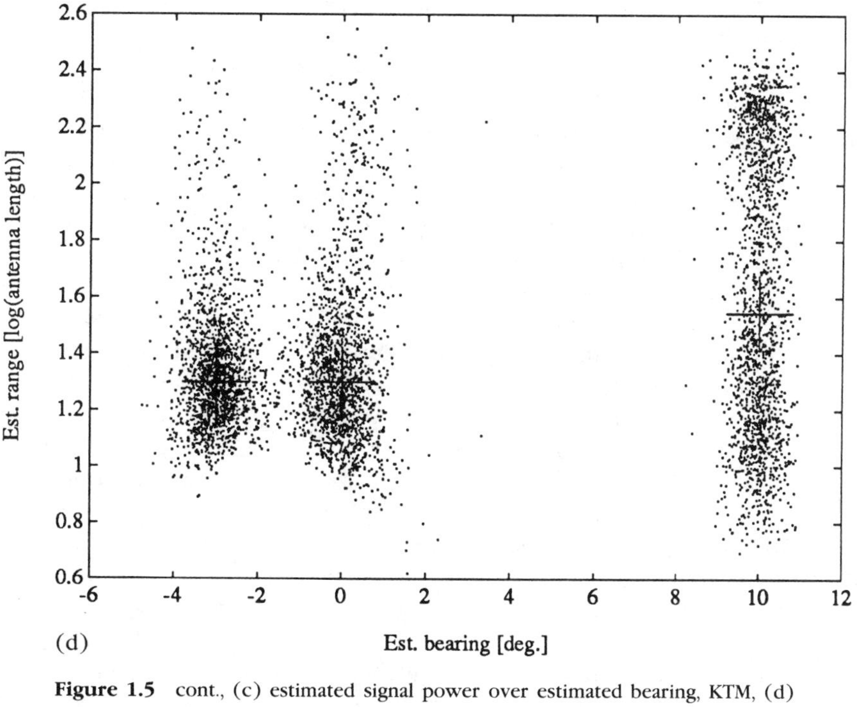

(d)

Figure 1.5 cont., (c) estimated signal power over estimated bearing, KTM, (d) estimated log (range/antenna length) over estimated bearing, AML.

mined using (1.124) and (1.125). The iterations for AML were executed as described in Section 1.4.3(c) with spectral estimates (1.131) and (1.132). The results are depicted as scatter diagrams in Fig. 1.5. Estimated power in terms of estimated SNR over estimated bearing is shown in Fig. 1.5(a) to (c) for MUSIC, KTM, and AML, respectively. Figure 1.5 (d) depicts estimated range in multiples of antenna length over estimated bearing for AML. The crosses indicate the model parameters of the sources.

We first observe in Fig. 1.5 that MUSIC cannot resolve sources 1 and 2. The number of snapshots is too small. KTM resolves the sources in the majority of cases; however, the estimates of the source parameters are biased. The third source is nearly unbiased estimated in both cases. The dispersion of KTM is larger than that of MUSIC as expected from the discussion in Section 1.3.1. AML resolves the sources in most cases and seems to estimate with lower bias and smaller dispersion than KTM. The outliers between sources 1 and 2 in Fig. 1.5(c) should be noticed. They can be the consequence of a property of the simple method used for computing the initial estimates. The method tends to put the first bearing estimate between sources 1 and 2 and the second in orthogonal direction right or left of the pair of sources. Sometimes, this type of error cannot be corrected by the iterations. Range estimation is known to be much more difficult than bearing estimation. This illustrates Fig. 1.5(d) for AML. The range estimates of the third, weak, and distant source are useless. The number of snapshots is too small. The range estimates of sources 1 and 2 start to cluster where they should. The corresponding results of KTM are even worse and not depicted.

Let us summarize some results of these and many other, similar experiments executed and statistically analyzed by Sandkühler [95] who also investigated estimates that use iterations (1.139) (called BML in the following) and Liggett's method (1.126). If a source is well separated from the other sources, the estimates of bearings and of signal spectral parameters are precise and stable. There are no essential differences, for example, between MUSIC and AML, except for range estimation. The estimates tend to be uncorrelated and not affected by the other sources. This property contrasts with estimations for sources close together, for example less than half a beamwidth. Estimations of signal spectral powers and bearings are correlated. We find significant influence by the neighbor sources. Best estimations were observed by the use of BML followed by AML and Liggett's method. For example, BML range estimates of source 3 start to cluster where they should as well in an experiment as earlier. However, AML is of particular interest because the effort for the numerical procedure we applied is not significantly greater than eigenvalue decomposition methods such as MUSIC combined with an automatic peak search. Sandkühler used the median deviation and Spearman's rank correlation coefficient for estimating the second moments of the parameter estimates and found, for his experiments, that these values can be approximately described by means of the Cramér-Rao lower bound if the SNR is not to low. The analysis of this bound also shows that the resolution of the sources is limited by the correlation of the estimates that depends on SNR and separation; see Messer and Schultheiss [126] for a similar investigation. The number of sources can be estimated by the MDL criterion. A possible overestimation of this number results in useful parameter estimates of the true sources and in spurious sources with very low SNR and high variance in bearings. AML has certain robustness properties. For example, uniformly distributed data in the time domain, suitably filtered and Fourier transformed with window length 32, did not appear to result in a significant reduction of either accuracy or stability of the AML estimates. Similarly, colored noise with smooth bearing spectra instead of sensor noise did not break down AML. Essentially, additional bias was only observed. This is the motivation to start the two-stage iterations described in connection to (1.160) with AML assuming only sensor noise, then correct the noise model and continue with AML and the prewhitened noise model and so on. This method, however, requires further research.

1.5 CONCLUDING REMARKS

Let us reexamine the main problem of array processing we discussed in this chapter: the estimation of the locations of sources and of their spectral properties, if a finite piece of array output is observed. In section 1.2, we investigated homogeneous wavefields as one possibility to describe arrivals of signals from farfield sources. We found that phased array beamforming followed by spectrum estimation of the beamsignal and computing the classical beamformer are suitable methods for obtaining an overview, or for surveillance purpose. Both methods are relatively robust; however, they perform poorly to resolve sources close together and to detect a weak source in the neighborhood of a strong interference. These problems were attacked by means of high-resolution diagrams and optimum beamformers in Section 1.3. Generally, these methods are successful if certain model assumptions are not violated. The high-resolution methods can separate closely spaced sources. The resulting estimates of the location and the spectral parameters are biased in some cases. Improvements require more careful modeling, where parametric methods are suitable. Parameter estimation by maximum likelihood in the frequency domain is a possibility; however, the global search in the parameter space of interest is prohibitive in most applications. Therefore, initial estimates from high-resolution methods that separate the sources are locally enhanced. These estimates can be precise and stable as discussed in the previous subsections. However, they can be unreliable if the model mismatches or the initial estimates are not precise enough. This is the reason why interactive systems should be used in practice, for example in sonar systems. The operator should apply several methods when a situation requires a more detailed analysis. In connection with more precise ranging, this is necessary all the more considering multipath ranging methods, and so on; see Rosenberger [127].

In this chapter, we have presented a tutorial view of array processing methods in the frequency domain for source location and spectrum estimation. We did not review work done, for example, on the effect of uncertainties in sensor positions and sensor characteristics on system performance, which is very important in sonar with towed arrays; see Rockah [128]. On the other hand, there are reasons to use parametric methods in the time domain. ARMA models and state space models in connection with signal subspace methods have been of interest. The interested reader may wish to refer to the work of Porat and Friedlander [129]; Su, Morf, and Nehorai [130] and [131]; Fuchs [132]; and Le Cadre [115].

ACKNOWLEDGMENT

D. Kraus's assistance in designing, programming, and executing the numerical experiments and U. Sandkühler's contributions to the results presented here are acknowledged.

REFERENCES

1. S. Haykin, ed., *Array Signal Processing* (Englewood Cliffs, N.J.; Prentice-Hall, 1985).
2. S. Haykin, ed., *Array Processing: Applications to Radar* (Stroudsburg, Pa.: Dowden, Hutchinsons, & Ross, 1980).
3. R. A. Monzingo and T. W. Miller, *Introduction to Adaptive Arrays* (New York: Wiley-Interscience, 1980).
4. J. E. Hudson, *Adaptive Array Principles* (Stevenage, U.K.: Peter Peregrinus, 1981).
5. A. B. Baggeroer, "Sonar Signal Processing," in *Applications of Digital Signal Processing,* ed. A. V. Oppenheim (Englewood Cliffs, N.J.; Prentice-Hall, 1978).
6. W. C. Knight, R. C. Pridham, and S. M. Kay, "Digital Signal Processing for Sonar," *Proc. IEEE,* **96** (1981), 1451–1506.
7. H. Urban, ed., *Adaptive Methods in Underwater Acoustics* (Dordrecht, Netherlands: Reidel, 1985).
8. Y. T. Chan, ed., *Underwater Acoustic Data Processing* (Dordrecht, Netherlands: Kluwer, 1988)
9. J. F. Böhme, "Array Processing," in *Les Houches,* Session XLV, 1985, *Signal Processing,* ed. J. L. Lacoume, T. S. Durrani, and R. Stora (Amsterdam; Elsevier, 1987), pp. 439–482.
10. F. S. Crawford, Jr., *Waves,* Berkeley physics course, Vol. 3, (New York, McGraw-Hill, 1968).
11. G. M. Wenz, "Acoustic Ambient Noise in the Ocean: Spectra and Sources," *J. Acoust. Soc. Am.,* **34**, (1962), 1936–1956.
12. J. M. Moura, "The Hybrid Algorithm: a Solution to Acquisition and Tracking," *J. Acoust. Soc. Am.,* **69** (1981), 1663–1672.
13. A. M. Yaglom, *An Introduction to the Theory of Stationary Random Functions* (Englewood, N.J.: Prentice-Hall, 1962).
14. J. Capon, "High-Resolution Frequency-Wavenumber Spectrum Analysis," *Proc. IEEE,* **57** (1969), 1408–1418.
15. D. R. Brillinger, *Time Series: Data Analysis and Theory,* expanded ed., (San Francisco, Holden-Day, 1981).

16. R. J. Adler, *The Geometry of Random Fields* (New York; Wiley-Interscience, 1981).
17. D. Middleton, "A Statistical Theory of Reverberation and Similar First-Order Scattered Fields," Pts. I–IV, *IEEE Trans. Inform. Theory,* **13** (1967), 372–414, **18** (1972), 35–90.
18. G. Linneberg (in German), "Über die diskrete Verarbeitung mehrdimensionaler Signale unter Verwendung von Wellendigitalfiltern," Dissertation, Fakultät für Elektrotechnik, Ruhr-Unversität Bochum, 1984.
19. J. F. Böhme, "Array Processing in Semi-homogeneous Random Fields," *Proc. GRETSI—Septième Coll. Traitement du Signal Appl.,* Nice, France, 1979, pp. 104/1–4.
20. H. Cox, "Resolving Power and Sensitivity to Mismatch of Optimum Array Processors," *J. Acoust. Soc. Am.,* **54** (1973), 771–785.
21. D. E. Dudgeon, "Fundamentals of Digital Array Processing," *Proc. IEEE* **65**, (1977), 898–904.
22. S. S. Reddi, "Multiple Source Location: A Digital Approach," *IEEE Trans. Aerosp. Electron. Syst.,* **15** (1979), 95–105.
23. G. Bienvenu and L. Kopp, "Optimality of High-Resolution Array Processing Using the Eigensystem Approach," *IEEE Trans. Acoust., Speech, Signal Processing,* **31** (1983), 1135–1248.
24. H. Mermoz, "Identification of Sources with Unknown Wavefronts," Techn. Rpt. 115–122, Naval Research Lab., Washington, D.C., March 1988.
25. V. F. Pisarenko, "On Estimation of Spectra by Means of Nonlinear Function of the Covariance Matrix," *Geophys. J. Roy. Astron. Soc.,* **28** (1972), 511–531.
26. V. F. Pisarenko, "The Retrieval of Harmonics from a Covariance Function," *Geophys. J. Roy. Astron. Soc.,* **33** (1973), 347–366.
27. G. Bienvenu, "Influence of the Spatial Coherence of the Background Noise on High-Resolution Passive Methods," *Proc. IEEE Int. Conf. Acoust., Speech, Signal Processing,* Washington, D. C., 1979, pp. 306–309.
28. A. Cantoni and L. C. Godera, "Resolving the Direction of the Sources in a Correlated Field Incidenting on an Array," *J. Acoust. Soc. Am.,* **67** (1980), 1247–1255.
29. R. O. Schmidt, "Multiple Emitter Location and Signal Parameter Estimation," Proc. RADC Spectrum Estimation Workshop, Rome, New York, 1979, pp. 243–258.
30. J. Ziegenbein, "Spectral Analysis Using the Karhunen-Loeve Transform," *Proc. IEEE Int. Conf. Acoust., Speech, Signal Processing,* Washington, D.C., 1979, pp. 182–185.
31. N. L. Owsley, "Spectral Signal Set Extraction," in *Signal Processing,* eds. J. W. R. Griffiths, P. L. Stocklin and C. van Schoonefeld, (New York: Academic Press, 1973), pp. 469–775.
32. W. S. Liggett, "Passive Sonar: Fitting Models to Multiple Time Series," in *Signal Processing,* eds. J. W. R. Griffiths, P. L. Stocklin, and C. van Schoonefeld (New York: Academic Press, 1973), pp. 327–345.
33. J. F. Böhme, "Source Parameter Estimation by Approximate Maximum Likelihood and Nonlinear Regression," *IEEE J. Oceanic Eng.,* **10** (1985), 206–212.
34. J. P. Burg, "Maximum Entropy Spectral Analysis," Proc. 37th Meeting Soc. Exploration Geophysicists, 1967; also in *Modern Spectrum Analysis,* ed. D. G. Childers, (New York; *IEEE Press,* 1978), pp. 34–41.

35. R. Kumaresan and D. W. Tufts, "Estimating the Angles of Arrivals of Multiple Plane Waves," *IEEE Trans. Aerosp. Electron. Syst.,* **19** (1983), 134–139.

36. U. Nickel, "Algebraic Formulation of Kumaresan-Tufts Superresolution Method, Showing Relation to ME and MUSIC Methods," *IEE Proc.,* **135**, Pt. F (1986), 7–10.

37. J. P. Burg, "The Relationship Between Maximum Entropy Spectra and Maximum Likelihood Spectra," *Geophys.,* **37** (1972), 375–376.

38. D. W. Tufts and R. Kumaresan, "Frequency Estimation of Multiple Sinusoids: Making Linear Predictions Like Maximum Likelihood," *Proc. IEEE,* **70** (1982), 975–989.

39. R. N. McDonough, "Application of the Maximum Likelihood Method and the Maximum-Entropy Method to Array Processing," in *Nonlinear Methods of Spectral Analysis,* ed. S. Haykin (Berlin: Springer, 1979), pp. 181–244.

40. D. H. Johnson, "The Application of Spectral Estimation Methods to Bearing Estimation Problems," *Proc. IEEE,* **70** (1982), 1018–1028.

41. S. W. Lang, "Spectral Estimation for Sensor Arrays," Ph.D. dissertation, MIT, Cambridge, Mass., 1981.

42. M. A. Lagunas, "The Variational Approach in Spectral Estimation," in *Signal Processing, III, Theory and Applications,* eds. I. J. Young et al., (Amsterdam: North Holland, 1986), pp. 307–314.

43. M. Wax, T. J. Shan, and T. Kailath, "Spatio-temporal Spectral Analysis by Eigenstructure Methods," *IEEE Trans. Acoust., Speech, Signal Processing,* **32** (1984), 817–827.

44. H. Wang and M. Kaveh, "Coherent Signal-subspace Processing for Detection and Estimation of Angles of Arrival of Multiple Wideband Sources," *IEEE Trans. Acoust., Speech, Signal Processing,* **33** (1985), 823–831.

45. T. J. Shan, M. Wax, and T. Kailath, "On Spatial Smoothing for Direction of Arrival Estimation of Coherent Signals," *IEEE Trans. Acoust., Speech, Signal Processing,* **33** (1985), 806–811.

46. R. Roy, A Paulraj, and T. Kailath, "ESPRIT—A Subspace Rotation Approach to Estimation of Parameters of Cisoids in Noise," *IEEE Trans. Acoust., Speech, Signal Processing,* **34** (1986), 1340–1342.

47. L. Kopp and G. Bienvenu, "Multiple Detection Using Eigenvalues When the Noise Spatial Coherency Is Partially Unknown," in [7], 385–391.

48. A. Paulraj and T. Kailath, "Eigenstructure Methods for Direction of Arrival Estimation in the Presence of Unknown Noise Fields," *IEEE Trans. Acoust., Speech, Signal Processing,* **34** (1986),13–20.

49. W. S. Liggett, "Passive Sonar Processing for Noise with Unknown Covariance Structure," *J. Acoust. Soc. Am.,* **51** (1972), 24–30.

50. H. A. d'Assumpcao, "Some New Signal Processors for Arrays of Sensors," *IEEE Trans. Inf. Theory,* **26** (1980), 441–453.

51. J. Capon and N. R. Goodman, "Probability Distribution for Estimators of the Frequency-Wavenumber Spectrum," *Proc. IEEE,* **58** (1970), 1785–1786.

52. A. B. Baggeroer, "Confidence Intervals for Regression (MEM) Spectral Estimates," *IEEE Trans. Inf. Theory,* **22** (1976), 534–545.

53. J. F. Böhme, "Remarks on the Statistical Behavior of Orthogonal Beamforming," *IEE Proc., vol.* **130**, Pt. F (1983), 256–260.

54. M. Kaveh and A. J. Barabell, "The Statistical Performance of the MUSIC and the Minimum-norm Algorithms in Resolving Plane Waves in Noise," *IEEE Trans. Acoust., Speech, Signal Processing,* **34** (1986), 331–341.

55. D. J. Jeffries and D. R. Farrier, "Asymptotic Results for Eigenvector Methods," *IEE Proc.,* **132**, Pt. F (1985), 589–594.

56. B. Porat and B. Friedlander, "Analysis of the Asymptotic Relative Efficiency of the MUSIC Algorithm," *IEEE Trans. Acoust., Speech, Signal Processing,* **36** (1988), 532–544.

57. R. Knötsch and J. F. Böhme, "Approximate Evaluation of Bounds for Peak Estimates," in *Mathematics in Signal Processing,* ed. T. S. Durrani et al. (Oxford: Clarendon Press, 1987), 133–145.

58. M. Wax and T. Kailath, "Determining the Number of Signals by Information Theoretic Criteria," *Proc. IEEE Acoust., Speech, Signal Processing,* Spectrum Estimation Workshop II, 1983, 192–196.

59. Y. Q. Yin and P. B. Krishnaiah, "On Some Nonparametric Methods for Detection of the Number of Signals," *IEEE Trans. Acoust., Speech, Signal Processing,* **35** (1987), 1533–1538.

60. F. Bryn, "Optimum Structures of Sonar Systems Employing Spatially Distributed Receiving Elements," *Proc. NATO ASI Signal Processing with Emphasis on Underwater Acoustics,* vol. 2, paper 30, The Hague, Netherlands, 1968.

61. B. Widrow, P. E. Mantey, L. J. Griffiths, and B. B. Goode, "Adaptive Antenna Systems," *Proc. IEEE,* **55** (1967), 2143–2159.

62. W. F. Gabriel, "Spectral Analysis and Adaptive Array Superresolution Techniques," *Proc. IEEE,* **68** (1980), 654–666.

63. O. L. Frost, III, "An Algorithm for Linearly Constrained Adaptive Array Processing," Proc. IEEE, 60 (1972), 926–935.

64. K. Krücker, "Rapid Interference Suppression Using a Kalman Filter," *IEE, Proc.* **130**, Pt. F (1983), 36–40.

65. J. C. McWhirter and T. J. Shepherd, "Least Squares Lattice Algorithm for Adaptive Channel Equalisation," *IEE Proc.,* **131**, Pt. F (1984).

66. R. Schreiber, "Implementation of Adaptive Array Algorithms," *IEEE Trans. Acoust., Speech, Signal Processing,* **34** (1986), 1038–1045.

67. B. Yang and J. F. Böhme, "Systolic Implementation of a General Adaptive Array Processing Algorithm," *Proc. IEEE Int. Conf. Acoust., Speech, Signal Processing,* New York, 1988, 2785–2788.

68. J. G. McWhirter and T. J. Shepherd, "An Efficient Systolic Array for MVDR Beamforming," *Proc. IEEE Int. Conf. on Systolic Arrays,* San Diego, California, 1988.

69. J. G. McWhirter and T. J. Shepherd, "A Systolic Array for Linearly Constrained Least-Squares Problems," *Proc. SPIE, Advanced Algorithms and Architectures for Signal Processing,* **696** (1986), 80–87.

70. B. Yang and J. F. Böhme, "A Multiple Constrained Adaptive Beamformer and a Systolic Implementation," in *Signal Processing* IV, *Theorie and Applications,* eds. J. L. Lacoume et al., (Amsterdam; Elsevier, 1988), 283–286.

71. U. Nickel, "Angle Estimation with Adaptive Arrays and Its Relation to Superresolution," *IEE Proc.,* **134**, Pt. H (1987), 77–82.

72. R. C. Davis, L. E. Brennan, and I. S. Reed, "Angle Estimation with Adaptive Arrays in External Noise Fields," *IEEE Trans. Aerosp. Electron. Syst.*, **12** (1976), 179–186.

73. I. Karasalo, "Estimating the Covariance Matrix by Signal Subspace Averaging," *IEEE Trans. Acoust., Speech, Signal Processing,* **34** (1986), 8–12.

74. M. D. Zoltowski, "On the Performance Analysis of the MVDR Beamformer in the Presence of Correlated Interference," *IEEE Trans. Acoust., Speech, Signal Processing,* **36** (1988), 945–947.

75. Y. Bresler, V. U. Reddi, and T. Kailath, "Optimum Beamforming for Coherent Signal and Interferences," *IEEE Trans. Acoust., Speech, Signal Processing,* **36** (1988), 833–843.

76. Y. Bresler and A. Macovski, "Exact Maximum Likelihood Parameter Estimation of Superimposed Exponential Signals in Noise," *IEEE Trans. Acoust., Speech, Signal Processing,* **34** (1986), 1081–1089.

77. I. J. Clarke, "High Discrimination Target Detection Algorithms and Estimation of Parameters," in [8].

78. V. H. McDonald and P. M. Schultheiss, "Optimum Passive Bearing Estimation in a Spatially Incoherent Noise Environment," *J. Acoust. Soc. Am.,* **46** (1969), 37–43.

79. W. J. Bangs and P.M. Schultheiss, "Space Time Processing for Optimal Parameter Estimation," in *Signal Processing,* eds. J. W. R. Griffiths et al. New York: Academic Press, 1973).

80. W. R. Hahn and S. A. Tretter, "Optimum Processing for Delay Vector Estimation in Passive Signal Arrays," *IEEE Trans. Inform. Theory,* **19** (1973), 608–614.

81. W. R. Hahn, "Optimum Signal Processing for Passive Sonar Range and Bearing Estimation," *J. Acoust. Soc. Am.,* **58** (1975), 201–207.

82. G. C. Carter, "Variance Bounds for Passively Locating an Acoustic Source with a Symmetric Line Array," *J. Acoust. Soc. Am.,* **62** (1977), 922–926.

83. P. M. Schultheiss and J. P. Ianniello, "Optimum Range and Bearing Estimation with Randomly Perturbated Arrays," *J. Acoust. Soc. Am.,* **68** (1980), 167–173.

84. M. J. Hinich and P. Shaman, "Parameter Estimation for an R-dimensional Plane Wave Observed with Additive Independent Gaussian Errors," *Ann. Math. Statist.,* **43** (1972), 153–169.

85. B. V. Hamon and E. J. Hannan, "Spectral Estimation of Time Delay for Dispersive and Non-dispersive Systems," *Appl. Statist.,* **23** (1974), 134–142.

86. E. J. Hannan, "Measuring the Velocity of a Signal," in *Perspectives in Probability and Statistics,* ed. J. Gani (New York: Academic Press, 1975), 227–237.

87. M. A. Cameron and E. J. Hannan, "Measuring the Properties of Plane Waves," *J. Int. Ass. Math. Geol.,* **10** (1978), 1–22.

88. J. F. Böhme, "On Parametric Methods for Array Processing," in *Signal Processing,* II, *Theory and Applications,* ed. H. W. Schüßler (Amsterdam: North-Holland, 1983), 637–644.

89. D. R. Brillinger, "Fourier Analysis of Stationary Processes," *Proc. IEEE,* **62** (1974), 1628–1643; also reprinted in [15].

90. D. R. Brillinger and P. R. Krishnaiah, eds., *Handbook of Statistics,* Vol. 3 (Amsterdam: Elsevier, 1983).

91. K. O. Dzhaparidze, *Parameter Estimation and Hypothesis Testing in Spectral Analysis* (Berlin: Springer, 1986).

92. H. Akaike, "A New Look at the Statistical Model Identification," *IEEE Trans. Autom. Control,* **19** (1974), 716–723.

93. J. Rissanen, "Modeling by Shortest Data Description," *Automatica,* **14** (1978), 465–471.

94. P. R. Krishnaiah, J. C. Lee, and T. C. Chang, "Likelihood Ratio Tests on Covariance Matrices and Mean Vectors of Complex Multivariate Normal Populations and Their Applications in Time Series," in [90], 439–476.

95. U. Sandkühler (in German), *Maximum-Likelihood Schätzer zur Analyse von Wellenfeldern* (Dr.-Ing. Dissertation, Fakultät für Elektrotechnik, Ruhr-Universität Bochum), Fortschr.-Ber., VDI Reihe 10 Nr. 75, VDI-Verlag, Düsseldorf, 1987.

96. G. Vezzosi, "Estimation of Phase Angles from the Cross-Spectral Matrix," *IEEE Trans. Acoust., Speech, and Signal Processing,* **34** (1986), 405–422.

97. P. Nicolas (in French), "Localisation des sources ponctuelles avec une antenne de géométrie inconnue," Thèse de 3ème cycle, Univ. Rennes I, Rennes, France, 1985.

98. C. R. Rao and J. Kleffe, "Estimation of Variance Components," in *Handbook of Statistics,* Vol. 1, ed. P. R. Krishnaiah (Amsterdam: North-Holland, 1980), 1–40.

99. J. F. Böhme, "Estimation of Spectral Parameters of Correlated Signals in Wavefields," *Signal Processing,* **11** (1986), 329–337.

100. J. P. Burg, D. G. Luenberger, and D. L. Wegner, "Estimation of Structurated Covariance Matrices," *Proc IEEE,* **70** (1982), 963–973.

101. Y. Bresler, "Maximum Likelihood Estimation of a Linearly Structurated Covariance with Application to Antenna Array Processing," *Proc. IEEE,* 4th Workshop on Spectrum Estimation and Modeling, Minneapolis, 1988, 172–175.

102. J. F. Böhme, "Separated Estimation of Wave Parameters and Spectral Parameters by Maximum Likelihood," *Proc. IEEE Int. Conf. Acoust., Speech, Signal Processing,* Tokyo, 1986, 2819–2822.

103. U Sandkühler and J. F. Böhme, "Accuracy of Maximum-Likelihood Estimates for Array Processing," *Proc. IEEE Intl. Conf. Acoust., Speech, and Signal Processing,* Dallas, 1987, 2015–2018.

104. Y. Bard, *Nonlinear Parameter Estimation* (New York: Academic Press 1974).

105. R. Fletcher, *Practical Methods of Optimization,* (New York: Wiley-Interscience, 1987).

106. M. J. Hinich, "Maximum Likelihood Estimation of the Position of a Radiating Source in a Waveguide," *J. Acoust. Soc. Am.,* **66** (1979), 480–483.

107. U. Nickel, "Superresolution by Spectral Line Fitting" in *Signal Processing,* II, *Theory and Applications,* ed. H. W. Schüßler (Amsterdam: North Holland, 1983).

108. M. Wax, "Detection and Estimation of Superimposed Signals," Ph.D. Thesis, Dept. of Electrical Engineering, Stanford University, Stanford, California, 1985.

109. I. Ziskind and M. Wax, "Maximum Likelihood Estimation via the Alternating Projection Maximization Algorithm," *IEEE Int. Conf. Acoust., Speech, Signal Processing,* Tokyo, 1987, 2280–2283.

110. M. Wax and I. Ziskind, "Detection of Fully Correlated Signals by the MDL Principle," *Proc. IEEE Int. Conf. Acoust., Speech, Signal Processing,* New York, 1988, 2777–2780.

111. R. H. Shumway, "Replicated Time-Series Regression: An Approach to Signal Estimation and Detection," in [90], 383–408.

112. R. Kumaresan, L. L. Scharf, and A. K. Shaw, "An Algorithm for Pole-Zero-Modelling and Spectral Analysis," *IEEE Trans. Acoust., Speech, Signal Processing,* **34** (1986), 637–640.

113. L. Kopp and G. Bienvenu (in French), "Détection par les valeurs propres de la matrice interspectrale: Adoption au bruit de fond," *Proc. GRETSI,* Nice, France, 1983.

114. R. Foka, H. Boucard, and H. Debart, "Factor Analysis and Estimation of Covariance Matrix," in [8].

115. J. P. Le Cadre (in French), "Contributions à l'utilisation des méthodes parametriques en traitment d'antennes," Thèse de Doctorat es sciences, USMG-INPG, Grenoble, France, 1987.

116. P. J. Thomson, "Signal Estimation Using an Array of Recorders," *Stochastic Process. and Appl.,* **13** (1982), 201–214.

117. D. Kraus (in German), "Modifikation von Maximum-Likelihood-Schätzern für Wellenparameter unter Berücksichtigung von Umgebungsrauschen," Diplomarbeit, Lehrstuhl für Signaltheorie, Ruhr-Universität, Bochum, 1987.

118. P. Forster and G. Vezzosi, "Approximate ML after Spheroidal Beamforming in the Narrow Band and Wide Band Cases," in *Signal Processing,* IV; *Theories and Applications,* eds. J. L. Lacoume et al., (Amsterdam: Elsevier, 1988), 307–310.

119. K. V. Mardia and R. J. Marshall, "Maximum Likelihood Estimation of Models for Residual Covariance in Spatial Regression," *Biometrica,* **71** (1984), 135–146.

120. J. F. Böhme and D. Kraus, "On Least Squares Methods for Direction of Arrival Estimation in the Presence of Unknown Noise Fields," *Proc. IEEE Int. Conf. Acoust., Speech, Signal Processing,* New York (1988), 2833–2836.

121. B. F. Cron and C. H. Sherman, "Spatial Correlation Function for Various Noise Models," *J. Acoust. Soc. Am.,* **34** (1962), 1732–1736.

122. J. T.-H. Lo, "New Maximum Likelihood Approach to Multiple Signal Estimation," *Proc. IEEE Intl. Conf. Acoust., Speech, Signal Processing,* New York, 1988, 2889–2892.

123. M. Feder and E. Weinstein, "Parameter Estimation of Superimposed Signals Using the EM Algorithm," *IEEE Trans. Acoust., Speech, and Signal Processing,* **36** (1988),477–489.

124. E. J. Hannan, "Non-Linear Time-Series Regression," *J. Appl. Probab.,* **8** (1971), 762–780.

125. D. R. Brillinger, "A Maximum Likelihood Approach to Frequency-Wavenumber Analysis," *IEEE Trans. Acoust., Speech, Signal Processing,* **33** (1985), 1076–1085.

126. H. Messer and P. M. Schultheiss, "Parameter Estimation for Two Sources with Overlapping Spectra Using an Arbitrary M Sensor Array," *Proc. IEEE Third ASSP Workshop on Spectrum Estimation and Modeling,* Boston, 1986, 149–152.

127. J. C. Rosenberger, "Passive Localization," in [8].

128. Y. Rockah, "Array Processing in the Presence of Uncertainty," Ph.D. dissertation, Yale University, New Haven, Connecticut, 1986.

129. B. Porat and B. Friedlander, "Estimation of Spatial and Spectral Parameters of Multiple Sources," *IEEE Trans. Inform. Theory,* **29** (1983), 412–425.

130. G. Su. and M. Morf, "The Signal Subspace Approach for Multiple Wide-Band Emitter Location," *IEEE Trans. Acoust., Speech, Signal Processing,* **31** (1983), 1502–1522.

131. A. Nehorai, G. Su, and M. Morf, "Estimation of Time Difference of Arrival by Pole Decomposition," *IEEE Trans. Acoust., Speech, Signal Processing,* **31** (1983), 1478–1492.

132. J. J. Fuchs, "State-Space Modeling and Estimation of Time Differences of Arrival", *IEEE Trans. Acoust., Speech, Signal Processing,* **34** (1986), 232–244.

2 Supervised Interpretation of Sampled Data Using Efficient Implementations of Higher-Rank Spectrum Estimation

Ira J. Clarke

2.1 INTRODUCTION

Information in a scenario under observation is, in general, sampled relatively sparsely in both time and space compared with the vast amount of detail potentially available. Numbers of sensors, sensor aperture, bandwidth, and sample rate are kept to the essential minimum to minimize sensor complexity and, in many cases, to reduce the amount of data to be processed or transmitted. The net result is that very little of the detailed information present in the input scenario is propagated through the sensor system to the processor.

Ultimately, the performance of any sensor system depends on a satisfactory choice of discriminants and sensors. Making this choice requires an adequate understanding of the laws of physics and of device technology; yet if we try to enhance discrimination by improvements to sensor design alone, we eventually hit the law of diminishing returns where cost outweighs advantage. There is always inevitable pressure to decrease both development and production costs.

Although lost information can never be recovered by processing alone, modern digital signal processing (DSP) technology now offers a new degree of freedom in the trade-off between sensor performance, the optimality of the processing stage, and robustness. It will be shown that we can, for example, now consider relaxing tolerances on sensor specifications by utilizing online adaptive self-calibration techniques. This approach has the potential to improve cost-effectiveness and to reduce development time scale.

There are two principal objectives for modern DSP:

1. To massage the data or image in such a way as to match better the specific, very narrow information bandwidth of a human observer, who makes the final commonsense decisions.
2. To perform fully automated data interpretation and decision taking, followed by the initiation of a conditioned reaction to the perceived stimuli at the sensor input.

In practice, excluding human beings from the loop is surprisingly difficult. The aim of much of the research in DSP over recent years has been that of automating ***reliable decision taking*** based on the limited evidence available through sensors. It is acknowledged that human observers are better suited to certain interpretative tasks because the brain is able to call on such a vast amount of previous experience, indirect collateral knowledge, and analogous situations and is able to reason in a logical manner. It is not easy to represent such knowledge in tractable mathematical form, and it is also surprisingly difficult to develop algorithms that utilize inexact and abstract knowledge in an efficient and consistent manner. The ***robustness*** of data processing and control algorithms is often crucial to the overall acceptability of a system design. A poor impression of the system is created if the signal processing system does not interpret data in a commonsense manner, when the conclusion is perfectly obvious to a skilled operator. Consequently, there is a continuing drive to develop robust algorithms able to extract as much information as possible from available sensor data.

2.1.1 Application-specific Algorithms

In practice, attempts to derive schemes for processing sensor data usually involve ad hoc sequences of system identification, spectrum estimation, mathematical transforms, data reduction, image enhancement, data fusion, inference, logical deduction, and parametric modeling techniques. We note that many of these operations require additional information, in the form of models, implicit and explicit assumptions, statistical relationships, direct constraints on the choice of solution, and sometimes as decisions from other sensors. The composite processor is highly specific to a target application and restricts the flexibility of the system in less than ideal practical conditions. There is a clear need to rationalize the approach to algorithm design and to develop near-optimal efficient and robust techniques, able to adapt to a wide range of practical conditions.

2.1.2 High Resolution

It has been shown by many researchers that, *in ideal conditions*, modern high-resolution algorithms have the potential to enhance greatly system identification, signal extraction, target acquisition, and tracking performance in radar, sonar, and many other sensor-based systems. For example, it is easily demonstrated by simulation that, in carefully controlled conditions, the resolving power of modern techniques, exemplified by the MUSIC algorithm [1], surpasses the bound set by the long-established Rayleigh resolution criterion. Improvements of an order of magnitude or more are readily attainable if the signal-to-noise ratio (SNR) is sufficiently large. There is also a significant improvement in the ability to discriminate weaker signals in the presence of sidelobe leakage from stronger signals and from jamming interference.

A basic question arises as to how such high-discrimination algorithms appear to recover detail information that has apparently been lost irretrievably within the sensing process. The lack of detail in sensor-derived data, due to aperture and sampling limitations, implies that, in most cases, we have an *underdetermined* inverse problem that has no unique mathematical solution. Clearly, on information theoretic grounds, it is impossible for detail that is not present in some form in the sensor data to be recovered by mathematical techniques alone. There is, at first, a suspicion that we are getting something for nothing, but it soon becomes clear that the algorithms in question depend on data-independent assumptions to constrain the solution space.

The so-called point target constraint, for example, is commonly applied in antenna array processing. Data are collected by simultaneously sampling the output ports of each of the sensors in the array. In the analysis, it is assumed that pointlike emitters lie in the field of view of the antenna. A large number of potential solutions, involving interpretation as diffuse emitters, are eliminated by this constraint alone. Attempts are then made to recognize the presence and location of individual pointlike emitters in an unknown scenario, by searching for characteristic templates or signatures in sensor output data.

The signatures are a function of the emitter, the propagation medium, the antenna array geometry, and the electrical characteristics of the sensors. It is assumed that several signature vectors are linearly combined in the sensing process with weights that are complex and depend directly on the varying relative phase and magnitude of the temporal waveforms of the individual emitters.

The data analysis problem is primarily the reverse of the sensing process, that of *decomposition* of data into, for example, spatial and temporal components. It is important to remember that there is no unique decomposition; some property or cost function for the decomposition must be defined. The support may be explicit in form or be provided implicitly via assumptions and algorithmic procedures. For example, in the radar case, we utilize steering vectors as *signatures.* Each vector is associated with a different value of a direction-of-arrival parameter. In general, components of a spatially oriented decomposition must relate directly to recognizable spatial signatures, as defined for the particular sensor system. The data can then be interpreted in terms of the spatial locations of emitters. By contrast, the orthonormal vectors, found as result of singular valued decomposition (SVD) or eigen decomposition of the data, serve only as a potentially useful means of separating signals from noise on the basis of magnitude. Such techniques are not directly appropriate to the task of resolving individual signals in a spatial or temporal context. We shall also see that SVD and eigen decomposition are both relatively inefficient at the task of noise suppression compared to optimal search methods based on matched filters that are accurately matched to predefined spatial or temporal features.

Numerous variants of modern high-resolution algorithms are currently available, each claiming that the unique point target interpretation deduced using that algorithm is *optimum* in some special way. We need therefore, to gain some insight into the fundamentals of these algorithms and into the reasons for the multitude of different solutions that appear to be based on the same basic premise of pointlike emitters. Several reasons for the variety of solutions are conceivable. We might, for example, suspect that the point assumption alone is an insufficient constraint on the span of the solution space, and that the problem is, as a consequence, still underdetermined or ambiguous. If that is the case, then each algorithm may be utilizing spurious information, taken perhaps from the stronger noise components, to solve the problem uniquely.

On the other hand, the point constraint might be more than sufficient and the problem becomes overdetermined in a mathematical sense. Yet, again, we would have the potential for several different solutions. Clearly, if an overdetermined situation does arise then, to select a single best solution, the algorithm must either exclude excess information or must form a balanced decision by suitably weighting all relevant information. It appears therefore that the differences between the various solutions offered by the algorithms might be due to the selection of different rules for determining the weight factors. The weights should, ideally, be based on Bayesian principles.

We also realize that various other *assumptions,* in addition to the point constraint, might be made when deriving an algorithm. These assumptions are equivalent to *importing* additional knowledge and therefore preferentially support different features in the solution. However, we shall see that the principal reason for the range of different solutions lies not only in the way the available information is weighted, but also in the shortcuts taken by the algorithms in searching for a *best* solution. For example, it is clear that several direction-of-arrival parameters are necessary in a mathematical model of a multiemitter scenario. These parameters, in general, cannot be found independently of each other, yet many existing algorithms attempt to do just that, with the implicit aim of reducing the computational demand.

The shortcut most commonly adopted by existing algorithms is based on an artificial constraint on the span of the search. It is arranged that the search is implemented as an inexpensive one-dimensional task rather than as a multidimensional problem. We shall see that the inherent shortcomings of this approach are the fundamental cause of classic resolution and sidelobe limitations. Both problems are, in principle, avoided by searching the entire span of permitted candidate solutions for a best match to incoming sensor data. If this full search reveals more than one equally satisfactory solution, then the problem is said to be ambiguous or underdetermined. Generally, because of hardware constraints, the system engineer insists that one solution only should be provided. It is then necessary to provide support in the form of a prior likelihood function that allows multiple solutions to be differentiated. The so-called *prior* is independent of the sensor data under analysis.

2.1.3 Field of Potential Application—The Problem

The task of defining the prior is not easy since in-service tasks for radar and sonar systems typically include the need to detect multiple emitters in a background comprising nonisotropic noise, various types of clutter, complex interference patterns, and multipath reception. In addition, sensor calibration errors may be present. Both the wanted emitters and background interference may be pointlike, textured, or diffuse in various ways, or may be nonstationary in several different respects, and may have a variety of statistical properties. The solution space is therefore extremely large and the prior is complex. If we are to contain the computational demand within manageable bounds, then some simplification is essential. The suggestions in this chapter relate to directing a search in a logical manner. The search pathway is partly data dependent and partly data independent (defined by the prior).

In some cases, support can be varied adaptively to match the prevailing circumstances. Knowledge of the operational environment can then be used to preclude regions of the solution space appropriate to other conditions. For example, the instrument function (sensor calibration) may be sensitive to temperature. It may be possible to measure temperature independently and direct the search for a satisfactory solution accordingly.

In many applications, parametric models are appropriate. Typical parameters include emitter location, Doppler frequency, and measures of statistical properties. In a nonstationary scenario, parameters related to rates of change may also be required. The number of emitters is, in general, also unknown and may vary with time. The concept of matching all permitted candidate solutions to incoming data, to find a best solution, is equally valid in the context of tracking applications. The task can, however, be eased significantly, if current estimates of track parameters are included in the prior. The range of potential application of parametric modeling and state estimation is exceptionally wide, in both civil and military contexts.

Unfortunately, the heuristic assumptions on which many existing algorithms depend are often mathematically so convenient to implement that they are difficult to modify. It is also often almost impossible to specify, in analytic terms, the sensitivity to minor deviations from those assumptions. We see that, because the range of allowed solutions is artificially restricted by the algorithm, the correct solution may be inaccessable. The corresponding operational flexibility of a system is then reduced. The validity of the point assumption, for example, depends in a complex way, not only on precise details of the scenario but also on the calibration accuracy and linearity of the sensor system.

The fundamental assertion is that, because the underlying problem of analyzing sensor data is underdetermined, it is necessary to predefine aspects of the preferred solution via a prior or via support that is *independent* of the sensor data. Clearly, since the aim is to force features, such as the point constraint, onto the eventual solution, the prior should be based on accurate knowledge of features of the actual scenario under observation. The engineer should therefore check, in the context of his or her particular problem, the validity of both implicit and explicit assumptions and other support built into a candidate algorithm. He must also check that algorithmic simplifications lead to acceptable results; hence the need for *application-specific* algorithms that are based on sound information theoretic concepts.

If the performance of a candidate algorithm is assessed in conditions that do not match algorithmic assumptions, then we should not be surprised if results are disappointing. For this reason adaptive and high-resolution algorithms are regarded by many engineers to lack robustness, to be computationally demanding, and to place unacceptable demands on signal-to-noise ratio.

We need, therefore, to examine the fundamentals of the algorithm design task and to develop basic techniques that are suited to exploiting the complex guiding rules and qualitative relationships essential to the optimal extraction and identification of poorly defined (marginal) signals. The problem is one of high discrimination rather than of high resolution per se. Clearly, we should first firmly establish the underlying causes of limitations in classical signal processing and determine to what extent improved performance is theoretically feasible. Although recent literature shows that considerable research effort has been committed to remedy deficiencies, a sound theoretical basis has not yet been established for developing *robust* application-specific high-discrimination DSP algorithms.

2.1.4 Novel Approach

A novel approach to tackling the high-discrimination problem is outlined in this chapter. The underlying concepts are based on maximum-likelihood (ML) and Bayesian principles. Although generally accepted as optimum, ML techniques are usually rejected on grounds of excessive expense in computational terms. The incremental multiparameter (IMP) algorithm suggested in Section 2.6 utilizes a novel *higher-rank spectrum estimation* procedure to obtain *near*-optimal high-discrimination performance and also promises to be exceptionally efficient and robust in tracking applications. The problem of computational load is addressed, in the main part, through an incremental directed search that is based on using the solution to a simpler problem as a starting point for the next stage. The computational demand is also reduced by carefully ordering the required matrix operations so that small matrices are involved at every stage. It is shown that the basic concepts can be extended to cover cases where collateral evidence is imprecise.

The IMP technique, by definition, is model based, but gives the engineer much freedom to apply a wide range of support to the analysis and interpretation of data. The engineer must actually ensure the validity of the models, but rigid constraints can be relaxed by increasing the number of variable parameters. The engineer should expect the final solution, rightly or wrongly, to directly reflect constraints applied by him. If, for example, he imposes a simple stationary point source model then he can hardly expect the solution to contain diffuse or moving emitters. In effect the engineer should simply get back features that he has asked for!

The design and development of cost-effective sensor systems, such as radar or sonar, are constrained by a variety of practical factors such as the laws of physics, size, power consumption, frequency allocation, and the availability of affordable technology suited to the task. Data processing software is becoming a major risk item and the primary cause of extended development and high cost in advanced systems. Robust well-understood generic algorithms are required that can be easily implemented and adapted to meet the specific requirements of real applications.

2.2 FUNDAMENTALS OF THE DESIGN TASK FOR DSP

Three distinct phases in the design of a high-performance signal processing system can be identified as follows:

1. Learning stage: *Define the cause, record the effect.*
   a. Define relationships between selected examples of known input conditions and the corresponding observable sensor output states.
   b. Record typical background noise and interference (another signature).
   c. Estimate the statistical uncertainty or variability of all measurements.
   d. Predefine appropriate system output actions (otherwise, DSP is pointless).
   e. Determine the relative costs of incorrect detections, decisions, and actions.
   f. Define the supporting rules, relationships, and models for generating the prior that is essential to solving the underdetermined data interpretation problem.
2. Algorithm definition: *The inverse problem—observe the effect, infer the cause.*
   a. Based on the prior, engineer an efficient robust algorithm able to deduce and/or estimate the probable input conditions given limited samples of sensor data.
   b. Specify decisions to be made and relate to the corresponding output action(s).

3. Hardware implementation: *Develop cost-effective computing hardware.*
   a. Satisfy the computational demands of the most suitable high-performance algorithm.
   b. If necessary, optimize partitioning between software and special-purpose hardware.

In addition, after completing the initial development phase, we generally need to test the system, in practical conditions, against theoretical predictions. We need to evaluate and refine the design to meet fully a specified performance. The final debugging task can be eased substantially if highly sophisticated simulations have been run previously and also by permanently building diagnosis facilities and test points into both software and hardware for monitoring and optimizing performance. It pays, at the outset, not to ignore the need for this critical and expensive stage of development. We see a clear need for a soundly based structured algorithm design methodology that can be readily understood and adapted to prevailing circumstances, at any stage during the operational life of the equipment.

In the remainder of this section, we review in greater detail, the objectives and rationale for each of three main interrelated stages of a system development. In later sections, we concentrate on the development of mathematical concepts for robust high-resolution, tracking, and state estimation algorithms.

2.2.1 Computational Hardware

It is well known that parallel computing systems require careful design. In a given application, architectures can be specifically developed to minimize memory access and communication requirements between the individual processors. Although remarkable improvements in efficiency and power consumption can be achieved through special-purpose hardware, it is at the expense of specialized and costly development effort. Economy of scale comes from multifunction software and mass-produced hardware that, in the DSP case, can perhaps only be achieved through a sound basic design methodology.

The mapping of an algorithm onto an archicture can, in principle, be a two-way compromise. It may, for example, be possible to realize the significant economy by optimizing minor features of an algorithm such that readily available, low-cost, or familiar hardware devices can be utilized.

In general, in any proposed algorithm, there are key functions that constitute the majority of the computational load. These include, for example, eigen decomposition, matrix inversion, and spectrum estimation. If calls for any of these stages can be minimized or the matrices can be reduced in size, then there is the potential for substantial cost savings. The algorithm, proposed in Section 2.6, for example, is based on higher-rank spectrum estimation and does not involve either eigen decomposition or a matrix inverse of the data covariance matrix. However, the number of points in a full higher-rank spectrum is extremely large, and we need to implement an efficient method of limiting the span of the search. It is clear that there are many interrelated, often subtle, factors that should be considered when developing algorithms and hardware for a specific application. If there is a need to compromise performance against computational demand, then great care must be exercised. We should avoid taking decisions based on suboptimal interpretations of the available sensor data without validation.

2.2.2 The Learning Stage

A question sometimes asked is: "Is a learning stage really necessary to the interpretation of data from a sensor system?" We are perhaps better able to give a reasoned answer by reversing the question and asking ourselves what conclusions might be drawn if we are simply presented with a string of numbers. We are told that the data represent samples from a sensor system. We can regard the device as a form of encoder, and we are asked to deduce the input. *No encoding key or other background information is provided.* Clearly, the task is impossible, we can only estimate certain statistical properties from the distribution of number values.

If we are then informed that the numbers are in sequence, then we can look for interesting patterns and may be able to express the data in simpler form, perhaps as linear combinations of simple sequences (decomposition). These patterns might be arbitrary vectors, such as eigenvectors, or could be basic mathematical functions, or familiar encoding keys. Examples range from simple sinusoids to complex speech waveforms.

In general, there are many different ways of decomposing data into arbitrary basis components. The best we might do then, without some prior knowledge of potential decoding keys, is to find the most economical representation of the numbers in some sense. If we find that the patterns conveniently fit a low-order mathematical or analytic model, then we conclude intuitively that the data are not a chance occurrence of random numbers but that there is an underlying cause for the structure in the data. We may be able to use this structure as a model to predict, interpolate, or compress the sampled data into fewer numbers or parameters. However, we are not much wiser since, in general, we cannot infer how the data were gathered nor can we interpret the model in physical terms. Clearly, a capability to predict is highly relevant in certain applications. Otherwise, the mathematical properties are of academic interest only. We conclude that prediction, in the absense of a prior, is based on copying repeating examples within the data provided. If, on the other hand, we find no preferred representation or model, then we might construe that there is no information hidden in the data. But can we be sure?

The main conclusion we draw from this discussion is that, to analyze data in useful terms, we must *link cause and effect* by examining detailed examples gathered during a learning phase. Given sensor data alone, we cannot make useful decisions about an input scenario or suggest a useful output action. Prior knowledge can take the form of previously identified signature vectors, data encoding formats, mathematical models, and operators. We must also include, as support, both quantitative and qualitative features of the input scenarios that go undetected by the primary sensors. We next consider how signature information can best be represented in usable form. The first step is to specify the cause-and-effect relationship either as analytic models or as the so-called array manifold.

2.2.3 The Basic Array Manifold

Definition

A simple example of the representation of a direct link between cause and effect is illustrated in Fig 2.1, which shows how a steering vector used in narrow-band antenna arrays is related to the physical sensor layout and the location of an emitter at a single point. A set of such relationships can be described by building up a library of discrete templates or signature vectors for a number of emitter locations. Clearly the discretization interval must be chosen to avoid undersampling, and sufficient vectors should be available to span the full field of view of the sensor system. The objective is to be able (subsequently) to infer which input condition prevails by matching the sensor output state to one of the reference vectors.

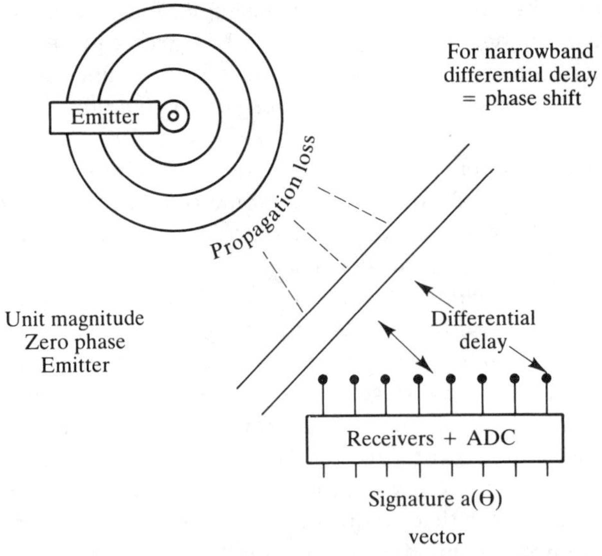

Figure 2.1 Concept of a signature vector for a basic sensor system comprising a narrow-band linear array of antennas, sampling a spatial wavefront. A different spatial frequency can be associated with each direction of arrival. In general, elements in the signature vector record propagation loss and delay to individual sensors in addition to the calibration of the individual receivers.

Individual signatures are indexed by a pointer. A selected value of the pointer (or parameter) is denoted θ. A basic signature is represented by a column vector, $\mathbf{a}(\theta)$, comprimising $\mathbf{n}_s$ complex-valued elements and corresponds to scenario comprising *a single emitter at unit reference magnitude* at a location indicated by θ. The library of such steering vectors, for individual emitters, is commonly referred to as the *array manifold*, denoted $\mathbf{a}(\boldsymbol{\Theta})$. $\boldsymbol{\Theta}$ defines the spatial field of view of the system. In many applications, the absolute phase and magnitude of an emitter waveform are not regarded as discriminants. The *relative* phases and magnitudes within the steering vector are relevant to the direction finding task. A signature vector,

a(θ), then defines a rank-one signal subspace. Although it is intended that the span of the array manifold defines the full permitted solution space, *it is important to note that the basic array manifold,* **a**($\boldsymbol{\Theta}$), *applies only to single emitter scenarios.* The case where a signature can be modeled by a single vector is termed the rank-one case.

In a typical radar application, a signature vector should extend over all available discriminants, such as direction of arrival, range, and Doppler. A vector should identify a single-resolution cell and could include, for example, a pulse compression waveform for the appropriate range bin. We shall see that, in the processing stage, we are required to generate a matched filter or weight vector appropriate to each signature vector. It is sufficient to represent the array manifold either as signature vectors or, equivalently in terms of information content, as the corresponding set of matched filters.

In the case of a narrow-band linear (or planar) array, using direction of arrival as the only discriminant, steering vectors can be reconstructed mathematically, using complex exponential functions sampled over the relevant *spatial* aperture. Each signature vector corresponds to a different value of spatial frequency. Frequency maps directly to direction of arrival, $\boldsymbol{\theta}$. In this case, because we have a simple analytic model, we do not necessarily need physically to store the array manifold in explicit form. The storage requirement can be traded against computational demand according to the relative cost in terms of hardware and access time. Although the underlying concept of an analytic model is directly equivalent to that of the array manifold, the sensors must be calibrated accurately and must be stable throughout the life of the equipment. For various practical reasons, we may prefer to use arbitrary sensor geometry or conformal antenna arrays. The signature vectors cannot then be expressed analytically. As a consequence, the calibration must be recorded experimentally and may have to be updated at regular intervals to compensate for changing environmental factors. In such cases, we lose the attractive features of frequency-domain spectrum analysis, obtained via the Fourier and inverse Fourier transforms; we must then generalize the processing algorithms to utilize arbitrary signatures and matched filters. Adequate means of accurate sensor calibration must be provided.

Estimation of the array manifold

The selection of suitable discriminants and the rejection of redundant information is, in general, key to the final performance of a system and requires very careful consideration. In a speech recognition task, for example, speaker-dependent features should be rejected. We note, as a further example, that in a spatial resolution problem, varying the absolute phase or magnitude of an emitter does not constitute a different spatial scenario. However, the magnitudes and statistical probability distribution functions (pdf), and temporal coherence properties of the emitters are, in certain cases, the relevant classifiers that discriminate an emitter from clutter, interference, and thermal sensor noise. The choice of discriminant is therefore highly application specific and may not be the same at all stages of processing within a system. We shall continue to illustrate the basic concepts of high-resolution matching algorithms by reference to the relatively simple specific example, shown in Fig. 2.1, of a sampled aperture antenna array in idealized conditions. Additive thermal sensor noise is assumed to be the significant source of interference. The mathematics generalizes easily to arbitrary multiple discriminants. In many cases, including radar and sonar, discriminants are processed sequentially, and we can illustrate the principles using the simple case of one discriminant, such as direction of arrival.

In the scenario depicted in Fig. 2.1, with one emitter located at θ, successive frames or snapshots of sensor output data, can be denoted by column vectors, $\mathbf{d}_{11}$, $\mathbf{d}_{21}, \ldots$, or as a matrix $\mathbf{D}$. $\mathbf{D}$ can be described in terms of the product of a single steering vector, $\mathbf{a}(\theta_1)$, and a series of real or complex scaling factors, $f_{11}, f_{21}, \ldots$ (representing the transmitted waveform of the emitter) plus additive measurement error

$$\mathbf{D} = [\mathbf{d}_1, \mathbf{d}_2, \ldots] = \mathbf{a}(\theta_1)\, [f_{11}, f_{21}, \ldots] + \text{observation noise} \qquad (2.1)$$

In the inverse problem, we are required to estimate the temporal waveform, $f_{11}, f_{21}, \ldots$ and to estimate a value for θ. In the learning phase, we must therefore estimate a full set of candidate spatial signature vectors, $\mathbf{a}(\boldsymbol{\Theta})$. Although in this case θ is a continuous variable, we shall regard the array manifold as a set of discrete vectors where the sampling grid at least satisfies the Nyquist sampling criterion. At intermediate points, interpolation between signature vectors can be performed by analytic curve fitting or by reference to physical models obtained by previous research.

In concept, a selected vector, $\mathbf{a}(\theta_1)$, can be estimated directly by experiment. A single emitter is set up at a known location, θ_1. A number of typical data vectors is then recorded in the form of a matrix $\mathbf{D}_1$. It can be shown that the principal left-hand singular vector $\mathbf{D}_1$ then defines a normalized version of $\mathbf{a}(\theta_1)$. A basic explanation of singular valued decomposition, as a mathematical tool for data reduction, is included in Section 2.5. The ratio of the corresponding singular value to the magnitude of the emitter is a measure of the propagation attenuation via the sensor and defines a suitable scaling factor for the vector, $\mathbf{a}(\theta_1)$.

A detailed experimental calibration of the sensor response can, in principle, be repeated at a sufficient number of emitter test sites to enable a representative set of signature vectors to be defined. The objective is to enable a signature vector to be regenerated subsequently, for a point emitter at any value of θ. Establishing a suitable analytic model, such as a complex expontential, can reduce the number of experiments required, but, for reasons of conceptual simplicity, analytic modeling of the basic array manifold will not be examined in this chapter.

In addition, internal state variable parameters or controls, within the sensor, should also be recorded as part of a signature. For example, adaptive feedback, receiver gain control, or a variety of illumination waveforms may be applied. In principle, internal state vectors can be monitored via additional elements in the signature vectors, $\mathbf{a}(\boldsymbol{\Theta})$, or via additional control parameters in a suitable analytic model.

2.2.4 Extended Array Manifold

Definition

In communication, sonar, and radar applications, an unknown input scenario may comprise several pointlike emitters simultaneously. The span of candidate solutions is now greatly increased, and we should, in principle, record an example of every distinct scenario. We introduce, at this stage, the concept of an *extended* array manifold comprising not only the single-emitter scenarios but also all permutations of possible multiemitter scenarios in the larger solution space. The proposal involves higher-rank or multimode signatures that each identify a distinct scenario comprising several pointlike emitters.

The foregoing concept is key to higher-rank or multimode spectrum estimation and to the IMP algorithm described in Section 2.6.

Solution space for the multiemitter case

If there are **r** emitters in the scenario, then the output data matrix, **D**, is assumed to be essentially a linear combination of the data matrices from the individual emitters. **D** is given by

$$\begin{aligned}\mathbf{D} = [\mathbf{d}_1, \mathbf{d}_2, \ldots] &= \sum_{i=1}^{\mathbf{r}} \mathbf{a}(\theta_i)\,[f_{1i}, f_{2i}, \ldots] + \text{observation errors} \\ &= \mathbf{A}_{nr}\,\mathbf{F}_{rk} + \text{observation errors (noise)}\end{aligned} \tag{2.2}$$

where $\mathbf{A}_{nr}$ is a column matrix comprising the r signature vectors and $\mathbf{F}_{rk}$ is a matrix comprising r row vectors representing the temporal signal waveforms with k sample points. The single principal left-hand singular vector of matrix **D** is no longer sufficient to model the data adequately, unless the emitters are fully coherent. In general, r principal singular vectors are now required to define the signal subspace of the data. The r singular vectors are orthonormal and define a rank r subspace. In principle, the subspace can be used to define a signature in the extended array manifold.

There is a *combinatorial explosion* in the number of separate higher-rank signatures required to define the subspaces for every possible permutation of r emitter positions, $\theta_1, \theta_2, \ldots, \theta_r$. Clearly, the number of permutations for even a small number of emitters is excessively large, and it is not practicable to estimate the required subspaces experimentally. However, although the r relevant steering vectors, $\mathbf{a}(\theta_1), \mathbf{a}(\theta_2), \ldots, \mathbf{a}(\theta_r)$, denoted by matrix $\mathbf{A}_{nr}$ are not orthonormal, these vectors occupy the same rank r subspace as defined in the foregoing by the r principal singular vectors. Therefore, we can *synthesize* algorithmically, the required signal subspace for any multiemitter scenario. We are required to measure and record only the basic array manifold vectors, characterized for single-emitter scenarios.

Rather than tediously evaluate singular vectors for all the permitted subspaces from $\mathbf{A}_{nr}$, it is convenient to regard the matrix $\mathbf{A}_{nr}$ as defining a candidate signature in the extended array manifold. A secondary advantage of using this idea is that the matrix $\mathbf{A}_{nr}$ directly includes the respective propagation attenuations for each emitter. The synthesis of a signature, at any required rank, can be performed, as and when required, during the processing stage. Constraints and rules defining permitted permutations, together with relevant prior likelihoods, form part of the prior.

As in the case of the basic array manifold, there is an assumption, implied by the use of normalized steering vectors, that the absolute phases and magnitudes of individual emitters are not useful as discriminants between emitters. It is also assumed that the sensor system is linear.

Extension to nonstationary problems

We have, so far, implicitly considered the case of stationary emitters. We note that it should be possible, in principle, to distinguish a stationary emitter from an emitter where θ changes progressively over the observation interval. Because of the additional discriminant now available, we must now define additional signatures, appropriate to each of the possible distinct solutions. In conceptual terms, the span of the solution space is now further enlarged, and the array manifold must again be extended. This extention can either be defined directly, via additional basis vectors, or indirectly, via a suitable analytic model. The prior must also be extended since, clearly, additional parameters, such as velocity, may lead to an increased incidence of ambiguity.

Sensor calibration

Sensor calibration is a potential problem in many practical systems. Drift, due to aging and environmental factors, leads to errors in the recorded basic steering vectors. The solution space can again be extended to include additional candidate signatures. In many cases, an analytic model can be envisaged, but the most significant contributions to sensor error must be included. Steering vectors in the extended solution space can synthesized by varying appropriate calibration parameters. The synthesis differs from the case of a nonstationary emitter since sensor calibration parameters apply generally to many emitters rather than to individual emitters.

We begin to realize that a conceptual basis for partial sensor calibration (system identification) and partial input identification is being proposed. Clearly, since the problem is likely to be underdetermined, sufficient prior knowledge must be available, in some form, to avoid ambiguities. Depending on the prior knowledge available at the time, the dominant task for the processor may be either that of input identification or that of sensor calibration.

In applications where sensor input and output is already known in detail, the objective becomes purely that of system identification. In general, models describing a sensor system have several degrees of freedom; it follows that the concept of an extended array manifold, with higher-rank or multimode signatures, is also relevant to system identification problems.

Preprocessing transforms

As an example of the application of prior knowledge to the partitioning of signal information from noise, we note that many sampled aperture systems have inherent geometric symmetry or regularity. Corresponding features are directly reflected in the underlying signatures that are required to be identified in the data. By contrast, typical noise components do not possess such symmetry.

Many existing algorithms therefore exploit symmetry as a relatively weak form of a priori knowledge. The aim is to reject antisymmetric components originating from interference. A typical case, for example, is backward-forward linear prediction. A simple processing operation of this kind can improve the marginal (weak signal) performance. The minimum SNR needed for detection or resolution can be reduced by several decibels. However, the discarded antisymmetric components may also contain potentially useful information about the calibration of the sensor system and, in principle, can help not only to identify the presence of errors but also to indicate the necessary corrective action. It is easy to overlook the value of discarded information.

2.2.5 Sensor Noise and Statistical Observation Error

It is inevitable that incoming data, obtained through a practical sensor, include thermal noise. Consequently, there is an information theoretic bound on the ability of an algorithm to differentiate or resolve similar candidate solutions. We have, so far, regarded the data analysis problem as one of decomposition into recognizable components, attributable to predefined cause-effect relationships specified in an extended array manifold. Clearly, in many applications, we should include thermal sensor noise in the category of an identifiable cause and can regard a record of typical background noise as an additional signature. In a practical problem, it may not be possible to define the noise exactly, and there may be range of candidate noise signatures, identified by one or more parameters.

Additive thermal sensor noise can be estimated experimentally for the case where no emitters are present in the input scenario. A large number of typical data vectors, denoted **R**, are recorded. In the case of Gaussian statistics, a covariance estimate is defined by computing the matrix product. $\mathbf{R}\,\mathbf{R}^{\dagger}$, where superscript † denotes the complex conjugate (Hermitian) transpose of a matrix or vector. A normalized estimate is obtained by averaging according to the number of noise vectors. The result can be regarded as sufficient to represent the properties of the thermal sensor noise. Other types of linearly additive interference can be modeled similarly. In the non-Gaussian case, additional statistical properties must be defined and estimated.

A signature for noise is generally of high rank or full rank. The primary discriminant is one of magnitude, but other statistical properties of both noise and emitters are also relevant. For example, the tracks of potential emitters in a candidate solution should possess predefined characteristics. If the number of emitters in the solution is increased, then one or more of the tracks is likely to break up into discontinuous points or segments. In conceptual terms, the extended array manifold must be constrained to exclude solutions that are not acceptable for any reason. Knowledge of permitted track behaviour should be available to the algorithm in usable form.

In general, the objective of processing is perceived as the allocation of all components in the available data to predefined causes. Residual information that cannot be allocated either to emitters or to thermal noise indicates additional, as yet unexplained, observation errors, incorrect models, interference, or an input scenario in a region of the solution space that has not been searched algorithmically. We should be prepared to recognize such cases and take appropriate action rather than, as is the tendency in many existing algorithms, blindly attribute the residue to sensor noise.

2.2.6 Output Actions

The required output actions should also be established in the learning or research stage. The requirements might include the control or update of visual displays, printers, audio systems, communication systems, or the many types of actuators used to control automatic systems. In practice, we may need to address estimation accuracy, resolvability, observability, and aspects of control theory. To minimize the cost of an incorrect decision, the supervising algorithm may also need to be aware of the consequences of alternative output actions. For example, we would not wish to initiate a war on the basis of a false detection of a nuclear missile. In other circumstances, the consequences of missed detection, for example, of a sea-skimming missile aimed at your ship, can also be disastrous. The nature of the problem, therefore, has impact on detection thresholds.

We need, therefore, to control detection and decision criteria knowing the cost of a mistake, and to include corroborative evidence from other sensor systems, higher-level discriminants (such as track behavior), and previous experience. In many processing schemes, the validation stage is separate from the initial processing. It is suggested that this approach is potentially inefficient.

2.2.7 The Inverse Problem

The first step, in a new development, is tentatively to propose a system comprising suitable sensors, processor hardware, and output devices. The next requirement is to define the operations to be performed by the processor. The final system design is evolved to meet operational specifications.

In principle, we could try to teach the proposed processor by presenting a series of *known* test input scenarios to the sensors and demanding that the output devices give the appropriate response by giving some suitable reward or penalty in terms of a cost function. We have seen that the number of scenarios needed might be extremely large and that some method of generalizing is required. It is the development of efficient methods of explicitly representing and implementing knowledge via a defined algorithm that distinguishes the concepts outlined in this chapter from those of neural networks and learning systems.

In conceptual terms, the development of a suitable processing algorithm requires that the essential properties of the forward transfer function between scenario and sensor output are known. Subsequently, given the sensor data alone, we must reverse this transform to infer an unknown input scenario. Our main task is the solution of this *inverse problem.*

We repeat that successful algorithmic interpretation of data depends on valid prior knowledge in the form of signatures, parametric models, rules, and predefined features. Algorithms cannot invent new solutions but are able only to synthesize candidate solutions based on models and rules. These models are generally controlled by a complex combination of many variable parameters. The principal difficulty is that, in general, the solution space is extremely large and the parameters cannot be estimated independently. The corresponding multidimensional likelihood surface is highly complex in shape.

We have seen that the forward transform is characterized by obtaining typical examples of the sensor output state for use as templates, fingerprints, or signatures in the extended array manifold. We should identify and eliminate features, such as rf carriers and speaker-dependent features, that are common to all candidate solutions in the required manifold. Similarly, interference should also be suppressed.

In principle, to solve the inverse problem by analytic means, we infer which input condition has been applied by *matching* sensor data to the remaining characteristics of each of the predefined signatures in turn. A search for the best match indicates a preferred solution. Clearly it is preferable that the signatures, as observed through the sensor, should be unique; otherwise, ambiguity will arise. If the differences between signatures are not discernible from sensor noise, then the inverse problem is said to be underdetermined or ambiguous. To select just one prefered solution, it is then necessary to refer to information in the prior likelihood function.

A unique or best solution can be obtained only if there is sufficient information, in total, for the problem to be overdetermined. For example, a subspace of given rank can, in general, be defined, within a reasonable tolerance, by several different linear combinations of basis vectors from the basic array manifold. To resolve the ambiguity, the prior should define explicitly the properties of the combination that is the most likely. The number of emitters in a scenario is usually an unknown parameter and must be estimated in solving the inverse problem. The prior might then indicate that the simplest *permissible* combination is the most likely. In this case, potential ambiguity is resolved by choosing that solution requiring the least acceptable number of emitters. This basic rule can be extended to resolve other ambiguities, for example, arising between nonstationary and stationary solutions. In line with the scientific approach to any problem, we may suggest that the prior should be strongly biased toward the simplest solution consistent with the available data.

Although there are many detailed algorithmic requirements in practical applications, two prominent objectives can be isolated:

1. Partition signal information from stocastic background components (noise and interference)
2. Isolation of deterministic components (signals) from each other and identification, by inference, of the most probable input scenario

The algorithm chosen to solve the inverse problem should not be overly sensitive to sensor calibration errors and should be adequately robust in less than ideal environmental conditions. Clearly, we cannot expect robustness if the underlying models do not have sufficient degrees of freedom in the specific application. The algorithm must also give appropriate weight to all available information whether originating from the prior or from the current sensor data.

The concept of simply *matching* all permitted candidate signatures to incoming data is not easily implemented in practice due to the large number of possible solutions. Ideally, the relative likelihood of every candidate solution must be evaluated using Bayesian principles. This approach is usually termed the maximum-likelihood method [2] and is regarded by most as excessively expensive in computational terms. The main emphasis of this chapter is on developing an understanding of near-optimal efficient methods of solution.

2.2.8 Implementation of the Prior

We have seen that, if the inverse problem for data alone is underdetermined, then we need to obtain additional support, independently of sensor data. Two basic options for implementation are suggested, either separately or in combination. In principle, a complete composite likelihood surface can be computed using both the sensor data and the prior. An exhaustive search for the global maximum of the resulting likelihood surface can then be initiated.

In general, the prior is initially defined independently of the sensor system. Clearly, the explicit evaluation of a full multidimensional surface for the prior, as a look-up table, is expensive in terms of storage requirement. It is easier, in general, to define the prior via a set of rules and models and to build these into an algorithm. A search pattern is developed that makes use of the prior to define a subset comprising the most likely solutions.

The strategy is to match sensor data to this limited number of individual candidate solutions and then, in order of decreasing prior likelihood, search a second set and third set until a satisfactory match is found. For example, simple solutions are usually given a greater prior likelihood than are complex candidates. It also follows, on grounds of computational efficiency, that simple candidate solutions should be tested before complex solutions. The search can then be terminated as soon as a satisfactory solution is obtained. Although a number of potentially valid solutions may exist, the first satisfactory match that is obtained between a candidate solution and the available data is adopted as the unique solution. Clearly, there is a potential reduction in computational load, particularly if the simpler solutions are easily tested.

As the complexity of a candidate solution increases, there is a combinatorial explosion in the span of the solution space. At a given level of complexity, there may be many candidate solutions that have equal prior likelihood. The search at the higher levels of complexity then becomes expensive. We also realize that the prior can be modified as a result of the first search, and again as a result of the second search. The complex solutions can then be reordered such that unlikely candidates are excluded from the search.

In general, deductions can be made progressively both about the scenario and the calibration of a sensor system; the modified prior is used to direct the search in the remainder of the analysis. In particular, since calibration parameters change relatively slowly, it is sensible to treat these parameters as part of the prior when searching for a best candidate input scenario. Clearly, we can conceive of an algorithm that alternates between system identification and high-resolution processing. The computational savings, implicit in a directed search of this type, lead to highly efficient algorithms such as IMP.

We should be aware that prior knowledge also has associated uncertainty. Rules defined in the prior may not be valid in all circumstances. In a number of algorithms, such as the Kalman filter, the number of degrees of freedom in a model is predetermined by the prior. In marginal cases, where SNR is insufficient, the inverse problem may then be underdetermined. In most cases, the resulting instability of the solution is unacceptable. Solutions obtained by rigid imposition of an inaccurate prior must be regarded as suboptimal. Since all uncertainty should be included in quantitive terms, we begin to realize that the necessary statistical information is best obtained by direct measurement rather than by heuristic, convenient, or simplified rules based on qualitive features.

A primary aim in algorithm development for specific applications is seen as that of the translation of the relevant prior into robust computationally efficient procedures for interpreting data in terms of cause and effect. Examples of how this objective might be achieved in practice are discussed, in a mathematical context, in Sections 2.3 through 2.7. The discussion is based on the joint concepts of an extended array manifold and of higher-rank spectrum estimation. The techniques are compared with existing methods.

2.3 THE POINT CONSTRAINT

In conceptual terms, the main conclusion with regard to solving the inverse problem is that sensor data should be *matched* to individual candidate signatures, defined in the extended array manifold as characteristic of the output of the sensor system to each and every known input scenario. Solutions outside the manifold are excluded. For a single-point emitter, it is well established that the matching process is best performed by computing optimum *matched filters.* Each filter is matched to one specific signature such that the response of the combined sensor and filter to an input emitter of unit reference magnitude is unity. The phase of the output also reflects the phase of the input emitter. If no interference is present and, if the sensor system is linear, then the output waveform from the filter should correspond exactly to that transmitted. The concept of a matched filter, as applied to scenarios with a single pointlike emitter, forms the basis for the classical spectrum estimation methods to be discussed in Section 2.4. The method is the basis of a search in many existing high-resolution algorithms (Section 2.5). Extending the concept of matched filtering to the signatures of a multiemitter scenario provides the basis for higher-rank spectrum estimation and for the IMP algorithm discussed in Section 2.6.

2.3.1 The Matched Filter for the Case of a Single Emitter

The narrow-band matched filter

For the narrow-band case, where time delays are small compared to the inverse of the signal bandwidth, it is well established that, for each rank-one signature vector, $\mathbf{a}(\theta)$, we can compute a matched filter, $\mathbf{a}^{\#}(\theta)$, such that

$$\mathbf{a}^{\#}(\theta)\mathbf{a}(\theta) = 1 \quad \text{(for each value of } \theta\text{)} \tag{2.3}$$

where $^{\#}$ denotes the Moore-Penrose pseudoinverse. The row vector, $\mathbf{a}^{\#}(\theta)$, is therefore given by

$$\mathbf{a}^{\#}(\theta) = [\mathbf{a}^{\dagger}(\theta)\,\mathbf{a}(\theta)]^{-1}\,\mathbf{a}^{\dagger}(\theta) \tag{2.4}$$

We see by inspection that equation (2.4) satisfies the requirement of unit response to a reference emitter at a phase and magnitude defined by the signature vector, $\mathbf{a}(\theta)$. The conjugation, implicit in the Hermitian transpose, denoted †, reverses the *phase shifts* introduced by propagation from the emitter to the output port of each sensor. The scaling factor $[\mathbf{a}^{\dagger}(\theta)\,\mathbf{a}(\theta)]^{-1}$ compensates for the propagation *attenuation.* The relative magnitudes of the individual elements in weight vector, $\mathbf{a}^{\dagger}(\theta)$, determine the relative weights given to each sensor in the matched filter. The full span of candidate vectors, defining matched filters for the single emitter case, is denoted $\mathbf{a}^{\#}(\Theta)$. The concept of signature vectors and corresponding matched filters generalizes to discriminants other than frequency and also to multiple discriminants.

The overall size of the resolution cell is reduced by including other discriminants in the signatures, with the principal advantage of improved discrimination between emitters (selectivity). However, the number of independent cells is increased, and the computational load may be high. In radar, for example, we might use, as discriminants, a pulse compression waveform in range, direction-of-arrival steering vectors in both planes, and Doppler waveforms to construct a large number of complex highly selective matched filters. At least one filter for each distinct resolution cell is required.

Leakage

The response of the matched filter is unity to the target signature only, but the response is not zero for most other candidate signatures. The reason is that there are, in general, similarities between signatures. Features that are common to another signature are not rejected by the filter. The effect is termed leakage. Consequently, in a multiemitter scenario, several other emitters may *leak* into the output of a selected matched filter. The output then comprises a linear combination of the emitter waveforms. The level of leakage can be predicted for a given filter, $\mathbf{a}^{\#}(\theta)$ by evaluating the function

$$\mathbf{P}(\theta, \boldsymbol{\Theta}) = \mathbf{a}^{\#}(\theta)\, \mathbf{a}(\boldsymbol{\Theta}) \tag{2.5}$$

$\mathbf{P}(\theta, \Theta)$ defines both the sidelobe response and the width of the main resolution cell (mainlobe) of a selected filter. The response is variously termed, according to application, as a point spread function, polar diagram, beam pattern, impulse response, or filter characteristic. Leakage is the cause of resolution and sidelobe limitations in the classical spectrum estimate. It should be noted that $\mathbf{P}(\theta, \Theta)$ is also related to the cross-correlation coefficient between a pair of signatures but is normalized differently.

Whitening transform

Leakage into the output of a matched filter also arises from *noise* components in the sensor data. The level of noise contaminating the output waveform depends primarily on the width of the mainlobe, but sidelobe leakage may also contribute a similar level of noise due to the greater span. The detectability of an emitter is degraded by noise. The objective of a whitening transform is to minimize leakage from noise and other interference.

The filter weight vector, $\mathbf{a}^{\#}(\theta)$, from equation (2.3), defines the weighting given to information provided by each of the sensors in an array, irrespective of the noise contribution from that sensor. Equation (2.3) defines the optimum matched filter only if the noise is distributed uniformly over the sensors and is uncorrelated. In general, an additional weighting function, related to the SNR of the available information at the sensors, must be included. If we change the design of a sensor system by including an arbitrary preprocessing transform, such as $\mathbf{W}_0$, then we can regard the *modified* sensor system in the same way. A signature in the *modified* system is then $\mathbf{W}_0^{\dagger}\mathbf{a}(\theta)$. Matched filters for the *modified* system are therefore defined by

$$[\mathbf{W}_0^{\dagger}\mathbf{a}(\theta)]^{\#}\, \mathbf{W}_0^{\dagger}\mathbf{a}(\theta) = 1 \quad \text{(for each value of } \theta) \tag{2.6}$$

Clearly, for maximum SNR, $\mathbf{W}_0$ should be chosen to minimize leakage. For example, noiselike interference may be generated by a jammer at a given location within the field of view of the sensor system. To reduce the effects of the interference, we should steer a null on that location via $\mathbf{W}_0$. The depth of the null should be inversely proportional to the magnitude of the interference. Unwanted additive noise contributions can be defined by a data matrix, denoted **R**. No emitters should be present, but all significant sources of interference must be included. If the statistics are Gaussian, then the optimum prewhitening transform is given by

$$\mathbf{W}_0^\dagger \, \mathbf{R} \, \mathbf{R}^\dagger \, \mathbf{W}_0 = \textbf{identity} \tag{2.7}$$

$\mathbf{W}_0$ can be evaluated as a pseudoinverse of $\mathbf{R}^\dagger$. If the whitening transform is included explicitly, as a separate stage, then the *modified* data become $\mathbf{W}_0^\dagger \, \mathbf{D}$ and the *modified* matched filters are given by

$$[\mathbf{W}_0^\dagger \mathbf{a}(\Theta)]^\# = [\mathbf{a}^\dagger(\Theta) \, \mathbf{W}_0 \, \mathbf{W}_0^\dagger \mathbf{a}(\Theta)]^{-1} \mathbf{a}^\dagger(\Theta) \, \mathbf{W}_0 \tag{2.8}$$

The noise is then defined by $\mathbf{W}_0^\dagger \, \mathbf{R}$ and the covariance is the identity matrix. (It is assumed that both data and noise vectors are normalized in proportion to the effective duration of the integration window.) We can easily deduce that the magnitude of the response to the *modified* filters to noise is unity (isotropic) for all θ. The reference matrix, **R**, must be obtained independently of **D**; otherwise, signals are suppressed or partially suppressed. In an active stationary problem, **R** can sometimes be obtained by switching off the active illuminating waveform. In passive applications, the background reference might be obtained, for example, from adjacent frequency bands or other sensors.

Adaptive cancellation

In certain applications, auxiliary sensors are deployed spatially to detect external interference. The reason, in general, is that external interference is not stationary and is not cooperative. Separate reference and data matrices, **R** and **D**, cannot be obtained easily. A suitable discriminant is required.

Adaptive sidelobe cancellers, for example, use a highly directional main antenna with several auxiliary wide-beam sensors. A whitening transform, $\mathbf{W}_d$, is computed as a time-varying pseudoinverse of incoming data from all sensors. This transform provides whitening in the spatial dimension. Matched filters, $[\mathbf{W}_d^\dagger \, \mathbf{a}(\Theta)]^\#$, if based on a steering vector utilizing all sensors, would then give an isotropic output as a function of θ. The length of the signature vectors, $/\mathbf{W}_0^\dagger \mathbf{a}(\Theta)/$, can, in principle, provide an indication of the location of pointlike interference as a function of θ. However, in many sidelobe cancellation systems, the output of the main antenna alone is used, on the basis that the signal information is concentrated in the output of the main sensor. The effect is that of an additional projection, through a subspace defined by a constraint vector denoted $\mathbf{Z}_1$. In the case in question, $\mathbf{Z}_1$ comprises $\mathbf{n}_s$ elements, one element is unity, and the remainder are zero. The nonzero element corresponds to the main sensor.

The product $[\mathbf{W}_d^\dagger \, \mathbf{Z}_1]^\# \, \mathbf{W}_d^\dagger \, \mathbf{D}$ defines the output waveform from a constrained canceller. The matrix $\mathbf{W}_d$ is evaluated using several weighted data vectors from **D**. Efficient numerical methods of recursive computation are available [3].

In an adaptive canceller for a multisensor antenna array with similar sensors, the constraint, $\mathbf{Z}_1$, can be replaced by a steering vector, $\mathbf{a}(\theta_c)$, where θ_c defines the direction of active illumination. Signal extraction is based on applying a second stage of filtering, for example, using range as a discriminant. It is assumed that the mark-to-space ratio of the desired signals is such that there is little mean contribution to $\mathbf{W}_d$ and, consequently, minimal signal suppression.

Multiple constraints can also be specified. $\mathbf{Z}_r$ defines a rank r subspace. The constraints provide a method of limited discrimination between interference and wanted information. Constraints are defined by prior knowledge. Some suitable discriminatory aspect of the signal information is necessary.

We realize that, in general, further processing, using nonspatial discriminants, is required to separate signals from a spatially whitened background. It is the algorithmic representation and implementation of suitable filters that exercise many design engineers.

For mathematical compactness in the remainder of the chapter, we shall, in general, assume that the whitening transform is included *implicitly,* where necessary, and do not consider further adaptive cancellation methods.

Signal extraction

The vector inner products $\mathbf{a}^{\#}(\hat{\theta})\,\mathbf{D}$ or $[\mathbf{W}_0^{\dagger}\mathbf{a}(\theta)]^{\#}\mathbf{W}_0^{\dagger}\,\mathbf{D}$ are the mathematical equivalent of applying a signal extraction filter to the data. The result is an estimate of the phase and magnitude of the emitter at each frame or snapshot. A suitable value of $\hat{\theta}$ must be estimated by algorithmic means.

Projection operator, signal blocking matrix, and modeling residue

The outer product of a selected candidate signature vector, $\mathbf{a}(\theta)$, and the corresponding matched filter vector, $\mathbf{a}^{\#}(\theta)$, form a square matrix, $\mathbf{a}(\theta)\,\mathbf{a}^{\#}(\theta)$, of size n_s by n_s. This matrix has one nonzero eigenvalue and is therefore rank one. Since the eigenvalue is unity, the operation defines projection through a rank-one subspace.

For example, given an estimate $\hat{\theta}$, the product $\mathbf{a}(\hat{\theta})\,\mathbf{a}^{\#}(\hat{\theta})\,\mathbf{D}$ defines a projection of the data vectors, $\mathbf{d}_1$, $\mathbf{d}_2$, . . . , through a candidate signal subspace defined by the selected signature. The resulting vectors form a matrix, $\hat{\mathbf{D}}$, that can be regarded as a rank-one estimate of the data matrix, $\mathbf{D}$. Clearly we can select a value of θ, from the span $\boldsymbol{\Theta}$, that optimizes the estimate of $\hat{\mathbf{D}}$ according to a cost function. The corresponding value, $\hat{\theta}$, is the required parameter estimate.

In least-mean-square (LMS) processing, the difference, $\mathbf{D} - \hat{\mathbf{D}}$ (modeling residue), is minimized as a function of θ. Weighted mean squares is used as the cost function. Clearly, the required residue can also be generated by a projection of $\mathbf{D}$ through the orthogonal subspace, $\mathbf{Q}_1$, defined by

$$\mathbf{Q}_1 = [\mathbf{identity} - \mathbf{a}(\theta)\,\mathbf{a}^{\#}(\theta)] \tag{2.9}$$

The subscript $_1$ denotes the rank of the missing subspace. The modeling residue, given by $\hat{\mathbf{Q}}_1\mathbf{D}$, can be minimized as required to give the estimate, $\hat{\theta}$, and the corresponding subspace matrix, $\hat{\mathbf{Q}}_1$.

Since the subspace defined by $\hat{\mathbf{Q}}_1$ is orthogonal to the signature vector, $\mathbf{a}(\hat{\theta})$, it can be regarded as a *signal blocking matrix*. It is useful to consider the Hermitian matrix, $\hat{\mathbf{Q}}_1$, in more detail. For example, if one emitter only is present in the input scenario, with a signature $\mathbf{a}(\theta)$, and if the parameter estimate $\hat{\theta}$, is sufficiently close to the true value, then the residue matrix, $\hat{\mathbf{Q}}_1\mathbf{D}$, should comprise noise only.

It is important to realize that $\hat{\mathbf{Q}}_1$ is an analogous transform to $\mathbf{W}_0$ and can be similarly applied as a preprocessing or weighting transform in a *modified* sensor system. In this case, $\hat{\mathbf{Q}}_1$ is a projection operator. In consequence, the *signal blocking transform* can be applied to data, to signatures, to *matched filters*, and to the background noise. The resulting effect is equivalent to a *modified* sensor system that is *not sensitive* to an emitter at the selected location, $\hat{\theta}$. There is, therefore, *no sidelobe leakage* from that emitter, in the matched filters defined for the *modified* system.

If the relative magnitudes of the *modified* signatures, $\hat{\mathbf{Q}}_1 a(\Theta)$, are evaluated, as a function of θ, then a null can be observed at $\hat{\theta}$. The null complements the size and shape to the corresponding resolution cell. The *modified* matched filters are defined by the pseudoinverse of each of the *modified* signatures and are defined by

$$[\hat{\mathbf{Q}}_1\ \mathbf{a}(\Theta)]^{\#} = [\mathbf{a}^{\dagger}(\Theta)\ \hat{\mathbf{Q}}_1\ \mathbf{a}(\Theta)]^{-1}\ \mathbf{a}^{\dagger}(\Theta)\ \hat{\mathbf{Q}}_1 \tag{2.10}$$

We note that $\hat{\mathbf{Q}}_1 = \mathbf{Q}_1^{\dagger} = \hat{\mathbf{Q}}_1\hat{\mathbf{Q}}_1$. The normalization terms, $[\mathbf{a}^{\dagger}(\Theta)\ \hat{\mathbf{Q}}_1\ \mathbf{a}(\Theta)]^{-1}$ compensate exactly for the apparent increase of sensor attenuation in the vicinity of the null. It can be seen that each of the matched filters, defined by equation 2.10, has unit response to the corresponding signature vector, $\mathbf{a}(\theta)$. At the precise center, $\hat{\theta}$, of the null the response is zero/zero (undefined) but is, in general, of no interest. If a second emitter is present in the input scenario, then that should appear in the output of the modified candidate matched filters *without leakage from the first emitter*. We can therefore, in principle, disregard the effects of the first emitter when searching for the filter that is best matched to the second emitter. The concept of the signal blocking transform is key to the efficient implementation of a directed search for multiple emitters in the IMP algorithm.

We note that features common to both the signatures of the first and the second emitter are suppressed by the signal blocking transform evaluated for either emitter. The corresponding signal extraction filters are matched only to the differences between the signatures. Since the magnitudes of the differences are, in general, less than the magnitude of either emitter, the SNR is reduced in both filters. This concept is important to the evaluation of a resolution limit.

Signal-to-noise ratio

The magnitude of the noise or interference present in the output of a matched filter can be predicted by applying the filter in question to the matrix, **R**. It is common practice to define the signal-to-noise power ratio by

$$\mathrm{SNR} = \frac{[\mathbf{a}^\dagger(\theta)\ \mathbf{a}(\theta)]^{-2}\ \mathbf{a}^\dagger(\theta)\ \mathbf{D}\ \mathbf{D}^\dagger\ \mathbf{a}(\theta)}{[\mathbf{a}^\dagger(\theta)\ \mathbf{a}(\theta)]^{-2}\ \mathbf{a}^\dagger(\theta)\ \mathbf{R}\ \mathbf{R}^\dagger\ \mathbf{a}(\theta)} \tag{2.11}$$

Clearly, the inverses in the numerator and the denominator can be canceled. The SNR for a *modified* filter can be evaluated similarly and simplified to

$$\begin{aligned}\mathrm{SNR} &= \frac{\mathbf{a}^\dagger(\theta)\ \hat{\mathbf{Q}}_1\ \mathbf{D}\ \mathbf{D}^\dagger\ \hat{\mathbf{Q}}_1\ \mathbf{a}(\theta)}{\mathbf{a}^\dagger(\theta)\ \hat{\mathbf{Q}}_1\ \mathbf{R}\ \mathbf{R}^\dagger\ \hat{\mathbf{Q}}_1\ \mathbf{a}(\theta)} \\ &= \frac{\mathbf{a}^\dagger(\theta)\ \hat{\mathbf{Q}}_1\ \mathbf{D}\ \mathbf{D}^\dagger\ \hat{\mathbf{Q}}_1\ \mathbf{a}(\theta)}{\mathbf{a}^\dagger(\theta)\ \hat{\mathbf{Q}}_1\ \mathbf{a}(\theta)}\end{aligned} \tag{2.12}$$

if whitened. If the signal blocking matrix cancels one of two signals, then equation 2.12 predicts the effective SNR of the other.

We have seen that, in a multiemitter scenario, an optimum signal extraction filter for one emitter should include nulls at the other emitter locations. The objective is to minimize leakage. We note that, if two emitters are close in spatial resolution terms, then the apparent signal power of one is reduced because of the proximity to the null steered on the other emitter. Because features common to both emitters are suppressed by $\mathbf{Q}_1$, the SNR is, in general, less than that predicted by equation 2.11. However, if the SNR of the remaining *difference* component is adequate, then the two emitters can, in principle, be resolved. Since the SNR of the difference components can be computed, a statistical bound on resolution can be evaluated. The bound bears no relationship to the Rayleigh resolution criterion.

There is, therefore, nothing theoretically suspect about the concept of *super-resolution;* the difficulty lies solely in the specification of suitable filters, correctly matched to the differences between the relevant signatures.

The wide-band matched filter

In the wide-band case, where phase shift is not constant over the signal band, the matched filter, $\mathbf{a}^{\#}(\theta)$, should reverse the relative *time delays* of signals arriving via each sensor, rather than the relative phase shifts. In conceptual terms, the required signature is achieved most readily by concatenating several adjacent column vectors from a test data matrix, to form a longer template vector that includes an appropriate aperture in the time delay domain. For processing expediency, redundancy in this longer vector should be removed. A suitable technique is based on projection, using singular vectors to define a subspace [4]. A brief description of singular value decomposition is included in Section 2.4. The analysis for the wide-band case then follows that of the narrow-band signatures. Similar considerations apply if other discriminants are available, used alone or in combination.

2.3.2 Matched Filters for the Extended Array Manifold

Multimode or higher-rank matched filter for the multiemitter case

For the proposed linear system, where the temporal waveforms (magnitudes and phases) of emitters are not discriminants, a higher-rank signature, $\mathbf{A}_{nr}$, comprises r signatures as column vectors, $\mathbf{a}(\theta_1)$, $\mathbf{a}(\theta_2)$, . . ., $\mathbf{a}(\theta_r)$, corresponding to r individual emitters each at unit reference magnitude. If we assume that the signatures are sufficiently dissimilar, then these vectors define a rank r subspace. We can compute a rank r matched filter, $\mathbf{A}_{nr}^{\#}$, such that

$$\mathbf{A}_{nr}^{\#}\mathbf{A}_{nr} = [\mathbf{A}_n^{\dagger r}\mathbf{A}_{nr}]^{-1}\,\mathbf{A}_n^{\dagger r}\mathbf{A}_{nr} = \textbf{identity} \tag{2.13}$$

A row of matrix $\mathbf{A}_{nr}^{\#}$ can be regarded as a weight vector or filter that is matched to a corresponding column of the signature matrix, $\mathbf{A}_{nr}$. The response to the other $r - 1$ signatures defined in $\mathbf{A}_{nr}$ is zero. The response is given directly by one row of the identity matrix. The matrix product $\mathbf{A}_{nr}^{\#}\,\mathbf{D}$ provides estimates, denoted $\hat{\mathbf{F}}_{kr}$, of the temporal waveforms (magnitudes and phases) of each of r emitters in the input scenario. The matrix $\mathbf{A}_{nr}^{\#}$ can therefore be regarded as multimode signal extraction filter for the r individual emitter waveforms. The *locations* of the point emitters are defined by the values of θ used to form $\mathbf{A}_{nr}$. The inverse problem reduces to that of estimating a suitable signature matrix, $\hat{\mathbf{A}}_{nr}$.

Signal blocking matrix for a higher-rank signature

Since the matrix $\hat{\mathbf{A}}_{nr}\,\hat{\mathbf{A}}_{nr}^{\#}$ should have r nonzero eigenvalues, each at unit magnitude, it defines a projection transform through a rank r subspace. As a generalization of the rank-one case, that is discussed in 2.3.1, we can similarly project data vectors, $\mathbf{D}$, through the rank r subspace defined by $\hat{\mathbf{A}}_{nr}\,\hat{\mathbf{A}}_{nr}^{\#}$. The result of the projection, denoted $\hat{\mathbf{D}}$, is defined by the matrix product

$$\hat{\mathbf{D}} = \hat{\mathbf{A}}_{nr}\,\hat{\mathbf{A}}_{nr}^{\#}\,\mathbf{D}$$

or, equivalently, by

$$\hat{\mathbf{D}} = \hat{\mathbf{A}}_{nr}\,\hat{\mathbf{F}}_{kr} \tag{2.14}$$

The projection can be regarded as a two-stage transform that is directly dependent on the point constraint. First, a matched filter (reverse) transform, $\hat{\mathbf{A}}_{nr}^{\#}$, is applied to estimate a pointlike version of the input scenario, and second, the corresponding cause to effect (forward) transform, $\hat{\mathbf{A}}_{nr}$, is applied. The latter step predicts the data matrix, $\hat{\mathbf{D}}$, corresponding to the proposed deconvolved scenario. The objective of the search process becomes that of finding a signal subspace projection that leaves **D** essentially unchanged. Equivalently, in a least-mean-square (LMS) context, the search algorithm could synthesize candidate *signal blocking matrices* or subspace projections, $\mathbf{Q}_r$, that generate the appropriate modeling residue, $\mathbf{D} - \hat{\mathbf{D}}$, defined by

$$\mathbf{Q}_r\,\mathbf{D} = [\textbf{identity} - \hat{\mathbf{A}}_{nr}\,\hat{\mathbf{A}}^{\#}_{nr}]\,\mathbf{D} \tag{2.15}$$

If the noise statistics are Gaussian, an SNR for each of the signals can be computed, in a manner similar to equation (2.12). A different signal blocking matrix, $\mathbf{Q}_{r-1}$, is required for each evaluation.

We observe from equation (2.13) that the pseudoinverse $\hat{\mathbf{A}}^{\#}_{nr}$ includes a square matrix $[\hat{\mathbf{A}}^{\dagger}_{nr}\,\hat{\mathbf{A}}_{nr}]$, of size r by r. Values in this matrix, denoted $\mathbf{P}_{rr}$, can be derived from the relevant point spread functions, $\mathbf{P}(\theta, \boldsymbol{\Theta})$, at $\theta = \theta_1, \theta_2, \ldots, \theta_r$. From an intuitive viewpoint, we note that $\mathbf{P}_{rr}$ serves two functions:

1. Diagonal elements of $\mathbf{P}_{rr}$ define the propagation attenuation for each emitter.
2. Off-diagonal elements of $\mathbf{P}_{rr}$ define the respective leakage terms (equation 2.5).

Since, in most practical cases, the size of the matrix $\mathbf{P}_{rr}$ is small, the inverse is easily computed. The inverse reverses the propagation attenuation and the effects of leakage. Also, since the cross-correlation between any pair of signatures is dependent only on the array manifold and is independent of the data, it is possible to interpolate the values required in matrix $\mathbf{P}_{rr}$ from *precomputed* point spread (leakage) functions. A suitable grid of sample points is required.

2.4 SPECTRUM ESTIMATION

2.4.1 The Classical Spectrum Estimate

Almost all existing data analysis algorithms are explicitly dependent on matching incoming data vectors to rank-one signatures defined by the basic or unextended array manifold. This search process is performed, in the case of complex exponentials, by classical spectrum estimation based on the inverse Fourier transform. In the generalized case of arbitrary signature vectors, we are required to compute appropriate matched filters.

It has been stated that the SNR for one candidate matched filter can be predicted directly (equation 2.11), given typical background noise vectors, **R**, and a batch of data vectors, **D**. It follows that the SNR can be similarly predicted for all candidate matched filters over the span defined by $\boldsymbol{\Theta}$. In conceptual terms, we simply present the data vectors, in turn, to the input ports of candidate filters and obtain a prediction of the average signal power, $S_0(\boldsymbol{\Theta})$ as a function of θ. We then similarly present reference vectors, defining the background noise (observation error), to a set of identical filters and predict the averaged output noise power for each filter, $N_0(\boldsymbol{\Theta})$. We denote the ratio, as a function of θ, as $\mathrm{SNR}_0(\boldsymbol{\Theta})$, given by

$$\mathrm{SNR}_0(\boldsymbol{\Theta}) = \frac{S_0(\boldsymbol{\Theta})}{N_0(\boldsymbol{\Theta})} = \frac{\mathbf{a}^\dagger(\boldsymbol{\Theta})\,\mathbf{D}\,\mathbf{D}^\dagger\,\mathbf{a}(\boldsymbol{\Theta})}{\mathbf{a}^\dagger(\boldsymbol{\Theta})\,\mathbf{R}\,\mathbf{R}^\dagger\,\mathbf{a}(\boldsymbol{\Theta})} \tag{2.16}$$

where the normalizing terms cancel and have been omitted. We also remember that, if the correct prewhitening has been applied, then $\mathbf{R}\,\mathbf{R}^\dagger$ should be an identity matrix. $N_0(\Theta)$ should be isotropic (unity). The evaluation of SNR then simplifies to

(2.17)

$$\mathrm{SNR}_0(\boldsymbol{\Theta}) = \mathbf{a}^{\#}(\boldsymbol{\Theta})\,\mathbf{D}\,\mathbf{D}^\dagger\,\mathbf{a}(\boldsymbol{\Theta})$$

$$= \sum_{k} \mathbf{d}_k^\dagger\,\mathbf{a}(\boldsymbol{\Theta})\,\mathbf{a}^{\#}(\boldsymbol{\Theta})\,\mathbf{d}_k \tag{2.17}$$

In the special case where the signatures are pure sinusoids, and the noise background is white, then $\mathrm{SNR}_0(\boldsymbol{\Theta})$ is the classical power spectrum density obtained via a discrete Fourier transform (DFT) (with padding). We refer to $\mathrm{SNR}_0(\boldsymbol{\Theta})$ as the classical spectrum estimate. Although spectrum estimation, in a strict classical sense applies only to complex exponentials, in this chapter, we also refer to the case where the signatures are arbitrary basis vectors as rank-one spectrum estimation. The operation can be regarded as the projection of data vectors through rank-one subspaces defined by the basic array manifold for single-emitter scenarios.

For Gaussian statistics, it can be shown that a value of $\mathrm{SNR}_0(\theta)$ is related monotonically to the likelihood of a given input scenario. The most likely solution to the inverse problem is therefore indicated by the filter with the highest SNR. The solution to the inverse problem, for a single signal, is found by a relatively simple one-dimensional *search* for the maximum value of $\mathrm{SNR}_0(\boldsymbol{\Theta})$. We denote the value of θ at the maximum SNR by $\hat{\theta}_1$. If one emitter only is present then the corresponding matched filter acts as the optimal signal extraction filter. An estimate of the emitter waveform is delivered as a function of time. If θ is a continuously variable parameter, then interpolation between sample points is necessary.

An alternative approach for finding $\hat{\theta}_1$ is that of minimizing the mean square of the residue, denoted $R_1(\boldsymbol{\Theta})$, that is *rejected* by each of the candidate matched filters. The latter approach is that of least mean squares. We note that the sum of the averaged residue power, $R_1(\boldsymbol{\Theta})$, and averaged signal power, $S_0(\boldsymbol{\Theta})$, corresponds to the total power of the data vectors. The two cost functions are therefore complementary.

A novel concept is that of estimating the ratio of residue to noise power, denoted as $\mathrm{SNR}_1(\boldsymbol{\Theta})$, and defined as the ratio of $R_1(\boldsymbol{\Theta})$ to the noise power rejected by each filter

$$\mathrm{SNR}_1(\boldsymbol{\Theta}) = \frac{R_1(\boldsymbol{\Theta})}{N_1(\boldsymbol{\Theta})} = \frac{\mathbf{a}^{\dagger}(\boldsymbol{\Theta})\ \mathbf{Q}_1\ \mathbf{D}\ \mathbf{D}^{\dagger}\ \mathbf{Q}_1\ \mathbf{a}(\boldsymbol{\Theta})}{\mathbf{a}^{\dagger}(\boldsymbol{\Theta})\ \mathbf{Q}_1\ \mathbf{R}\ \mathbf{R}^{\dagger}\ \mathbf{Q}_1\ \mathbf{a}(\boldsymbol{\Theta})} \tag{2.18}$$

where normalizing terms have again been omitted. The notation SNR is used because the spectrum is primarily regarded as that of a *modified* system that is not sensitive to an emitter at $\hat{\theta}_1$. The concept of the spectrum of the *residue* can be generalized to multiple emitters by replacing $\mathbf{Q}_1$ by $\mathbf{Q}_r$. If the sensor system includes an accurate whitening transform, then

$$\mathrm{SNR}_r(\boldsymbol{\Theta}) = [\mathbf{Q}_r\ \mathbf{a}(\boldsymbol{\Theta})]^{\#}\ \mathbf{D}\ \mathbf{D}^{\dagger}\ \mathbf{Q}_r\ \mathbf{a}(\boldsymbol{\Theta}) \tag{2.19}$$

The subscripts again refer to the rank of the missing subspace in the projection defined by $\mathbf{Q}_r$. Equation (2.17) is satisfied if $\mathbf{Q}_0$ is defined as an identity matrix.

If no emitters are present, then $\mathrm{SNR}_0(\boldsymbol{\Theta})$ should be statistically isotropic, being simply the ratio of filtered incoming samples of noise to the corresponding samples of reference noise. A detection threshold can then be set, in the conventional manner, to distinguish between the signal and no-signal case. Similarly, if r signals are present and $\mathbf{Q}_r$ is estimated accurately, then, since the *modified* sensor system is not sensitive to any of the signals, $\mathrm{SNR}_r(\boldsymbol{\Theta})$ should be essentially isotropic. If a signal is present in the residue then a value, $\mathrm{SNR}_r(\theta_{r+1})$, is directly related to the likelihood that $\mathbf{a}(\theta_{r+1})$ corresponds to a candidate $r + 1$th emitter. The most likely solution is therefore indicated by searching for the maximum SNR as a function of θ_{r+1}. For the highest probability of detection, the contribution from leakage in the both numerator and denominator of the computation of SNR should be minimized. In order that leakage does not cause bias on the maximum of $\mathrm{SNR}_r(\boldsymbol{\Theta})$, we require that the noise contribution is isotropic as a function of θ.

The foregoing requirement continues to be satisfied by applying the whitening transform, $\mathbf{W}_0$, chosen such that a covariance estimate of typical noise background data becomes an identity matrix. The *modified* covariance, $\mathbf{Q}_r\ \mathbf{W}_0\ \mathbf{R}\ \mathbf{R}^{\dagger}\ \mathbf{W}_0\ \mathbf{Q}_r$, is a rank-reduced identity matrix.

For practical reasons of storage and computation it is usual practice to restrict the span of the candidate signatures and matched filters in the array manifold to the simple case of single-emitter scenarios, where the sensor data are expected to be essentially rank one. We note that *each point* in the classical power spectrum can be regarded as a measure of the degree of correlation between data and a *single* normalized rank-one signature. A similar operation can be undertaken for a higher-rank signature corresponding to a multiemitter scenario.

2.4.2 Higher-Rank Spectrum Estimation

Definition

The novel concept of a higher-rank spectrum is directly related to the estimation of SNR for a candidate multiemitter matched filter. It is assumed that the higher-rank signature can be synthesized from r rank-one basis signatures, denoted $\mathbf{A}_{nr}$. A rank r spectrum estimate is a projection of data vectors, $\mathbf{d}_1, \mathbf{d}_2, \ldots, \mathbf{d}_k$ through a rank r subspace, defined by $\mathbf{A}_{nr}\,\mathbf{A}_{nr}^{\#}$,

$$\begin{aligned} S_r(\Theta_1, \Theta_2, \ldots, \Theta_r) &= Tr(\mathbf{A}_{nr}^{\#}\,\mathbf{D}\,\mathbf{D}^{\dagger}\,\mathbf{A}_{nr}) \\ &= \sum_k \mathbf{d}_k^{\dagger}\,\mathbf{A}_{nr}\,\mathbf{A}_{nr}^{\#}\,\mathbf{d}_k \end{aligned} \tag{2.20}$$

In principle, a point in the extended spectrum is evaluated for every candidate higher-rank signature as defined in the extended array manifold. The result can be regarded as a high-dimension surface. The spectrum applies directly, for example, to a linear or planar antenna array, where the signatures at a given rank, r, are defined by linear combinations of r complex sinusoids. We can evaluate the surface, not only for all permutations of r direction-of-arrival parameters, but we can also vary r as a parameter. The availability of these additional degrees of freedom contrasts with the classical power spectrum estimate that is limited to the one dimension defined by the basic array manifold. The higher-rank spectrum estimate closely resembles a maximum-likelihood surface for a rank r model. The noise, $N_r(\Theta_1, \Theta_2, \ldots, \Theta_r)$, associated with each point in the signal subspace, is essentially isotropic, as required for unbiased estimates. The level is r units of power.

A residue, $R_r(\Theta_1, \Theta_2, \ldots, \Theta_r)$, can also be evaluated by subtraction of $S_r(\Theta_1, \Theta_2, \ldots, \Theta_r)$ from the total power in the data. Alternatively, a complementary subspace, $\mathbf{Q}_r$, that is orthogonal to $\mathbf{A}_{nr}^{\#}\,\mathbf{A}_{nr}$, can be computed. The noise associated with the subspace of the residue is isotropic at $n - r$ units of power. The value of $R_r(\Theta_1, \Theta_2, \ldots, \Theta_r)$ approaches $n - r$ when all signals are accurately blocked by a suitable choice of $\theta_1, \theta_2, \ldots, \theta_r$.

The location of the global minimum of the spectrum of the residue, at a given rank, represents a best solution, in the LMS sense and approximates to the equivalent ML solution. Estimates of directions of arrival, $\hat{\theta}_1, \hat{\theta}_2, \ldots, \hat{\theta}_r$ are directly available from the location of the minimum. At the global minimum, a mean ratio of residue power to noise, denoted RNR(r), can be defined by

$$\begin{aligned} \mathrm{RNR}(r) &= \frac{1}{n-r} R_r(\hat{\theta}_1, \hat{\theta}_2, \ldots, \hat{\theta}_r) \\ &= \frac{1}{n-r} \sum_k \mathbf{d}_k^{\dagger}\,\mathbf{d}_k - \mathbf{d}_k^{\dagger}\,\hat{\mathbf{A}}_{nr}\,\hat{\mathbf{A}}_{nr}^{\#}\,\mathbf{d}_k \end{aligned} \tag{2.21}$$

As r is increased, the value of RNR(r) decreases toward a constant value of unity. In concept, a simple threshold, based on a chi-squared test, enables a lower bound on $\hat{r}$

to be estimated, where $\hat{r}$ is an estimate of the number of emitters in the scenario. In Section 2.6, more sensitive tests are suggested. Inevitable residual leakage from sensor noise implies that estimates are statistical, but since the leakage is isotropic (and is minimized for other emitters), parameter estimates are essentially unbiased.

The concepts of higher-rank spectrum estimation are easily generalized to the case of arbitrary geometry, to discriminants other than frequency, and to self-calibration. An extended higher-rank spectrum can be obtained, in principle, by evaluating $S_r(\boldsymbol{\Theta}_1, \boldsymbol{\Theta}_2, \ldots, \boldsymbol{\Theta}_r)$ for all permutations of appropriate control parameters, including calibration variables and rates of change of $\hat{\boldsymbol{\theta}}$ as necessary.

There is a daunting computational problem associated with higher-rank spectrum estimation, caused by the inevitable combinatorial explosion in the number of candidate higher-rank matched filters as the number of parameters is increased.

Search philosophy

In the majority of radar and sonar applications, the data rate is such that a *full* multiparameter search of an extended spectrum is well beyond technological feasibility at present. The prospect of near optimality in the multiparameter case can be retained only by resorting to an efficient computational compromise. The objective, in algorithmic implementations such as IMP, is to minimize performance degradation. The question is how might we tackle the compromise from a rational viewpoint.

Higher-rank spectrum estimation differs from ML estimation in one significant respect, notably that of null depth. For full optimality, null depths should each be just sufficient to suppress the mean leakage for the corresponding emitter to that of whitened background noise. Instead, a higher-rank spectrum is generated by steering zeros that are intended to suppress completely the potential leakage from the relevant emitters. Since magnitude and statistics of possible signals are unknown at the outset, the number of parameters required to generate a higher-rank spectrum is significantly reduced by steering zeros. The additional parameters can be estimated subsequently if a fully optimal solution is essential. Nulls rather than zeros can be steered in a second phase of the search. By this means, the ML search task has, in effect, been partitioned into two stages. Estimates from the first constrained search are used as a starting point for a local (unimodal) search in the second phase.

The idea of first solving a simpler problem leads to the concept of a *directed search*. The philosophy is based on setting relatively unimportant parameters to default values. The concept depends also on utilizing the rules that define the prior. For example, the objective, in many cases, is to assess the evidence that indicates how many emitters are present. In general, the prior dictates that the simplest solution that does not conflict with other evidence is the most likely. We, therefore, first propose the simplest possible solution comprising one emitter, then proceed to two, and so on, until the evidence for the next emitter, as measured by the SNR of the residue, is below a preset threshold. The advantages are that candidate solutions of unnecessarily high rank are not evaluated and parameter estimates from a lower-rank solution act as a starting point in the search at the next higher rank. We do not need to compute the full higher-rank spectrum at any stage (except the relatively simple rank-one case).

The directed search approach provides the basis for novel, computationally efficient, multistage algorithms discussed in Section 2.6.

2.5 STANDARD REPRESENTATION OF HIGH-DISCRIMINATION ALGORITHMS BASED ON A RANK-ONE SPECTRUM

2.5.1 Aperture Weighting

Historically, aperture taper functions are perhaps the earliest example of a transform used to modify sidelobe leakage effects. This basic method therefore represents the first real application of a high-discrimination concept. The main disadvantage is that a desired reduction in sidelobe leakage must be traded for poorer mainbeam resolution (or vice versa). SNR is also degraded because the whitening is modified by the aperture weighting transform, denoted $\mathbf{W}_a$. The power spectrum is given by

$$\mathrm{SNR}(\boldsymbol{\Theta}) = \frac{\mathbf{a}^{\dagger}(\boldsymbol{\Theta})\,\mathbf{W}_a\,\mathbf{W}_a^{\dagger}\,\mathbf{D}\,\mathbf{D}^{\dagger}\,\mathbf{W}_a\,\mathbf{W}_a^{\dagger}\,\mathbf{a}(\boldsymbol{\Theta})}{\mathbf{a}^{\dagger}(\boldsymbol{\Theta})\,\mathbf{W}_a\,\mathbf{W}_a^{\dagger}\,\mathbf{R}\,\mathbf{R}^{\dagger}\,\mathbf{W}_a\,\mathbf{W}_a^{\dagger}\,\mathbf{a}(\boldsymbol{\Theta})} \tag{2.22}$$

where $\mathbf{W}_a\,\mathbf{W}_a^{\dagger}$ is a diagonal matrix. It is commonly assumed that the denominator is isotropic and that it can therefore be omitted. In general, any pair of rank-one signature vectors corresponding to arbitrary emitter locations are likely to have features in common; in other words, they are unlikely to be orthogonal. It is the common features that give rise to leakage. Any fixed preprocessing transform, such as an aperture weighting function, when applied to all vectors in the array manifold, leads to a *modified* sensor system with a different level of correlation between each pair of signatures. Although a limited number of zeros can be introduced, the locations are not dependent on the external scenario and therefore do not minimize leakage from other emitters if present.

Aperture weighting or shading, enhances discrimination in some respect, but if several emitters are present simultaneously, then leakage causes problems of bias, poor resolution, and poor discrimination that cannot be eliminated by nonadaptive preprocessing transforms, in a classical or rank-one spectrum.

2.5.2 Adaptive Processing Techniques

Conceptual basis

Considerable research and development effort has been expended on designing data dependent transforms. The techniques are, in most cases, based on eigen decomposition of a data covariance estimate, $\mathbf{D}\,\mathbf{D}^{\dagger}$, or on the inverse of the same matrix. One version of the algorithms can, for example, be regarded as applying a data dependent transform, $\mathbf{W}_d$, such that $\mathbf{W}_d^\dagger \mathbf{D}\,\mathbf{D}^\dagger \mathbf{W}_d$ is an identity matrix. We see that the *modified* data covariance estimate is, in effect, whitened rather than the background noise. Equation (2.22) then becomes, for the *adaptively modified* system,

$$\begin{aligned}\mathrm{SNR}(\boldsymbol{\Theta}) &= \frac{\mathbf{a}^\dagger(\boldsymbol{\Theta})\,\mathbf{W}_d\,\mathbf{W}_d^\dagger\,\mathbf{D}\,\mathbf{D}^\dagger\,\mathbf{W}_d\,\mathbf{W}_d^\dagger\,\mathbf{a}(\boldsymbol{\Theta})}{\mathbf{a}^\dagger(\boldsymbol{\Theta})\,\mathbf{W}_d\,\mathbf{W}_d^\dagger\,\mathbf{R}\,\mathbf{R}^\dagger\,\mathbf{W}_d\,\mathbf{W}_d^\dagger\,\mathbf{a}(\boldsymbol{\Theta})} \\ &= \frac{\mathbf{a}^\dagger(\boldsymbol{\Theta})\,[\mathbf{D}\,\mathbf{D}^\dagger]^{-1}\,\mathbf{a}(\boldsymbol{\Theta})}{\mathbf{a}^\dagger(\boldsymbol{\Theta})\,[\mathbf{D}\,\mathbf{D}^\dagger]^{-2}\,\mathbf{a}(\boldsymbol{\Theta})} \qquad \text{if } \mathbf{R}\,\mathbf{R}^\dagger \text{ is an identity matrix}\end{aligned} \tag{2.23}$$

If the signal magnitudes are large relative to noise then the matrix $\mathbf{W}_d\,\mathbf{W}_d^\dagger$ is essentially orthogonal to the signal subspace. The weighting, $\mathbf{W}_d\,\mathbf{W}_d^\dagger$, has some similarity to the signal blocking matrix, $\mathbf{Q}_r$, defined in Section 2.3 but with nulls rather than zeros. In the term $\mathbf{W}_d^\dagger\,\mathbf{R}\,\mathbf{R}^\dagger\,\mathbf{W}_d$, the weighting suppresses components that match emitters that are present in $\mathbf{D}\,\mathbf{D}^\dagger$. Since similar nulls are required in a maximum-likelihood approach, it might be expected that leakage is at a minimum in the matched filters of the *adaptively modified* system. It should be noted, however, that the target signature for each filter is also attenuated by $\mathbf{W}_d$.

The objective, in solving the inverse problem, becomes that of interpreting the weighting matrix, $\mathbf{W}_d$, in terms of a recognizable input scenario. In principle, the task can be tackled either by a search of higher-rank candidate solutions defined in the extended array manifold or, indirectly, by first computing a pseudoinverse of $\mathbf{D}$ and then interpreting that inverse via a rank-one spectrum estimate. The direct method leads to the computional expense of the multidimensional maximum-likelihood search, and it is therefore attractive to examine the indirect method. It is assumed that the principal maxima or minima in a suitable spectrum indicate the estimated locations of several pointlike emitters.

Equation (2.23) provides a suitable spectrum but other variants have been originated, each claiming to be optimum in some respect. The derivations are based on mathematical assumptions. A number of algorithms are of the form

$$\mathrm{SNR}(\boldsymbol{\Theta}) = \frac{\mathbf{a}^{\#}(\boldsymbol{\Theta})\,[\mathbf{D}\,\mathbf{D}^\dagger]^{1-\rho}\,\mathbf{a}(\boldsymbol{\Theta})}{\mathbf{a}^{\#}(\boldsymbol{\Theta})\,[\mathbf{D}\,\mathbf{D}^\dagger]^{-\rho}\,\mathbf{a}(\boldsymbol{\Theta})} \tag{2.24}$$

where ρ usually denotes an integer. Noninteger exponents can be implemented using eigen decomposition of $\mathbf{D}\,\mathbf{D}^\dagger$ defined by

$$\mathbf{D}\,\mathbf{D}^\dagger = \mathbf{U}\,\boldsymbol{\lambda}\,\mathbf{U}^\dagger \tag{2.25}$$

Since $\mathbf{D}\,\mathbf{D}^\dagger$ is Hermitian, $\mathbf{U}$ defines a set of orthonormal column vectors. Eigenvalues, denoted λ, are proportional to the power of each eigencomponent in $\mathbf{D}\,\mathbf{D}^\dagger$. Implementation of equation (2.24), using noninteger values of ρ, is defined by

$$[\mathbf{D}\,\mathbf{D}^\dagger]^{-\rho} = \mathbf{U}\,\boldsymbol{\lambda}^{-\rho}\,\mathbf{U}^\dagger \tag{2.26}$$

In equation (2.24), it is again assumed that $\mathbf{R}\,\mathbf{R}^\dagger$ is the identity matrix (whitened noise). $\mathbf{W}_d\,\mathbf{W}_d^\dagger$ is given by $[\mathbf{D}\,\mathbf{D}^\dagger]^{-\rho/2}$. It can be observed that if ρ is zero then equation (2.24) defines the unmodified system, and if the value of ρ is set to two, then the spectrum becomes that of equation (2.23).

Figure 2.2(a) shows spectra for a single emitter at values of ρ from zero to two in steps of 0.5. The number of data vectors used in the simulation is 20. The sensor system is a uniformly spaced linear array comprising 16 elements. The horizontal axis is spatial frequency measured in cycles within the full aperture. It can be observed that, as ρ is increased, the mainlobe is reduced in width and becomes relatively sharp. It can also be seen that the SNR at the maximum is markedly reduced. The effects are due to a null in the denominator at the location of the emitter. The background noise is therefore no longer white. The depth of the null depends on emitter power, on r, and on the statistical differences between the sample of noise in the numerator and in the reference, $\mathbf{R}\,\mathbf{R}^\dagger$. The *modified* signal power of a pointlike emitter, as measured in the numerator, is directly attenuated by the null. Since noise is not pointlike, the level of noise leakage, as measured in the denominator, is therefore not attenuated by the same factor.

It is clear from equation (2.23) that the noise term, in the whitened case, is given by $\mathbf{a}^\dagger(\Theta)\,\mathbf{W}_d\,\mathbf{W}_d^\dagger\,\mathbf{W}_d\,\mathbf{W}_d^\dagger\,\mathbf{a}(\Theta)$ rather than the naive assumption given by $\mathbf{a}^\dagger(\Theta)$ $\mathbf{W}_d\,\mathbf{W}_d^\dagger\,\mathbf{a}(\Theta)$. The additional terms have no effect on a selected value, SNR(θ), only if $\mathbf{a}(\theta)$ is an eigenvector of $\mathbf{W}_d\,\mathbf{W}_d^\dagger$ and therefore of $\mathbf{D}\,\mathbf{D}^\dagger$. Since, in a practical case, the principal eigenvector includes some contribution from noise, the steering vector of an emitter, $\mathbf{a}(\theta)$, is not exactly orthogonal to the so-called noise subspace defined by the remaining eigenvectors. A small proportion of emitter power therefore appears in the noise subspace. We note that eigenvalues in the noise subspace are each close to unity. The noise subspace is therefore little changed by computing an inverse, such as $\mathbf{W}_d\,\mathbf{W}_d^\dagger$. It follows that the leakage of signal into the noise term, $\mathbf{a}^\dagger(\theta)\,\mathbf{U}\,\boldsymbol{\lambda}^{-\rho}$ $\mathbf{U}^\dagger\,\mathbf{a}(\theta)$, is relatively independent of ρ. Since the principal eigenvalue is relatively much greater, in general, the corresponding eigencomponent (primarily signal) is attenuated significantly by an inverse. The loss in SNR that is observed in Fig. 2.2(a) can therefore be explained in qualitative terms. The existence of the signal leakage can be displayed by excluding signal subspace component(s) from $\mathbf{W}_d$. The resulting effect on the spectrum is shown in Fig. 2.2(b), at the same values of ρ. The level of leakage demonstrates the effectiveness or otherwise of eigen-based partitioning in this specific case.

The case where ρ is set to unity in equation (2.24) provides the basis for an algorithm due to Capon [5]. It can easily be deduced that the numerator is isotropic (unity). This approach is sometimes termed the maximum-likelihood method but must be distinguished from the multidimensional search that is more correctly called maximum likelihood [2].

Standard format

A number of high-resolution algorithms also use the rank-one spectrum as a search technique. An additional weighting of the information in the covariance estimate, $\mathbf{D}\,\mathbf{D}^\dagger$, is applied via the numerator of equation (2.23). The weighting can either take the form of an additional filter stage (projection matrix) or of a nonlinear weighting of the eigencomponents of $\mathbf{D}\,\mathbf{D}^\dagger$ (or both).

A standard mathematical representation and implementation of such algorithms is feasible, based on weighting the left-hand singular vectors [6]. The weighting, denoted by a transform, $\mathbf{W}_\alpha$, is dependent on the singular values of the available data. Left-hand singular vectors of the data matrix, $\mathbf{D}$, are identical to the eigenvectors, $\mathbf{U}$, of $\mathbf{D}\,\mathbf{D}^\dagger$. Nonzero singular values, denoted $\boldsymbol{\alpha}$, are defined by $\boldsymbol{\alpha}^2 = \lambda$. A *modified* rank-one spectrum is evaluated using

$$L(\Theta) = \mathbf{a}^{\#}(\Theta)\,\mathbf{U}\,\mathbf{W}_\alpha\,\mathbf{W}_\alpha^\dagger\,\mathbf{U}^\dagger\,\mathbf{a}(\Theta) \tag{2.27}$$

or the inverse of this function.

TABLE 2.1. Standard format for algorithms based on a rank-one spectrum.

Algorithm (for arbitrary vectors)	Weight	Algorithm (for complex exponentials)	Weight
Classical	$\boldsymbol{\alpha}$		
Capon (MLM) [5]	$\boldsymbol{\alpha}^{-1}$	Burg [7]	$\boldsymbol{\alpha}^{-2}\mathbf{U}^\dagger\mathbf{Z}_1$
MUSIC [1]	$\mathbf{P}_n$	Kumaresan and Tufts [8]	$\mathbf{P}_n\mathbf{U}^\dagger\mathbf{Z}_1$
Principal components [9]	$\mathbf{P}_s$		
Johnson and DeGraaf [9]	$\boldsymbol{\alpha}^{-1}\,\mathbf{P}_n$		
Standard form [6]	$\boldsymbol{\alpha}^{-\nu}\,\mathbf{P}$	Standard form	$\boldsymbol{\alpha}^{-\nu}\,\mathbf{P}\mathbf{U}^\dagger$

Table 2.1 lists a number of different examples of values for $\mathbf{W}_\alpha$. The exponent ν is, in principle, any real number. The algorithms in the left-hand column are applicable to arbitrary signatures, including analysis in terms of frequency. $\mathbf{P}_s$ and $\mathbf{P}_n$ define rank-reduced identity operators that select the singular vectors of the signal and noise subspace, respectively. The basis for this partitioning is the magnitude of the singular values. Each value is tested against a threshold that is preset in relation to the noise background. Since the signal and noise subspaces are complementary, the spectra corresponding to the principal components and MUSIC algorithms are also complementary. The principal components method is therefore regarded as an alternative implementation of MUSIC. Note that the algorithm due to Capon [5] can also be cast

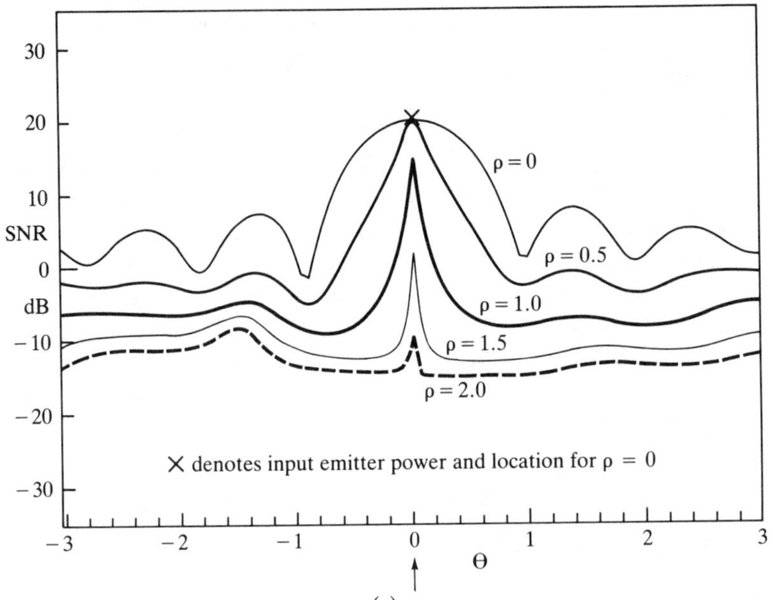

(a)

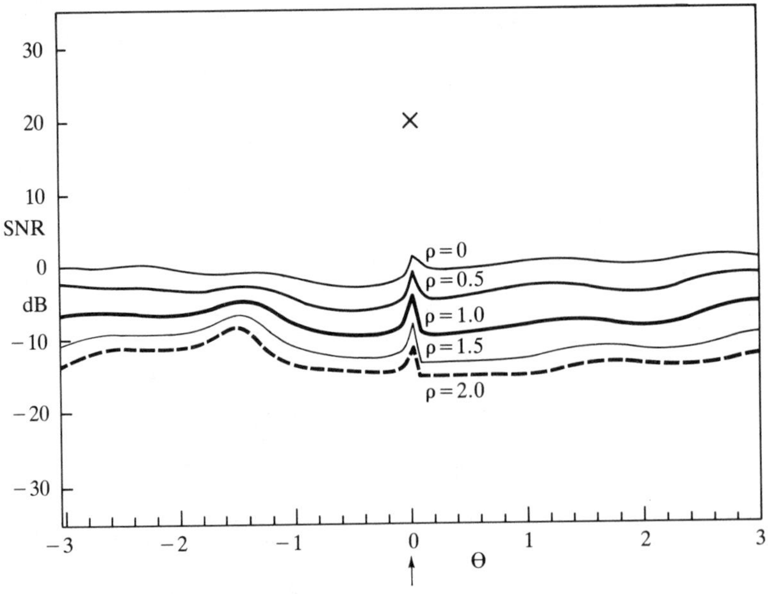

(b)

Figure 2.2 (a) Rank-one spectra, using equation (2.24), are shown for a single emitter, as a function of ρ. The horizontal scale represents spatial frequency measured in cycles in the aperture of a linear array of 16 elements. Zero decibels on the vertical SNR scale is referenced to the noise power per degree of freedom in the unmodified system ($\rho = 0$). The spectra illustrate the progressive reduction in the width of the mainlobe and associated reduction in SNR as ρ is increased from 0 to 2. (b) Similar spectra are shown but for the case where the principal eigencomponent has been excluded from the data and from the signatures by a projection operation. The SNR indicates the level of signal not in the so-called signal subspace. It can be

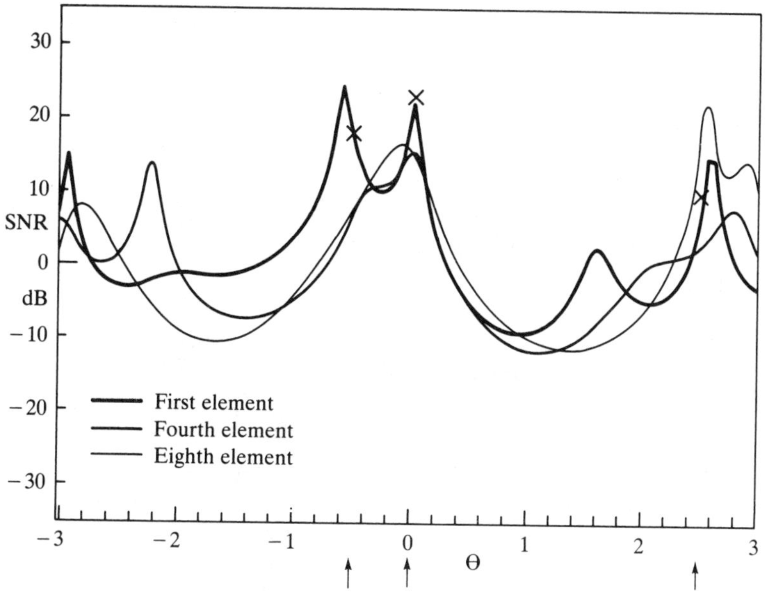

(c)

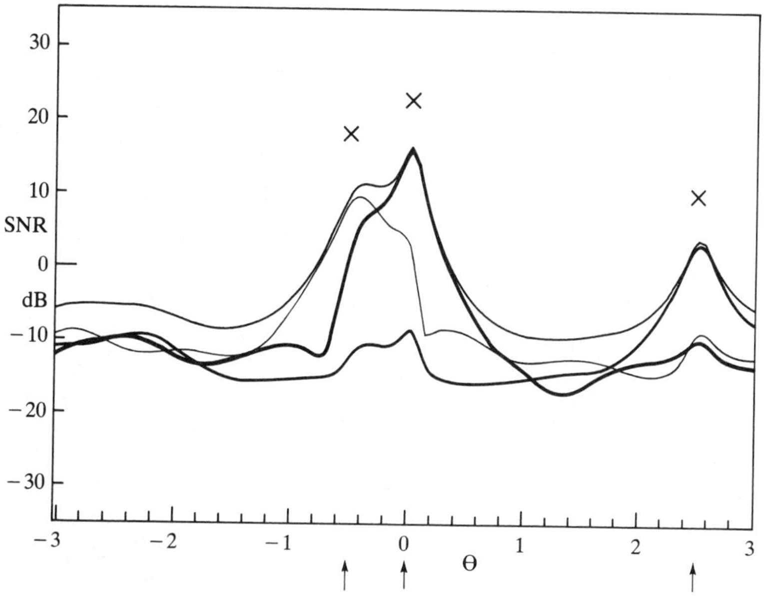

(d)

seen that the SNR is essentially independent of ρ. (c) Rank-one spectra are shown using equation (2.28) for the case where $\nu = 1$. The indicated element is the only nonzero element in the constraint vector **Z**. It can be seen that the highest resolution is obtained using the first (or last) element. Inner elements are less likely to generate spurious maxima. The number of elements is 16. (d) Rank-one spectra for the individual emitters are shown using the data-dependent weighting of equation (2.23). The solid line shows the composite spectrum as the sum of the individual spectra. The conclusion to be drawn is that leakage, though reduced, remains an inherent limitation of all high-resolution algorithms that are based on rank-one spectrum estimation.

in the standard form shown. There is some debate on whether the algorithm due to Burg [7] should use $\nu = 1$ or $\nu = 2$ in practice. Clearly, noninteger values of ν should provide a progressive trade-off between SNR and resolution. Other forms of weighting are also feasible.

In general, the weighting given to the vectors in the signal subspace is close to zero and the weight given to the noise (or residue) subspace vectors is close to unity. The weighting matrix, $\mathbf{W}_\alpha \mathbf{W}_\alpha^\dagger$, therefore has some similarity, in effect, to a signal blocking transform, $\mathbf{Q}_r$. Relatively small differences in the weighting give rise to significantly different resolution performance. The matrix, $\mathbf{U}^\dagger$, can be regarded as a transform forming a stage of the *modified* system. The *modified* data then become simply the matrix of singular values. It is instructive to examine the power density spectrum for each singular component individually. These spectra can subsequently be summed according to the weighting needed to implement any of the published algorithms or to meet a specific criterion defined in the prior.

In the right-hand column, $\mathbf{Z}_1$ is a constraint vector, comprising zero values excepting the first element only. For the general case, where $\mathbf{Z}$ is any set of vectors, equation (2.27) then becomes

$$L(\Theta) = \frac{\mathbf{a}^\dagger(\Theta)\,\mathbf{U}\,\mathbf{W}_\alpha^\dagger\,\mathbf{W}_\alpha\,\mathbf{U}^\dagger\,\mathbf{Z}\,\mathbf{Z}^\dagger\,\mathbf{U}\,\mathbf{W}_\alpha^\dagger\,\mathbf{W}_\alpha\,\mathbf{U}^\dagger\,\mathbf{a}(\Theta)}{\mathbf{a}^\dagger(\Theta)\,\mathbf{Z}\,\mathbf{Z}^\dagger\,\mathbf{a}(\Theta)} \tag{2.28}$$

The denominator ensures that the spectrum is isotropic if no emitters are present in the scenario. The constraint(s) defined in $\mathbf{Z}$ enhance selected information and are, in general, used to extract the information most relevant to discriminating between emitters. In some implementations, the computational load can also be reduced.

The algorithms in the right-hand column are applicable to analysis in terms of complex exponentials. In this case, the vector $\mathbf{Z}_1^\dagger\,\mathbf{U}\,\mathbf{W}_\alpha^\dagger\,\mathbf{W}_\alpha\,\mathbf{U}^\dagger$ can also be analyzed in the z-plane [10] rather than by searching for maxima or minima. The locations of poles and zeros are used instead as estimates of the frequencies of pointlike emitters. $\mathbf{Z}_1^\dagger$ has the effect of selecting the first element from the vector, $\mathbf{U}\,\mathbf{W}_\alpha^\dagger\,\mathbf{W}_\alpha\,\mathbf{U}^\dagger\,\mathbf{a}(\theta)$. An element from this vector defines the residue for the corresponding sensor, at a given value of θ. It is instructive to examine individual spectra, one for each sensor by selecting the appropriate element in $\mathbf{Z}$. The effect, at $\nu = 1$, is shown in Fig. 2.2(c). It is also possible to devise a similar algorithm in the signal subspace, but emitters must be located by finding points in the spectrum closest to unit magnitude (zero phase). Improved resolution is obtained at the expense of a reduced SNR and increased ambiguity due to rejection of some of the available information.

The missing entries in Table 2.1 represent other possible variants. Various ratios or products of two or more spectra are also feasible. In many cases the inverse of a spectrum can be examined if preferred, but the parameter estimates are unaffected. All the algorithms can be applied to a fixed block of unprocessed data vectors, with uniform weighting over the sensor aperture. Alternatively, in the case of analysis in terms of frequency, a sliding rectangularly weighted window or subaperture can be applied to one or more data vectors. The effect is that a greater number of data vectors with fewer elements are available. The objective is to increase the rank of the data covariance estimate. Recursive weighting can also be applied within the window and may be applied in both forward and backward directions over the aperture. Some originators specified one choice of operator in the relevant publication. In general, the operator chosen should relate to prior knowledge about the scenario. For example, if emitters are not expected to be pointlike due to distortion in the propagation medium, then the diffusing effect leads to a loss of performance. Additional false emitters would be included in the interpretation. The effect can be reduced by choosing a subaperture appropriate to the coherence distance in the prevailing conditions.

2.5.3 Performance

Performance differences are to be expected between algorithms that employ heuristic and indirect techniques for including the prior in a rank-one spectrum. It is often difficult to quantify, in analytic terms, the consequences of convenient assumptions on the desired features of the solution. Although error is perhaps minimized in some intuitive sense, it is, in general, hard to quantify the trade-off between, for example, probability of resolution, false alarm rate, and bias on parameter estimates. It is often necessary to resort to Monte Carlo techniques to compare performance in relative terms [4, 9].

A specific example serves to illustrate that there is an inherent limitation in superresolution techniques that are based on a modified rank-one spectrum. We have seen (equation 2.17) that a rank-one spectrum can be regarded as the summation of spectra for individual data vectors because regrouping of terms in a matrix product has no effect on the result. The principle applies, in particular, to the summation in the term $\mathbf{D}\,\mathbf{D}^{\dagger}$ in the numerator of equation (2.23). In concept, if several emitters are activated individually, on a mutually exclusive basis then the summation can be split into groups that each correspond to one emitter. We can assume that the covariance estimate that is used to compute $\mathbf{W}_d$, includes all available data vectors. It follows that a power density spectrum for one emitter alone can be evaluated using the data vectors that correspond to that emitter. The result is similar to a scaled point spread function but defined for the *modified* system that includes $\mathbf{W}_d$. This point spread function predicts the level of leakage to be expected from all other candidate locations including the remaining emitters not yet included in the numerator. In principle, individual spectra yields the composite spectrum that is equivalent to analyzing the complete matrix, $\mathbf{D}\,\mathbf{D}^{\dagger}$ in one step. The principle is illustrated in Fig. 2.2(d) for the Capon algorithm (three emitters).

Since it appears that we can regard any rank-one spectrum as a summation of weighted point spread functions, it can be argued that there is a direct analogy between the limitations of high-resolution spectra and the classical (unmodified) spectrum. In the case of high-resolution algorithms, the *modified* point spread functions are relatively sharp and narrow. This sharpness is directly related to the depth of null in the denominator (measured from zero) and therefore depends, through $\mathbf{W}_d$, on the SNR of the relevant emitter. Although, in the *modified* system, SNR is also reduced as a direct consequence of the transform, $\mathbf{W}_d$, the sharp reduction in width of the mainlobe leads to a potential resolving power that is far in excess of that indicated by the Rayleigh resolution criterion for the unmodified system. Although the recognized problems of bias, the masking of weaker components, and resolution limitations are reduced, leakage is not eliminated. In general, performance trade-offs are required between loss of SNR and improved discrimination between emitters.

The reasons for applying adaptive and nonadaptive transforms are primarily concerned with preserving, emphasizing, or suppressing information according to the perceived relative importance of that information to a rank-one spectrum estimate. The aim is to reduce the leakage that arises if the data originate from a combination of several emitters, but the primary cause of suboptimal resolution remains that leakage.

In principle, several different weighting functions (algorithms) can be applied independently to the same batch of data vectors, some to improve resolution and others to reduce sidelobe level. The resulting spectra could perhaps be compared and interpreted in such a way as to maximize the probability of resolution without increasing the false alarm rate.

The foregoing conceptual basis for high-resolution algorithms leads to the intuitive evaluation of performance limits for candidate algorithms in a target application. It should, however, be noted that, in the specific scenario used as an illustration, the emitters were mutually orthogonal. Because the eigenvalues of a covariance matrix, $\mathbf{D}\,\mathbf{D}^{\dagger}$, depend both on the spatial correlation between steering vectors and on the temporal correlation between emitter waveforms, null depth is, in general, less than for the idealized case of orthogonal temporal waveforms. If multipath is present or if the waveforms are of short duration and therefore unlikely to be orthogonal, then it follows that resolution is degraded. The relevant cross-product terms must be included as additional components in the composite spectrum. Cross-product terms between signal waveforms and noise should be considered.

The parameter estimates, obtained via any high-discrimination technique, can be used to compute corresponding signal extraction filters, each with nulls at all but one emitter location. We have seen in Section 2.3 that we can also evaluate the level of *unexplained* modeling residue for any higher-rank signal model by generating a signal blocking matrix. The effectiveness of the eigen-based partitioning algorithms can therefore be observed by evaluating a spectrum of the residue subspace using equations (2.18) or (2.19). Although there are many algorithm variants based on a rank-one spectrum, the originators failed to suggest that the residual error can be computed and corrections made to the proposed solution by iterative means. The degree of confidence to be placed on the selected solution is therefore not quantified and corrective actions for refining parameter estimates are not suggested.

2.6 THE IMP ALGORITHM

2.6.1 Description

Basic concept

In principle, a maximum-likelihood (ML) solution to a parametric modeling problem is obtained by searching for the global maximum of a relative likelihood function. The function is, in general, a multidimensional surface covering the complete span of permitted solutions. A combinatorial explosion in the span of the solution space occurs as the number of variable parameters increases.

Collateral knowledge can similarly be represented as a likelihood surface, but derived in the absence of sensor data. Hard constraints, for example, set up out of bounds regions that do not require to be searched; the computational load can then be reduced by many orders of magnitude. Clearly hard constraints must be used *with care* if there is any possibility that a required solution may arise in a zone forbidden by the search pattern. In addition, for a real problem, there may also be complex probabilistic interrelationships arising in the prior through abstract collateral knowledge. The full likelihood surface, including priors, is therefore often complex in shape and difficult to compute.

We have seen that classical methods suffer resolution problems because a *single* parameter only is varied. In principle, all parameters should be varied to cover the full solution space of the multiemitter model. ML concepts are not used widely for the simple reason that the computational cost associated with a full search of all permitted solutions is prohibitive. In practice, simplifications are essential. For example, we might devise a simple approximation to the ML surface where the global maximum is close to that of the true surface. An initial solution is then found more easily and refined as necessary. For example, the IMP algorithm initially eliminates parameters, such as signal magnitude, that have little effect on estimates of the main parameters. The surface utilized by IMP is that defined by the higher-rank spectrum estimate introduced in Section 2.4.

Major computational savings are also achieved by avoiding an exhaustive search. The aim is to compute as little as possible of the likelihood surface, but the danger in this approach is that the search reveals a submaximum. The IMP algorithm adopts the idea of a *data-dependent* directed search based on examining cross-sections through the approximate ML surface. This approach contrasts with the random point-by-point search associated with synthetic annealing techniques.

The directed search technique is based on first solving a simpler related problem with fewer parameters. The number of parameters is incremented in an order of significance chosen to maximize the SNR for the least number of decomposition components.

The problem

For the purpose of discussion, we propose to illustrate the IMP concept using a common radar problem where the actual number of emitters, r, present in incoming sensor data is unknown. We are required to *count and isolate individual* components and to estimate corresponding directions of arrival $\hat{\theta}_1, \hat{\theta}_2, \ldots, \hat{\theta}_r$. It may be necessary to track the location of each emitter.

We assume that each emitter can be identified by a unique signature observed at the output of an antenna array. Each signature is indexed to previously observed input conditions via the array manifold. For simplicity in the discussion, we can, without loss of generality, associate each candidate signature with one value of a single pointer or parameter, θ, sampled on a grid of at least n_S points satisfying the Nyquist criterion. In many radar problems, θ is a continuously variable parameter (such as direction of arrival) and must be interpolated to achieve adequate accuracy. Incoming data vectors are complex linear combinations of r such signatures. The estimate of the number of emitters is denoted $\hat{r}$.

The incremental sequence

The initial step in the IMP algorithm is to test the SNR of the permitted solutions for single emitters defined by the basic array manifold ($\hat{r} = 1$). The best rank-one solution is selected, as a function of θ, using the rank-one spectrum estimate defined by equations (2.16) or (2.17). The resulting parameter estimate, $\hat{\theta}_1$, is taken as indicating the approximate location of the strongest emitter. The next step is to examine a *selection* of possible two signature solutions ($\hat{r} = 2$), estimate $\hat{\theta}_2$, and then proceed, stage by stage, $\hat{r} = 3, 4$, and so on. Each search uses the parameter estimates from the previous stage to generate a signal blocking projection matrix, $\mathbf{Q}_r$. The search is limited to one dimension

$$\mathrm{SNR}_{r-1}(\Theta_r) = [\mathbf{Q}_{r-1}\, \mathbf{a}(\Theta_r)]^{\#}\, \mathbf{D}\, \mathbf{D}^{\dagger}\, \mathbf{Q}_{r-1}\, \mathbf{a}(\Theta_r) \tag{2.29}$$

This equation relates directly to equation (2.19) and can be recast in several different forms to achieve optimum computational economy. The function predicts the SNR of a possible next signal component in the residue remaining after suppressing the larger emitters identified in the previous stage. Signatures from the basic array manifold are utilized to define *modified* candidate matched filters as a function of θ.

At the stage $\hat{r} = 2$, one zero is steered, via a signal blocking matrix, $\mathbf{Q}_1$. This matrix is set up to be orthogonal to the signature for the first estimate of the largest emitter. In effect, the sensor is *modified* such that it no longer responds to the signature tentatively assumed to be at $\hat{\theta}_1$. The SNR is predicted for each *modified* filter in turn by directly applying the filters to both incoming data and to the noise background. The ratio function, $\mathrm{SNR}_1(\Theta_2)$, indicates the relative likelihood of the candidate solutions as a function of θ_2. The maximum indicates the most likely location of the second emitter, $\hat{\theta}_2$, given the current estimate of $\hat{\theta}_1$.

The iterative step

To implement the one-dimensional search for $\hat{\theta}_2$, an estimate of $\hat{\theta}_1$ was needed. However, if the true number of signal components is two (or more), then the single-emitter assumption underpinning the classical matched filter approach of equation (2.18) is violated. In general, any attempt to estimate $\hat{\theta}_1$ independently of $\hat{\theta}_2, \ldots, \hat{\theta}_r$ gives a biased result because the filter matched to the true θ_1 signature also responds partially to leakage components from (as yet) unidentified emitters. Since the bias is caused primarily by signals weaker than the first, the degree of bias is generally less than the width of the mainlobe of the filter. In typical applications, $\hat{\theta}_1$ is therefore a reasonable trial point for steering the one zero required to estimate $\hat{\theta}_2$. The estimate $\hat{\theta}_2$ is taken as the location of the maximum of $\text{SNR}_1(\Theta_2)$, but, because $\hat{\theta}_1$ is not yet accurate, $\hat{\theta}_2$ is also biased. The estimation task becomes that of a standard unimodal optimization problem at rank two and, in the IMP algorithm, is performed iteratively.

When values for both $\hat{\theta}_1$ and $\hat{\theta}_2$ are available, then a single zero is steered via $\mathbf{Q}_1$ alternately $\hat{\theta}_1$ and $\hat{\theta}_2$. The zero, when located at $\hat{\theta}_2$, suppresses a major proportion of the leakage arising from the second emitter at the true θ_2, bias on a reestimate of $\hat{\theta}_1$ is therefore reduced. Further swapping of the zero between $\hat{\theta}_1$ and $\hat{\theta}_2$ leads iteratively to better estimates of both $\hat{\theta}_1$ and $\hat{\theta}_2$. The single zero, when correctly positioned at either θ_1 or θ_2, suppresses the other component and eliminates the primary source of bias. Estimates are still affected by unpredicted leakage from stocastic noise, additional emitters, and other interference. *Before proceeding to the rank-three stage, it is essential to complete sufficient iterations to satisfy a suitable convergence criterion.* Sidelobe leakage from both emitters, due to residual modeling error, *must* be less than any possible third signal to avoid a spurious detection in the third stage.

At stage $\hat{r} = 3$, two zeros are steered, at $\hat{\theta}_1$ and $\hat{\theta}_2$. $\text{SNR}_2(\Theta_3)$ is computed and $\hat{\theta}_3$ is estimated. Again, IMP solves, iteratively, the unimodal optimization problem, now with three variables.

Clearly, we can continue to increment $\hat{r}$ in like manner until, in principle, we reach stage $n_s - 1$, where n_s is the number of independent degrees of freedom in signatures.

Convergence

We note that, at stage $\hat{r} - 1$, the $\hat{r}$th parameter, $\hat{\theta}_r$, is assumed to be unknown. The first step, at stage $\hat{r}$, is a one-dimensional search of $\text{SNR}_{r-1}(\Theta_r)$, over the full span of possible values. The objective is to estimate both the SNR and location of the next largest signal component. The search problem is, in general, not unimodal, in the sense that the function has many submaxima. After finding the principal maximum, a simple detection threshold can be applied. We anticipate that the largest component in the residue has been identified and, therefore, that removal at stage $\hat{r} + 1$, via $\mathbf{Q}_{r+1}$, would minimize the power of remaining residue in an LMS sense. However, the solution is not yet exact. The outstanding iterative optimization task is, in most cases, unimodal and converges, with few iterations, to the true global maximum at rank $\hat{r}$.

At each iteration, $\hat{r} - 1$ of the $\hat{r}$ parameters are held at fixed values. The search function employed by the IMP algorithm, can be regarded as generating a *cross-section* through a rank $\hat{r}$ spectrum estimate. The initial estimate of $\hat{\theta}_r$ is not at the global maximum of that surface unless the other $\hat{r} - 1$ estimates are accurate. If one estimate is improved, then the remainder can also be improved by a cyclic permutation. Unimodal gradient climbing reestimation of each of the $\hat{r}$ parameters, in turn, leads progressively to the required maximum. A convergence criterion can be devised, based on detecting the degree of improvement at each iteration. The sidelobe and mainlobe leakage in the matched filters then becomes of secondary importance, and, in principle, we no longer require aperture weighting functions or other forms of data manipulation to reduce leakage. The task is, instead, accomplished by matrix $\mathbf{Q}_r$.

The three-point method of interpolation

At stage $\hat{r}$, *reestimates* of parameters should be close to the present values. A *local* evaluation of $\text{SNR}_{r-1}(\theta_i)$ for the ith parameter at three points, $\hat{\theta}_i$ and $\hat{\theta}_i \pm \delta$, is normally sufficient. An updated estimate of θ_i can then be interpolated by fitting a second-order curve through the three points and solving analytically for the maximum. The outer two *guard* points should be chosen to lie within the mainlobe of the central matched filter response. Ideally the shape of the mainlobe from the point spread function could be utilized in the analytic curve fit, but, in practice, it is often adequate to match the logarithm of $\text{SNR}_{r-1}(\theta_i)$ to a quadratic function. The three-point method avoids the computational expense and stability of gradient climbing routines that are, in effect, based on two points only.

Estimation of model order or rank

The *final* task, equivalent to that of detection, is to estimate the number of identifiable signal components, $\hat{r}$, are present in the data. Trial solutions at $\hat{r} = 0$ through incrementally to $n_s - 1$ solutions are potentially available. The approach taken in the IMP algorithm [12] contrasts to conventional methods, where the parameter estimation stage generally follows detection.

On grounds of computational economy, it is clear that we would prefer not to evaluate ML solutions at greater rank than the minimum needed model signals adequately in the incoming data. We note that if $\hat{r}$ is at or above the correct rank, then the function $\text{SNR}_r(\Theta_{r+1})$ relates to noise components only and should be essentially isotropic. An upper limit on $\hat{r}$ can be established. A simple detection threshold can be used to determine if the function is sufficiently isotropic. If the detection criterion is not met, then there is little point in proceeding to higher-rank models. Tracking at the current model order, $\hat{r}$, can be initiated or, alternatively, if no more data are available, then the current solution is accepted. Additional computational expense is then avoided. Introduction of undesirable noise into the signal model is also minimized.

Other tests for estimating model order, $\hat{r}$, are feasible. One possible method is based on estimating the residue-to-noise power ratio, RNR(r), at each of various ranks. Since RNR is a measure of the mean residue per degree of freedom (equation 2.21), a chi-squared test of significance can be applied. Alternatively, examination of RNR as a function of $\hat{r}$ should reveal a plateau region close to unit level where no signal information exists. A low-order function can be fitted to the plateau region. An upper value of $\hat{r}$ is estimated at the point where the RNR deviates from the plateau by exceeding a preset threshold. In the same way the RNR can be plotted as a function of $\hat{r}$, $\text{SNR}_r(\hat{\theta}_{r+1})$ can also be plotted as a function of $\hat{r}$. In marginal detection conditions, due to the lower level of noise leakage in selective (rank-one) matched filter, it is to be expected that the plateau should be reached at a higher value of $\hat{r}$. Better resolution is therefore expected. The method proposed in IMP, that of a preset detection threshold on the effective SNR in a *modified* matched filter, is significantly more efficient but does require knowledge of the relevant statistical parameters of the noise. We can then better achieve the lowest probability of false detections and the highest probability of true detections. Clearly, methods for estimating $\hat{r}$ must rely on prior knowledge of characteristics that enable signals to be distinguished from noise. Estimates of $\hat{r}$ can be validated by applying signal identification algorithms that, for example, examine track behavior.

2.6.2 Operation of the IMP Algorithm

A higher-rank spectrum can be regarded as a topological multiparameter surface approximating to the likelihood surface in most respects. In the case of pointlike emitters, this surface comprises intersecting ridges in a multidimensional hyperspace. The signal blocking matrix, **Q**, removes one or more ridges from the surface by projection. The rank-one spectrum of the residue, defined in equation (2.29), is used as an essential estimation tool in IMP and represents a one-dimensional cross-section of the resulting surface. The underlying algorithmic objective of the IMP algorithm is to remove progressively all ridges by blocking contributions from each emitter on an incremental basis (deflation) until the residual surface, due to noise, is essentially isotropic. Figure 2.3 illustrates the concept for two emitters.

In the IMP algorithm, decomposition components are found progressively in decreasing order of magnitude, using rank-one spectra. Figure 2.4 shows a sequential decomposition of simulated data from a 16-element linear array (20 frames) in a nonisotropic background that has been whitened. Successive plots represent the spectrum of the modeling residue as the rank of the model is incremented. The maximum in the spectrum at each stage is regarded as indicating the most likely location of the next decomposition component. The new component is then extracted from the data and suppressed in subsequent stages. All six emitters in the simulation have been located in this way. The final spectrum (hatched) uses the

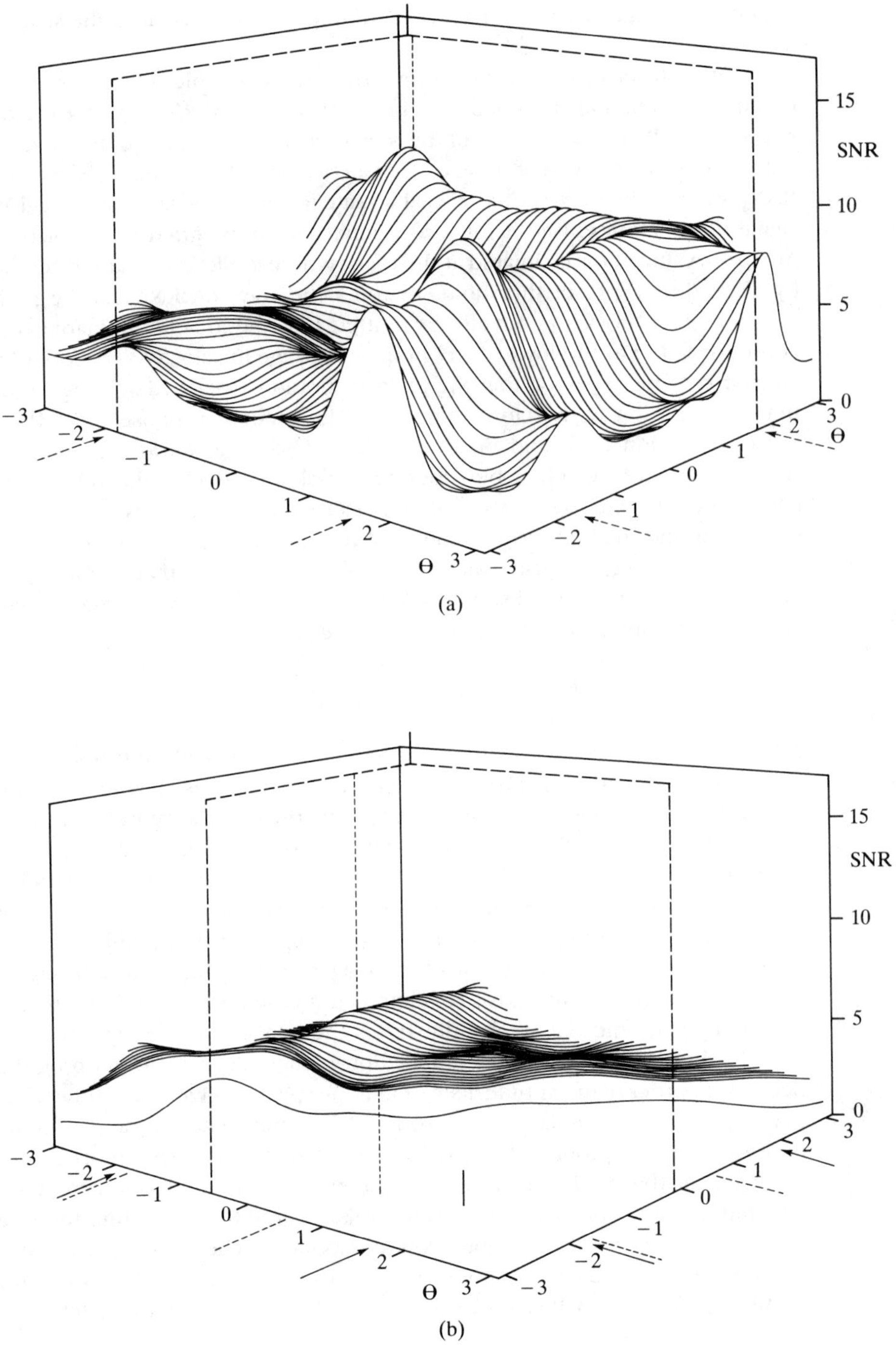
15
SNR
10
5
0
−3
−2
−1
0
1
2
Θ 3
3
2
Θ
1
0
−1
−2
−3
(a)
15
SNR
10
5
0
−3
−2
−1
0
1
2
Θ 3
3
2
1
0
−1
−2
−3
(b)

Figure 2.3

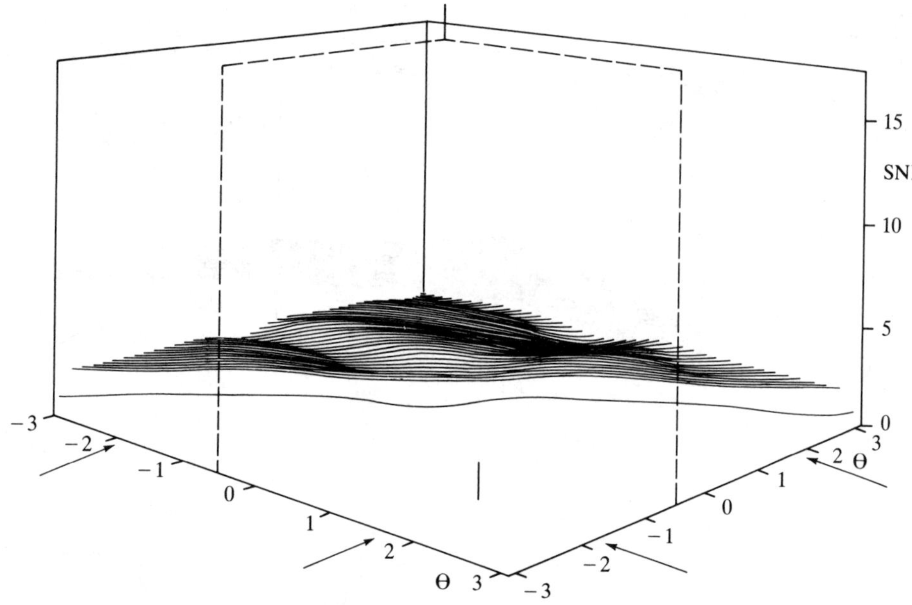

(c)

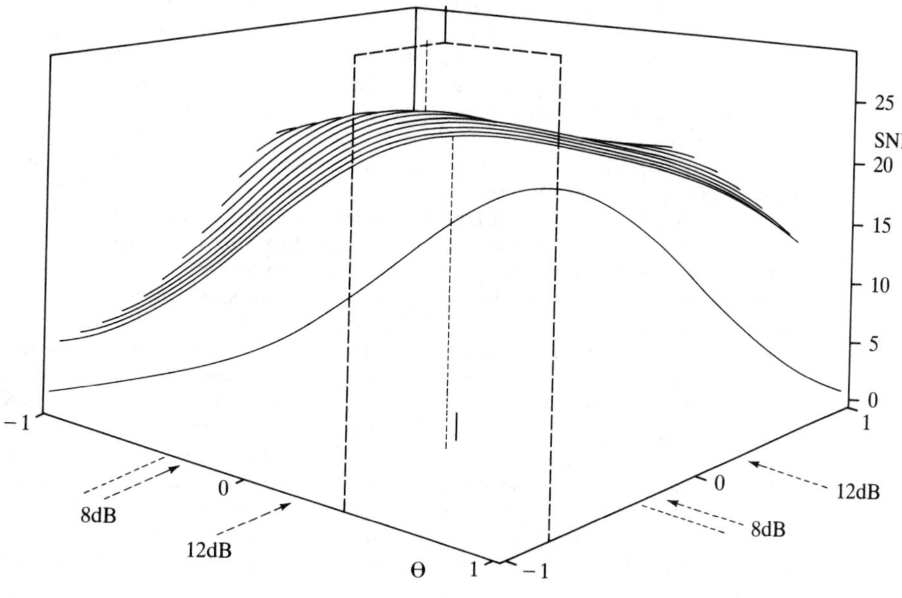

(d)

Figure 2.3 continued

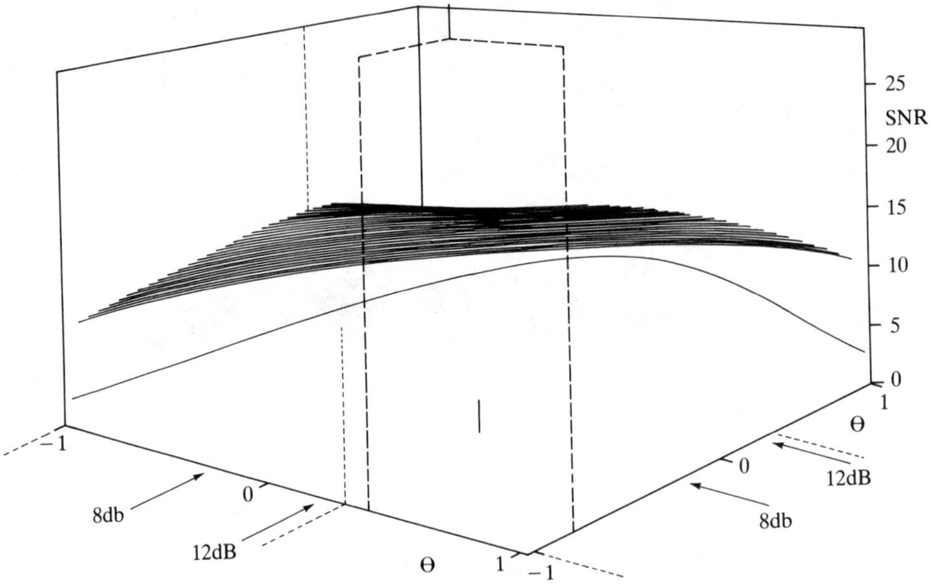

(e)

Figure 2.3 continued (a) An example of a rank-two spectrum for a simulated 8-element linear array. The vertical axis represents the spectral power density as a ratio (SNR). The scale maximum is normalized to the total power in the available data relative to the noise power per degree of freedom. The total noise power is therefore 8 units. The horizontal axes define the two identical spatial-frequency coordinates required for rank-two spectrum estimation. The span, in each case, is ± 3 cycles measured in the aperture of the sensor (note that zero is broadside). Since interchanging the two frequencies has no effect on the defined subspace, symmetry about the diagonal can be observed. Two emitters, of 5 dB and 8 dB (SNR is 3.2 and 6.3, respectively), and separated by 3.5 cycles, were used as an input scenario. The locations are arrowed. Two pairs of intersecting ridges are visible, each associated with one of the emitters. The location of either of the two global maxima provides estimates (dashed) of the spatial frequencies for both emitters. The error, due to noise, can be observed. (b) The same scenario is shown, but where the 8-dB emitter has been canceled by a signal blocking matrix. The ridges for the 5-dB emitter remain. The front section of the spectrum has been omitted to reveal the corresponding rank-one spectrum on the diagonal. It can be seen that the maximum of this spectrum indicates the location of the 5-dB emitter. The global maximum of the rank-two spectrum provides a similar estimate but also includes an unnecessary estimate for the strongest noise component. (c) The same scenario again is shown but where both emitters have been canceled. It can be seen that the rank-two spectrum of the residual noise is essentially isotropic. (d) An example of a rank-two spectrum showing the superresolution of two closely spaced emitters. The input scenario comprises 8- and 12-dB emitters at 0.5 cycle separation. The dashed lines indicate the location of the global maximum. The associated location errors can be observed by comparison with the arrowed input locations. (e) The same scenario is repeated but where the smaller emitter has been blocked. The rank-one spectrum (along the diagonal) for the remaining emitter shows little bias. The significantly lower value of SNR should be observed.

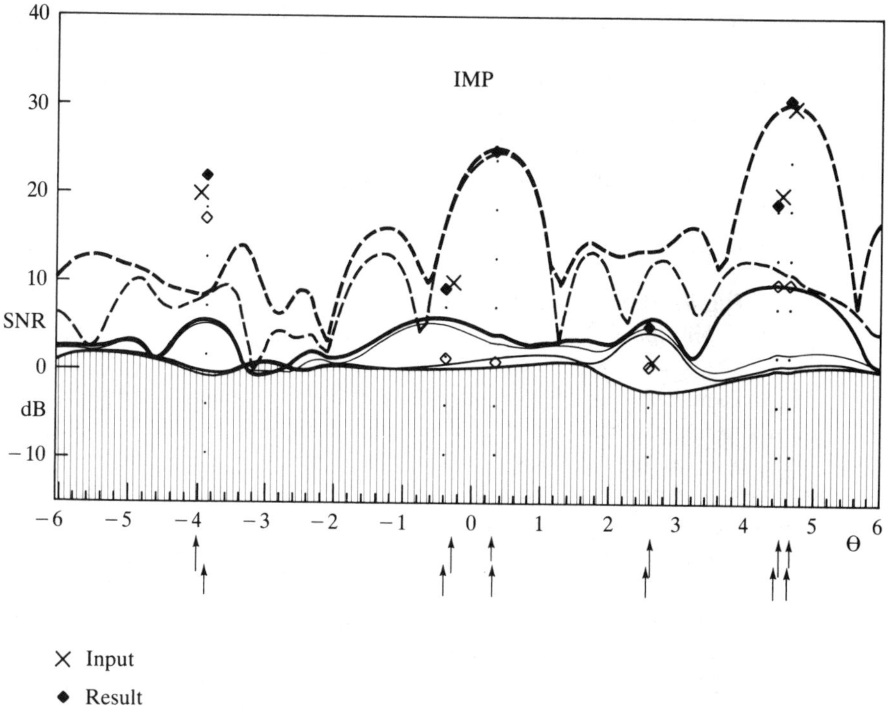

Figure 2.4 Rank-one spectra evaluated in the IMP algorithm and used to estimate progressively the locations of the six emitters arrowed and marked by X. The horizontal axis represents the field of view, Θ, spanning 12 cycles measured over the aperture of a 16-element linear antenna array. Zero is broadside. The vertical axis is power spectral density measured in the form of SNR (decibel scale). As each signal is identified in turn, the corresponding signal subspace is removed via a signal blocking matrix until the residue is essentially isotropic.

corresponding signal blocking matrix, $\mathbf{Q}_6$, to check, against a threshold, that dominant spectral components do not remain.

The filled diamond symbols indicate estimates of emitter power using signal extraction filters defined at the final model order. The open diamond symbols similarly indicate estimated background interference. The left-hand emitter (first) is competing against a high level of background noise, the second and third are fully correlated (identical waveforms), the fourth is very weak, the fifth and sixth are spatially very close. By comparison, three emitters only (the first, third, and sixth) were located in the MUSIC spectrum estimate.

2.6.3 Performance of IMP

An attractive feature of the IMP algorithm is the simple method of linearly combining selected points in an image or classical spectrum. An SNR function for the residue can be generated in terms of predefined point spread functions evaluated from the basic array manifold. The technique leads to novel methods of maximizing SNR for the largest, as yet unmodeled, signal component and allows the IMP algorithm to achieve a near optimum probability of detection in a noise background. The minimum SNR required to resolve two closely spaced marginal emitters can be improved by 10–15 dB relative to the MUSIC algorithm [13]. A significantly improved resolution of correlated or multipath signal components is also possible.

2.6.4 The Relationship of the IMP Solution to Maximum Likelihood

The maximum of a likelihood surface, that has been constrained to $\hat{r}$ decomposition components, can be termed a rank $\hat{r}$ ML solution. The solution comprises (1) estimates of $\hat{r}$ parameter values, (2) $\hat{r}$ matched filters, (3) corresponding estimates of the $\hat{r}$ largest identifiable *signal magnitudes* and waveforms (obtained by passing data through the matched filters), and (4) an unmodeled residue. The residue is the minimum, as measured by a suitable cost function, that can be achieved by jointly varying the $\hat{r}$ parameters, but other state variable parameters may also have been varied.

The solution achieved via the IMP algorithm corresponds to the global maximum of a higher-rank spectrum estimation at rank $\hat{r}$. It can be shown that the IMP approach is equivalent to generating, mathematically, a rank $\hat{r}$ signal subspace projection that is independent of the magnitudes of the decomposition components. The signal blocking matrix, $\mathbf{Q}_r$, is used by the IMP algorithm to define a subspace for the residue. $\mathbf{Q}_r$ is orthogonal to the signal subspace and therefore locates zeros to suppress leakage from r candidate emitters. The equivalent matrix, in a fully optimum LMS algorithm, is a whitening transform, $\mathbf{W}_r\,\mathbf{W}_r^\dagger$. This transform locates a null rather than a zero on each emitter. Leakage is then suppressed to the nominal level of background noise rather than to zero. The two algorithms should give near identical results if the signal magnitudes are large, but the difference becomes noticeable if one or more of the mean signal magnitudes is comparable to noise. Since a shallow null has less effect on the SNR of an adjacent emitter than a deeper null or zero, the marginal performance of an ML method is expected to be superior. If the additional computation is deemed cost-effective, then a candidate whitening transform, $\mathbf{W}_r$, can be evaluated using the rank $\hat{r}$ IMP estimates of the signal waveforms, $\hat{\mathbf{F}}_{rk}$, and the higher-rank signature, $\mathbf{A}_{nr}$ (or $\hat{\mathbf{D}}$)

$$\begin{aligned} \mathbf{W}_r\,\mathbf{W}_r^\dagger &= [\mathbf{A}_{nr}\hat{\mathbf{F}}_{rk}\,\hat{\mathbf{F}}_{rk}^\dagger\,\mathbf{A}_{nr}^\dagger + \mathbf{Q}_r\,\mathbf{R}\,\mathbf{R}^\dagger\,\mathbf{Q}_r]^{-1} \\ &= \mathbf{Q}_r + [\hat{\mathbf{D}}_r\,\hat{\mathbf{D}}_r^\dagger]^{\#} \qquad \text{if } \mathbf{R}\,\mathbf{R}^\dagger \text{ is an identity} \end{aligned} \tag{2.30}$$

The residue (to be minimized) is then $\mathbf{W}_r^\dagger\mathbf{D}$ rather than $\mathbf{Q}_r\,\mathbf{D}$. In the initial searches up to rank $\hat{r}$ and in the iterative optimization steps, equation (2.29) can be utilized, but, to implement fully optimum LMS, the final optimization should be modified to include equation (2.30). However, the basic IMP algorithm has an advantage in terms of robustness if non-Gaussian or impulsive signals are present. Clearly, if the pdf's of the signals are not Gaussian or are unknown, then likelihoods cannot be properly evaluated by an LMS cost function.

2.7 TECHNIQUES FOR ROBUST PROCESSING

It is commonly assumed, with little regard to the effect on algorithmic robustness, that there is an exact link between cause and effect. For example, it is assumed that the signatures represented in the array manifold are accurate. An allowance for perturbations in sensor calibration is rarely included in theoretical predictions. For this reason, the performance of a high-resolution algorithm in a practical environment does not, in general, live up to naive expectations. An active area of research is the development of robust DSP algorithms intended to withstand deficiencies in the knowledge base. Several basic questions need to be addressed:

1. How can the prior be better represented in usable form?
2. Can we design robust parametric modeling algorithms based on a deeper understanding of the information handling aspects?
3. Can we predict a performance limit?
4. Can we reduce the computational demand to an acceptable level?

The basic mechanism embodied in data analysis algorithms is the weighing up of the total evidence for and against a given hypothesis or decision. The fundamental algorithmic design problem is one of choosing appropriate weight factors. The weighting should depend on the following:

1. The relative degree of uncertainty of the factors in the input evidence
2. The degree of relevance of each factor to the possible interpretations
3. The prior preference for each candidate solution
4. The predicted consequences of the action taken as a result of the decision

Maximizing the net level of confidence in a decision is equivalent, in the Gaussian case, to maximizing the SNR of a matched filter. However, to maximize the robustness of the interpretation, we must include all potential contributions to uncertainty. On the assumption that the statistics are Gaussian there is a concept termed total least squares (TLS) that applies if all sources of observation uncertainty are included. A number of algorithms have been adapted based on this concept.

The concept of total least squares can be incorporated directly into the IMP algorithm to improve robustness. Signal-to-noise ratio is evaluated for each of the candidate solutions using knowledge of the statistical parameters associated with *all* observations including sensor data, the array manifold, and the prior. Additional whitening transforms are applied as necessary.

A first step is to arrange, as far as possible, that the individual factors influencing a decision contribute an equal measure of uncertainty in each of the possible outcomes, in a manner analogous to whitening. This basic idea is, in most cases, difficult to implement if the level of uncertainty depends on the magnitude of a signal yet to be isolated by analysis. An iterative approach is, in principle, the best practical alternative. It should be noted that estimates of signal waveforms, $\hat{\mathbf{F}}_{rk}$, can be derived at each stage of the IMP algorithm. The level of uncertainty associated with each component can then be assessed. If the modeling residue, associated with a candidate solution, can be allocated, in full, to predefined causes of observation error, then that solution should be acceptable on the basis of the available evidence. Analysis of unexplained residue can be used to direct the search toward a better solution.

The IMP algorithm generates separately optimized maximum-likelihood solutions at each model order, $\hat{r}$, and therefore presents a multiple-hypothesis situation. The choice of a suitable model order (estimated number of emitters) is critical to the robustness of the solution and should depend both on the prior and on the SNR of the largest component in the residue. It is suggested in the description of IMP that model order is most easily selected by simply applying a threshold to the spectrum, $\text{SNR}_r(\Theta_{r+1})$, for the residue at each candidate rank. Intuitively, it can be seen that if the signal blocking projection matrix, $\mathbf{Q}_r$, is not sufficiently orthogonal to the larger components due, for example, to calibration errors or nonstationary locations of emitters, then residual leakage may give rise to artifacts in the spectrum of the residue. In the next stage, these maxima compete with legitimate components in the spectrum and may cause detections at spurious maxima. Gross tracking errors and instability result if the unimodal optimization is misdirected, but robustness can be improved simply by increasing the threshold at the expense of a lower probability of detection. For example, adaptive control of the threshold can be introduced when it is expected that a null or zero at a fixed location is unlikely to suppress an emitter fully.

In the descriptio… of the IMP algorithm (Section 2.6), the solution was constrained to be a linear combination of signatures, each uniquely identified by the single indexing parameter, θ. However, if the array manifold is extended to include other parameters, then the lost performance can, in principle, be recovered. Other parameters are defined by the extended array manifold and can be introduced into the directed search procedure of the IMP algorithm. The objective remains that of minimizing the residue $\mathbf{D} - \hat{\mathbf{D}}$ with an appropriate cost function. The additional parameters are initially set to default values defined by the prior.

The objective, in a robust version of IMP, is to introduce parameters one by one, in order of importance to the solution. The idea is based on the underlying preference for the simplest solution. At a given stage, the algorithm must assess which additional parameter is the most relevant. This is achieved by a search that establishes the effect of varying each candidate parameter separately but over the full span. The incrementing of the model order $\hat{r}$ is regarded as one of the possibilities but is implemented only if other parameters have a lesser effect on reducing the RNR. One additional parameter is selected, and the default value is dropped in favor of the best value from the search. A local (unimodal) optimization of all parameters in the new model is performed. Research is required, in each specific application, to ensure that parameters are chosen that promote fast convergence toward the best solution. Making the correct choice of analytic model is important.

Clearly, in nonstationary problems, tracks of the estimated values of individual parameters can be validated against predicted bounds on the behavior expected. Validation of tract behavior can be made at several values of $\hat{r}$ and the best option selected. Individual emitters can, in principle, be included or excluded from the model on the basis of frame-to-frame behavior using, for example, modulations of magnitude as a discriminant. If emitters are known to fade or blink rapidly, then continuous tracking of the magnitudes of the signal waveforms potentially allows the vectors necessary to a higher-rank signature, $\mathbf{A}_{nr}$, to be selected and deselected in a highly responsive manner. The approach contrasts with current adaptive procedures, where state variables (including model order) are heavily filtered. Improved stability obtained by flexible control of model order should enable relevant parameters to be tracked robustly. The early stages of IMP can be bypassed if initial or default parameter values are available via the prior but, to ensure stability, care must be taken that the estimates are not misleading.

If track behavior constraints are in the form of an analytic model, then the model, with current values of track parameters, can be used to predict values of θ in the next batch of data. The prior is modified accordingly. In principle, it is also possible to update track parameters directly, using the approach to parameter estimation adopted in IMP. The constraint that parameters must be stationary within a block of data can then be relaxed (at low computational expense). The effective integration period can be extended without causing lag in the response. The technique is analogous to that of the extended Kalman filter with similar potential performance advantages in marginal scenarios. In addition, the integration period can be controlled adaptively, and the restrictive constraints of fixed model order and slow track acquisition are alleviated.

2.8 CONCLUSIONS

In general, the output data from a sensor system require expert interpretation. Automating this task is essentially a complex signal processing problem based on prior knowledge. Although the performance of digital processing technology improves each year, the development of suitable algorithms has not been advancing as rapidly.

Many modern high-resolution algorithms, such as MUSIC [1], can be categorized in a common format. A combination of different weighting transforms is, in general, applied to data before generating a modified form of a classical rank-one spectrum. The detection of multiple emitters is obtained by searching for maxima or minima in a one-dimensional search. The corresponding locations provide the necessary estimates of discriminatory parameters. The validity of the corresponding decomposition and modeling residues are not cross-checked in these algorithms. It is concluded that noniterative algorithms of this type are inherently suboptimal, due to residual leakage due to cross-correlation effects that cannot be suppressed in a single rank-one estimation stage. A better insight into the basis for each algorithm is obtained by examination of the weighting given to available evidence in relation to the underlying information theory.

An important conclusion, from an information theoretic viewpoint, is that there is nothing theoretically suspect about *superresolution;* the difficulty lies solely in the specification of a suitable prior and of filters, correctly matched to the differences between the relevant signatures of individual emitters. The span of permitted solutions can be defined in terms of an extended array manifold. There is, however, a daunting computational problem associated with the estimation of a higher-rank spectrum or of a likelihood function. The problem is caused by the inevitable combinatorial explosion in the number of candidate higher-rank matched filters as the number of parameters is increased.

Applying the principles of maximum likelihood to the inverse problem leads to optimal solutions, if a full search is undertaken, but with an excessively heavy computational penalty. The development of computationally efficient robust high-performance DSP algorithms based on sound information theoretic principles is highly desirable. The aim should be to minimize computational load by avoiding the search of a large proportion of the solution space that is permitted in the maximum-likelihood approach. Simulations of the IMP (incremental multiparameter) algorithm demonstrate that a directed search is a much undervalued method of exploiting prior knowledge by comparison with competitive techniques such as synthetic annealing.

The IMP data analysis algorithm is iterative, uses a directed search, and takes a very different approach to that of other high-resolution, parametric modeling, state estimation, and tracking algorithms intended for similar applications. The explicit inclusion of support to augment sensor data in a direct and flexible manner leads to stable robust solutions to problems that cannot be solved by current methods. Efficient versions of IMP operate on data in the image, focal plane, spectral, or beam-space domains. Data vectors from sampled apertures can be transformed into the appropriate (whitened) domain by preprocessing. The sensor coupled with the preprocessor can be regarded as a single module, that is, as a *modified* sensor system generating spectral domain data directly. By contrast, methods that are based on eigen decomposition (or matrix inversion) of a covariance matrix are computationally expensive, limit the applicability of many existing algorithms, such as the MUSIC algorithm, to near stationary problems, and also rely on a high degree of decorrelation in the temporal or other domain. A robust version of IMP, based on an extended array manifold, has potential applications in a wide range of data interpretation tasks.

In summary, the IMP algorithm proceeds by first proposing a simple (low-rank) deconvolved model of the scenario. This proposed interpretion is then convolved with the appropriate point spread functions for the *modified* sensor system. The result is a prediction of the sensor data that would correspond to the proposed scenario. These data are compared with actual data. The comparison is performed in the beamspace, focal plane, or image domain and is expressed as a weighted difference, modeling error, or residue. The residue, expressed in SNR form, is examined algorithmically to provide an estimated best correction to the currently proposed deconvolved version of the scenario. The cycle is repeated iteratively, progressively refining the estimate of the scenario and incrementing the rank of the model as necessary. When the interpretation is sufficiently complex for the predicted data to match the input data to within observation error, then the algorithm is stopped. The final choice is therefore the least complex solution permitted by the available data.

Each correction or update to the deconvolved scenario requires the inclusion of support in the form of constraints on expected features. A point target constraint is applied in the basic IMP algorithm, but other features such as edges and lines or tracks are feasible. At each stage, spatial positions and other parameters are adjusted iteratively to minimize the modeling residue. Aperture weighting is, in principle, unnecessary as a means of controlling the sidelobes of the point spread functions because the residue is evaluated by projection through a subspace where the expected degree of leakage is built into the model.

If we can regard the scientific approach as the art of oversimplication of natural laws followed by progressive refinement by more detailed research then the IMP algorithm has a justifiable foundation. The philosophy, illustrated in the IMP algorithm, should lead to application-specific variants that can be applied to a wide range of problems.

The algorithm is applicable to sampled aperture systems, including both planar and focal plane antenna arrays, large or small, of arbitrary configuration with or without built-in beamformers, pulse compression filters, or Doppler channels. For example, the number of elements in sampled data from a radar may be very large. Eigen-based computation is then excessively expensive. The signatures used in the IMP algorithm may be extended across several discriminants, such as angle of arrival, range, and Doppler, enhancing both the resolution and detection of marginal signals. Since eigen decomposition is not required in the IMP algorithm, data vectors can be analyzed individually. Cross-correlation between signals, due for example, to multipath can also be addressed. By contrast with eigen-based techniques, the IMP algorithm, implemented in efficient form, adds little to the computational demand over that needed to form the basic image.

Adaptive control of model order (rank), achieved by monitoring the effective SNR of individual components, ensures that stability is not affected by unnecessary noise as signals fade. Performance can be expected to be close to that of optimal ML methods but without the excessive computational demand. The principal difference lies in the depth of null used to suppress leakage in a multiemitter scenario. However, if the additional expense is deemed cost-effective, then the solution obtained via the IMP method can be used to provide initial estimates both of emitter location and of mean magnitude for a relatively inexpensive unimodal search in a localized region of the ML surface.

It is also concluded that the IMP concept, used as parametric modeling or state estimation algorithm, leads to novel methods of tracking. The technique, outlined in this chapter, has the potential to alleviate widely recognized design problems of the ubiquitous Kalman filter, notably poor acquisition, instability, slow response, and fixed model order. Rapid simultaneous variation of many interrelated parameters can, in principle, be monitored. The method has the potential to open up a wide range of new applications for parametric modeling algorithms. A technique for direct track extraction illustrates that support can be applied in a single optimized step. This approach contrasts with the practice of separate stages for plot extraction and track formation.

In general, the robustness of the IMP high-resolution algorithm, in nonideal conditions, is better than competitive eigen-based techniques [13]. However, if the support imposed by the engineer is seriously in error, then alternative algorithms *using little or no support* might, at first, *appear* to perform better than IMP because incorrect features are not being forced onto the solution. The flexibility of the IMP concept encourages the system designer to optimize the use of support in a way that matches the confidence in that support. In particular, the IMP approach can be extended to include TLS. The task of online self-calibration of the sensor system is also simplified. Invariant features of the solution can be used to best effect. Analysis is alternated between signal parameter estimation and system identification. There is a clear limit, defined by information theory, beyond which the performance is degraded.

Given adequate parametric models, implementation of the algorithm in software is a straightforward design task that does not suffer the common problems associated with designing fixed-rank gradient-based algorithms such as the Kalman filter. Although the computational demand of IMP is relatively low, the high performance potential opens up a wide range of novel but highly demanding applications. Certain applications are likely to require algorithmically engineered specialized hardware in the form of novel devices and architectures. The key task is to define a satisfactory extended array manifold, with suitable analytic models and discriminants, for the intended specific application.

Potential applications range from data interpretation in small conformal sensor arrays to large multielement dispersed antennae on land or sea or in space. Sensors may be active or passive and include radar, sonar, seismic, acoustic, thermal, and optical. System identification, self-calibration, state estimation, data fusion, and tracking are included.

REFERENCES

1. R. O. Schmidt, "Multiple Emitter Location and Spectral Parameter Estimation," *Proc. RADC Spectrum Estimation* workshop, (1979), 243–258.
2. J. M. Mendel, *Lessons in Digital Estimation Theory* (Englewood Cliffs, N.J.: Prentice-Hall, 1987).
3. J. G. McWhirter and T. J. Shepherd, "Efficient Minimum Variance Distortionless Response Processing Using a Systolic Array," *SPIE Proc.,* Vol. 975–39, *Advanced Algorithms and Architectures for Signal Processing,* III, San Diego, California (1988).
4. J. L. Mather, "A Monte Carlo Performance Analysis of Accelerated SVD-Based High-Discrimination Algorithms," memorandum no. 4083, RSRE, Malvern, England.
5. J. Capon, R. J. Greenfield, and R. J. Kolker "Multidimensional Maximum Likelihood Spectra," *Geophysics* **37**(2) (1972), 375–376.
6. I. J. Clarke, "Robustness of Eigen-Based Analysis Techniques Versus Iterative Adaptation," *RADAR 87,* IEE Conference Publication No. 281 (1987), 84–88.
7. J. P. Burg, "The Relationship Between Maximum Entropy Spectra and Maximum Likelihood Spectra," *Geophysics* **37**(2) (1972), 375–376.
8. R. Kumaresan and D. W. Tufts, "Estimating Angles of Arrival of Multiple Plane Waves," *IEEE Trans.* AES-19(1) (1983), 134–139.
9. A. J. Barabell, J. Capon, D. F. DeLong, J. R. Johnson, and K. D. Senne, "Performance Comparison of Superresolution Array Processing Algorithms," MIT Lincoln Lab. Project report TST-72 (May 1984).
10. A. J. Barabell, "Improving the Performance of Eigenstructure-based Direction-Finding Algorithms," *IEEE Proc. Int. Conf. Acoustics and Speech Signal Processing* **1,** (1983), 336–339.
11. U. Nickel, "Angle Estimation with Adaptive Arrays and Its Relation to Super-resolution," *IEE Proc. H,* **134** (1987), 77–82.
12. I. J. Clarke, "High-Discrimination Detection Bound and Model Order Control," *SPIE Proc.* Vol. 975–33, *Advanced Algorithms and Architectures for Signal Processing,* III, San Diego, California (1988).
13. J. L. Mather, "Performance of the IMP Array Processing Algorithm — First Results," memorandum no. 4291, RSRE, Malvern, England (1989).

3

Maximum Likelihood for Angle-of-Arrival Estimation in Multipath

S. Haykin, V. Kezys, and *E. Vertatschitsch*

3.1. INTRODUCTION

Multipath is a phenomenon that arises when a radiating source lies in close proximity to a reflecting surface [1]. This radiating source may, for example, represent a low-flying target that is illuminated by the transmitting antenna of a low-angle tracking radar. The task of such a radar system is to track the movement of the target, which becomes particularly acute when the target lies inside a small fraction of the receiving antenna beamwidth. In this situation, the antenna hunts up and down, sometimes pointing upward correctly at the target and then pointing erroneously downward at the image of the target. Indeed, the low-angle tracking radar problem remains to be an engineering challenge [2]. The purpose of this chapter is to show that the solution to this problem may well lie in the use of a multiparameter array system.

Specifically, we present a study of the classical maximum-likelihood principle for estimating the angle of arrival of a target in the presence of multipath. The study is supported by real-life data obtained using a 32-element sampled-aperture radar system. Two particular system configurations with different amounts of available information are considered:

1. *Uniformly spaced antenna elements.* In this part of the study, we explore the role of *prior information* represented by a priori knowledge of the antenna height above the reflecting surface and the target range from the radar. Here, we also explore the potential benefits that may be derived from the use of *multiple transmit frequencies,* a technique that expands on the pool of information available for processing.
2. *Nonuniformly spaced antenna elements.* In this case, we explore the benefits obtained by using an array configuration known as *nonredundant arrays* [3].

In both cases, the model describing the received signal is assumed to be *specular.* In other words, the *diffuse* component arising from surface irregularities is neglected.

The body of the chapter is organized as follows. In Section 3.2 we briefly review the characteristics of multipath as experienced in a *low-angle tracking radar environment.* In Section 3.3 we describe the essentials of a 32-element sampled-aperture radar system used in the study. In Section 3.4 we focus on the *maximum-likelihood (ML) estimation* procedure for a uniformly spaced array antenna. In particular, we describe the ML algorithm, assuming a specular multipath environment (i.e., ignoring the effect of diffuse multipath). We present experimental results on the use of a uniformly spaced receiving array antenna system that includes the benefits derived from the use of (1) prior information and (2) multifrequency illumination. In Section 3.5 we present experiment results on the use of a nonuniformly spaced array antenna. In Section 3.6 we present conclusions drawn from our experimental study of the low-angle tracking radar problem.

3.2. MULTIPATH

Consider the idealized *flat-earth* geometry depicted in Fig. 3.1. In the situation described here, *diffuse multipath* arising from surface irregularities is ignored. According to this figure, the incident field at the receiving antenna consists of two basic components:

1. *A free-space component* that propagates from the *target* to the receiving antenna along a direct path; this component is also referred to as the *direct ray* (signal).
2. *A specular multipath component* that results from surface reflections; it reaches the receiving antenna along an indirect path that points to the *image* of the target.

The important point to note is that both the free-space and specular multipath components are well defined in terms of the amplitude, phase difference, and direction

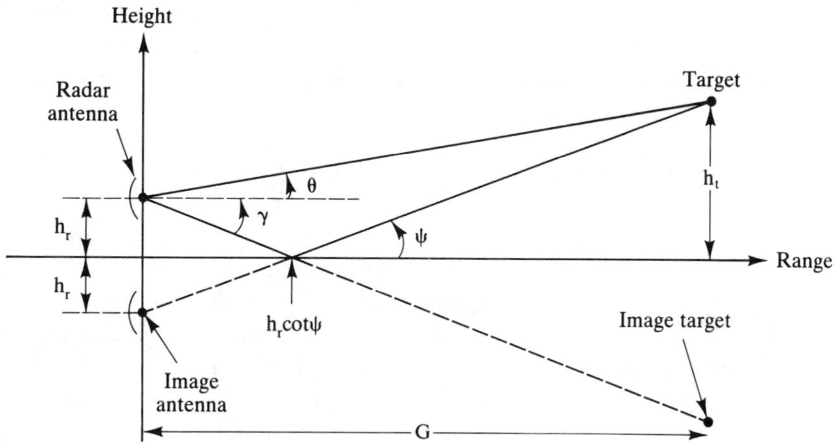

Figure 3.1 Geometry of flat-earth reflection.

when the geometry, reflection process, and propagation effects are known exactly. Under this condition, we may view the multipath model of Fig. 3.1 as *deterministic.*

Let θ denote the *elevation angle* of the target, and $A(\theta)$ denote the *voltage gain pattern* of the receiving antenna measured at angle θ. We may then express the free-space component at the receiving antenna as

$$a_F = IA(\theta), \tag{3.1}$$

where I is the *free-space field intensity* at the receiving antenna. To describe the specular multipath component, let ψ denote the *grazing angle* for specular reflection and $A(-\psi)$ denote the antenna voltage gain pattern at angle $-\psi$ (i.e., the elevation angle to the reflection point). We may thus express the specular multipath component as

$$a_S = \rho_0 IA(-\psi)\exp(-j\alpha), \tag{3.2}$$

where ρ_0 is the magnitude of the *reflection coefficient* for the surface at grazing angle ψ. The angle $-\alpha$ is the total phase shift of the specular multipath component with respect to the free-space component; it is defined by

$$\alpha = \frac{2\pi\delta}{\lambda} + \phi, \tag{3.3}$$

where λ is the *radar wavelength,* δ is the *path-length difference* between the free-space and specular multipath components, and ϕ is the phase angle of the surface reflection coefficient. For a flat reflecting surface, we find from the geometry of Fig. 3.1 that the path-length difference δ is approximately given by

$$\delta \simeq \frac{2h_r h_t}{G}, \tag{3.4}$$

where h_r is the *receiving antenna height*, h_t is the *target height*, and G is the *ground range of the target*; for the definition of these terms, see Fig. 3.1. For the more general case of a curved reflecting surface, the reader is referred to Blake [4].

3.3. EXPERIMENTAL SAMPLED-APERTURE RADAR SYSTEM

In this section we briefly describe an experimental radar system used to collect real-life data for testing the maximum-likelihood estimation of elevation angle of a target embedded in a multipath environment. The experimental system was implemented in two phases. In *Phase I* of the implementation, a single fixed radar frequency was employed. Data collected with this version of the system were used to study the effect of multipath on maximum-likelihood estimation for nonuniform arrays; see Section 3.5. In *Phase II* of the implementation, two transmit frequencies were used simultaneously; one frequency was *fixed* while the other was *agile.* Data collected with the second version of the system were used to study the effect of multipath on maximum-likelihood estimation for uniform arrays; see Section 3.4.

In Phase I of the system implementation, the transmit frequency was fixed at 9.81 GHz, providing a free-space wavelength of approximately 3.05 cm. The transmitter consists of a free-running 5-MHz double-oven, crystal-controlled oscillator that is used to phaselock a 9.81-GHz source. The signal is amplified to approximately 10 W out of a *traveling-wave tube amplifier* (*TWTA*) and then transmitted through a 10-dB gain horn.

The 5-MHz crystal oscillator, also found in the receiver, provides very-low-phase noise. The short-term-phase noise was characterized by an *Allan variance* of 5×10^{-12} in 1 second. After a 24-hour burn-in, the long-term drift of the oscillator is less than 3 parts in 10^{10} per day.

The receiver consists of a 32-element *sampled aperture.* Each channel of the receiver consists of a 10-dB gain horn, followed by a 10-dB directional coupler. A test signal may be injected into the system through this coupler when the transmitter is shut down; this is done for *calibrating* the system. The signal is then mixed down to 45 MHz and amplified. The path is split and mixed down to *inphase* and *quadrature* baseband signals that have frequencies of 15.625 Hz. After further amplification and low-pass filtering (cutoff at 31.25 Hz), it is sampled at 125 Hz, or eight samples per cycle.

The low-frequency signals are all digitally generated, synchronous with a computer system clock. The baseband frequency, filter bandwidths, and sampling signals are all under computer control and can be varied as experimental conditions require. The IF (intermediate frequency) amplification is also varied through software control.

Prior to data collections, with the transmitter on, the 5-MHz oscillator at the receiver is fine-tuned such that the receiver operates with 0.1 Hz of the transmitted signal at X-band. For all practical purposes, we may describe the receiver as *coherent* for the duration of the data collection, usually less than 10 seconds. For long-term data collection, provision was made for continuous adjustment of the 5-MHz oscillator. When the test signal is applied instead, the system is truly coherent. Since the electronic gain and phase shift will vary slightly with frequency, the system is operated at the identical frequencies for which it is calibrated.

The vertical linear array consists of thirty-two 10-dB gain horns oriented for horizontal polarization. The structure is machined such that the spacing between horns is 5.715 ± 0.010 cm. A similar tolerance is used for the remaining two horizontal dimensions. The electrical phase error with respect to neighboring elements is less than 1°. With the interelement spacing larger than $\lambda/2$, the unambiguous field of view of the receiving array is approximately ±15.5°. In terms of the normalized parameters, where the spacing between elements is considered equal to 1 unit, the span of wavenumber is $\pm\pi$, with $+\pi$ corresponding to a physical elevation angle of 15.5°. Since the aperture is 1.77 meters, the beamwidth in physical angle is approximately 1°.

The experimental multipath data were collected at a site located on the mouth of Dorcas Bay that opens onto the eastern end of Lake Huron. The transmitter was situated at a distance of 4.61 km from the receiver, both being within 10 meters of the water's edge (and occasionally, during storms, within the bay when the water level rose). The site is described in Fig. 3.2. Alternative transmitter heights were selectable through a radio link. The center of the receiving array was at 8.6 meters above the water surface.

For Phase II of the experimental system, the system was modified to include *frequency diversity*. In particular, two receiver channels were included behind each of the 32 elements of the array antenna. One channel was designed to operate at a *fixed frequency* of 10.2 GHz. The second channel was *agile*, with its operating frequency selectable from 8.05 to 12.34 GHz in 30-MHz steps. Moreover, the transmitting horn was placed on a vertically moving carriage structure. This setup permitted the height of the transmitter to be continuously varied between 2 and 20 meters under computer control from the receiving site.

3.4. MAXIMUM LIKELIHOOD APPLIED TO A UNIFORM ARRAY

The *signal received* at the nth element of a vertical linear array due to a single target in the presence of specular multipath may be modeled as

$$r_n(f) = G_n(f)s_n(f) + w_n(f), \qquad (3.5)$$

where $s_n(f)$ is the *uncorrupted signal* defined by

$$s_n(f) = a_1(f)\exp[jk_1(f)x_n] + a_2(f)\exp[jk_2(f)x_n] \qquad (3.6)$$

and the *wavenumbers* $k_1(f)$ and $k_2(f)$ are defined by

$$k_m(f) = \frac{2\pi fd}{c}\sin\theta_m, \qquad m = 1, 2. \qquad (3.7)$$

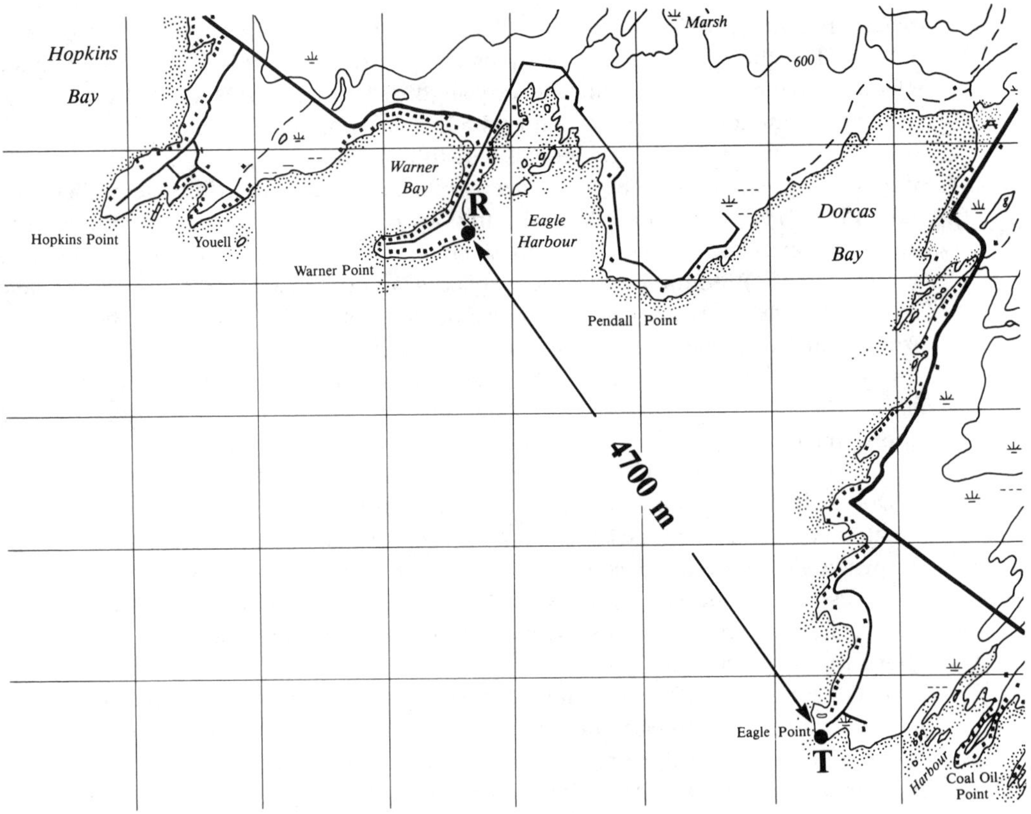

Figure 3.2 Plan view of the Dorcas Bay site near Tobermory, Ontario, showing the locations of the transmitter (T) and sampled aperture receiver (R).

The parameter f denotes the *operating frequency*; c denotes the *speed of light*; θ_1 and θ_2 are the *elevation angles* of the direct and specular rays, respectively; and a_1 and a_2 are their *complex amplitudes*. Note that according to the definition of (3.7), the wavenumber $k_m(f)$ is measured in radians. The multiplicative factor $G_n(f)$ denotes residual *calibration error*. Last, the term w_n represents a sample of *independent additive white Gaussian noise*.

The value x_n refers to the normalized location of the nth element of the receiving array antenna. In the case of a linear array with N uniformly spaced elements, we have

$$x_n = \left(n - \frac{N-1}{2}\right), \qquad n = 1, 2, \ldots, N. \tag{3.8}$$

On the other hand, in the case of a nonuniform array (e.g., nonredundant array), x_n takes on a corresponding set of integer values that are nonuniformly spaced. In this section, we use the specular multipath model of (3.5) to (3.8) to formulate the maximum-likelihood estimation problem for the case of a uniform array. In the next section, we consider the corresponding case of a nonredundant array.

The separation between the direct and specular rays is given by the absolute value, $|k_1 - k_2|$. Given that this separation may possibly be less than the beamwidth of the aperture of the receiving array, the requirement is to use the classical maximum-likelihood principle [5] to estimate the wavenumber k_1 of the direct ray.

In the model described by (3.5) to (3.7), the operating frequency f is shown as a variable radar parameter. The reason for so doing is the potential for increased accuracy of estimation as predicted by calculations based on the *Cramér-Rao bound*; for details, see [6]. The variability of the operating frequency f, however, only applies to the material presented in this section.

3.4.1 Estimation Algorithm

A basic premise behind the maximum-likelihood estimation strategy used herein is the incorporation of prior information. Let $\mathbf{r}$ denote the received vector with elements $r_n(f)$. The vector $\mathbf{r}$ is a concatenation of M observations, with each observation being N elements long and corresponding to a specific operating frequency. Thus the dimension of vector $\mathbf{r}$ is MN. The uncorrupted signal vector $\mathbf{s}$, also of dimension MN, is defined in a similar fashion. Without prior information, the *conditional probability density function* of the received vector $\mathbf{r}$ may be formulated as

$$p(\mathbf{r}|\mathbf{a}, k_1, k_2) = \frac{1}{\pi^N|\mathbf{R}|} \exp[-(\mathbf{r} - \mathbf{s})^\dagger \mathbf{R}^{-1}(\mathbf{r} - \mathbf{s})], \tag{3.9}$$

where $\mathbf{R}$ is the *noise/error covariance matrix*, $|\mathbf{R}|$ is its *determinant*, and $\mathbf{R}^{-1}$ is its *inverse*. The superscript $\dagger$ denotes *Hermitian transposition*. The fixed wavenumber k_m, $m = 1, 2$, is written in (3.9) in place of $k_m(f_O)$, where f_O is an arbitrary measurement frequency. By contrast, the vectors $\mathbf{r}$, $\mathbf{s}$, and $\mathbf{a}$ are all dependent on the frequency f. The vector $\mathbf{a}$ is a $2M$-by-1 vector with its mth elements denoted by $a_m(f_l)$, where $m = 1, 2$ and $l = 1, 2, \ldots M$.

In much of the array-processing literature, the noise covariance matrix is assumed to represent an additive identically distributed and independent process. In such a situation, we may write

$$\mathbf{R} = \sigma_W^2 \mathbf{I} \tag{3.10}$$

where σ_W^2 is the additive noise variance and $\mathbf{I}$ is the N-by-N identity matrix. However, through the process of using a physical sampled aperture, it has been found that a significant amount of measurement error may arise due to the unavoidable presence of residual calibration errors. In other words, in (3.5) we have

$$G_n(f) \neq 1, \qquad \text{for all } n \text{ and } f. \tag{3.11}$$

Thus, in general, the measurement error may be expressed as

$$\begin{aligned}\epsilon_n(f) &= r_n(f) - s_n(f) \\ &= w_n(f) + [G_n(f) - 1]s_n(f).\end{aligned} \tag{3.12}$$

Let $N(\mu, \sigma)$ denote *a Gaussian probability density function* with mean μ and standard deviation σ (i.e., variance σ^2). Then, for small, independent, and Gaussian-distributed residual gain and phase errors, we may express their respective distributions as $N(O, \sigma_g)$ and $N(O, \sigma_p)$. Accordingly, the combined noise/error covariance matrix may be written as

$$\mathbf{R} = \mathbf{q}\mathbf{I}, \tag{3.13}$$

where $\mathbf{q}$ is an MN-by-1 vector with its elements defined by

$$q_n(f) = \sigma_w^2 + (\sigma_g^2 + \sigma_p^2)|s_n(f)|^2, \qquad n = 1, 2, \ldots, MN. \tag{3.14}$$

Under these assumptions, the covariance matrix $\mathbf{R}$ remains diagonal; however, unlike the idealized noise-only case of (3.10), the covariance matrix of (3.13) has *unequal diagonal elements.*

The use of (3.14) requires knowledge of the uncorrupted elemental signal $s_n(f)$. Since the estimator is not privy to such a knowledge, we will simplify the calculation of $q_n(f)$, albeit at the cost of some degradation in estimation accuracy, by using the received elemental signal $r_n(f)$ in place of the uncorrupted elemental signal $s_n(f)$.

As mentioned previously, a specular multipath environment encountered in the use of a low-angle tracking radar system poses a challenge in signal processing difficulty; this is particularly so when the highly correlated direct and specular rays are closely spaced. Nevertheless, such an environment does provide for useful relationships between unknown parameters of the system through the reflection geometry; see Fig. 3.1. Specifically, assuming a flat-earth model and knowledge of the receiving array height above the reflecting surface, the phase difference (measured at the *center of the array*) between the specular and direct rays at an operating frequency f may be expressed as

$$\begin{aligned}\Delta\psi(f) &= \arg\left[\frac{a_2(f)}{a_1(f)}\right] \\ &\simeq -\frac{2\pi f}{c}\delta + \arg[\rho(f)],\end{aligned} \tag{3.15}$$

where δ is the difference between the direct and specular path lengths, defined approximately by (3.4). The frequency-dependent parameter $\rho(f)$ is the *complex reflection coefficient* of the reflecting surface. It is assumed that $\rho(f)$ has a constant phase with respect to the frequency f over the frequency band of interest. Under this condition, (3.15) reveals that the phase difference $\Delta\psi(f)$ may be viewed as a linear function of the frequency f. This linear dependence is described by an unknown offset $\Delta\psi(f_1)$ and slope β, as shown by

$$\Delta\psi(f) = \Delta\psi(f_1) + \beta(f - f_1). \tag{3.16}$$

The slope β is itself defined by

$$\beta = -\frac{h_r(k_1 - k_2)}{df_0}, \tag{3.17}$$

where d is the array element spacing.

The prior information described here may be used to formulate equality constraints on some of the phase differences. The complexity of the estimator is thereby reduced. However, when the prior information such as array height is inexact, this would lead to a suboptimum solution. Specifically, given a value for h_r and at least two frequency measurements, we may consider the prior information as represented by the conditional Gaussian probability density function:

$$\begin{aligned} p[\Delta\psi(f_1), \Delta\psi(f_2)] &= p[\Delta\psi(f_2)|\Delta\psi(f_1)]\,p[\Delta\psi(f_1)] \\ &= N[\overline{\Delta\psi(f_2)}, \sigma_{\Delta\psi}(f_2)]\,p[\Delta\psi(f_1)], \end{aligned} \tag{3.18}$$

where

$$\sigma_{\Delta\psi}(f_2) = \frac{2\pi}{df_0}(f - f_1)\sigma_{hr}(k_1 - k_2), \tag{3.19}$$

where $\overline{\Delta\psi(f)}$ is the mean value as in (3.16) and σ_{hr} represents the inaccuracy in our knowledge of h_r. Here we have constrained the phase difference at frequency f_i, $i > 2$, without loss of generality.

Likewise, given the array height and the target range, the wavenumber k_2 may be expressed as

$$k_2 \simeq -\frac{4\pi df_0 h_r}{cG} - k_1. \tag{3.20}$$

The corresponding a priori probability density function may be expressed as

$$\begin{aligned} p(k_1, k_2) &= p(k_2|k_1)p(k_1) \\ &= N(\overline{k_2}, \sigma_{k_2})p(k_1), \end{aligned} \tag{3.21}$$

where

$$\sigma_{k_2}^2 = \left[\frac{4\pi df_0\sigma_{hr}}{cG}\right]^2 + \left[\frac{4\pi df_0 h_r\sigma_G}{cG^2}\right]^2 \tag{3.22}$$

and $\overline{k_2}$ is the mean value as in (3.20) and σ_G represents the inaccuracy in our knowledge of the range G. In the case of both G and h_r, a distinction must be made between the actual variances and those that are assumed and used in the estimator.

Combining (3.9), (3.16), (3.19), (3.20), and (3.21), taking the natural logorithm, and ignoring those terms that are independent of the unknown parameters, the overall *objective function* to be maximized assumes the form

$$\ell(\mathbf{a}, k_1, k_2) = -(\mathbf{r} - \mathbf{s})^\dagger \mathbf{R}^{-1}(\mathbf{r} - \mathbf{s}) - \frac{(k_2 - \overline{k_2})^2}{2\sigma_{k2}^2} - \frac{[\Delta\psi(f_2) - \overline{\Delta\psi(f_2)}]^2}{2\sigma_{\Delta\psi}^2(f_2)}. \tag{3.23}$$

The use of an exhaustive search to locate the global maximum of the log-likelihood function of (3.23) with respect to the unknown complex amplitude **a** and the wavenumbers k_1 and k_2 is impractical. We say this because, even for two transmit frequencies, the number of unknown parameters to be estimated is 10, none of which appears linearly.

A simpler strategy, however, is to note that in the first two terms of (3.23), the vector **a** appears *linearly* in **s** and is therefore *separable* [6]. The strategy is then to search the feasible domain of k_1 and k_2 (i.e., $k_1 > k_2$) using a *coarse grid* for only the two terms so as to find a point that lies close to the global maximum. One drawback to the grid approach is that, by leaving out the last term, the a priori information due to the phase difference slope does not contribute to reducing the *probability of outliers* and hence the threshold *signal-to-noise ratio* (*SNR*). We proceed from the point located previously and maximize the log-likelihood (objective) function with respect to all parameters. For this optimization step in the computation we use a modified *Fletcher's quasi-Newton algorithm* [7] with equality constraints implicitly placed on $\Delta\psi(f_l)$ for $l > 2$. The optimization strategy described here is quite efficient since it uses gradients of the objective function expressed explicitly.

3.4.2 Experimental Results

For the uniform array part of the study, we used data that were collected with version II of the experimental system described in Section 3.2. In this mode of operation, the system uses a pair of transmit frequencies, one fixed at 10.2 GHz and the other adjustable over the band 8.05 to 12.34 GHz in 30-MHz steps. The data were collected at the Lake Huron site (identified in Section 3.2) over several days. The measurements performed were accompanied by calibration runs.

Using this data base, the results plotted in Fig. 3.3 were obtained. In this figure, we show the phase difference (measured in radians) at the center of the array plotted versus the operating frequency f (measured in gigahertz). This figure clearly illustrates the linear relationship between the phase difference $\Delta\psi$ and frequency f, as predicted by (3.16). The phase differences plotted in Fig. 3.3 were estimated from single snapshots at each of the adjustable frequencies from 8.05 to 12.34 GHz for a target height of 15.59 meters; this height corresponds to a 0.41 beamwidth separation at 10.2 GHz. The slope of a linear fit in the experimentally derived phase differences was −1.11 rad/GHz. Theory, based on (3.16), predicts a value of −1.18 rad/GHz for the slope. The closeness of the experimental to theoretical result establishes confidence in the validity of the model used to derive, in part, the log-likelihood function of (3.23).

Figures 3.4 and 3.5 show the angle-of-arrival maximum-likelihood estimates for each of 64 snapshots. For Fig. 3.4(a), only one frequency (12.34 GHz) is used, and no

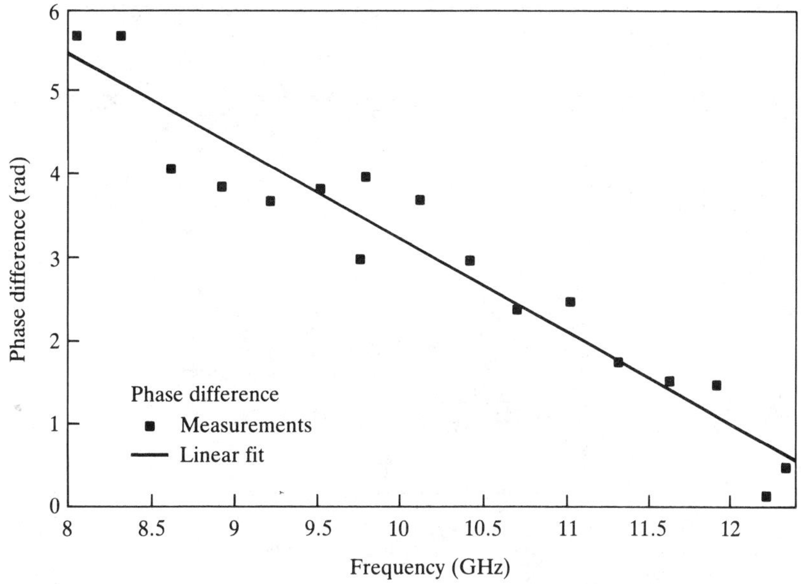

Figure 3.3 Direct-specular phase difference 0.4 beamwidth separation at 10.2-GHz slope of fit = −1.11 rad/GHz.

prior information is assumed. The height of the transmitter for this experiment was 15.59 meters (0.41 beamwidth separation at 10.2 GHz). A decrease in the variability of the estimates for this array geometry can be seen in Fig. 3.4(b) where four different transmit frequencies (9.52, 10.72, 11.92, and 12.34 GHz) are used and the height of the receiving array is assumed known with an accuracy of 0.1 meter.

The results of a more difficult multipath environment are plotted in Fig. 3.5. For this experiment the transmitter height was 9.59 meters, resulting in a 0.233 beamwidth separation at 9.52 GHz. With only one transmit frequency (9.52 GHz) and no prior information used in the estimation, we get the results shown in Fig. 3.5(a). Here we clearly see that the estimator fails to resolve the direct and specular rays for many of the snapshots used in the computation. When, however, for the same geometry as in Fig. 3.5(a), we use four different transmit frequencies (9.52, 10.72, 11.32, and 11.92 GHz) and also use knowledge of the receiver height and target range with an assumed accuracy of 0.1 meter and 100 meters, respectively, we get the results shown in Fig. 3.5(b). This figure, compared to Fig. 3.5(a), clearly demonstrates a significantly improved estimator performance that results from the combined use of prior information and four different transmit frequencies.

Note that each snapshot of data used for Figs. 3.4 and 3.5 corresponds to the transmission of two frequencies, one fixed and the other adjustable. This means that the experimental results plotted in Figs. 3.4(b) and 3.5(b) involve the use of three snapshots for each point (sample) shown there.

The results shown in Figs. 3.4 and 3.5 assume a flat-earth geometry. With a target range of 4.61 kilometers that was used in the collection of the experimental

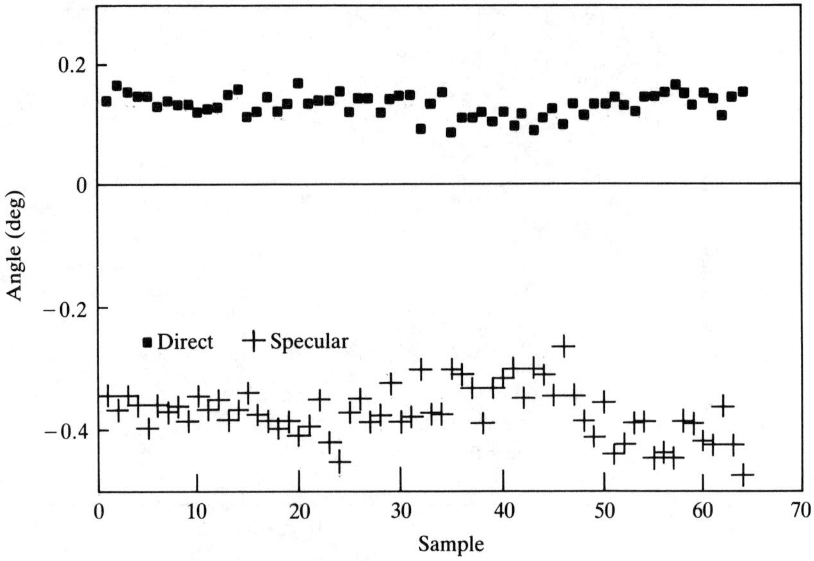

(a)

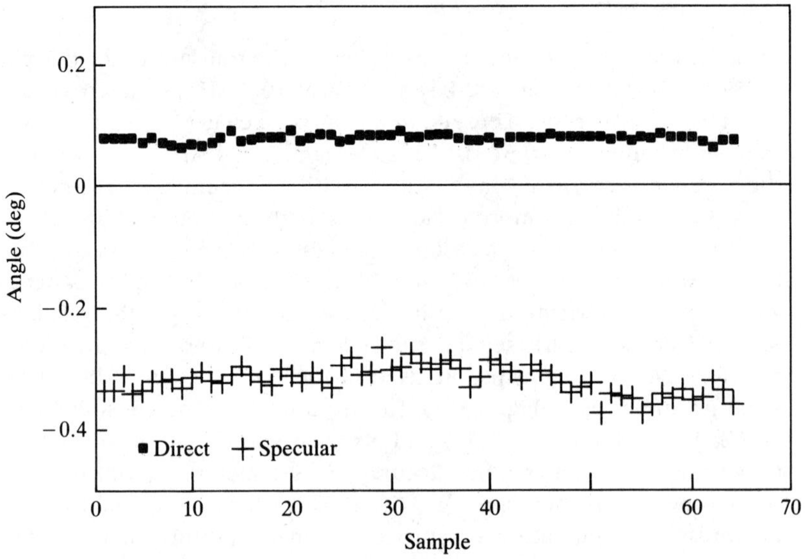

(b)

Figure 3.4 Angle-of-arrival estimates 0.41 beamwidth separation at 12.34 GHz. (a) 1 freq., no a priori assumptions; (b) 4 freq., σ_{br} assumed 0.1 meter.

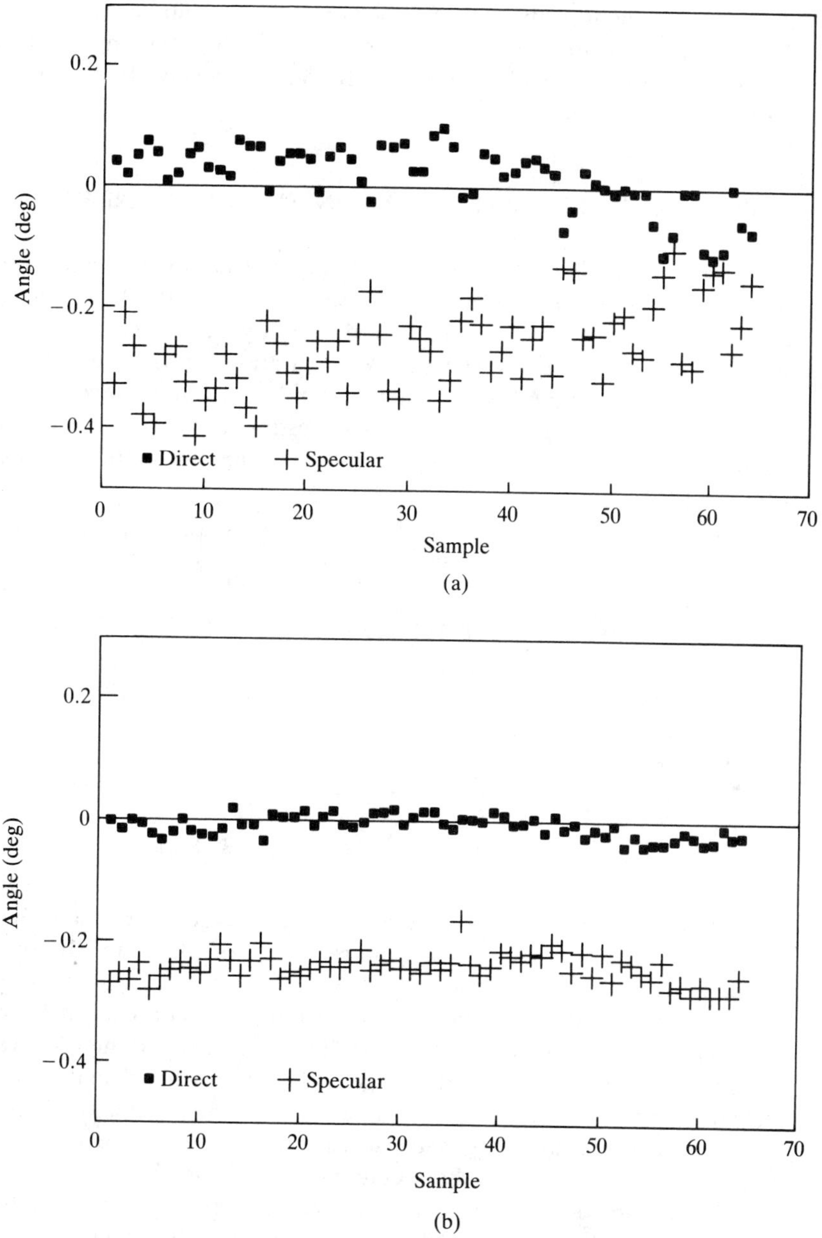

Figure 3.5 Angle-of-arrival estimates 0.233 beamwidth separation at 9.52 Ghz. (a) 1 freq., no a priori assumptions; (b) 4 freq., σ_{br}, σ_{Rt} assumed 0.1, 100 meters.

data for these figures, the reflecting water surface exhibits a noticeable curvature. The results with this data base may therefore be further improved in estimation accuracy by using a mathematical model for the received signal that supposes a curved-earth geometry.

3.5. MAXIMUM LIKELIHOOD APPLIED TO NONREDUNDANT ARRAY

In this section, we consider the problem of two plane waves incident on a linear, but not necessarily uniform, array using a single "snapshot" of the data (at only one frequency). The N-array locations are defined by a set of values x_n that have been normalized to the smallest spacing between array sensors, denoted by d, and the signal received by the nth sensor, denoted by r_n. Given the received signal vector $\mathbf{r}$ with element r_n, the maximum-likelihood estimates for the two wavenumbers are those values of k_1 and k_2 that maximize the log-likelihood function (objective function)

$$\ell(k_1, k_2) = \frac{|D_1|^2 + |D_2|^2 - 2 \cdot \mathrm{Re}[\rho_{12} D_1^* D_2 / N]}{N^2 - |\rho_{12}|^2} \tag{3.24}$$

where Re denotes the "real part of" and

$$D_m = \sum_{n=1}^{N} z_n \cdot \exp(-jk_m x_n), \qquad m = 1, 2, \tag{3.25}$$

and

$$\rho_{12} = \sum_{n=1}^{N} \exp[j(k_1 - k_2)x_n]. \tag{3.26}$$

Note that the likelihood function of (3.24), except for a multiplying factor, is a reformulation of the first term on the right side of (3.23) for the special case of a single frequency.

The ML estimate is found by performing a two-dimensional search in wavenumber space for pairs of values k_1, k_2. The search is two-dimensional since no prior information is assumed about the range of the target other than it is located in the far field of the receiving aperture. Therefore, the wavenumber and relative phase of the specular reflection from the water surface cannot be completely described by the target wavenumber even if the receiver height is known.

This estimation applies to an arbitrary array spacing; however, the nonuniform array considered in this text has the property that the array spacings are all multiples of d. Therefore, the set (x_n) is a set of integers. This fact permits us to take advantage of a number of algorithms in the maximization of the objective function such as using a padded FFTs (fast Fourier transforms) to evaluate $D(k)$ initially at a large number of potential locations from which all pairs (k_1, k_2) are taken in a course grid search to approximately locate the global maximum.

3.5.1 Experimental Results

For this second part of the study, data were collected with Phase I of the experimental radar system described in Section 3.3. The data collection were made under various water roughness conditions. For the purpose of our present discussion, the different surface roughness levels are referred to as "smooth," "chop," and "rough." The respective total peak to trough waveheights were approximately 0.25 m, 1 m, and 2.5 m. The data base spanned approximately 2 seconds, for which 256 snapshots were collected.

From the 32-element array, we may select subsets corresponding to a smaller number of elements. Consider, for example, a subset consisting of seven elements. Table 3.1 gives the configurations for this number of elements (i.e., 7). The first row of Table 3.1 refers to a *uniform configuration*, in which the elements of the array are all equispaced. The second row of the table refers to a *non*redundant configuration, defined as an array that minimizes the number of holes of the coarray under the constraint that there be no redundance. A *coarray* is described by marking the values of the separation that exist for each pairing of elements in the array as an interferometer. A *nonredundant array* has the unique property that it provides the densest packing of the coarray of all possible zero-redundancy arrays. The nonredundant array increases the aperture size compared to a uniformly spaced array of the same number of elements without reducing the scan coverage capability; see Vertatschitsch and Haykin [3] for details. Returning to the example of a 7-element array, we see that its nonredundant version spans a total of 26 (uniformly spaced) elements, as indicated in row 2 of Table 3.1. There are a total of 7 such arrays, each one obtained by "sliding" the first subset along the length of the full array. Since the array is not symmetrical, we can identify another 7 arrays using the inverted representation. These two configurations will be referred to as NR7a and NR7b. For comparison purposes, we use the 26-element uniform array that spans the same aperture as NR7 and therefore find 7 such subarrays in the 32-element set. We also compare the results obtained by using a 14-element uniform array U14 of which there are 19.

The target was known to be located above the local horizon of the receiving aperture (and the reflection from the water surface was below the local horizon). The determination of the local horizon requires an approximate knowledge of the receiving aperture height above the reflecting surface. This reduced the search space and therefore also the computational effort; the condition can, in general, be removed. No additional prior information about the target or reflection was used in the estimation.

TABLE 3.1 Array Structures

No. of Sensors	Location							Array Property
7	0	1	2	3	4	5	6	Uniform
7	0	1	4	10	18	23	25	Nonredundant

Each estimate is compared snapshot by snapshot to the corresponding result obtained by using the 32-element uniform array at the same time instant. This will remove the effect of tower sway and wind loading from influencing the estimation errors. We call the 32-element estimates the "true" values for a given snapshot, and determine the mean-squared error for each array structure over the 256 snapshots. The arrays U26, NR7a, and NR7b all have beamwidths defined as 0.25 units of wavenumber. The array U14 has a beamwidth of 0.45. Since the phase difference at the midpoint of the array under investigation may greatly influence the performance as shown in the simulations of the previous section, we keep such subarray separate and determine the errors independently.

The two sets of results presented in Figs. 3.6 and 3.7 show the MSE (mean-squared error) for maximum-likelihood estimation for different array configurations and different angular separations between the direct and specular rays.

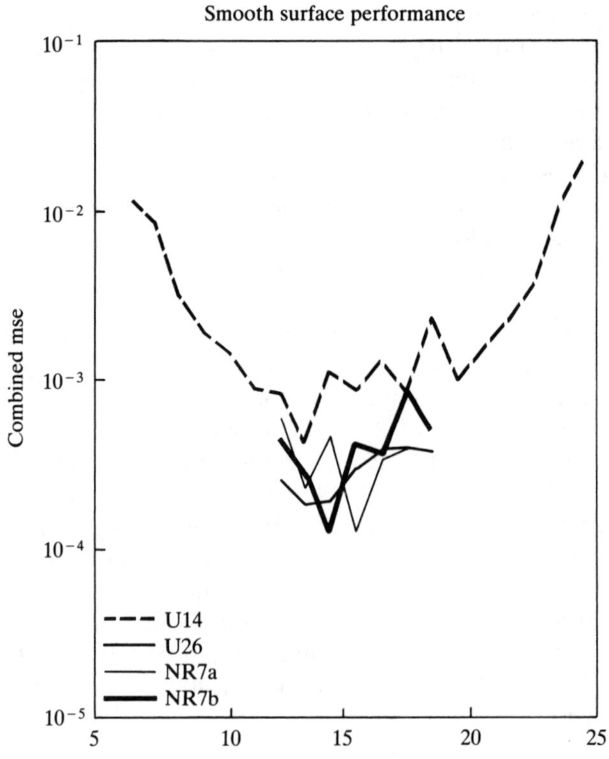

(a)

Figure 3.6 Comparison of 14- and 26-element uniform with 7-element nonredundant subarrays using the full 32-element array as a reference on the "smooth" surface with a transmitter height of 19.0 meters (i.e., largest separation between direct and specular rays).

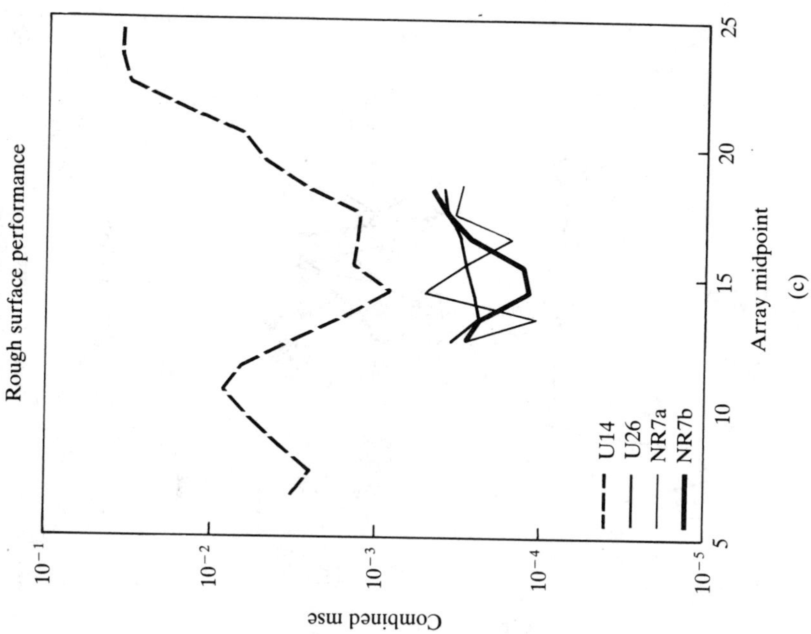

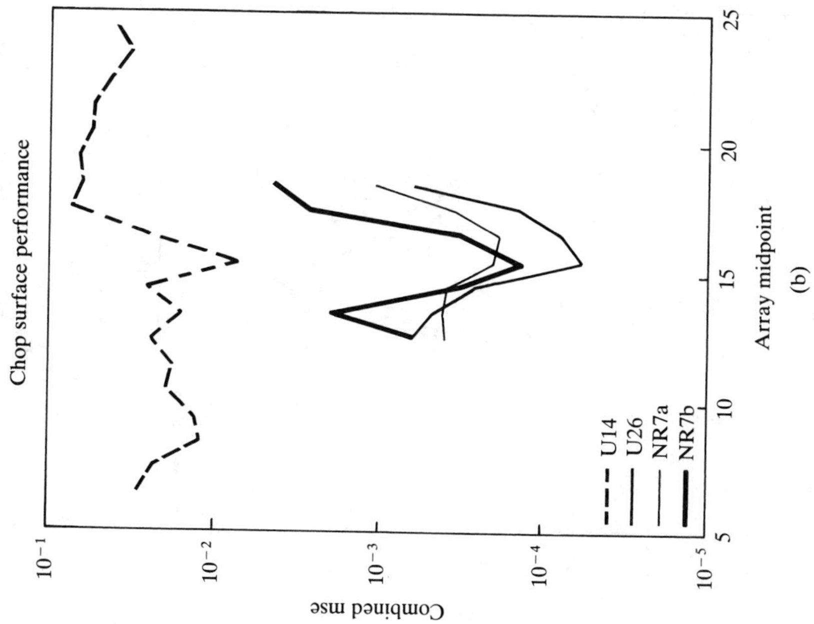

Figure 3.6 continued

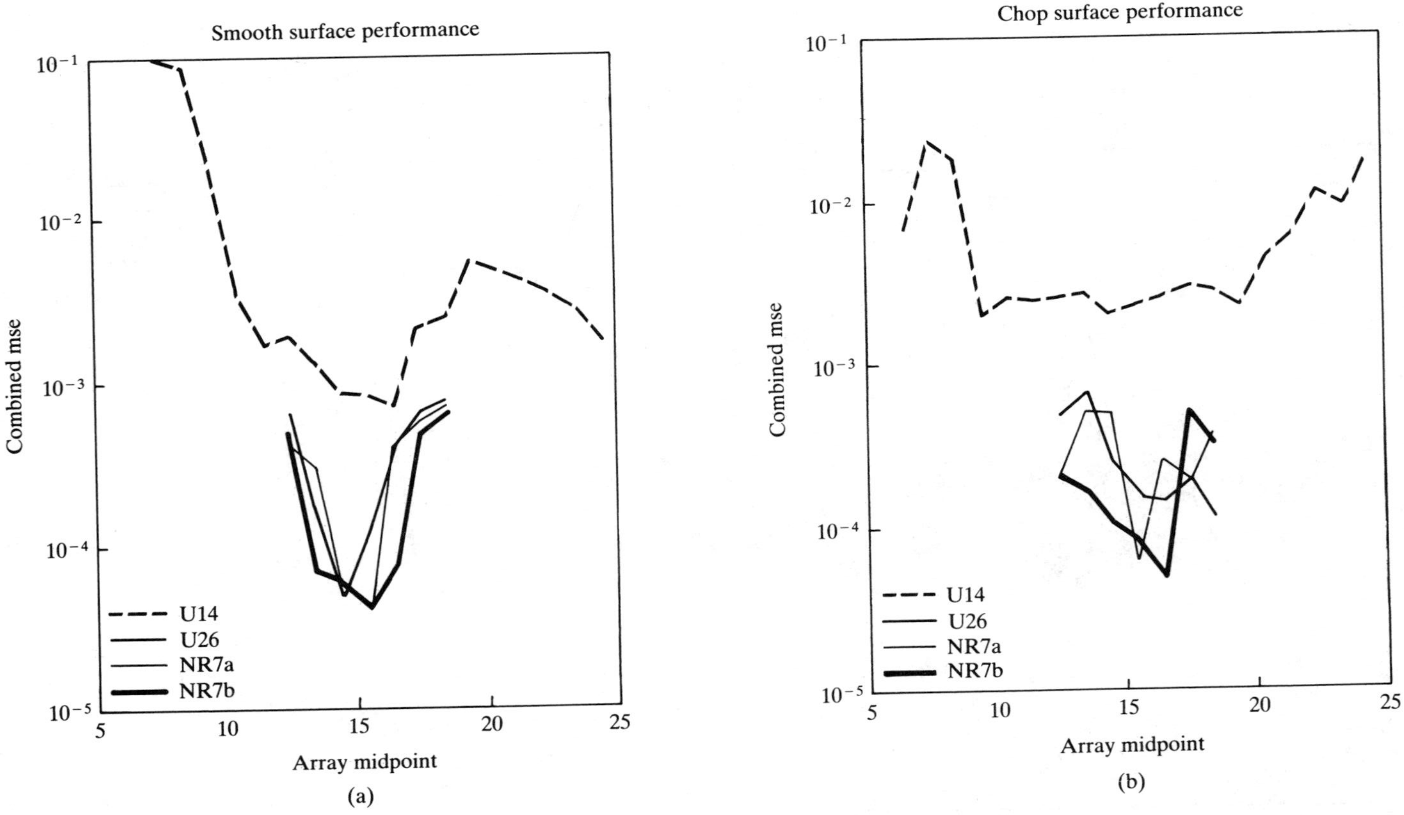

Figure 3.7 Comparison of 14- and 26-element uniform with 7-element nonredundant subarrays using the full 32-element array as a reference on the "smooth" surface with a transmitter height of 16.6 meters (smallest separation between direct and specular rays).

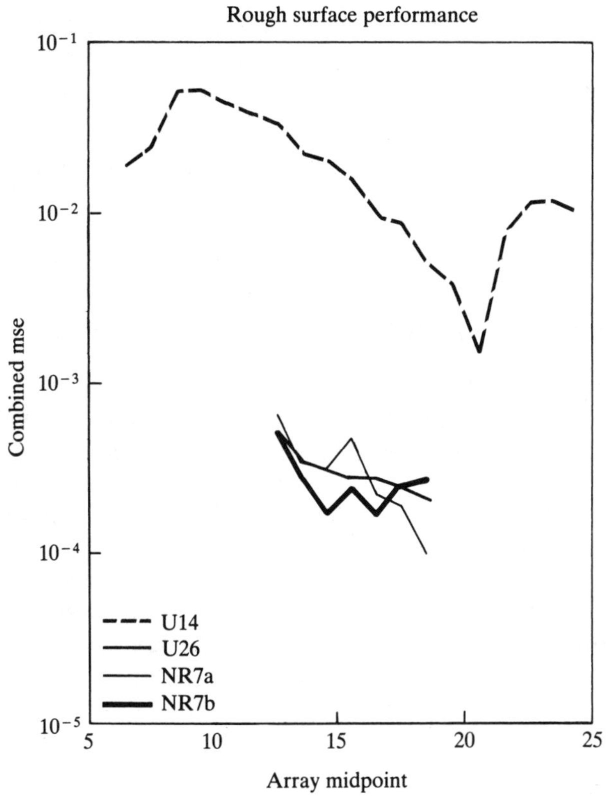

(c)

Figure 3.7 continued

The two transmitter heights at the time Phase I of this experiment was run yield target to specular image separations in wavenumber of approximately 0.090 and 0.078 units, which are well within half the beamwidth of the 32-element receiving aperture. Figure 3.6 demonstrates the performance for each subarray as compared to the 32-element full array for the three different surface conditions when the uppermost transmitter at a height of 19.0 meters was used. The horizontal axis corresponds to the location of the midpoint of the subaperture array measured with respect to the top of the 32-element full array (whose midpoint would be 15.5 on the same axis). The nonredundant array performance generally oscillates about the 26-element uniform array performance of all three surface conditions. In all the nonredundant subarrays considered, the performance was better than that of the 14-element uniform array when compared with the same midpoint. In 37 of the 42 comparisons, the improvement was greater than 3 dB. It is significant to note that the 7-element uniform array could not estimate the target angle of arrival in these experiments, and therefore no results are presented. This is probably due to the calibration errors, which in essence limit the effective signal-to-noise ratio obtained.

TABLE 3.2 MSE in Wavenumber Comparisons

Array	Smooth	Chop	Rough	
U14	4.3	74.0	8.5	$\times 10^{-4}$
NR7a	1.3	1.8	1.1	$\times 10^{-4}$
NR7b	1.3	1.4	1.2	$\times 10^{-4}$

A comparison of the minimum MSE is presented in Table 3.2 for the uniform 14-element array and the two sets of nonredundant arrays. This comparison is made independent of the subarray center and we find the improvement was 5.2 dB on the smooth surface, 16.1 dB for the choppy sea state, and 8.5 dB on the rough surface data.

For the smaller separation, with the lowermost transmitter height of 16.6 meters, the absolute performance of the 32-element array was not as accurate as for the larger separation. However, we can still examine the performance of the subarrays with respect to the 32-element array. Figure 3.7 presents the results of this examination. The results are similar to those observed in Fig. 3.6. The degradation of the 14-element uniform array is significantly larger than it was for the larger separation. That is, the NR7 and U26 arrays provide results that are in much closer agreement with the 32-element estimates than does U14.

The results of these experiments were not limited by the thermal noise but rather by calibration accuracy. This is the accuracy with which the relative phase and amplitude of each channel can be evaluated as well as the DC (direct current) offsets. It should be pointed out that the MLE (maximum-likelihood estimate) based upon equal additive noise powers at each sensor is not necessarily optimum in a calibration accuracy-limited environment. The most general MLE, in which an accurate assessment of the calibration error statistics would be required, may be significantly more complex in terms of implementation. Factors such as gain errors, which may not be Gaussian distributed and so on, greatly complicate the final expression. It is, however, clear that for the obtained calibration accuracy, the nonredundant arrays perform significantly better than uniform arrays consisting of even twice as many elements.

3.6. CONCLUSIONS

The following important conclusions are drawn from the study, in the context of superresolution elevation estimation in the presence of multipath:

1. The use of multiple frequencies for illuminating a multipath environment makes it possible to expand on the pool of information available on the environment, thereby improving the accuracy of elevation estimation.
2. The incorporation of prior information into a maximum-likelihood estimate provides another method for improving the accuracy of elevation estimation in a multipath environment.

3. Combining the use of prior information and a multiplicity of transmit frequencies in the maximum-likelihood algorithm provides a powerful method for angle-of-arrival estimation in multipath.
4. The Rayleigh limit on the resolution capability of an array antenna has indeed been overcome in a fairly reliable fashion, using the strategy described under point 3.
5. A nonuniformly spaced array, with a judicious choice of the placement of array elements, can provide a significant performance improvement over a uniformly spaced array antenna with the same number of elements and the same scan coverage capability.
6. Ultimately, however, array calibration errors (arising from system imperfections) and diffuse multipath (arising from the reflecting surface irregularities) may impose practical limits on the accuracy of angle-of-arrival estimation in a multipath environment.

REFERENCES

1. D. K. Barton, "Low-Angle Radar Tracking," *Proc. IEEE*, **62**, no. 8 (June 1974), 687–704.
2. S. Haykin, "Where Do We Stand On High Resolution for Low-Angle Tracking Radar?" CRL Report #205, McMaster University. The substance of this report was presented at the IEEE Acoustics, Speech, and Signal Processing Workshop on Spectrum Estimation and Modeling, Minneapolis, Minnesota, August 1988.
3. E. Vertatschitsch and S. Haykin, "Nonredundant Arrays," *Proc. IEEE*, **74**, no. 1 (1986), 217.
4. L. V. Blake, *Radar Range-Performance Analysis* (Lexington, Mass.: Lexington Books, 1980).
5. H. L. Van Trees, *Detection, Estimation, and Modulation Theory*, Part 1 (New York: John Wiley, 1969).
6. V. Kezys and S. Haykin, "Multi-Frequency Angle-of-Arrival Estimation: An Experimental Evaluation," in *Advanced Algorithms and Architectures for Signal Processing*, Vol. III, ed. F. T. Luk, *Proc. SPIE*, **975** (1988), 93–100.
7. J. W. Bandler and N. Sinha, "FLOPT V-A Program for Minimax Optimization Using the Accelerated Least-Path Algorithm," S.O.C. Report, Faculty of Engineering, McMaster University, Hamilton, Canada, February 1980.

4

Maximum-Likelihood Methods for Toeplitz Covariance Estimation and Radar Imaging

Michael I. Miller, Daniel R. Fuhrmann, Joseph A. O'Sullivan, and Donald L. Snyder

4.1. INTRODUCTION

In the spectrum estimation problem, we observe a wide-sense-stationary process over a finite interval and wish to estimate the frequency, or spectral, distribution of its power. From a theoretical and intellectual standpoint, this times-series problem is interesting and it motivates much of the material in the first half of this chapter. In our more application-oriented research, the spectrum estimation problem arises in the narrowband *direction-finding (DF)* problem for uniform linear sensor arrays [1] and in *delay-Doppler radar imaging* [2].

In the DF application, each sensor responds to plane waves that arrive from many discrete directions or a continuum of directions. The phase relationships among sensors are analogous to the phase relationships among points in a time series, and the problem of estimating the spatial distribution of incident power is analogous to estimating the power spectral density of a time series. The fundamental differences between the time-series problem and direction finding are as follows: (1) in

DF, we observe multiple independent samples of a random vector instead of just one, and (2) in DF there is a nonlinear mapping between the physical *direction-of-arrival (DOA)* and the *electrical angle* (the phase shift between adjacent sensors). Depending on the wavelength and the intersensor spacing, the range of electrical angles may be less than or greater than 2π; in the latter case, there are ambiguities in the DOA. In the delay-Doppler radar imaging problem, a single measured signal is modeled as the sum of additive noise and a superposition integral of the product of target reflectances and the transmitted signal. The reflectances are assumed to be samples from a Gaussian process. The imaging problem is to estimate the spectrum of the reflectance process in additive noise [2].

Let us return for the moment to the generic time-series problem that is of broad interest to the digital signal-processing community. Here the data consist of a finite number of samples of a discrete-time process; in some cases, the data consists of a finite set of autocovariance values. A host of covariance and spectrum estimators for this problem can be found in Kay and Marple [3], for example, or the recent textbooks by Kay [4] or Marple [5]. One approach that we consider philosophically sound is that based on Burg's *maximum entropy method* [6]. Assume that one is given a finite set of autocovariance values, then the maximum entropy density subject to second-order moment constraints is a multivariate Gaussian. Burg derived the maximum entropy density of an infinite length time series subject to the contraints imposed by given autocovariances and the fact that the time series is wide-sense stationary. The maximum entropy power density spectrum estimate is of the same analytical form as that resulting from an autoregressive model [6]–[9].*

For the covariance estimation problem we address, the data consist of samples of the time series rather than its autocovariance function. The entropy method generates the functional form of the power spectral density when given covariance constraints, but does not provide a method for estimating the power spectrum when given data. To apply the entropy method, investigators have generated second-order covariance statistics from the data, from which the maximum entropy spectra may be generated [6], [10]. As just one example, an average of lag products weighted with a triangular window has been used [6]. For this choice of covariance constraints, the maximum entropy method generates a spectrum consistent with the autoregressive model, with coefficients determined by the triangularly weighted lag products.

For the problem in which data samples are given, we have adopted a statistically valid methodology for estimating the parameters that enter into the density function for the observations. We invoke a *Gaussian model,* based on the consideration that it is the maximum entropy density for second-order moment constraints. Wide-sense stationarity implies that the covariance matrix of the observations (which we now characterize as forming a random vector) is a *Toeplitz matrix*, that is, constant

*See also the discussion on the maximum entropy method presented in Chapters 1 and 7.

along the diagonals. We adopt the *method of maximum likelihood* to estimate the parameters. Hence the problem that we address may be stated as follows. *Given one or more samples of a random vector, find the maximum-likelihood estimate of the covariance matrix under the Gaussian model, subject to the constraint that the covariance matrix is Toeplitz.*

As an aside, we find it curious that this problem did not appear in the engineering literature prior to 1982 when the paper of Burg, Luenberger, and Wenger [11] appeared. In the flurry of activity in the 20-odd years of "modern spectrum estimation," much work has been based on correlation estimates; however, little or no attention has been paid to the fact that "law of large numbers"–type correlation estimators are decidedly suboptimal. This may be due to a number of factors. In some problems, the data actually *do* consist of covariance samples; in others, the sheer volume of data removed the need for sophisticated statistical methods. Some investigators may have been put off by the perceived intractability of the problem, while others may be reluctant to rely on the Gaussian model. Our point of view is that this particular problem needs to be solved so that its usefulness may be evaluated in the context of real-world applications.

In the first half of this chapter we examine this fundamental covariance estimation problem. In Section 4.2, we review the basic results of Burg et al. [11] and consider the question of the existence of positive definite solutions. In Section 4.3 we derive the EM (expectation maximization) algorithm for the maximization of the log likelihood; we have proven that this algorithm generates a sequence of estimates that increase the likelihood and that have stable limit points that satisfy the necessary conditions for a maximizer (see Miller and Snyder [12]). Section 4.4 discusses a fast method for implementing the EM algorithm, and in Section 4.5 we show simulation results that demonstrate the superiority of the *maximum-likelihood (ML) method* over conventional biased and unbiased estimators.

The second half of this chapter, Section 4.6, is devoted to a new approach for high-resolution delay-Doppler radar imaging. At first glance the radar problem seems quite different from ordinary spectrum estimation, primarily because the desired parameters (a set of powers in the delay-Doppler plane) enter into the density of the observations via a fairly complicated linear transformation on the reflectivity process. We show that, by modeling the target reflectance at each point on the target as a sample of a Gaussian process with covariance determined by the powers at each Doppler frequency in the return signal, the imaging problem is fundamentally a spectrum estimation problem and therefore has very similar structure to the Toeplitz covariance estimation problem. This allows us to propose an iterative EM algorithm for its solution.

In Section 4.7 we finish the chapter by stating the necessary conditions for the existence of unique spectrum estimates satisfying the maximum-likelihood conditions, along with a brief statement of our convergence results on the iterative EM algorithms.

4.2. PROBLEM FORMULATION AND EXISTENCE OF POSITIVE-DEFINITE SOLUTIONS

Let $\mathbf{y}_G(1), \ldots, \mathbf{y}_G(M)$ be M independent realizations of a $G \times 1$ complex random vector with a *zero-mean multivariate complex Gaussian distribution and unknown Toeplitz covariance* $\mathbf{K}_G$. (G is our mnemonic for "given.") The ML covariance estimate $\hat{\mathbf{K}}_G$ is the positive definite Toeplitz matrix maximizing the likelihood function of the data, given by

$$L(\mathbf{K}_G; \mathbf{y}_G(k), 1 \le k \le M) = \prod_{k=1}^{M} \pi^{-G}(\det \mathbf{K}_G)^{-1}\exp[-\mathbf{y}_G^{\dagger}(k)\mathbf{K}_G^{-1}\mathbf{y}_G(k)]. \tag{4.1}$$

As first proven by Burg and others [11], for the real case, which is easily extended to the complex case, the necessary condition for the *maximum-likelihood estimate (MLE)* of $\mathbf{K}_G$ is given by the following trace condition.

$$\mathrm{tr}[(\hat{\mathbf{K}}_G^{-1}\mathbf{S}\hat{\mathbf{K}}_G^{-1} - \hat{\mathbf{K}}_G^{-1})\delta\mathbf{K}_G] = 0, \tag{4.2}$$

where $\mathbf{S}$ is the sample covariance matrix given by

$$\mathbf{S} = \frac{1}{M}\sum_{k=1}^{M} \mathbf{y}_G(k)\mathbf{y}_G^{\dagger}(k). \tag{4.3}$$

In (4.2), $\hat{\mathbf{K}}_G$ is constrained to be Toeplitz, and $\delta\mathbf{K}_G$ represents *any allowable variation in the constraint space.* If $\hat{\mathbf{K}}_G$ is an allowable variation, one necessary condition that the solution must satisfy is

$$\mathrm{tr}[\hat{\mathbf{K}}_G^{-1}\mathbf{S}] = G. \tag{4.4}$$

When this constraint is embedded in the likelihood equation of (4.1), the exponential term becomes a constant, and the variational problem reduces to

$$\hat{\mathbf{K}}_G = \underset{\mathbf{K}_G}{\mathrm{argmin}} \det \mathbf{K}_G \tag{4.5}$$

subject to the constraint that $\hat{\mathbf{K}}_G$ is a positive definite Toeplitz matrix that satisfies (4.4). Thus the maximum-likelihood problem can be interpreted as a problem of finding a minimum entropy Gaussian density subject to these two constraints.

The first order of business in attacking this difficult variational problem is to determine whether or not a solution exists. For simplicity, we now explore the existence of positive-definite solutions assuming real Gaussian processes, thereby more closely following Burg's original work. These results on existence of solutions may be found in greater detail in [13].

For the time-series problem in which $M = 1$, and hence $\mathbf{S}$ is of rank 1, it is clear that (4.3) may not have a maximizer in the set of positive-definite Toeplitz matrices and that our chosen parameter space may in fact be too large for this optimization problem to produce meaningful results. Upon further investigation, it was found that certain data vectors can lead to positive definite maximizers, but that the probability or measure of the set of such vectors was exceedingly small. This is stated more precisely in the following theorem.

Theorem 1. Let $\mathbf{y}_G$ be a $G \times 1$ real, zero-mean Gaussian random vector with arbitrary circulant covariance. Let A be the set of all vectors $\mathbf{x}$ such that the log-likelihood $L(\mathbf{K}_G; \mathbf{x})$ is unbounded above over the set of positive-definite Toeplitz matrices $\mathbf{K}_G$. Then,

$$P(\mathbf{y}_G \in A) > 1 - \frac{1}{2^{G-2}}. \tag{4.6}$$

Appendix A contains a basic outline of the proof of Theorem 1. For a more complete proof, see [13].

The consequence of Theorem 1 is that, even for random vectors with perfectly conditioned covariance ($\sigma^2\mathbf{I}$ is circulant), the probability of observing a sample vector that leads to a positive-definite solution is virtually zero. In this sense, the ML problem is *ill-posed* in that a single observation is insufficient to produce a meaningful estimate of $\mathbf{k}_G$.

In Theorem 1, we consider only the case in which the true underlying covariance is circulant. In [13] we extend this result to consider Gaussian densities with arbitrary Toeplitz covariance; the result is not as strong, although the measure of the complement of set A in (4.6) still goes to zero exponentially with G. Given the "closeness" of nonsingular Toeplitz matrices to circulant matrices in an asymptotic sense [14], we consider Theorem 1 to be a very good indication that the probability that a single observation will yield a positive definite solution is extremely small for values of G of interest.

The difficulty of maximizing the likelihood over the set of Toeplitz matrices is analogous to a larger class of likelihood problems addressed by Grenander [15] for which the maximizer may not exist. Grenander notes that in ML problems such as these, the parameter space may be too large. He proposes maximizing the likelihood over a constrained subset and then relaxing the constraint with sample size by allowing the subset to grow. Under the condition that the modified constraint space grows sufficiently slowly, consistent estimates are produced. This is Grenander's *method of sieves.*

This is precisely the approach taken here, to constrain the space of allowable Toeplitz covariances to those with circulant extension. This new constraint set arises naturally in the EM algorithm to be derived in the next section; as we shall see, it also leads to estimates that are guaranteed to be positive definite.

Our new constraint set is the set of nonnegative definite $G \times G$ Toeplitz matrices that have a nonnegative definite $N \times N$ circulant extension. By this we mean that $\mathbf{K}_G$ can be written as the upper-left $G \times G$ block of a nonnegative definite $N \times N$ circulant matrix. It may not be immediately obvious that this is a restriction at all; however, a further characterization of this set will make this clear.

An $N \times N$ circulant matrix $\mathbf{K}_N$ is diagonalized by the discrete Fourier transform (DFT). Thus, a nonnegative definite symmetric circulant matrix can be written in the form

$$\mathbf{K}_N = \mathbf{W}^{\dagger}\mathbf{\Sigma}\mathbf{W}, \tag{4.7}$$

where $\mathbf{W}$ is the normalized DFT matrix and $\mathbf{\Sigma}$ is the diagonal matrix of nonnegative eigenvalues. Any $\mathbf{K}_G$ that is the upper left $G \times G$ block of some circulant $\mathbf{K}_N$ can be written

$$\mathbf{K}_G = \mathbf{W}_G^{\dagger}\mathbf{\Sigma}\mathbf{W}_G, \tag{4.8}$$

where $\mathbf{W}_G$ is the $N \times G$ submatrix made up of the first G columns of $\mathbf{W}$.

The constrained MLE is found by performing the maximization of the likelihood over the set of Toeplitz matrixes with this restriction. The necessary trace conditions for this maximizer are identical to those of (4.2), but the variations are performed over the constraint set described by (4.8). As stated in the following theorem, the constrained maximizer is guaranteed to exist and will be nonsingular.

Theorem 2. Let $\mathbf{y}_G$ be a $G \times 1$ real, zero-mean Gaussian random vector with nonsingular covariance. Let A be the set of vectors $\mathbf{x}$ such that the log-likelihood $L(\mathbf{K}_G; \mathbf{x})$ is unbounded above over the set of positive definite $\mathbf{K}_G$ with nonnegative definite circulant extension to period N. Then,

$$P(\mathbf{y}_G \in A) = 0. \tag{4.9}$$

Proof. See Appendix B.

For the original problem stated by Burg, Luenberger, and Wenger [11], it was assumed that there were multiple data vectors $\mathbf{y}_G(i)$, $i = 1, \ldots, M$, with $M \geq G$ so that the sample covariance matrix $\mathbf{S}$ is full rank. For the time-series problem, it seems unnatural to us to introduce multiple independent vector samples into the problem since a single observation vector supports longer lags. By performing the maximization over the set of Toeplitz matrices with circulant extension, we have assured the existence of the MLE without assuming multiple vector samples.

4.3. EXPECTATION-MAXIMIZATION ALGORITHM FOR MAXIMIZING LIKELIHOOD

Because the necessary conditions of (4.2) are fairly complex, we have been unable to solve for the maximizing $\hat{\mathbf{K}}_G$ explicitly. Instead, we must rely on indirect, iterative techniques for maximizing the log likelihood. Burg and others [11] do provide such an algorithm that they term the *inverse iteration algorithm*; this algorithm is somewhat awkward in that each iterate must be tested to determine whether or not it is positive definite and whether or not the likelihood has increased. Our experience with the *EM* (expectation-maximization) algorithm of Dempster, Laird, and Rubin

[16] in radionuclide imaging [12] led us to consider it as an alternative approach to this optimization problem. For completeness, we include here a brief description of the EM algorithm.

The EM algorithm is an iterative method for determining the MLE of some parameter vector, call it $\boldsymbol{\phi}$, from some observed data, call them $\mathbf{X}$. Using the terminology of Dempster and others, $\mathbf{X}$ is termed the "incomplete data." The incomplete data take values in a sample space $\overline{\mathbf{X}} = \{\text{all possible values of } \mathbf{X}\}$ and have a density $g(\boldsymbol{\phi}; \mathbf{X})$ over their sample space. It is often difficult to maximize $g(\boldsymbol{\phi}; \mathbf{X})$ with respect to ϕ. To circumvent this difficulty, $\overline{X}$ is embedded in a large space $\overline{Z}$ in which some hypothetical data Z, termed the "complete data," takes values. A many-to-one mapping from $\overline{\mathbf{Z}}$ to $\overline{\mathbf{X}}$, defined by some function H such that $\mathbf{X} = H(\mathbf{Z})$ is assumed, as is a density $f(\boldsymbol{\phi}; \mathbf{Z})$ of $\mathbf{Z}$ over $\overline{Z}$. The densities of the incomplete and complete data are related according to

$$g(\boldsymbol{\phi}; \mathbf{X}) = \int_{\{\mathbf{Z};\, \mathbf{X}=H(\mathbf{Z})\}} f(\boldsymbol{\phi}; \mathbf{Z})\, dZ.$$

There are two steps for each iteration in the EM algorithm, an E step and M step. In the E (for expectation) step on the $p + 1$ iteration, the conditional expectation of the complete-data likelihood function $E\{\log f(\phi; \mathbf{Z})|\mathbf{X}, \phi^{(p)}\}$ is determined, where $\phi^{(p)}$ is the estimate of ϕ determined at iteration p. In the M (for maximization) step, the conditional expectation of the complete-data log likelihood is maximized with respect to ϕ, yielding the $\phi^{(p+1)}$ iterate. As proven by Dempster and others, the sequence of log-likelihood functions of the incomplete-data $\log g(\phi^{(p)}; \mathbf{X})$, $\log g(\phi^{p+1)}; \mathbf{X})$, . . . , will be nondecreasing.

We now propose the EM algorithm for generating Toeplitz covariances that satisfy the necessary maximizer conditions of the likelihood of (4.2). We first state the basic algorithm for generating the sequence of covariance estimates. We then show that this is an instance of an EM algorithm, and thereby is monotonic in the likelihood.

Proposition 1: Define the sequence of $G \times G$ covariance estimates $\mathbf{K}_G^{(p)}$ with the n, l entry given by the following iteration:

$$K_G^{(p+1)}(n-l) = \frac{1}{N}\sum_{m=0}^{N-1} E\{y\,(m) y^*[(m+n-l)_{\text{mod } N}]|\mathbf{y}_G, \mathbf{K}_G^{(p)}\}, \tag{4.10}$$

where $(\ )_{\text{mod } N}$ denotes modulo N and the asterisk denotes complex conjugation. Then, the set of iterates $\{\mathbf{K}_G^{(p)}; p \geq 1\}$ is an instance of an expectation-maximization algorithm and the sequence of likelihoods

$$L(\mathbf{K}_G^{(p)}; \mathbf{y}_G) = \pi^{-G}\,(\det \mathbf{K}_G^{(p)})^{-1} \exp[-\mathbf{y}_G^{\dagger} \mathbf{K}_G^{(p)-1} \mathbf{y}_G \tag{4.11}$$

is nondecreasing.

Demonstrating that (4.10) is an instance of an EM algorithm implies that the sequence is monotonic in the likelihood. That (4.10) defines an EM sequence follows from the fact that the complete data correspond to the N-dimensional vector $\mathbf{y}_N$ consisting of the given $\mathbf{y}_G$ augmented by the $(N-G)$-dimensional vector $\mathbf{y}_A = [y(G) \cdots y(N-1)]^T$; that is, $\mathbf{y}_N = [\mathbf{y}_G^T \mathbf{y}_A^T]^T$. The incomplete data is the observed vector $\mathbf{y}_G$, with the many-to-one function $\mathbf{H}$ mapping $\mathbf{y}_N$ to $\mathbf{y}_G$ ignoring all elements corresponding to the augmented vector of length $N-G$; that is, $\mathbf{y}_G = \mathbf{H}\mathbf{y}_N$, with $\mathbf{H}$ a $G \times N$ matrix with unit entries along the diagonal, and zero otherwise. The complete-data likelihood is given by

$$L(\mathbf{K}_N; \mathbf{y}_N) = \pi^{-N} (\det \mathbf{K}_N)^{-1} \exp[-\mathbf{y}_N^\dagger \mathbf{K}_N^{-1} \mathbf{y}_N]. \tag{4.12}$$

Using the Fourier transform matrix $\mathbf{W}$ from the orthogonal decomposition of $\mathbf{K}_N$ yields the data in the uncorrelated rotated coordinates $\mathbf{c}_N = \mathbf{W}\mathbf{y}_N$ with $\mathbf{c}_N = [c(0) \cdots c(N-1)]^T$. The estimation problem in the rotated data becomes one of estimating the eigenvalues $\sigma(m)$ and the diagonal matrix $\mathbf{\Sigma} = \text{diag}[\sigma(0), \ldots, \sigma(N-1)]$ corresponding to the spectral power at discrete frequencies $2\pi m/N$, for $m = 0, \ldots, N-1$. Rewriting the log likelihood in the rotated coordinates and discarding terms not a function of the parameters yields the following simple log likelihood to be maximized:

$$-\sum_{m=0}^{N-1} \log \sigma(m) - \sum_{m=0}^{N-1} \frac{|c(m)|^2}{\sigma(m)}. \tag{4.13}$$

The E-step of the EM algorithm involves taking the conditional mean of the complete-data log-likelihood; the M-step performs the maximization over the eigenvalues $\sigma(m)$.

The maximizing spectral coefficient on iteration $p+1$ is obtained by evaluating the conditional expectation with respect to the data and power spectral coefficients from the previous iteration, yielding the following estimates:

$$\sigma(m)^{(p+1)} = E\{|c(m)|^2 | \mathbf{\Sigma}^{(p)}, \mathbf{y}_G\}, \qquad \text{for } m = 0, 1, \ldots, N-1, \tag{4.14}$$

with

$$\mathbf{\Sigma}^{(p+1)} = \text{diag}[\sigma(0)^{(p+1)}, \ldots, \sigma(N-1)^{(p+1)}]. \tag{4.15}$$

The covariances at iteration $p+1$ are obtained by simply transforming back to the original coordinates yielding

$$K_N^{(p+1)}(n-l) = \frac{1}{N}\sum_{m=0}^{N-1} E\{y(m)y^*[(m+n-l)_{\text{mod } N}] | \mathbf{y}_G, \mathbf{K}_N^{(p)}\}, \tag{4.16}$$

demonstrating the proposition.

That the iteration makes intuitive sense can be seen as the constrained ML procedure requires the augmentation of the lag products via generation of conditional mean estimates of the missing lags. The algorithm fills in the lag products that are missing due to the finite window length. At the convergence point of the iteration, the covariance estimate are those values that equal the average of the conditional mean of the lag products.

We have proven previously (see Miller and Snyder [12]) that all the limit points of the EM sequence are stable and satisfy the necessary maximizer conditions. This we explore in more detail in Section 4.7.

4.4. FAST IMPLEMENTATION STRATEGY

While (4.16) captures the simple beauty of the EM algorithm, an efficient implementation can be built using the spectral representation of the covariances, as follows. The expectation of the covariance lags in (4.16) corresponds to off-diagonal entries in the conditional mean matrix $E\{\mathbf{y}_N\mathbf{y}_N^\dagger \mid \mathbf{y}_G, \mathbf{K}_G^{(p)}\}$, where $\mathbf{y}_N$ is the full period of the periodic process. Only $\mathbf{y}_G$ is measured so that the lag products comprising the full period of the process must be estimated. Using the rotated orthogonal spectrum coordinates as $\mathbf{c}_N = \mathbf{W}\mathbf{y}_N$, where $\mathbf{W}$ is the $N \times N$ matrix consisting of the orthonormal discrete Fourier transform columns, the conditional lag product matrix needed for (4.16) becomes

$$E\{\mathbf{y}_N\mathbf{y}_N^\dagger \mid \mathbf{y}_G, \mathbf{K}_G^{(p)}\} = \mathbf{W}^\dagger E\{\mathbf{c}_N\mathbf{c}_N^\dagger \mid \mathbf{y}_G, \mathbf{K}_G^{(p)}\}\mathbf{W}. \tag{4.17}$$

The conditional mean estimation problem reduces to finding the conditional covariance of the rotated data $\mathbf{c}_N$ in the orthogonal coordinates defined by $\mathbf{W}$. Denoting the correlation $E\{\mathbf{c}_N\mathbf{y}_G^\dagger\}$ of $\mathbf{c}_N$ and $\mathbf{y}_G$ as $\mathbf{K}_{cy}$ and using the standard expressions for the conditional mean and variance of Gaussian random variables from Rhodes [17], we obtain

$$E\{\mathbf{c}_N\mathbf{c}_N^\dagger \mid \mathbf{y}_G\} = \mathbf{K}_{cy}\mathbf{K}_G^{-1}\mathbf{y}_G\mathbf{y}_G^\dagger\mathbf{K}_G^{-1}\mathbf{K}_{cy}^\dagger + \Sigma - \mathbf{K}_{cy}\mathbf{K}_G^{-1}\mathbf{K}_{cy}^\dagger, \tag{4.18}$$

where $\Sigma = E\{\mathbf{c}_N\mathbf{c}_N^\dagger\}$. Since $\mathbf{y}_N = \mathbf{W}^\dagger\mathbf{c}_N$, then $\mathbf{y}_N^\dagger = \mathbf{c}_N^\dagger\mathbf{W}$. Substituting this into the defining relation for $\mathbf{K}_{cy}$, we find

$$\mathbf{K}_{cy} = E\{\mathbf{c}_N\mathbf{c}_N^\dagger\mathbf{W}_G\} = \Sigma\mathbf{W}_G, \tag{4.19}$$

where $\mathbf{W}_G$ is the $N \times G$ matrix of the leftmost G columns of $\mathbf{W}$. We now write (4.18) on iteration $p + 1$ as

$$\begin{aligned} E\{\mathbf{c}_N\mathbf{c}_N^\dagger \mid \Sigma^{(p)}, \mathbf{y}_G\} &= \Sigma^{(p)}\mathbf{W}_G\mathbf{K}_G^{(p)-1}\mathbf{y}_G\mathbf{y}_G^\dagger\mathbf{K}_G^{(p)-1}\mathbf{W}_g^\dagger\Sigma^{(p)} \\ &\quad + \Sigma^{(p)} - \Sigma^{(p)}\mathbf{W}_G\mathbf{K}_G^{(p)-1}\mathbf{W}_G^\dagger\Sigma^{(p)}. \end{aligned} \tag{4.20}$$

The diagonal elements of (4.20) form the new estimate of Σ, and the new estimate of $\mathbf{K}_G$ is given by

$$\mathbf{K}_G^{(p)} = \mathbf{W}_G^\dagger\Sigma^{(p)}\mathbf{W}_G. \tag{4.21}$$

There are further computational savings to be found in the evaluation of the first and third terms in (4.20). The first term is the set of squared magnitudes of the elements of $\Sigma \mathbf{W}_G \mathbf{K}_G^{-1} \mathbf{y}_G$. This term could be computed using the Levinson-Durbin algorithm (G^2 multiplications) followed by one N-point *fast Fourier transform* (*FFT*) ($2N \log N$ multiplications). However, the third term requires the explicit computation of $\mathbf{K}_G^{-1}$ via Trench's algorithm ($\frac{7}{4}G^2$) and thus the use of the Levinson-Durbin algorithm is unnecessary.

The diagonal elements of $\mathbf{W}_G \mathbf{K}_G^{-1} \mathbf{W}_G^\dagger$ can be found using one FFT as follows. Define

$$\mathbf{Q} = \begin{bmatrix} \mathbf{K}_G^{-1} & 0 \\ 0 & 0 \end{bmatrix}. \tag{4.22}$$

Then

$$\mathbf{W}_G \mathbf{K}_G^{-1} \mathbf{W}_G^\dagger = \mathbf{W} \mathbf{Q} \mathbf{W}^\dagger = \mathbf{P}, \tag{4.23}$$

with the (i, i) element of $\mathbf{P}$ given by

$$p(i, i) = \sum_{k=0}^{N-1} \sum_{l=0}^{N-1} e^{-j2\pi ik/N} q(k, l) e^{+j2\pi il/N} \tag{4.24}$$

$$= \sum_{k=0}^{N-1} \sum_{l=0}^{N-1} e^{-j2\pi i(k - l)/N} q(k, l). \tag{4.25}$$

With the substitution

$$n = (k - l)_{\mathrm{mod}\, N}, \tag{4.26}$$

we have

$$\sum_{n=0}^{N=1} e^{-j2\pi n/N} \sum_{k=0}^{N-1} q[k, (k - n)_{\mathrm{mod}\, N}]. \tag{4.27}$$

The sum over k represents the sum along the nth subdiagonal $\mathbf{Q}$ for n positive, and the nth superdiagonal for n negative. The desired diagonal elements of $\mathbf{W}_G \mathbf{K}_G^{-1} \mathbf{W}_G^\dagger$ are then found by taking the FFT of this sequence of diagonal sums, as indicated in (4.27).

Since $\mathbf{K}_G^{-1}$ was computed explicitly, $\mathbf{K}_G^{-1} \mathbf{y}_G$ can be computed with $G^2/2$ multiplications by exploiting the *persymmetry* of $\mathbf{K}_G^{-1}$.

For the problem in which there are multiple independent vector samples $\mathbf{y}_G(1) \cdots \mathbf{y}_G(M)$, there are two approaches to computing the first term of (4.20). The first is to compute $\Sigma \mathbf{W}_G \mathbf{K}_G^{-1} \mathbf{y}_G(i)$ for $i = 1, \ldots, M$. The second method is to form

$$\mathbf{S} = \frac{1}{M} \sum_{k=1}^{M} \mathbf{y}_G(k) \mathbf{y}_G^\dagger(k), \tag{4.28}$$

at the first iteration, then compute $\Sigma \mathbf{W}_G \mathbf{K}_G^{-1} \mathbf{S} \mathbf{K}_G^{-1} \mathbf{W}_G^{\dagger} \Sigma$ using straightforward matrix multiplications, along with the trick of computing the diagonal elements of $\mathbf{W}_G \mathbf{Q} \mathbf{W}_G^{\dagger}$ with one FFT. The relative merits of the two approaches depend on the values of G, N, and M. Method 1 requires $M[(G^2/2) + 2N \log N)]$ multiplications, whereas method 2 requires $(G^3 + 2N \log N)$ multiplications.

These computational tricks actually reveal a hidden interpretation of the EM algorithm in terms of the gradient of the cost function. The computation of (4.20) can be thought of as comprising the following five steps: (1) compute the unconstrained gradient of the cost function $(\mathbf{K}_G^{-1} \mathbf{S} \mathbf{K}_G^{-1} - \mathbf{K}_G^{-1})$; (2) project this unconstrained gradient onto the space of Toeplitz matrices by averaging along the diagonals; (3) multiply the elements of this Toeplitz gradient by a triangular window [this is done when averaging along the diagonals of $\mathbf{Q}$ in (4.22)]; (4) transform to the frequency domain; and (5) update the spectrum by this windowed, transformed gradient, weighted by the squares of the spectral values from the previous iteration.

4.5. SIMULATION RESULTS

We now explore the performance of the method. We have generated periodic Gaussian processes with period $N = 32$. Shown in Fig. 4.1 is the process covariance (left column) and spectrum (right column) of the zero-mean Gaussian process used for the simulations.

Shown in Fig. 4.2 are the results of applying the ML algorithm to $G = 16$ pieces of one realization of the 32-length process. Plotted via the narrow bars in both the top and bottom rows of Fig. 4.2 are the maximum-likelihood estimates given the full 32 pieces of the period of the process. The MLE given the full period of the process $\mathbf{y}_{32}$ corresponds to the conventional periodogram [12] and is given by

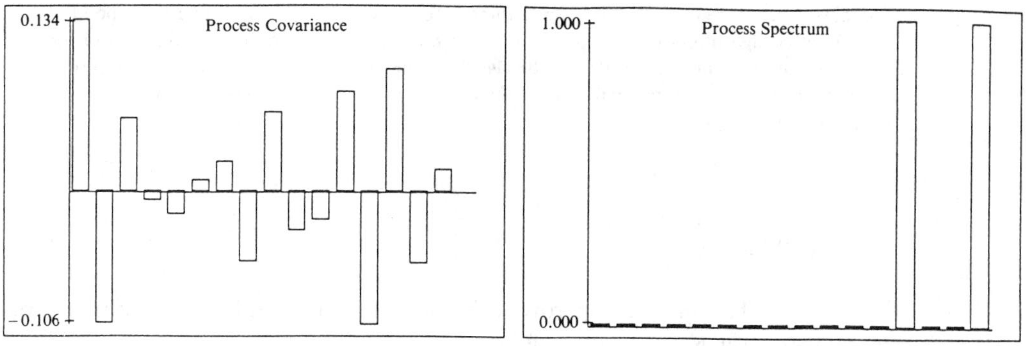

Figure 4.1 Left column shows the process covariances and right column the spectrum of zero-mean Gaussian 32-periodic process used in the simulations. Only the first 16 covariances and spectral components are shown.

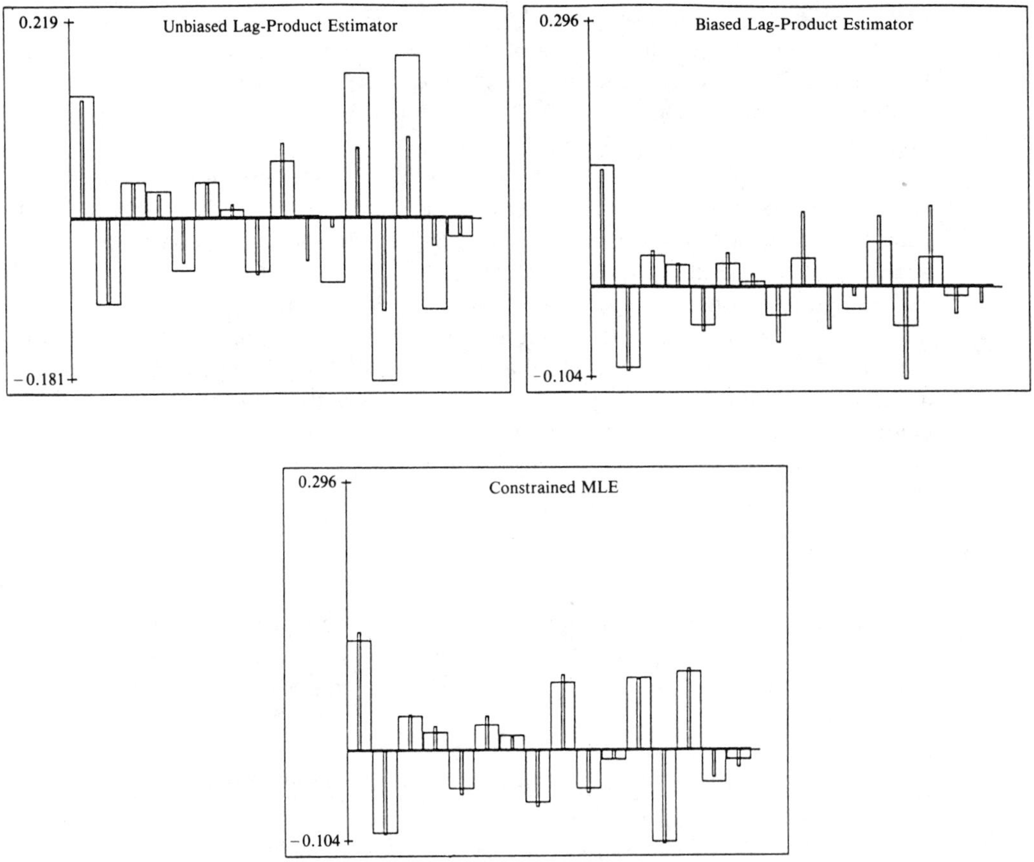

Figure 4.2 Top row shows the unbiased covariance estimator (left column, wide bars) and biased covariance estimator (right column, wide bars), given $G=16$ pieces of the 32-length process. Bottom row (wide bars) shows the MLE corresponding to the convergence point of the iterative algorithm of (4.10) generated from the single realization of $G=16$ pieces of the process, with the initial condition for the algorithm the unbiased lag estimator of the top row. Superimposed in both rows, depicted via the narrow bars is the MLE given all 32 pieces of the full period of the process.

$$\hat{K}(\tau) = \frac{1}{32}\sum_{m=0}^{31} y(m)y^*((m+\tau)_{\text{mod } 32}), \qquad \text{for } 0 \le \tau \le 31. \tag{4.29}$$

The wide bars in the top row, left column correspond to the conventional unbiased estimator $\hat{K}_u(\tau)$ given as follows:

$$\hat{K}_u(\tau) = \frac{1}{G-\tau-1}\sum_{m=0}^{G-1-\tau} y(m)y^*(m+\tau), \qquad \text{for } 0 \le \tau \le G-1. \tag{4.30}$$

Note how for the long lags ($\tau \approx 15$), since there are so few data points, the estimator has high variance as manifest by the large swings around the narrow bars.

The wide bars in the top row, right column correspond to the conventional biased estimator $\hat{K}_b(\tau)$ given as follows:

$$\hat{K}_b(\tau) = \frac{1}{N} \sum_{m=0}^{G-1-\tau} y(m)y^*(m+\tau), \qquad \text{for } 0 \leq \tau \leq G-1. \tag{4.31}$$

Note how for the long lags ($\tau \approx 15$), since there are so few data points, the estimator is biased low.

Shown in the bottom row via the wide bars are the covariance estimates corresponding to the convergence point of the iterative ML algorithm. The initial condition for the algorithm was the biased lag product estimator of the top row, right column. Notice how the algorithm has filled in the missing lag products (those outside of the data window). The MLE corresponds almost exactly to the estimator generated with all 32 pieces of the process.

Shown in Fig. 4.3 are the ensemble results generated from 1000 realizations of the periodic process, from which mean-squared error statistics were derived. The iteration was allowed to continue until a stable point was generated (within the floating point precision of the matrix operations). The simulation was used to study the performance of the method versus varying numbers of data points, with $N = 32$ and $G = 16, 12, 8, 4$. The performance of the iterative ML estimator of (4.10) was examined by using G pieces of the full period of the process and generating estimates of the first G covariances. For the performance comparison the following three different estimators were generated: (1) the complete-data MLE of (4.29) based on all 32 samples of the process; (2) the unbiased estimator of (4.30) consisting of the first $G < N$ samples of the process; and (3) the MLE corresponding to the convergence point of the algorithm of (4.10) given $G < N$ pieces of the process.

Plotted via the solid lines in Fig. 4.3 are the *mean-squared error* (*MSE*) results for the covariance lags $G = 16, 12, 8, 4$ for the ML algorithm (left column) and, for comparison, the conventional unbiased estimator (right column). The MSE was found by summing the variance and the square of the bias for each covariance lag. Since the estimators for any G can only estimate the first G covariances, the solid lines only extend to the $G - 1$ lag. Plotted via the dashed lines are the MSEs for the complete-data estimators, for which all 32 pieces of the full period of the process were assumed given. We emphasize that the complete-data estimator is the "best one can do."

The left panel shows the major result of the new maximum-likelihood method. The remarkable result is that, using only 16 of 32 pieces of data, the MLE performs as well as the full 32-data estimator. That is, even for lags that are large (≈ 15), where only one lag product is available, the MLE does as well as the complete-data estimator did with the full 32 lag products. This is consistent with Fig. 4.2, in which the missing data have been fairly well estimated via the conditional expectations of (2) for $G =$ 16 data points. The other excellent result is that the ML algorithm is fairly robust

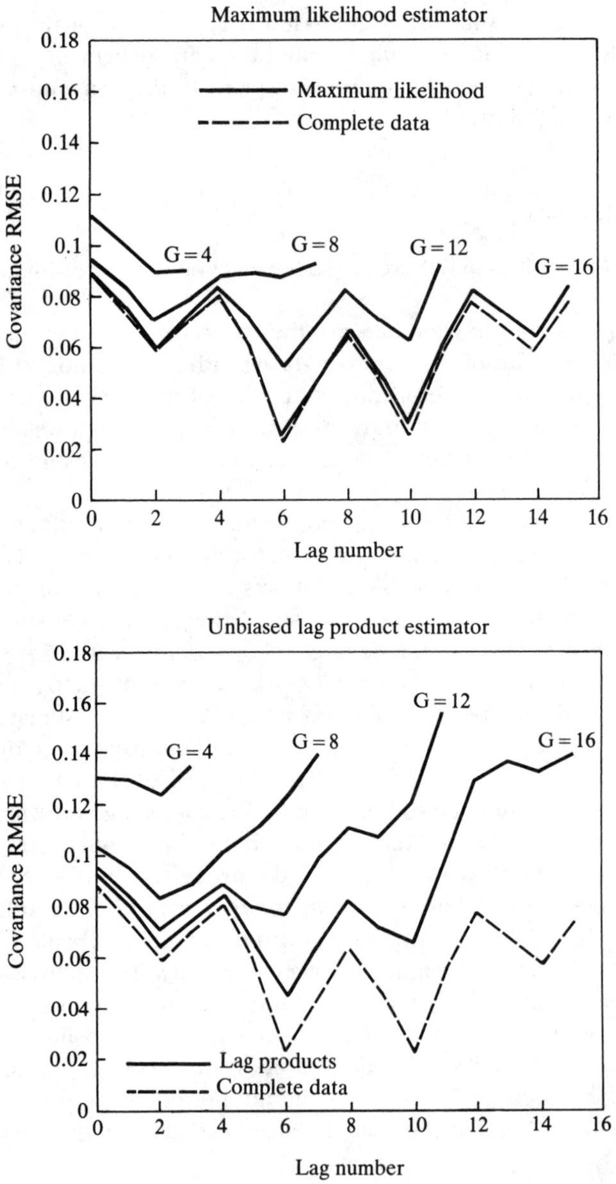

Figure 4.3 Left panel shows the mean-squared error performance versus covariance lag (solid lines) for the ML algorithm generated from G=16, 12, 8, 4 pieces of the 32-length process. Right panel shows the mean-squared error for the unbiased lag product estimator. Plotted via the dashed lines are the mean-squared errors for the MLE given all 32 pieces of the N = 32 periodic process.

with respect to the total number of lags available. While the performance decreases as G tends to zero, the algorithm continues to estimate the lags without showing an unstable decrease in performance.

As seen in the right panel, the conventional unbiased estimator suffers from the major difficulty that at the edge of the data window its variance grows. For all choices of G, the ML algorithm on the left has superior performance.

4.6. DELAY-DOPPLER RADAR IMAGING

This section presents recent results we have obtained [2] in *high-resolution delay-Doppler spotlight-mode radar imaging.* In this problem, a two-dimensional reflecting target is illuminated by some radar signal propagating toward the target. The objective of the imaging problem is to form an image of the target by processing the reflected radar signal.

We will derive a model for the target and for the received data. The received data, when sampled, form a discrete-time zero-mean complex Gaussian process whose second-order properties depend in a complicated way on the desired parameters that characterize the target. The estimation of these parameters has much in common with the estimation of Toeplitz covariance matrices, discussed previously. We will show, however, that this problem differs significantly in the following ways:

1. There exists a sequence of random vectors $\mathbf{y}_G(1) \cdots \mathbf{y}_G(K)$ that are independent and that have unrelated Toeplitz covariances.
2. The underlying discrete spectrum for each $\mathbf{y}_G(k)$ is constrained to have several zero values.
3. The $\mathbf{y}_G(k)$'s are not directly observable. The received data consists of a linear combination of these vectors, with the transformation determined by the transmitted signal.
4. There is additive noise in the received data.

This section proceeds as follows. The model describing the reflectivity process is given first. Since the processing is performed digitally, the discrete form of this model is examined in detail. Next, the manner in which the transmitted signal interacts with the reflectivity to form the radar return is presented. The imaging problem is then stated as a maximum-likelihood covariance estimation problem. A necessary condition for the maximum-likelihood solution is obtained, and an EM algorithm approach to solving for the maximum is taken.

4.6.1 The Reflectivity Process Model

Radar systems used to produce high-resolution images of reflecting targets illuminate the target with a series of pulses, resulting in an estimation of the reflectivity function based on the series of observed return echoes. It is convenient to think of the target as being composed of a discrete set of "patches" in two dimensions; each patch corresponds to 1 pixel in the desired image. It is assumed that the radar transmitter and receiver are colocated and the target is rotating relative to the transmitter. In this way, each patch is indexed by a unique pair of numbers: its distance from the transmitter and its velocity relative to the transmitter. This is depicted in Fig. 4.4.

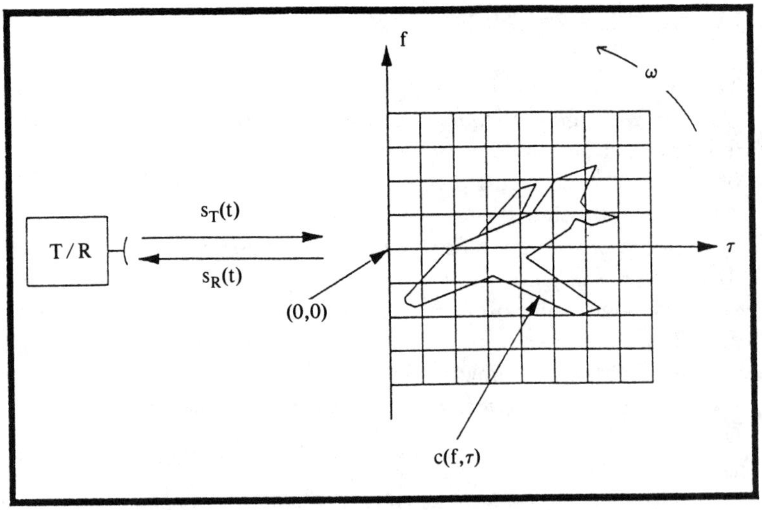

Figure 4.4 Plot of a radar transmitter/receiver transmitting signal $s_T(t)$ and receiving signal $s_R(t)$. Relative motion between the transmitter/receiver and the target results in an apparent rotation about $(0, 0)$. A reflecting path on the target at range x and cross-range y introduces a delay τ, Doppler shift f, and reflectivity gain $c(f, \tau)$.

Each patch of the target reflects RF (radio frequency) electromagnetic energy. If it were possible to illuminate a single patch with a plane wave signal $s(t)$ (measured at the target), then the reflected signal (also measured at the target) is $cs(t)$, where c is the complex reflectivity. We model the reflectivity c as a zero-mean complex Gaussian random variable with independent, equidistributed real and imaginary parts; that is, c has density

$$p(c) = \frac{1}{\pi\sigma^2}\exp\left[\frac{-|c|^2}{\sigma^2}\right]. \tag{4.32}$$

The variance, or power σ, of this reflectivity is the parameter of interest; a larger power means a brighter patch. The imaging problem is to estimate the target reflectivity power as a function of patch location in range and cross-range coordinates. Since it is assumed that the target surface is rough compared to the RF wavelength, the reflectivities for all the patches are mutually independent.

If it were possible to illuminate a single patch, it would have two effects on the reflected radar signal. First, the range coordinate (along the line of sight) determines the two-way delay between transmitted and received signal. The cross-range coordinate (perpendicular to the line of sight) determines the Doppler frequency shift, since the directed velocity relative to the transmitter varies linearly with the distance from the center of rotation for rigid bodies. These relationships are summarized by the expression

$$s_R(t) = c(f, \tau)s_T(t - \tau)e^{j2\pi f(t - \tau/2)}, \tag{4.33}$$

where $s_T(t)$ and $s_R(t)$ are the envelopes measured at the radar of the transmitted and received signals, respectively, and $c(f, \tau)$ is the reflectivity of the target patch that introduces the delay τ and the Doppler shift f.

4.6.2 Spotlight-Mode Measurement Model

The fundamental difficulty in radar imaging at microwave frequencies is that, at the resolution available, it is *not* possible to illuminate patches small enough to form images at a useful resolution. In spotlight-mode radar, the entire target is illuminated by one signal, and the target reflectivity as a function of delay and Doppler is determined by processing the return signal. This return signal is the superposition of reflections from all the patches on the target. From (4.33), we see that the return signal can be modeled by the following expression:

$$s_R(t) = \sum_f \sum_\tau c(f, \tau) s_T(t - \tau) e^{j2\pi f(t - \tau/2)}. \tag{4.34}$$

The variables f and τ in (4.34) take on values from a discrete set. Assume that the resolution widths of each patch are Δf and $\Delta\tau$ in Doppler and delay coordinates, respectively. Assume further that there are I_R bins in the range direction (the target is of length $I_R\, \Delta\tau/2$ in delay coordinates), there are I_{CR} bins in the cross-range direction (the target is of width $I_{CR}\, \Delta f$ in Doppler coordinates), and the sampling period of the radar return signal is Δt seconds. Then (4.34) can be rewritten as

$$s_R(n\, \Delta t) = \sum_{k=1}^{I_R} \sum_{m=-(I_{CR}-1)/2}^{m=+(I_{CR}-1)/2} c(m, k) s(n\, \Delta t - k\, \Delta\tau) e^{j2\pi nm\, \Delta f\, \Delta t}, \tag{4.35}$$

where

$$c(m, k) \triangleq c(m\, \Delta f, k\, \Delta\tau) e^{-j2\pi m\, \Delta f\, \Delta\tau/2}. \tag{4.36}$$

Because of the uniform phase of the probability density for $c(f, \tau)$, $c(m, k)$ is statistically indistinguishable from $c(m\, \Delta f, k\, \Delta\tau)$. The set of $c(m, k)$'s form a set of mutually independent zero-mean Gaussian random variables on a two-dimensional grid, and we wish to estimate the variance of each.

There are sampling issues associated with the selection of Δf, $\Delta\tau$, and Δt. Some of these issues are addressed in our paper [2]. The simplest of these issues relates the sampling rate of the received signal Δt to the achievable Doppler resolution Δf. To achieve I_{CR} samples of the target region with resolution Δf and no aliasing, a necessary condition is that Δt be less than $1/(I_{CR}\, \Delta f)$. Typically, we take G samples of $s_R(t)$ and we have $I_{CR} < G < N$, where N is the periodic extension and is introduced shortly.

It is convenient now to introduce the variable $y(n, k)$ defined by

$$y(n, k) = \sum_m c(m, k) e^{j2\pi mn\, \Delta f\, \Delta t}. \tag{4.37}$$

For each $k, y(n, k)$ is a zero-mean Gaussian random variable formed by taking a linear combination of DFT columns, with the coefficients being independent random variables. This is the characterization of a random vector with a Toeplitz covariance matrix; from the finite sum in (4.37), we know that this covariance must be a member of the class of Toeplitz matrices with circulant extensions. The estimation of the covariance of y is equivalent to the estimation of the variances of c, and thus we have established the connection between this radar-imaging problem and the Toeplitz covariance estimation problem discussed in the earlier sections of this chapter.

As can be seen from (4.35), the y variables are not directly observable. Combining (4.35) and (4.37) we have

$$s_R(n\,\Delta t) = \sum_k s_T(n\,\Delta t - k\,\Delta\tau)y(n, k). \tag{4.38}$$

There is also an additive noise component that must be considered. Define the sampled output of the radar receiver to be

$$r(n) = s_R(n\,\Delta t) + w(n), \tag{4.39}$$

where the $w(n)$ are independent zero-mean Gaussian random variables with variance N_0.

4.6.3 Maximum-Likelihood Problem Formulation

Having discretized all the variables in the problem, it is now helpful to restate the estimation problem using matrix-vector notation. Define the vector $\mathbf{y}_G(k)$ by

$$\mathbf{y}_G(k) = [y(0, k) \;\cdots\; y(G-1, k)]^T, \tag{4.40}$$

and then define

$$\mathbf{y} = [\mathbf{y}_G^T(1) \;\cdots\; \mathbf{y}_G^T(I_R)]. \tag{4.41}$$

Let $\mathbf{K}_G(k)$ be the Toeplitz covariance matrix of $\mathbf{y}_G(k)$, and let $\mathbf{K}_G$ be the covariance matrix of $\mathbf{y}$. By the independence of the $\mathbf{y}_G(k)$'s, $\mathbf{K}_G$ is a block diagonal matrix. The observation vector is

$$\mathbf{r} = [r(1) \;\cdots\; r(G)]^T, \tag{4.42}$$

and the noise vector is

$$\mathbf{w} = [w(1) \;\cdots\; w(G)]^T. \tag{4.43}$$

Define the signal matrix $\mathbf{S}_T^\dagger$ as the $G \times GI_R$ matrix composed of I_R $G \times G$ submatrices each of which is diagonal:

$$\mathbf{S}_T^\dagger = [\mathbf{S}_1\ \mathbf{S}_2\ \mathbf{S}_3 \;\cdots\; \mathbf{S}_{I_R}], \tag{4.44}$$

where

$$\mathbf{S}_k = diag\{s_T(-k\Delta\tau)s_T(\Delta t - k\Delta\tau)s_T(2\Delta t - k\Delta\tau)\cdots$$

$$s_T[(G-1)\Delta t - k\Delta\tau]\}.$$

Then combining (4.38) and (4.39) yields the following vector form:

$$\mathbf{r} = \mathbf{S}_T^{\dagger}\mathbf{y} + \mathbf{w} = \sum_{k=1}^{I_R} \mathbf{S}_k\mathbf{y}_G(k) + \mathbf{w}. \tag{4.45}$$

The measurement vector $\mathbf{r}$ is a zero-mean Gaussian random variable with covariance

$$\mathbf{K}_R = \mathbf{S}_T^{\dagger}\mathbf{K}_G\,\mathbf{S}_T + N_0\mathbf{I}. \tag{4.46}$$

The likelihood of the radar return data becomes

$$L(\mathbf{K}_R;\mathbf{r}) = \pi^{-G}(\det\mathbf{K}_R)^{-1}\exp[-\mathbf{r}^{\dagger}\mathbf{K}_R^{-1}\mathbf{r}]. \tag{4.47}$$

As discussed in Section 4.2, a necessary condition for the maximizer $\hat{\mathbf{K}}_R$ is that it satisfy the following trace condition:

$$\mathrm{tr}[(\hat{\mathbf{K}}_R^{-1}\mathbf{r}\mathbf{r}^{\dagger}\hat{\mathbf{K}}_R^{-1} - \hat{\mathbf{K}}_R^{-1})\delta\mathbf{K}_R] = 0. \tag{4.48}$$

The matrix $\delta\mathbf{K}_R$ is a variational matrix that takes values in all possible additive variations of the matrix $\hat{\mathbf{K}}_R$. Elements in the set of possible $\mathbf{K}_R$ are described by (4.46). If we rewrite the trace condition in terms of $\hat{\mathbf{K}}_G$, then we get

$$\mathrm{tr}[\mathbf{S}_T(\mathbf{S}_T^{\dagger}\hat{\mathbf{K}}_G\mathbf{S}_T + N_0\mathbf{I})^{-1}(\mathbf{r}\mathbf{r}^{\dagger} - \mathbf{S}_T^{\dagger}\hat{\mathbf{K}}_G\mathbf{S}_T - N_0\mathbf{I})(\mathbf{S}_T^{\dagger}\hat{\mathbf{K}}_G\mathbf{S}_T + N_0\mathbf{I})^{-1}\mathbf{S}_T^{\dagger}\delta\mathbf{K}_G] = 0. \tag{4.49}$$

The relationship between the radar problem and the Toeplitz constrained covariance problem studied in the first half of this chapter can be seen by comparing the necessary maximizer conditions of (4.49) to those stated in (4.2). The outstanding similarity is that the allowable variations $\delta\mathbf{K}_G$ are constrained to be in the Toeplitz class. The three major differences illustrated via (4.49) between the radar problem and the generic spectrum estimation problem are as follows: (1) the radar signal matrix $\mathbf{S}_T$ weights $\mathbf{K}_G$ throughout; (2) the additive noise is manifest via the matrix $N_0\mathbf{I}$; and (3) $\mathbf{K}_G$ is a block diagonal matrix with Toeplitz blocks. These differences notwithstanding, the iterative EM algorithm that we propose for the solution of (4.49) takes an almost identical form to that already presented.

4.6.4 Expectation-Maximization Algorithm

We proceed by proposing the following iterative algorithm for generating the maximizers that satisfy the necessary conditions of (4.49), and then we demonstrate that it is an instance of an EM algorithm.

Proposition 2: Define the set of sequences of $G \times G$ covariance estimates $\{\mathbf{K}_G^{(p)}(k);\ 1 \le k \le I_R\}$ with n, l entries of $\mathbf{K}_G^{(p)}(k)$ given by the following iteration:

$$K_G^{(p+1)}(n-l, k) = \frac{1}{N} \sum_{m=0}^{N-1} E\{y(m,k)y^* [(m+n-l)_{\mathrm{mod}\,N}, k] | \mathbf{r}, \mathbf{K}_G^{(p)}(k);\ 1 \le k \le I_R\}. \tag{4.50}$$

Then, the set of iterates $\{\mathbf{K}_G^{(p)}(k);\ p \ge 1\}$ is an instance of an EM algorithm. Forming the block diagonal matrix $\mathbf{K}_G^{(p)}$ from the I_R Toeplitz estimates $\mathbf{K}_G^{(p)}(k)$ at each delay k, then the sequence of likelihoods $L(\mathbf{S}_T^{\dagger}\mathbf{K}_G^{(p)}\mathbf{S}_T + N_0\mathbf{I};\ \mathbf{r})$ of (4.47) is nondecreasing.

We proceed as in Section 4.3, leaving out details which have already been included. Define the set of complete-data vectors as $\{\mathbf{y}_N(k);\ 1 \le k \le I_R\}$, with the complete data for delay k corresponding to the periodic extension of the data vector $\mathbf{y}_G(k)$ given by

$$\mathbf{y}_N(k) = [\mathbf{y}_G(k)^{\mathrm{T}} \mathbf{y}_A(k)^{\mathrm{T}}]^{\mathrm{T}}, \tag{4.51}$$

where $\mathbf{y}_A(k)$ is an $(N-G) \times 1$ vector that augments $\mathbf{y}_G(k)$ to obtain a full period of the periodic process at range k. The crucial part of the model now comes into play. Since the processes are uncorrelated across range bins k, the complete-data likelihood becomes

$$L[\mathbf{K}_N(k),\ 1 \le k \le I_R;\ \mathbf{y}_N(k),\ 1 \le k \le I_R] = \prod_{k=1}^{I_R} L[\mathbf{K}_N(k);\ \mathbf{y}_N(k)], \tag{4.52}$$

with each $\mathbf{K}_N(k)$ a circulant Toeplitz matrix corresponding to range bin k. The likelihood on the left in (4.52) is the likelihood of the complete data for all range bins taken together. The likelihoods on the right in (4.52) are each associated with one delay. Now defining the rotated coordinates $\mathbf{c}_N(k) = \mathbf{W}\mathbf{y}_N(k)$, we use the fact that the Gaussian variates $c(m,k)$ have nonzero variance $\sigma(m,k) > 0$ for $-(I_{CR}-1)/2 \le m \le (I_{CR}-1)/2$, and are zero otherwise. Writing the complete-data log likelihood over the rotated coordinates yields the following complete-data log likelihood to be maximized:

$$-\sum_{k=1}^{I_R} \sum_{m=-(I_{CR}-1)/2}^{(I_{CR}-1)/2} [\log \sigma(m,k) + \frac{|c(m,k)|^2}{\sigma(m,k)}]. \tag{4.53}$$

Taking the conditional expectation on the E-step and maximizing yields the following estimates at iteration $p+1$,

$$\sigma^{(p+1)}(m,k) = E\{|c(m,k)|^2 | \mathbf{r}, \Sigma^{(p)}(k); 1 \le k \le I_R\}, \tag{4.54}$$

with

$$\Sigma^{(p+1)}(k) = diag[\sigma^{(p)}(0, k), \sigma^{(p)}(1, k), \ldots, \sigma^{(p)}(N-1, k)]. \quad (4.55)$$

Note that for $(I_{CR} - 1)/2 < m < N - (I_{CR} - 1)/2$ there holds $\sigma(m, k) = 0$. The estimate of the covariance $K_N^{(p+1)}(n - l, k)$ at range bin k is found by transforming back to the original coordinates yielding

$$K_N^{(p+1)}(n - l, k) = \frac{1}{N} \sum_{m=0}^{N-1} E\{y(m, k) \; y^*[(m + n - l)_{\text{mod } N}, k] | \mathbf{r}, \mathbf{K}_N^{(p)}(k), 1 \le k \le I_R\}, \quad (4.56)$$

demonstrating Proposition 2.

4.6.5 Implementation Strategy

The equation developed in Proposition 2 demonstrates the similarity between the EM algorithm for the radar problem and the EM algorithm developed earlier in this chapter. It should be anticipated that the implementation strategy for the radar problem would be similar.

The algorithm is implemented using (4.54) in its matrix form. First, we define vector $\mathbf{c}_N(k)$ as before:

$$\mathbf{c}_N(k) = [c(0, k) \; c(1, k) \; c(2, k) \; \cdots \; c(N-1, k)]^T. \quad (4.57)$$

Now using the fact that $\mathbf{y}_G(k) = \mathbf{W}_G^{\dagger}\mathbf{c}_N(k)$, with $\mathbf{W}_G$ the first G columns of $\mathbf{W}$ from (4.17–4.19), and substituting into (4.45) yields the following model for $\mathbf{r}$:

$$\mathbf{r} = \sum_{k-1}^{I_R} \Gamma^{\dagger}(k)\mathbf{c}_N(k) + \mathbf{w}, \quad (4.58)$$

where

$$\Gamma^{\dagger}(k) = \mathbf{S}_k \mathbf{W}_G^{\dagger}. \quad (4.59)$$

The vectors $\mathbf{c}_N(k)$ can be seen to have covariance $\Sigma(k)$ from (4.55), so at step $p + 1$, we have to compute $E\{\mathbf{c}_N(k)\mathbf{c}_N^{\dagger}(k) \mid \mathbf{r}, \Sigma^{(p)}(k), 1 \le k \le I_R\}$. Following Rhodes [16] again,

$$E\{\mathbf{c}_N(k)\mathbf{c}_N^{\dagger}(k) \mid \mathbf{r}, \Sigma(k), 1 \le k \le I_R\} = \mathbf{K}_{cr}(k)\mathbf{K}_R^{-1}\mathbf{r}\mathbf{r}^{\dagger}\mathbf{K}_R^{-1}\mathbf{K}_{cr}(k)^{\dagger} + \Sigma(k) - \mathbf{K}_{cr}(k)\mathbf{K}_R^{-1}\mathbf{K}_{cr}^{\dagger}(k), \quad (4.60)$$

where $\mathbf{K}_{cr}(k) = \Sigma(k)\Gamma(k)$ is the correlation between $\mathbf{c}_N(k)$ and $\mathbf{r}$. The covariance matrix for $\mathbf{r}$ may be written in terms of the $\Gamma(k)$'s as

$$\mathbf{K}_R = \sum_{k=1}^{I_R} \Gamma^\dagger(k)\Sigma(k)\Gamma(k) + N_0\mathbf{I}. \tag{4.61}$$

Written for iteration $p+1$, (4.60) becomes

$$\begin{aligned} E\{\mathbf{c}_N(k)\mathbf{c}_N^\dagger(k) \mid \mathbf{r}, \Sigma^{(p+1)}(k),\ 1 \le k \le I_R\} \\ = \Sigma^{(p)}(k)\Gamma(k)\mathbf{K}_R^{(p)-1}\mathbf{r}\mathbf{r}^\dagger\mathbf{K}_R^{(p)-1}\Gamma^\dagger(k)\Sigma^{(p)} \\ + \Sigma^{(p)}(k) - \Sigma^{(p)}(k)\Gamma(k)\mathbf{K}_R^{(p)-1}\Gamma^\dagger\Sigma^{(p)}(k). \end{aligned} \tag{4.62}$$

The diagonal elements of (4.62) form the new estimate for $\Sigma(k)$.

The matrices $\Gamma(k)$ are computed once off-line and then are used at each iteration to find $\mathbf{K}_R$. The matrix $\mathbf{K}_R$ is Hermitian symmetric, but not necessarily Toeplitz, and its inverse must be computed at each iteration. The matrix $\Gamma(k)$ is not necessarily composed of columns from orthogonal DFT matrices. Thus the algorithms proposed in Section 4.4 may not be applicable to the computations indicated in (4.62).

4.7. UNIQUENESS OF SPECTRUM ESTIMATES AND CONVERGENCE OF THE ALGORITHMS

This section addresses the question of the uniqueness of maximum-likelihood spectrum estimates as well as the convergence of the EM algorithm for Toeplitz covariance estimation of (4.10). In what directly follows we explore the necessary conditions for the spectrum associated with the maximum-likelihood Toeplitz estimates of the covariances to be unique. Then we explore the convergence of the algorithm.

4.7.1 Uniqueness of the Spectrum Estimates

First, we examine the uniqueness of the spectrum estimates for both the radar imaging problem and the Toeplitz estimation problem from the preceding sections. For the radar imaging problem, define the matrix Γ to be

$$\Gamma^\dagger = [\Gamma^\dagger(1)\ \Gamma^\dagger(2)\ \cdots\ \Gamma^\dagger(I_R)], \tag{4.63}$$

where $\Gamma(k)$ is given by (4.59). Note, for the Toeplitz covariance problem of Sections 4.1–4.3, the matrix $\Gamma(k)$ is simply W_G, the first G columns of the DFT matrix. Also, define the matrix Σ for the radar imaging problem to be the block diagonal matrix whose kth block is $\Sigma(k)$ from (4.55). The integer M stands for the rank of Σ; this is N if we are discussing the original Toeplitz problem and $I_R I_{CR}$ if we are discussing the radar problem. Let γ_l denote the row of Γ corresponding to the lth nonzero entry of Σ. There are M such rows, each of which has G entries.

The approach is based upon the Cramér-Rao lower bounds on the variance of estimates. The Cramér-Rao lower bounds are obtained by inverting the *Fisher information matrix*. When the Fisher information matrix is singular, these bounds are infinite, thereby implying that the maximum-likelihood spectrum estimates are nonunique.

Theorem 3. The $M \times M$ Fisher information matrix for estimating the nonzero entries of Σ given data r has m, n entry given by the magnitude squared of

$$\gamma_m^\dagger K_R^{-1} \gamma_n. \tag{4.64}$$

Proof. The Fisher information matrix is simply the negative of the expected value of the second derivative of the log-likelihood function, and as proven in the appendix of [2] has entries given by (4.64).

Theorem 4. Suppose the Fisher information matrix determined by the entries of (4.64) is singular and that the matrix K_R is positive definite. Then there does not exist a Σ that is positive definite that yields a unique maximum of the log likelihood.

Proof. Define the $M \times G^2$ matrix F to have kth row given by $\gamma_k \bigcirc \gamma_k^*$, where $\bigcirc$ denotes the Kronecker product. The Fisher information matrix may be rewritten using F as

$$F(K_R^{-1} \bigcirc K_R^{-1*})F^\dagger. \tag{4.65}$$

Since the rank of the matrix $K_R^{-1} \bigcirc K_R^{-1*}$ equals G^2, and by the form of the matrix, the Fisher information matrix is singular if and only if the matrix F has rank less than M, which is true if and only if there exists a real vector s such that $F^\dagger S = 0$. Such an s exists if and only if there exists a real diagonal matrix D ($D = diag(s)$) such that

$$\Gamma^\dagger(\Sigma + \alpha D)\Gamma = \Gamma^\dagger \Gamma \Sigma \Gamma \tag{4.66}$$

for all real α. If Σ is positive definite and maximizes the log likelihood, then there exists a β such that for all $0 \le \alpha \le \beta$, the matrix $\Sigma + \alpha D$ is nonnegative definite and yields the same covariance matrix and hence the same value for the log likelihood.

Corollary 1. For the spectrum estimation problem from the previous sections, there does not exist a positive definite Σ that yields a unique maximum of the log likelihood if $N > 2G - 1$.

Proof. Examine the matrix F constructed earlier to see that its rank must be less than or equal to $2G - 1$.

Note that this theorem does not imply that the Toeplitz covariances generated by the algorithm are not unique. It relates only to the uniqueness of the spectrum associated with the Toeplitz covariance. There may in fact be many Σ that yield the same estimate for the Toeplitz covariance matrix (for M large this is precisely the case). For problems such as in radar imaging where the image array are spectral coefficients across different range bins, the uniqueness conditions of the spectrum from Theorem 3 are extremely important.

4.7.2 Convergence of the Iterative Algorithms

Proving convergence of the EM algorithm of (4.10) to the set of MLEs of the Toeplitz constrained covariance involves three parts: (1) showing that the likelihood is finite over the closed set of bounded, positive semidefinite covariance matrices and attains its bound in the set; (2) showing that all limit points of the sequences of (4.10) are stable and that these stable points satisfy the necessary maximizer conditions; and (3) showing that every sequence converges. We have proven (1) and (2) in [12] for the straight Toeplitz covariance estimation problem of Sections 4.1–4.3. We sketch briefly here the ideas of the proof.

Part 1 relies on the fact that over the constrained set of Toeplitz matrices with circulant extensions the Gaussian density has value zero for all singular matrices (this is precisely Theorem 2 of this chapter). This then assures that algorithms with increasing likelihood such as (4.10) will not converge to singularity, the iteration sequence is contained in a compact set of positive semidefinite matrices, and that the log likelihood is upper semicontinuous and finite from above over the compact set. Then it follows that the likelihood is both bounded above and attains its maximum.

Part 2 is proven by showing that the algorithm produces a monotonic sequence of likelihoods, that the map defined by (4.10) is continuous over the set of positive-definite iterates, and the set of limit points of the subsequences of the algorithm are all stable. It is then straightforward to show that the limit points satisfy the maximizer conditions.

For Part 3 we must show that the full sequence converges. This still remains an open research question.

4.8. CONCLUSIONS

In this chapter we have proposed a constrained ML approach for generating Toeplitz covariance estimates, for which the constraint set is the set of positive-definite Toeplitz matrices with periodic extensions. By imposing the periodic extension constraint on the maximization, we have proven that it ensures the existence of a nonsingular estimator. It also provides a natural starting point for the EM algorithm, for which we show that when the observed data is of length less than the assumed period of the process, conditional mean estimates of the lag products that are outside of the data collection window must be generated, from which the Toeplitz covariances are obtained. Given an estimate of the Toeplitz covariance, the lag products must be reestimated, thereby giving rise to a new estimate of the Toeplitz covariance. This is precisely the EM algorithm for the constrained estimation problem. We have also shown, via simulations, that estimates generated using the ML procedure have excellent mean-squared error properties when compared with conventional lag-product estimators.

We have also developed a new approach for high-resolution delay-Doppler radar imaging of diffuse targets. By modeling the target reflectance at each point on the target as a sample of a Gaussian process with covariance determined by the target's scattering function, the imaging problem is fundamentally a spectrum estimation problem and therefore has very similar structure to the Toeplitz covariance estimation problem. This has allowed us to propose an iterative EM algorithm for its solution, involving the estimation of a series of Toeplitz covariances at each delay.

APPENDIX A. PROOF OF THEOREM 1

In [13] it is shown that, if there exists a singular Toeplitz matrix whose range space contains $\mathbf{y}_G$, then the likelihood function will be unbounded. Therefore, we must show that the union of the range spaces of all singular Toeplitz matrices is a set of measure greater than $1 - 1/2^{G-2}$.

As a result of a theorem of Caratheodory [18], it is possible to write any $G \times G$ nonnegative definite Toeplitz matrix in a unique decomposition of the form

$$\mathbf{T} = \sum_{k=1}^{G} c_k \gamma_k \gamma_k^\dagger + \sigma^2 \mathbf{I}, \tag{4.A.1}$$

where $c_k > 0$, $M < G$, $\sigma^2 \geq 0$, and

$$\gamma_k = [1 \; e^{j\omega_k} \; e^{2j\omega k} \; \cdots \; e^{(G-1)j\omega_k}]^T. \tag{4.A.2}$$

The frequency ω_k can take on any value in the range $[\pi, -\pi]$.

A singular nonnegative definite Toeplitz matrix can be written in the form

$$\mathbf{T}_0 = \sum_{k=1}^{G} c_k \gamma_k \gamma_k^\dagger. \tag{4.A.3}$$

The γ_k's are linearly independent; hence, a vector $\mathbf{x}$ will lie in the range of some $\mathbf{T}_0$ if it can be written as the superposition of $(G - 1)$ complex exponentials of arbitrary frequency and complex amplitude. For real $\mathbf{x}$ and $\mathbf{T}$, the condition is that $\mathbf{x}$ be expressible as the sum of $(N - 1)/2$ (N odd) or $N/2$ (N even) real sinusoids of arbitrary amplitude, frequency, and phase. For G even, one of the frequencies must be 0 or π, and for G odd, it may be that two of the frequencies are 0 and π and there are $(G - 3)/2$ other sinusoids. It follows that an equivalent condition on $\mathbf{x}$ is that there exists at least one polynomial of the form

$$B(z) = \sum_{i=0}^{G-1} b(i) z^i \tag{4.A.4}$$

such that $B(z)$ has $G - 1$ simple roots on the unit circle and $\mathbf{x}$ is orthogonal to the coefficient vector $\mathbf{b}$.

We abbreviate the remainder of the proof and refer the reader to [13]. There it is shown that, in the spaces of symmetric and antisymmetric vectors, there exist *orthants* which contain coefficient vectors that form polynomials with $G - 1$ simple roots on the unit circle. Any vector that is not a member of these orthants is orthogonal to at least one vector in the orthant, and any vector within the orthant cannot be orthogonal to any other member of the orthant. These orthants have small measure, $O(2^{-G})$, and it is this measure that leads to the bound in the theorem statement.

APPENDIX B. PROOF OF THEOREM 2

Any Toeplitz matrix in the class with nonnegative definite circulant extensions to period N can be written in the form

$$\mathbf{T} = \sum_{k=0}^{N-1} \sigma_k^2 \gamma_k \gamma_k^\dagger, \tag{4.B.1}$$

with the $G \times 1$ columns γ_k having the form given in (4.A.2), and

$$\omega_k = \frac{2\pi k}{N}. \tag{4.B.2}$$

Note the similarity to the decomposition (4.A.1) of an arbitrary Toeplitz matrix; here the frequencies ω_k are fixed, but there are N terms in the summation.

Since any G of the γ_k's in (4.B.1) are linearly independent, any singular $\mathbf{T}_0$ in this class must be written with less than G terms in the summation; that is,

$$\mathbf{T}_0 = \sum_{k=1}^{G-1} \sigma_k^2 \gamma_k \gamma_k^\dagger \tag{4.B.3}$$

where the frequencies ω_k are chosen from the set

$$\Omega_N = \frac{2\pi i}{N}, \qquad i = 0, \ldots, N-1. \tag{4.B.4}$$

For an observation vector $\mathbf{x}$ to lie in the range space of some $\mathbf{T}_0$, it must belong to *span*$(\gamma_1 \cdots \gamma_{G-1})$ for some set of $G - 1$ frequencies from the set Ω_N. For any particular set of $G - 1$ frequencies, the probability that $\mathbf{x}$ lies in the subspace spanned by $(\gamma_1 \cdots \gamma_{G-1})$ is zero because the probability of a reduced-dimension subspace is always zero when the true covariance is full rank. Furthermore, there is a finite number, namely $\binom{N}{G-1}$, of such sets of frequencies from Ω_N, and hence the probability that $\mathbf{x}$ lies in the subspace spanned by any of the corresponding γ's is zero.

There are some technicalities involved in showing that this is sufficient to prove that the likelihood cannot be unbounded above over the restricted class of Toeplitz matrices. Again the reader is referred to [13] for more details.

ACKNOWLEDGMENTS

We express our appreciation to Michael J. Turmon for his contributions to the computational efficiency and performance section results. Michael I. Miller was supported by the National Science Foundation under a Presidential Young Investigator Award (Grant Number ECE-8552518); Donald L. Snyder and Joseph O'Sullivan were supported by O.N.R. Contract N00014-96-K-370 and the National Science Foundation under grant MIP-8722463.

REFERENCES

1. D. H. Johnson, "The Application of Spectral Estimation Methods to Bearing Estimation Problems," *Proc. IEEE*, **70**, no. 9 (1982), 1018–1028.
2. D. L. Snyder, J. A. O'Sullivan, and M. I. Miller, "The Use of Maximum-Likelihood Estimation for Forming Images of Diffuse Radar-Targets from Delay-Doppler Data," *IEEE Transactions on Information Theory* (1989).
3. S. Kay and L. Marple, Jr., "Spectrum Analysis—A Modern Perspective," *Proc. IEEE*, **69**, no. 11 (1981), 1380–1418.
4. S. Kay, *Modern Spectral Estimation* (Englewood Cliffs, N.J.: Prentice-Hall, 1987).
5. L. Marple, Jr., *Digital Spectral Analysis and Applications* (Englewood Cliffs, N.J.: Prentice-Hall, 1987).
6. J. P. Burg, *Maximum Entropy Spectral Analysis*, Univ. Microfilms No. 75-25, Stanford University, Stanford, California, 1975.
7. E. T. Jaynes, "On the Rationale of Maximum-Entropy Methods," *Proc. IEEE*, **70** (September 1982), 939–952.
8. A. van den Bos, *IEEE Trans. Information Theory*, **17** (1971), 493–494.
9. J. E. Shore, "Minimum Cross-Entropy Spectral Analysis," *IEEE Trans. Acoustics, Speech, and Signal Processing*, **29**, no. 2 (April 1981), 230–237.
10. D. G. Childers, ed., *Modern Spectrum Analysis* (New York: IEEE Press, 1978).
11. J. P. Burg, D. G. Luenberger, and D. L. Wenger, "Estimation of Structured Covariance Matrices," *Proc. IEEE*, **70**, no. 9 (September 1982), 963–974.
12. M. I. Miller and D. L. Snyder, "The Role of Likelihood and Entropy in Incomplete-Data Problems: Applications to Estimating Point-Process Intensities and Toeplitz Constrained Covariances," *Proc. IEEE* (July 1987), 892–907.
13. D. R. Fuhrmann and M. I. Miller, "On the Existence of Positive Definite Maximum Likelihood Estimates of Structured Covariance Matrices," *IEEE Trans. Information Theory* (July 1988).
14. R. M. Gray, *Toeplitz and Circulant Matrices II*, Stanford University Information Systems Laboratory Technical Report No. 6504-1, 1977.

15. U. Grenander, *Abstract Inference* (New York: John Wiley, 1981).

16. A. D. Dempster, N. M. Laird, and D. B. Rubin, "Maximum Likelihood from Incomplete Data via the EM Algorithm," *J. Roy. Statis. Soc.*, **B39** (1977), 1–37.

17. I. B. Rhodes, "A Tutorial Introduction to Estimation and Filtering," *IEEE Trans. Automatic Control*, **16**, no. 6 (1971), 688–706.

18. U. Grenander and G. Szego, *Toeplitz Forms and Their Applications* (Berkeley: University of California Press, 1958).

5

Threshold Properties of Narrow-Band Signal-Subspace Array Processing Methods

M. Kaveh and H. Wang

5.1. INTRODUCTION

Signal-subspace processing refers to estimation procedures that are formulated in the space spanned by the eigenvectors corresponding to the *principal eigenvalues* of an estimate of the covariance matrix of the observed data. Therefore, the accuracy with which the number of principal eigenvalues are determined and the quality of the resulting estimated subspace are crucial factors in the success of a particular estimator of this type.

This chapter explores some statistical aspects of the performance of signal-subspace methods for the estimation of the frequencies of multiple complex sinusoids corrupted by uncorrelated noise and observed over several independent realizations of the process. The specific application that is considered is the estimation of the *angles of arrival* (*AOA*) of narrow-band plane waves. Hence, throughout the chapter we use the model of an array of sensors measuring plane waves arriving from its far field. The characterization of the performance of the methods will range from the estimation of the number of plane waves (called *detection* here) and a general assessment of the quality of estimated signal subspaces to an analysis of the properties of two specific algorithms MUSIC [1] and minimum-norm (min-norm) [2] for resolving closely spaced plane waves. We shall use some unconventional performance measures and criteria that are, however, reasonable and are amenable to asymptotic theoretical analyses that reveal the threshold behavior of MUSIC and min-norm as functions of physically meaningful source, array, and noise parameters.

Theoretical characterization of the performance of nonlinear parameter estimators, such as the ones considered here, is at best a difficult task. The *Cramér-Rao lower bound* (*CRLB*) to the variance of unbiased estimates of the parameters [3] is often the sole theoretical measure that is used to indicate the best possible performance of estimators. Unfortunately, this bound does not directly characterize the low *signal-to-noise ratio* (*SNR*) limitations of the estimators. Thus, the performance of algorithms is usually obtained by numerical simulations.

For the types of problems considered here, performance measures other than the variance of the estimators may be the primary factors in determining the onset of the threshold regions. These measures often possess direct, albeit nontransparent, connections to the variability of the estimated parameters. One such measure is the probability of resolution of (angularly) closely spaced plane waves observed in noise. Figure 5.1 shows an example of such a resolution probability (sources are resolved if two peaks of a MUSIC "spectrum" are located within a beamwidth of the true angles) as a function of signal-to-noise ratio. Figure 5.2 is a plot of the CRLB and the sample standard deviation of one of the estimated angles for the same example as in Fig. 5.1. It is clear that once the two sources are resolved with a high probability, the standard deviation of the estimated angles are also close to the CRLB. Thus, the resolution threshold SNR may loosely be viewed as a reasonably tight lower bound for the traditional mean-squared error or variance threshold SNR. Therefore, in this chapter,

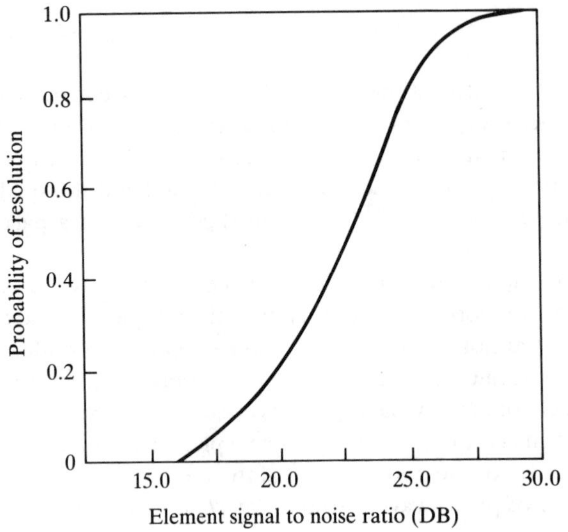

Figure 5.1 Example of the probability of resolution versus element signal-to-noise ratio for the MUSIC algorithm.

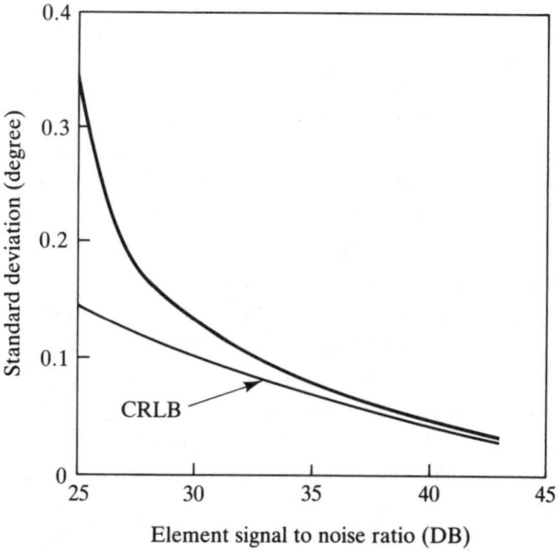

Figure 5.2 Standard deviation of the MUSIC angle estimate and the corresponding Cramér-Rao lower bound.

we will use performance measures such as the *probability of correctly estimating the number of sources*, *quality of the estimated signal-subspace*, and the *resolution thresholds of specific algorithms to characterize signal-subspace processing methods.*

The organization of the chapter is as follows. In Section 5.2, the signal and sensor array models are presented, followed by a brief exposition of the MUSIC and min-norm estimators. Sections 5.3 and 5.4 contain the asymptotic statistics of the eigenvalues and eigenvectors of the sample covariance matrix of the data and expressions for the eigenvalues and eigenvectors of the covariance matrix as functions of signal, noise, and array parameters for one- and two-source situations. Detection criteria and detection performance are addressed in Section 5.5. In Section 5.6 the quality of the estimated signal subspaces are evaluated through the introduction of a measure of closeness between the actual and estimated subspaces. Sections 5.7 and 5.8 present derivations of resolution thresholds for the MUSIC and min-norm algorithms. It should be emphasized that the material discussed in Sections 5.3, 5.5, and 5.6 is applicable to situations that are substantially more general than the sinusoid in noise model. In Section 5.9 we present a review of new theoretical results that have been recently published on the subject matter of this chapter.

5.2. MODEL AND ALGORITHMS

In this section we first formulate a vector-space model for the signals that are presented to the estimator by the L-coherent receiver elements of an array of sensors. We then briefly review the general approach to signal-subspace angle-of-arrival estimation and summarize two popular algorithms for carrying out such estimation.

5.2.1 The Model

Let us assume that d signals, $s_i(t)$, $i = 1, \ldots, d$, are incident on the array from sources in its far field. The array geometry is arbitrary but known to the processor, and the impinging waves on which the signals are carried arrive at the array from distinctly different directions. The ensuing analyses of the algorithms will require a tractable statistical characterization of the signals and the receiver noise. It is, therefore, assumed that the $s_i(t)$ are sample functions from stationary, zero-mean complex Gaussian processes with center frequency f_0 and bandwidth B_x. We begin by introducing the second-order statistics of the source signals, receiver noise, and the received signals that are needed in the development of the algorithms.

Depending on the context in which it is used, we shall use the $FT[\ ]$ to denote the continuous-time or the discrete-time Fourier transform of the argument; indicate complex conjugation, transposition, and Hermitian transposition by superscripts $*$, T, and $\dagger$, respectively; and use $E\{X\}$ and $\overline{X}$ interchangeably to denote statistical expectation of the random variable X.

Define a source signal vector by

$$\mathbf{s}(t) = [s_1(t), s_2(t), \ldots, s_d(t)]^T. \tag{5.1}$$

The source spectral density matrix, $\mathbf{P}_s(f)$, is given by [4]

$$\mathbf{P}_s(f) = FT[E\{\mathbf{s}(t+\tau)\mathbf{s}^H(t)\}]. \tag{5.2}$$

$\mathbf{P}_s(f)$ is in general unknown to the receiver and may be singular as in the cases of specular multipath or active jamming.

Let $\mathbf{x}(t)$ be an L vector formed from the outputs of the sensors. According to the commonly used linear model, the combination of the propagation characteristics of the source signals and the sampling of the wavefronts by the sensor array results in the following model for $\mathbf{x}(t)$:

$$\mathbf{x}(t) = \mathbf{a}(t)*\mathbf{s}(t) + \mathbf{n}(t), \tag{5.3}$$

where $\mathbf{a}(t)$ denotes the $L \times d$ *propagation/array impulse response matrix,* referenced in time and space to the reception by one of the sensor elements. The ith column of $\mathbf{a}(t)$, $\mathbf{a}_i(t)$, represents the propagation and reception of the ith source signal by the array, $\mathbf{n}(t)$ is the receiver noise vector that is assumed to be stationary, complex Gaussian with zero mean, and $*$ is the convolution operator. The *cross-spectral density matrix* (*CSDM*) for the received signal vector is then given by

$$\mathbf{P}_x(f) = \mathbf{A}(f)\mathbf{P}_s(f)\mathbf{A}(f)^\dagger + \sigma_n^2(f)\mathbf{P}_n(f), \tag{5.4}$$

where $\mathbf{A}(f)$ is the Fourier transform of $\mathbf{a}(t)$, $\sigma_n^2(f)$ is an unknown constant, and $\mathbf{P}_n(f)$ is the normalized receiver noise CSDM, which is assumed to be known. In this chapter we assume $\mathbf{P}_n(f) = \mathbf{I}$.

Let us choose element one as reference, and denote the difference between the arrival times of the ith wavefront at the first and the jth sensors by τ_{ij}. The most common model represents the jth element of $\mathbf{a}_i(t)$ by a pure time delay of τ_{ij} seconds. Thus,

$$\mathbf{a}_i(t) = \begin{bmatrix} \delta(t) \\ \delta(t - \tau_{i2}) \\ \vdots \\ \delta(t - \tau_{iL}) \end{bmatrix} \quad \text{and} \quad \mathbf{A}_i(f) = \begin{bmatrix} 1 \\ e^{-j2\pi f \tau_{i2}} \\ \vdots \\ e^{-j2\pi f \tau_{iL}} \end{bmatrix},$$

where $\delta(t)$ is the Dirac delta function. For planar wavefronts generated by sources in the far field of the array, τ_{ij} is determined by the relative location of the jth sensor, direction of propagation of the ith wave, and the velocity of propagation in the medium. Therefore, in the neighborhood of any frequency f, $P_x(f)$ embodies the information on the directions of arrival of the plane waves as functionally manifested by $\mathbf{A}(f)$. Thus, $\mathbf{A}(f)$ and $\mathbf{A}_i(f)$ are, respectively, referred to as the ***direction matrix*** and the ith ***direction vector at frequency f***.

In what follows we are primarily interested in the performances of algorithms as a function of the temporal and spatial characteristics of sources, the observation time, and the array aperture. To standardize these results with respect to array geometry and noise, we shall use a linear, uniformly spaced array of $L > d$ elements with sources that are in the plane of the array and a spatially white noise, wherever in the chapter that an array model is called for. Otherwise, the results of the chapter are general. Figure 5.3 shows a uniform linear array with a vector indicating the direction of propagation for the ith wave. The relative delays are given by

$$\tau_{ij} = \frac{(1 - j)D_e \sin \theta_i}{C},$$

where C is the velocity of propagation. We shall refer to

$$\omega_i(f) = \frac{2\pi f D_e \sin \theta_i}{C}$$

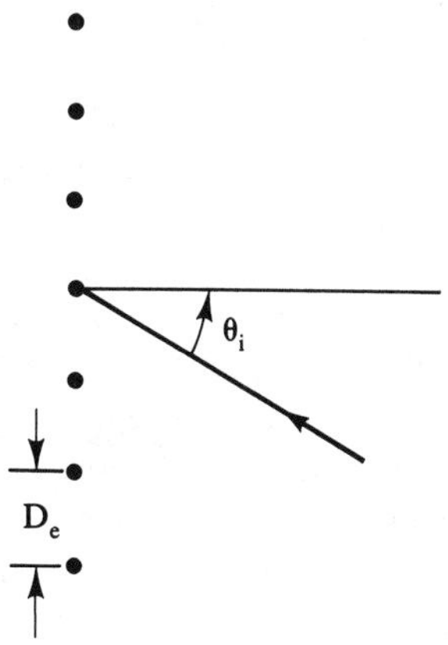

Figure 5.3 A linear, uniformly spaced sensor array with one ray of an impinging plane wave.

as the *i*th spatial frequency at a temporal frequency of f Hz. The direction matrix may now be written as

$$\mathbf{A}(f) = \begin{bmatrix} 1 & 1 & \cdots & 1 \\ e^{j\omega_1(f)} & e^{j\omega_2(f)} & \vdots & e^{j\omega_d(f)} \\ e^{j2\omega_1(f)} & e^{j2\omega_2(f)} & \vdots & e^{j2\omega_d(f)} \\ \vdots & \vdots & \vdots & \vdots \\ e^{j(L-1)\omega_1(f)} & e^{j(L-1)\omega_2(f)} & \cdots & e^{j(L-1)\omega_d(f)} \end{bmatrix}. \tag{5.5}$$

In principle, it is possible to formulate an optimum estimator for θ_i directly based on the observed spatiotemporal data vector $\mathbf{x}(t)$. Because of the complexity of the resulting algorithm, however, this is rarely done in practice. Instead, the data are usually transformed into the frequency domain to generate an estimate of $\mathbf{P}_x(f)$. Using these reduced data, the spatial frequencies or angles are estimated based on the postulated model in (5.4).

It is common in AOA estimation problems to divide the measurement time into K disjoint intervals, each of length ΔT seconds. We will assume that ΔT is much greater than the maximum propagation delay across the array. Let us denote the measured data over the kth time interval by $\mathbf{x}_k(t)$. When $\mathbf{x}(t)$ is narrow-band, that is, $B_x \ll f_0$, we let $\Delta T = 1/B_x$. If $\mathbf{x}(t)$ is wide-band, we shall assume that ΔT is large enough, so that

$$\frac{1}{\Delta T} \ll f_0 - \frac{B_x}{2}. \tag{5.6}$$

Let $f_j = f_0 \pm \frac{j}{\Delta T}$, j an integer, and $\mathbf{X}_k(f_j)$ be the Fourier transform of $\mathbf{x}_k(t)$ at frequency f_j. It is well known that $\mathbf{X}_k(f_j)$ for different k and/or j are pairwise asymptotically (large ΔT) normally distributed and independent [4], [5]. Furthermore,

$$\mathbf{R}(f_j) \triangleq E\{\mathbf{X}_k(f_j)\mathbf{X}_k^{\dagger}(f_j)\} \simeq \frac{1}{\Delta T}\mathbf{P}_x(f_j). \tag{5.7}$$

$\mathbf{R}(f_j)$ will be called the *spatial correlation matrix* at frequency f_j. Clearly, interest in the frequency components is confined to those within the bandwidth of the signal. Hence in the narrow-band case, only the component at f_0 is used. For wide-band sources, (5.6) guarantees that $\mathbf{X}_k(f_j)$, $k = 1, \ldots, K$, for $f_0 - B_x/2 \le f_j \le f_0 + B_x/2$, represents spatiotemporal samples of a *narrow-band* plane wave at frequency f_j.

In this chapter we are concerned with the performance of narrow-band direction-finding algorithms. This implies that $\mathbf{X}_k(f_j)$, $k = 1, \ldots, K$, for only one f_j is used in formulating an estimator. Consequently, from now on we omit the argument f_j in denoting the frequency-dependent parameters, vectors, and matrices. In particular, the observed data shall be modeled as K independent samples (snapshots) of zero-mean complex Gaussian vectors $\mathbf{X}_k$, $k = 1, \ldots, K$. A class of AOA estimation algorithms and some performance analyses for the wide-band model, where the data for all f_j in the signal bandwidth are used, can be found in [6] and [7]. A different approach to solving the same problem is given in [8].

5.2.2 Narrow-Band Signal-Subspace Methods

Following the notation established earlier for narrowband signals, and the assumed noise and array model, the spatial covariance matrix **R**, and its maximum-likelihood estimate $\hat{\mathbf{R}}$ are given by

$$\mathbf{R} = \mathbf{R}_0 + \sigma_n^2\mathbf{I}, \tag{5.8}$$

where

$$\mathbf{R}_0 = \mathbf{A}\mathbf{Q}_0\mathbf{A}^\dagger \tag{5.9}$$

is a nonsingular signal-only spatial covariance matrix, with the ijth element of $\mathbf{Q}_0$ given by $E\{s_i(t)s^*_j(t)\}$, and

$$\hat{\mathbf{R}} = \frac{1}{K}\sum_{k=1}^{K} \mathbf{X}_k\mathbf{X}_k^\dagger. \tag{5.10}$$

$\hat{\mathbf{R}}$ is the statistic on which the narrow-band signal-subspace AOA estimators are based. In the remainder of the chapter, "^" appearing over a variable shall indicate the estimate of the variable based on $\hat{\mathbf{R}}$.

To describe the general approach to signal-subspace processing, we begin by decomposing **R** in terms of its eigenvalues, λ_i, and orthonormal eigenvectors, e_i. Thus,

$$\mathbf{R} = \sum_{i=1}^{L} \lambda_i\mathbf{e}_i\mathbf{e}_i^\dagger, \tag{5.11}$$

where $\lambda_1 > \lambda_2 > \cdots > \lambda_d > \lambda_{d+1} = \cdots = \lambda_L = \sigma_n^2$. $\lambda_i(\mathbf{e}_i)$, $i = 1, \ldots, d$, are known as the signal-subspace eigenvalues (eigenvectors) and $\lambda_i(\mathbf{e}_i)$, $i = d+1, \ldots, L$, are called the noise-subspace eigenvalues (eigenvectors). Define the matrices $\mathbf{E}_s$ and $\mathbf{E}_N$ as

$$\mathbf{E}_s \underline{\underline{\Delta}} [\mathbf{e}_1, \mathbf{e}_2, \ldots, \mathbf{e}_d]$$

and

$$\mathbf{E}_N \underline{\underline{\Delta}} [\mathbf{e}_{d+1}, \mathbf{e}_{d+2}, \ldots, \mathbf{e}_L].$$

The column spans of $\mathbf{E}_s$ and $\mathbf{E}_N$ are the signal- and noise-subspaces, respectively.

Two observations follow immediately.

1. The span of $\mathbf{E}_s$ = column span of **A**. So,

$$\|\mathbf{A}_i^\dagger\mathbf{E}_s\|^2 = \sum_{j=1}^{d} |\mathbf{A}_i^\dagger\mathbf{e}_j|^2 = L, \qquad i = 1, \ldots, d, \tag{5.12}$$

where $\|\mathbf{A}_i\|$ denotes the Euclidean norm of the vector $\mathbf{A}_i$.

2. $\mathbf{A}_i$, $i = 1, \ldots, d$ are orthogonal to the span of $\mathbf{E}_N$. That is,

$$\|\mathbf{A}_i^\dagger \mathbf{E}_N\| = 0, \qquad i = 1, \ldots, d. \tag{5.13}$$

The deterministic principles of signal-subspace processing for angle (or frequency) determination can now be stated. For the linear uniform array, it is as follows:

> Given $\mathbf{R}$ and d, find an appropriate vector or set of vectors belonging to the span of $\mathbf{E}_N$. Project $\mathbf{V}(\omega) \triangleq 1/\sqrt{L}\,[1, e^{j\omega}, \ldots, e^{j(L-1)\omega}]^T$ on the noise-subspace vector(s) for all ω. The d values of ω for which the projection(s) are zero are the desired spatial frequencies.

What is clear from the foregoing statement is that any vector $\in$ span $\mathbf{E}_N$ will do the job. The practical estimation problem, however, arises from the fact that only an estimate, $\hat{\mathbf{R}}$, of $\mathbf{R}$ is available. In general $\hat{\mathbf{E}}_s \neq \mathbf{E}_s$, $\hat{\mathbf{E}}_N \neq \mathbf{E}_N$, and $\mathbf{A}_i$ is not orthogonal to $\hat{\mathbf{E}}_N$. Under these conditions it is not at all clear what vector(s) in the *estimated noise subspace* should be used to form the projections. Consequently, several signal-subspace frequency estimators have been proposed in the literature. They require the following steps:

1. Based on $\hat{\mathbf{R}}$, estimate the number of sources. We call this stage of processing the detection stage and denote the estimated number by $\hat{d}$. This issue is addressed in Section 5.5.
2. Determine $\hat{\mathbf{E}}_N$ or alternatively $\hat{\mathbf{E}}_s$.
3. Form a statistic, $\hat{D}(\omega)$, called the ***null spectrum,*** from the projection(s) of $\mathbf{V}(\omega)$ onto appropriately selected vector(s) in the span of $\hat{\mathbf{E}}_N$ or $\hat{\mathbf{E}}_s$. The location of the $\hat{d}$ smallest minima of $D(\omega)$ are the *estimated spatial frequencies.*

Various choices for $\hat{D}(\omega)$ may be found in [1], [2], and [9]–[15]. In Sections 5.7 and 5.8 we investigate the resolving properties of two popular methods: MUSIC [1] and minimum-norm [2]. In the following, we present the expressions for $D(\omega)$ for these two methods, for the plane waves-in-uncorrelated-noise model.

$D(\omega)$ for MUSIC may be expressed in terms of signal- or noise-subspace eigenvectors. These follow directly from (5.12) and (5.13) as

$$D(\omega) = 1 - \mathbf{V}^\dagger(\omega)\left(\sum_{i=1}^{d} \mathbf{e}_i \mathbf{e}_i^\dagger\right)\mathbf{V}(\omega) \tag{5.14}$$

and

$$D(\omega) = \mathbf{V}^\dagger(\omega)\left(\sum_{i=d+1}^{L} \mathbf{e}_i \mathbf{e}_i^\dagger\right)\mathbf{V}(\omega). \tag{5.15}$$

It is clear that the MUSIC estimator uses a *uniform average* of the norms of the projections of $\mathbf{V}(\omega)$ onto the estimated signal- or noise-subspace eigenvectors.

The min-norm $D(\omega)$ uses a vector $\mathbf{h}$ with one of its elements set to unity that is entirely in the noise subspace and that has the minimum Euclidean norm. Then

$$D(\omega) = |\mathbf{V}^{\dagger}(\omega)\mathbf{h}|^2 \tag{5.16}$$

In the following we outline the derivation of a useful formula for $\mathbf{h}$ with its first element set to unity, in terms of signal-subspace quantities. The noise-subspace formula can be found in [2].

Since $\mathbf{h}$ is in the noise-subspace, it must satisfy

$$\mathbf{E}_s^{\dagger}\mathbf{h} = \mathbf{0} \tag{5.17}$$

Partition $\mathbf{E}_s$ and $\mathbf{h}$ as

$$\mathbf{E}_s = \begin{bmatrix} \mathbf{g}^T \\ \mathbf{E}_s' \end{bmatrix} \quad \text{and} \quad \mathbf{h} = \begin{bmatrix} 1 \\ \mathbf{h}' \end{bmatrix}$$

where $\mathbf{g}$ is a vector of the first elements of $\mathbf{e}_i$, $i = 1, \ldots, d$. Substituting for $\mathbf{E}_s$ in (5.17), we have

$$\mathbf{E}_s'^{\dagger}\,\mathbf{h}' = -\,\mathbf{g}^*. \tag{5.18}$$

The minimum-norm solution of (5.18) is given by [16]:

$$\mathbf{h}' = -\,\mathbf{E}_s'(\mathbf{E}_s'^{\dagger}\mathbf{E}_s')^{-1}\mathbf{g}^*. \tag{5.19}$$

But

$$\mathbf{E}_s'^{\dagger}\mathbf{E}_s' = \mathbf{I} - \mathbf{g}^*\mathbf{g}^T. \tag{5.20}$$

Using the *matrix inversion lemma* in (5.20), $(\mathbf{E}_s'^{\dagger}\mathbf{E}_s')^{-1}$ can be expressed as

$$(\mathbf{E}_s'^{\dagger}E_s')^{-1} = I + \frac{\mathbf{g}^*\mathbf{g}^T}{1 - \mathbf{g}^T\mathbf{g}^*}, \tag{5.21}$$

resulting in

$$\mathbf{h}' = \frac{-\,\mathbf{E}_s'\mathbf{g}^*}{1 - \mathbf{g}^{\dagger}\mathbf{g}}. \tag{5.22}$$

It is finally worth noting that neither MUSIC nor min-norm requires uniform linear sampling of the aperture. A more general situation simply requires an appropriate specification for $\mathbf{V}(\omega)$.

5.3. STATISTICS OF THE ESTIMATED EIGENVALUES AND EIGENVECTORS

In this section the asymptotic (large K) first- and second-order statistics of the eigenvalues of $\hat{\mathbf{R}}$ and the normalized eigenvector estimates corresponding to the distinct eigenvalues of $\mathbf{R}$ will be presented. Let $\hat{\lambda}_i > \hat{\lambda}_{i+1}$ be the ordered eigenvalues of $\hat{\mathbf{R}}$.

The asymptotic statistics of $\hat{\lambda}_i$, $i = 1, \ldots, d$, and the associated eigenvectors are based on the work of Anderson [17] and its extensions by Gupta [18] and Brillinger [4]. The derivations in these references, however, are for the unnormalized eigenvector estimates. In the following, Wilkinson's perturbation theory [19] is used in conjunction with the results in [4] to derive the statistics for the *normalized* eigenvector estimates (see also [20] and [21]). Computational formulas for the distributions of all the eigenvalues of $\hat{\mathbf{R}}$ are reviewed in [22]. In the derivation of the probability of overdetermining the number of sources, we need the asymptotic mean, variance, and the probability density function of $\hat{\lambda}_{d+1}$, which is, to a very good approximation, the largest eigenvalue of a *central complex Wishart-distributed matrix with its average proportional to the identity matrix.* An algorithm for obtaining the percentage points for this eigenvalue is given in [23]. Later in this section we present empirical models for the asymptotic statistics of $\hat{\lambda}_{d+1}$ and compare them with the computational results of [23]. We now proceed with a summary of the statistics of the d largest eigenvalues of $\hat{\mathbf{R}}$ and their associated normalized eigenvectors.

5.3.1 Statistics of $\hat{\lambda}_i$ and $\hat{\mathbf{e}}_i$, $i = 1, \ldots, d$

It is well known that $\hat{\mathbf{R}}$ has a complex Wishart distribution with elements that are asymptotically jointly normal. Furthermore, $\hat{\lambda}_i$ are asymptotically normal and independent of $\hat{\mathbf{e}}_j$, $i, j = 1, \ldots, d$, and

$$E\{\hat{\lambda}_i\} = \lambda_i + o(K^{-1}) \tag{5.23}$$

$$\text{cov}\,\{\hat{\lambda}_i, \hat{\lambda}_j\} = \delta_{ij}\lambda_i^2/K + o(K^{-2}) \tag{5.24}$$

where $\text{cov}\,\{a, b\} = E\{ab^*\} - E\{a\}E\{b^*\}$ and δ_{ij} is the Kronecker delta.

$\hat{\mathbf{R}}$ may be represented in terms of a random perturbation to $\mathbf{R}$. Denoting the perturbation factor by p, $0 < p \ll 1$, $\hat{\mathbf{R}}$ is given by

$$\hat{\mathbf{R}} = \mathbf{R} + p\mathbf{B}, \tag{5.25}$$

where $\mathbf{B} = (\hat{\mathbf{R}} - \mathbf{R})/p$ is a Hermitian, zero-mean random matrix with elements that are jointly normal. The following result is needed later. Let χ_i, $i = 1, 2, 3, 4$, be four complex vectors with appropriate dimensions. Using the expansion of the fourth-order moment of jointly normal random variables, and the asymptotic statistics of $\hat{\mathbf{R}}$, it can easily be shown that (see [4, p. 114])

$$E\{(\chi_1^\dagger\,\mathbf{B}\,\chi_2)(\chi_3^\dagger\,\mathbf{B}\,\chi_4)\} = \frac{(\chi_1^\dagger\,\mathbf{R}\,\chi_3)(\chi_4^\dagger\,\mathbf{R}\,\chi_2)}{Kp^2} \tag{5.26}$$

Let $\tilde{\mathbf{e}}_i$ denote the unnormalized eigenvector given in a perturbation expansion given by

$$\tilde{\mathbf{e}}_i = \mathbf{e}_i + \sum_{\substack{j=1 \\ j \neq i}}^{L} \left(\sum_{k=1}^{\infty} t_{kj}^{(i)}\, p^k \right) \mathbf{e}_j, \tag{5.27}$$

where $t_{kj}^{(i)}$, $k = 1, 2, \ldots$, are the coefficients of the perturbation expansion of $\tilde{\mathbf{e}}_i$ along $\mathbf{e}_j$. The following steps develop the lowest order perturbation expression for the normalized estimate of the ith eigenvector. Using the orthonormality of $\mathbf{e}_i$ and keeping the term with the lowest order of p, we have

$$\|\tilde{\mathbf{e}}_i\|^2 \simeq 1 + \sum_{\substack{j=1 \\ j \neq i}}^{L} |t_{1j}^{(i)}|^2 p^2, \tag{5.28}$$

$$\|\tilde{\mathbf{e}}_i\|^{-1} \approx 1 - \frac{1}{2} \sum_{\substack{j=1 \\ j \neq i}}^{L} |t_{1j}^{(i)}|^2 p^2, \tag{5.29}$$

$$\hat{\mathbf{e}}_i = \tilde{\mathbf{e}}_i \, \|\tilde{\mathbf{e}}_i\|^{-1} \approx \Big(1 - \frac{1}{2} \sum_{\substack{j=1 \\ j \neq i}}^{L} |t_{1j}^{(i)}|^2 p^2\Big)\mathbf{e}_i + \sum_{\substack{j=1 \\ j \neq i}}^{L} t_{1j}^{(i)} \, p\mathbf{e}_j + \sum_{\substack{j=1 \\ j \neq i}}^{L} t_{2j}^{(i)} p^2 \mathbf{e}_j. \tag{5.30}$$

The expressions for $t_{kj}^{(i)}$ are obtained by substitution of (5.30) into the perturbed equation for the eigenvalue problem. Then

$$t_{1j}^{(i)} = \frac{\mathbf{e}_j^\dagger \, \mathbf{B} \, \mathbf{e}_i}{\lambda_j - \lambda_i}, \qquad j \neq i \tag{5.31}$$

and

$$t_{2j}^{(i)} = \frac{(\mathbf{e}_i^\dagger \, \mathbf{B} \, \mathbf{e}_i)(\mathbf{e}_j^\dagger \, \mathbf{B} \, \mathbf{e}_i)}{(\lambda_j - \lambda_i)^2} - \sum_{\substack{k=1 \\ k \neq i}}^{L} \frac{(\mathbf{e}_j^\dagger \, \mathbf{B} \, \mathbf{e}_k)(\mathbf{e}_k^\dagger \, \mathbf{B} \, \mathbf{e}_i)}{(\lambda_j - \lambda_i)(\lambda_k - \lambda_i)}. \tag{5.32}$$

The expected value of $\hat{\mathbf{e}}_i$ in (5.30) only involves $E\{|t_{1j}^{(i)}|^2\}$ and $E\{t_{2j}^{(i)}\}$, as $E\{t_{1j}^{(i)}\} = 0$ from (5.31). Using (5.26) and the fact that $\mathbf{e}_i$ are eigenvectors of $\mathbf{R}$ (i.e., $\mathbf{e}_i^\dagger \mathbf{R} \mathbf{e}_j = \lambda_i \delta_{ij}$), it follows that

$$E\{|t_{1j}^{(i)}|^2\} = \frac{\lambda_i \lambda_j}{(\lambda_i - \lambda_j)^2 \, kp^2}, \qquad i \neq j \tag{5.33}$$

$$E\{t_{2j}^{(i)}\} = 0. \tag{5.34}$$

Thus,

$$E\{\hat{\mathbf{e}}_i\} \approx \mathbf{e}_i - \frac{1}{2} \sum_{\substack{j=1 \\ j \neq i}}^{L} \frac{\lambda_i \lambda_j}{(\lambda_i - \lambda_j)^2 K} \mathbf{e}_i. \tag{5.35}$$

The approximation of order K^{-1} to the central second-order moments of $\hat{\mathbf{e}}_i$ are the same as those given by Brillinger [4, p. 454] for $\tilde{\mathbf{e}}_i$. This can be seen from the following. Using (5.29) in the definition of $\hat{\mathbf{e}}_i$, we obtain

$$\text{cov}\,\{\hat{\mathbf{e}}_i, \hat{\mathbf{e}}_j\} \approx \text{cov}\,\{\tilde{\mathbf{e}}_i, \tilde{\mathbf{e}}_j\}$$

$$-\frac{1}{2}\left[\operatorname{cov}\left\{\sum_{\substack{k=1\\k\neq i}}^{L}|t_{1k}^{(i)}|^2\,p^2\,\tilde{\mathbf{e}}_i,\ \tilde{\mathbf{e}}_j\right\}\right.$$

$$\left.+\operatorname{cov}\left\{\tilde{\mathbf{e}}_i,\ \sum_{\substack{k=1\\k\neq j}}^{L}|t_{1k}^{(i)}|^2\,p^2\tilde{\mathbf{e}}_j\right\}\right]$$

$$+\frac{1}{4}\operatorname{cov}\left\{\sum_{\substack{k=1\\k\neq i}}^{L}|t_{1k}^{(i)}|^2\,\tilde{\mathbf{e}}_i,\ \sum_{\substack{k=1\\k\neq j}}^{L}|t_{1k}^{(i)}|^2\,\tilde{\mathbf{e}}_j\right\}p^4, \tag{5.36}$$

where

$$\operatorname{cov}\{\chi_i, \chi_j\} = E\{\chi_i\,\chi_j^\dagger\} - E\{\chi_i\}\,E\{\chi_j^\dagger\}.$$

Careful examination of the last three terms in the foregoing equation shows that they involve third- or higher-order moments of elements of the zero-mean complex Guassian matrix $\mathbf{B}$. Thus, these terms contribute zero (for odd orders) or quantities of order less than or equal to K^{-2}. It follows that

$$\operatorname{cov}\{\hat{\mathbf{e}}_i, \hat{\mathbf{e}}_j\} = \operatorname{cov}\{\tilde{\mathbf{e}}_i, \tilde{\mathbf{e}}_j\} + o(K^{-2}). \tag{5.37}$$

Thus the covariance expressions from [4, p. 343] may be used for the normalized eigenvectors. Under an order K^{-1} approximation, these are given by

$$\operatorname{cov}\{\hat{\mathbf{e}}_i, \hat{\mathbf{e}}_j\} \approx \sum_{\substack{k=1\\k\neq i}}^{L}\frac{\lambda_i\lambda_k}{(\lambda_i-\lambda_k)^2K}\,\mathbf{e}_k\mathbf{e}_k^\dagger\delta_{ij} \tag{5.38}$$

$$\operatorname{cov}\{\hat{\mathbf{e}}_i, \hat{\mathbf{e}}_j^*\} \simeq -\frac{\lambda_i\lambda_j}{(\lambda_i-\lambda_j)^2K}\,\mathbf{e}_j\mathbf{e}_i^{\mathrm{T}}(1-\delta_{ij}). \tag{5.39}$$

It is useful to make the asymptotic statistics of the errors in estimating $\mathbf{e}_i$, $i = 1, \ldots, d$ explicit. Therefore, let $\hat{\mathbf{e}}_i = \mathbf{e}_i + \boldsymbol{\eta}_i$. The order K^{-1} approximations of the first- and second-order statistics of $\boldsymbol{\eta}_i$ follow directly from (5.35), (5.38), and (5.39) to be

$$E\{\boldsymbol{\eta}_i\} \simeq -\frac{\lambda_i}{2K}\sum_{\substack{k=1\\k\neq i}}^{L}\frac{\lambda_k}{(\lambda_i-\lambda_k)^2}\,\mathbf{e}_i \triangleq c_i\mathbf{e}_i \tag{5.40}$$

$$E\{\boldsymbol{\eta}_i\boldsymbol{\eta}_j^\dagger\} \simeq \frac{\lambda_i}{K}\sum_{\substack{k=1\\k\neq i}}^{\mathrm{L}}\frac{\lambda_k}{(\lambda_i-\lambda_k)^2}\,\mathbf{e}_k\mathbf{e}_k^\dagger\,\delta_{ij}, \tag{5.41}$$

$$E\{\boldsymbol{\eta}_i\boldsymbol{\eta}_j^T\} \simeq -\frac{\lambda_i\lambda_j}{K(\lambda_i-\lambda_j)^2}\,\mathbf{e}_j\mathbf{e}_i^{\mathrm{T}}(1-\delta_{ij}). \tag{5.42}$$

It is also worth noting that (5.40) is consistent with (5.41) and the normalization constraint on $\|\hat{\mathbf{e}}_i\|$ and $\|\mathbf{e}_i\|$, as can be seen in the following. The constraint results in

$$\|\hat{\mathbf{e}}_i\|^2 = 1 + \mathbf{e}_i^\dagger \boldsymbol{\eta}_i + \boldsymbol{\eta}_i^\dagger \mathbf{e}_i + \boldsymbol{\eta}_i^\dagger \boldsymbol{\eta}_i = 1, \tag{5.43}$$

which gives

$$2\,\mathrm{Re}[\mathbf{e}_i^\dagger \boldsymbol{\eta}_i] = -\boldsymbol{\eta}_i^\dagger \boldsymbol{\eta}_i. \tag{5.44}$$

Taking the expectation of (5.44) and retaining terms of order K^{-1} results in

$$2\,\mathrm{Re}\,[\mathbf{e}_i^\dagger E\{\boldsymbol{\eta}_i\}] \simeq -\frac{\lambda_i}{K} \sum_{\substack{k=1 \\ k \neq i}}^{L} \frac{\lambda_k}{(\lambda_i - \lambda_k)^2}\, \mathbf{e}_k^\dagger \mathbf{e}_k = \frac{-\lambda_i}{K} \sum_{\substack{k=1 \\ k \neq i}}^{L} \frac{\lambda_k}{(\lambda_i - \lambda_k)^2}. \tag{5.45}$$

Clearly, the expression for $E\{\boldsymbol{\eta}_i\}$ in (5.40) satisfies the constraint in (5.45).

Empirical Statistics of $\hat{\lambda}_{d+1}$

In Section 5.5 we show that the probability of overdetermining the number of sources is dictated by the statistics of $\hat{\lambda}_{d+1}$. Asymptotically, the statistics of $\hat{\lambda}_1, \ldots, \hat{\lambda}_d$ and $\hat{\lambda}_{d+1}, \ldots, \hat{\lambda}_L$ decouple. So, in this sense the behavior of $\hat{\lambda}_{d+1}$ is the same as that of the largest eigenvalue of a central complex Wishart-distributed matrix with its average proportional to the identity matrix. A computational scheme for obtaining the (nonasymptotic) percentage points of this eigenvalue is given in [23]. However, we are interested in the mean and standard deviation of $\hat{\lambda}_{d+1}$ as well as a closed-form expression for its probability density function. Consequently we proceed to approximate the density of $\hat{\lambda}_{d+1}$ by a Gaussian with mean and variance given by their sample values obtained through 5000 trials [24]. In the following we give some examples of these results and compare the empirical percentage points with those given in [23].

Figure 5.4 shows plots of the probability distribution for $\hat{\lambda}_{d+1}$ obtained empirically, from a Gaussian model with mean and variance given by their sample estimates and by the algorithm given in [23], when $L - d = 14$ and $K = 100$. The Gaussian model and the histogram for this case are shown in Fig. 5.5. The computations given in [23] for values of $L - d > 8$ require double-precision arithmetic on a large mainframe computer. Another example is shown in Fig. 5.6, in which the theoretical asymptotic density for $L - d = 2$, which is derived from the joint density of $\hat{\lambda}_{d+1}, \ldots, \hat{\lambda}_L$ as given in [18], is compared with the empirical model.

The preceding examples and many others that we have investigated demonstrate that the Gaussian model is adequate for the characterization of the order determination schemes as carried out in Section 5.5. Figures 5.7 and 5.8 give examples of the asymptotic mean and the standard deviation of $\hat{\lambda}_{d+1}$. Based on our empirical results, we model the mean of λ_{d+1} by $\{1 + [b(L - d, K)]/\sqrt{K}\}\sigma_n^2$ and the variance of $\hat{\lambda}_{d+1}$ by $[c(L - d, K)/K]\sigma_n^4$. $b(1 - d, K)$ is on the order of $2\sqrt{L - d}$ and $0 < c\,(L - d, K) < 1$.

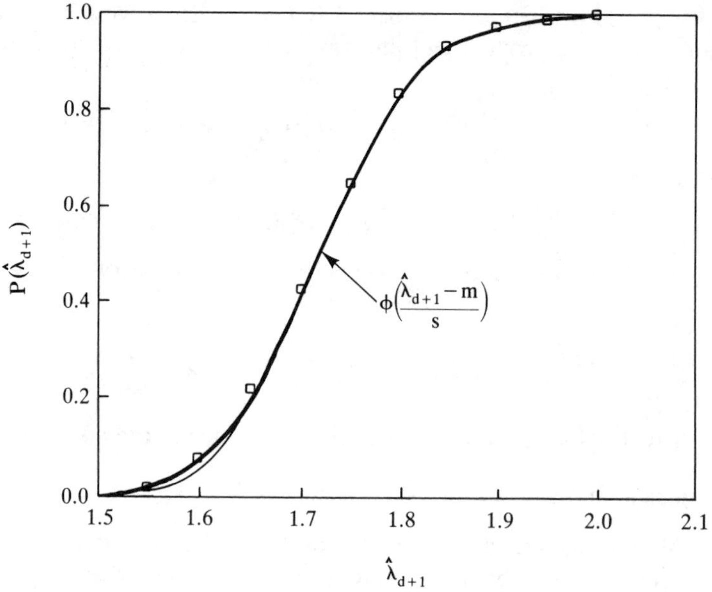

Figure 5.4 Empirical, Gaussian model, and computed exact (□) probability distributions for $\hat{\lambda}_{d+1}$; $L = 16$, $d = 2$, $K = 100$.

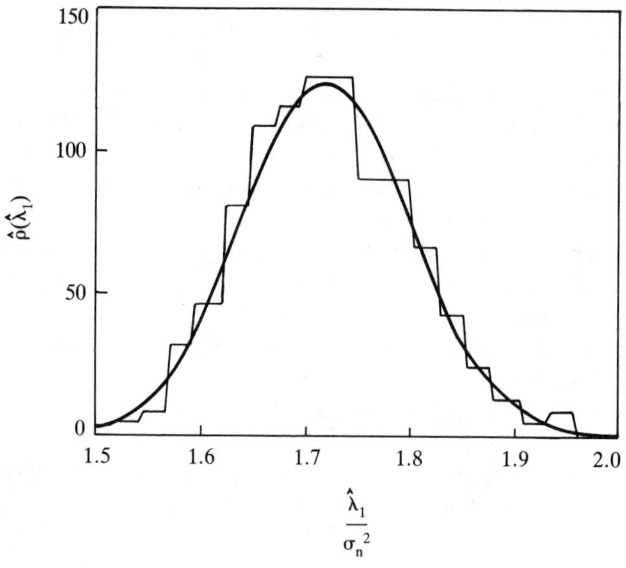

Figure 5.5 The histogram and the Gaussian model for the asymptotic probability density function of $\hat{\lambda}_{d+1}$; $L = 16$, $d = 2$, $K = 100$.

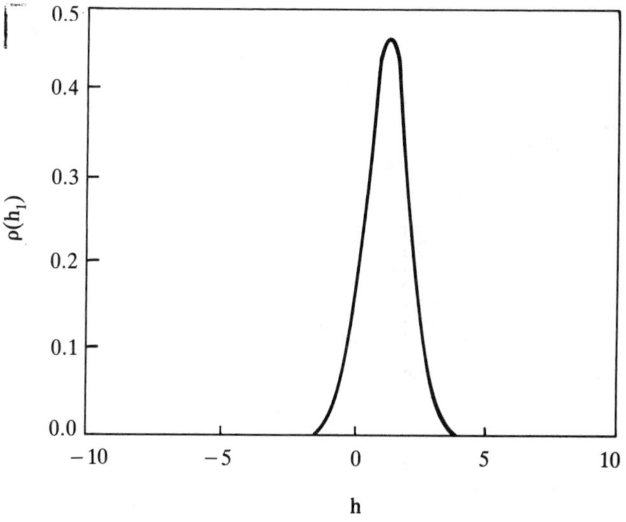

Figure 5.6 The theoretical and the Gaussian model for the asymptotic probability density function of h_1, where $h_1 = \sqrt{K}\,(\hat{\lambda}_{d+1} - \sigma_n^2)$; $L = 4$, $d = 2$.

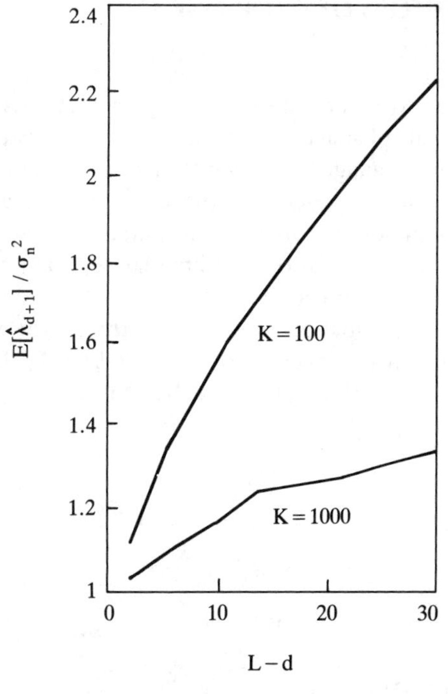

Figure 5.7 Simulation results for the asymptotic mean of $\hat{\lambda}_{d+1}$.

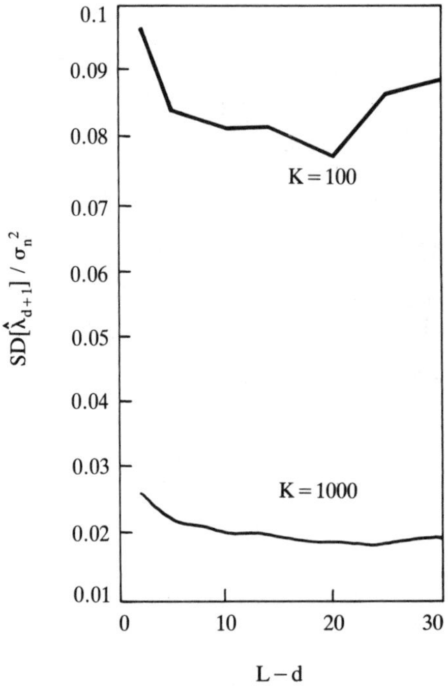

Figure 5.8 Simulation results for the asymptotic standard deviation of $\hat{\lambda}_{d+1}$.

5.4. EIGENVALUES AND EIGENVECTORS OF R FOR ONE AND TWO SOURCES

Analytical evaluation of the performance of signal-subspace methods requires closed-form expressions for the signal-subspace eigenvalues and eigenvectors of **R**. These, however, may be obtained in a manageable and informative form only for $d = 1$ and 2. In this section we derive the expression for these eigenvalues and eigenvectors and explicitly show their dependence on physical parameters such as signal and noise powers, array length, and correlation and angular separation of the sources. This development is similar to that given in [25].

Let $\lambda_i' = \lambda_i - \sigma_n^2$ be the ith signal-only eigenvalue, that is, the ith eigenvalue of $\mathbf{R}_0 \cdot \mathbf{e}_i i$, $i = 1, \ldots, d$, is the ith eigenvector of **R** as well as $\mathbf{R}_0$. Let $\mathbf{V}(\omega_i)$ be the normalized direction vector of the ith source that has an average power of P_i. For $d = 1$, it is clear that

$$\lambda_i' = LP_i, \tag{5.46}$$

and

$$\mathbf{e}_i = \mathbf{V}(\omega_i). \tag{5.47}$$

For the two-signal case we have from (5.9)

$$\mathbf{R}_0 = \mathbf{A}\begin{bmatrix} P_1 & \sqrt{P_1P_2}\psi^* \\ \sqrt{P_1P_2}\psi & P_2 \end{bmatrix}\mathbf{A}^H, \tag{5.48}$$

where the source correlation coefficient, ψ, is given by

$$\psi = \frac{E[s_1^*(t)\, s_2(t)]}{\sqrt{P_1 P_2}}.$$

Evaluation of the nonzero eigenvalues and the associated eigenvectors of $\mathbf{R}_0$ is simplified considerably by noting that each eigenvector is a linear combination of the two direction vectors. Let $\mathbf{e}_i'$ be the ith unnormalized eigenvector given by

$$\mathbf{e}_i' = \mathbf{A}_1 + \kappa_i \mathbf{A}_2, \qquad i = 1, 2,$$

where κ_i is an appropriate constant. Since $\mathbf{e}_i'$ is proportional to $\mathbf{e}_i$, we have

$$\mathbf{R}_0 \mathbf{e}_i' = \lambda_i' \mathbf{e}_i'. \tag{5.49}$$

Substituting the expressions for $\mathbf{R}_0$ and $\mathbf{e}_i'$ in (5.49) results in

$$\begin{aligned}(LP_1 + P_1\kappa_i L\phi + \sqrt{P_1P_2}\,\psi^* L\phi^* + \kappa_i\sqrt{P_1P_2}\,\psi^* L)\mathbf{A}_1 + (\sqrt{P_1P_2}\,\psi L \\ + \kappa_i\sqrt{P_1P_2}\,\psi L\phi + P_2 L\phi^* + P_2\kappa_i L)\mathbf{A}_2 = \lambda_i'\mathbf{A}_1 + \lambda'_i\kappa_i A_2,\ i = 1, 2,\end{aligned} \tag{5.50}$$

where $\phi = \mathbf{V}^H(\omega_1)\mathbf{V}(\omega_2)$ is the cosine of the angle between the two signal vectors.

The eigenvalues can be obtained by equating the coefficients of $\mathbf{A}_1$ and $\mathbf{A}_2$ (5.50), for any i. Thus, suppressing the index i we have

$$\begin{aligned} LP_1 + \sqrt{P_1P_2}\,L\psi^*\phi^* + (LP_1\phi + \sqrt{P_1P_2}\,L\psi^*)\kappa &= \lambda' \\ L\phi^* P_2 + \sqrt{P_1P_2}\,L\psi + (LP_2 + \sqrt{P_1P_2}\,L\psi\phi)\kappa &= \lambda'\kappa. \end{aligned} \tag{5.51}$$

Solving the preceding set of equations for λ' and κ yields

$$\lambda_{1,2}' = \frac{L}{2}\,[P_1 + P_2 + 2\sqrt{P_1P_2}\,\mathrm{Re}(\phi\psi)] \times \left[1 \overset{+}{(-)} \sqrt{1 - \frac{4P_1P_2(1 - |\phi|^2)(1 - |\psi|^2)}{(P_1 + P_2 + 2\sqrt{P_1P_2}\,\mathrm{Re}(\phi\psi)^2}}\,\right], \tag{5.52}$$

and

$$\kappa_i = \frac{\lambda_i' - L(P_1 + \sqrt{P_1P_2}\,\psi^*\phi^*)}{L(P_1\phi + \sqrt{P_1P_2}\,\psi^*)}, \qquad i = 1, 2. \tag{5.53}$$

In the performance analyses of this chapter we will be interested in situations involving two equipowered, uncorrelated sources with fraction of a beamwidth angular separations. In the following we obtain the resulting simple expressions for λ_i' and $\mathbf{e}_i$, $i = 1, 2$.

Let $P_1 = P_2$, $\omega_d = (\omega_1 - \omega_2)/2$, $\psi = 0$, and $\Delta^2 = L^2\omega_d^2/3$. Furthermore, define centered normalized direction vectors $\mathbf{U}i = e^{-j[(L-1)\omega i/2]}$. Then

$$\phi = \frac{1}{L}\, e^{-j(L-1)\omega_d}\, \frac{\sin(L\omega_d)}{\sin(\omega_d)}. \tag{5.54}$$

For $\Delta^2 \ll 1$ and $L^2 \gg 1$, ϕ may be expanded as

$$\phi = e^{-j(L-1)\omega_d}\left(1 - \frac{\Delta^2}{2} + 3\frac{\Delta^4}{40} - \cdots\right). \tag{5.55}$$

The eigenvalues and eigenvectors follow easily as

$$\lambda'_{1,(2)} \simeq P_1 L[1 \overset{+}{\underset{(-)}{}} |\phi|] \simeq P_1 L[1 \overset{+}{\underset{(-)}{}} (1 - \frac{\Delta^2}{2} + \frac{3\Delta^4}{40})] \tag{5.56}$$

and

$$\mathbf{e}_{1,(2)} = \frac{\mathbf{U}_1 \overset{+}{\underset{(-)}{}} \mathbf{U}_2}{\sqrt{2[1 \overset{+}{\underset{(-)}{}} |\phi|]}}. \tag{5.57}$$

Once again, substituting the approximate value of $|\phi|$ in (5.57) we obtain

$$\mathbf{e}_1 \simeq \frac{1}{2}\left(1 + \frac{\Delta^2}{8} - \frac{3\Delta^4}{160}\right)(\mathbf{U}_1 + \mathbf{U}_2),$$

$$\mathbf{e}_2 \simeq \frac{1}{\Delta}\left(1 + \frac{3\Delta^2}{40}\right)(\mathbf{U}_1 - \mathbf{U}_2). \tag{5.58}$$

5.5. DETECTION: DETERMINATION OF NUMBER OF SOURCES

In Section 5.2 it was shown that the $L - d$ smallest eigenvalues of $\mathbf{R}$ are equal. Thus, the problem of determining the number of sources, given the observation samples X_k, $k = 1, 2, \ldots, K$, or the estimated spatial covariance matrix, can be treated as that of testing the multiplicity of the smallest eigenvalues of $\hat{\mathbf{R}}$. A statistical multiple-hypothesis testing procedure, known as the *Lawley-Bartlett test* for factor analysis and principal component analysis in statistics, has been used by many investigators. However, this approach has a practical limitation in that it requires a proper set of threshold levels for dependent sequential testing. Reference [26] contains a detailed discussion of this approach in the context of direction finding.

An alternative approach is provided by applying the general model-fitting theory to order determination. Using an information theoretic criterion Akaike [27] developed a method, called *an information criterion (AIC)*,* for parametric model fitting to observed data. Based on the point of view of the *minimum* (shortest) *data description* (*MDL*), Rissanen [28] also developed a similar method. The direct application of these techniques to the problem of determining the number of sources with the model given in Section 5.2 would be very difficult, mainly due to the complicated nature of likelihood maximization over all the parameters specifying the model.

Wax and Kailath [29] applied AIC and MDL methods to the problem of determining the number of sources. A pre-overparameterization of the model in terms of

*AIC is also referred to as the *minimum information-theoretic criterion estimation* (*MAICE*).

the eigenvalues and eigenvectors of **R** was used to reduce the computational complexity of the methods. As we may expect, the overparameterization causes some degradation of the detection performances, especially in the wide-band signal situations [6]. For the narrow-band model treated here, the methods give the estimate of the number of sources, $\hat{d}$, by minimizing the following function over d,

$$\Lambda(d, p, K) \triangleq K(L - d) \log\left[\frac{a(d)}{q(d)}\right] + p(d, K), \tag{5.59}$$

where

$$a(d) \triangleq \frac{1}{L - d} \sum_{i = d + 1}^{L} \hat{\lambda}_i \tag{5.60}$$

$$q(d) \triangleq \left(\prod_{i = d + 1}^{L} \hat{\lambda}_i \right)^{1/(L - d)} \tag{5.61}$$

and $p(d, K)$ is a *penalty function* for the overdetermination of the number of sources with the overparametrized model and log denotes the natural logarithm. This function is given by

$$p(d, K) = d(2L - d) \tag{5.62}$$

for the AIC method and

$$p(d, K) = \frac{1}{2} d(2L - d) \log K \tag{5.63}$$

for the MDL method. It is easy to see that the MDL method places a heavier penalty on the overdetermination of the number of sources.

In the following we examine the detection performance of methods that estimate d by minimizing Λ in (5.59) in general and the AIC and MDL estimators in particular. We pay special attention to the probabilities of underestimating and overestimating the number of sources for the case of up to two closely spaced sources [30], [24]. The basic analytical methods used here can also be used to examine the detection performance for more general scenarios. The one-source case is examined in [30].

Let H_2 denote the hypothesis that the number of sources is $d = 2$. We derive the probability of underestimating the number of sources given H_2, defined by

$$P_M(2) = P\{(\hat{d} < 2) | H_2\}, \tag{5.64}$$

and the probability of overestimating the number of sources given H_2, defined by

$$P_F(2) = P\{(\hat{d} > 2) | H_2\}. \tag{5.65}$$

Since Λ is a smooth function of d and convex, as observed experimentally, we have

$$P_M(2) = P\{\Lambda(1, p, K) < \Lambda(2, p, K) | H_2\} \tag{5.66}$$

and

$$P_F(2) = P\{\Lambda(3, p, K) < \Lambda(2, p, K)|H_2\}. \tag{5.67}$$

By the definitions of $a(d)$ and $q(d)$ given by (5.60) and (5.61), respectively, it follows that

$$a(1) = \frac{1}{L-1}\hat{\lambda}_2 + \frac{L-2}{L-1}a(2) \tag{5.68}$$

and

$$[q(1)]^{L-1} = \hat{\lambda}_2\ [q(2)]^{L-2}. \tag{5.69}$$

Thus we have

$$\begin{aligned}
(L-1)\log\left[\frac{a(1)}{q(1)}\right] &= \log\frac{[a(1)]^{L-1}}{[q(1)]^{L-1}} \\
&= \log\frac{\left[\frac{1}{L-1}\hat{\lambda}_2 + \frac{L-2}{L-1}a(2)\right]^{L-1}}{\hat{\lambda}_2[q(2)]^{L-2}} \\
&= \log\left\{\frac{[a(2)]^{L-2}}{[q(2)]^{L-2}}\cdot\frac{a(2)\left[\frac{L-2}{L-1}+\frac{\hat{\lambda}_2}{(L-1)a(2)}\right]^{L-1}}{\hat{\lambda}_2}\right\} \\
&= (L-2)\log\frac{a(2)}{q(2)} + \log Q_M
\end{aligned} \tag{5.70}$$

where

$$Q_M \triangleq \left\{\frac{L-2}{L-1} + \frac{1}{L-1}[\hat{\lambda}_2/a(2)]\right\}^{L-1}/[\hat{\lambda}_2/a(2)] \tag{5.71}$$

Similarly we have

$$(L-3)\log\frac{a(3)}{q(3)} = (L-2)\log\frac{a(2)}{q(2)} + \log Q_F \tag{5.72}$$

where

$$Q_F \triangleq [\frac{L-2}{L-3} - \frac{1}{L-3}[\hat{\lambda}_3/a(2)]]^{L-3}[\hat{\lambda}_3/a(2)]. \tag{5.73}$$

Using (5.70) and (5.71) we can now express $P_M(2)$ and $P_F(2)$ in terms of Q_M, Q_F, and the chosen penalty function $p(d, K)$ as

$$\begin{aligned}
P_M(2) &= P\{\Lambda(1, p, K) < \Lambda(2, p, K)|H_2\} \\
&= P\{\log Q_M < \frac{p(2, K) - p(1, K)}{K}|H_2\}
\end{aligned} \tag{5.74}$$

and

$$P_F(2) = P\{\Lambda(3, p, K) < \Lambda(2, p, K) | H_2\}$$
$$= P\{-\log Q_F > \frac{p(3, K) - p(2, K)}{K} | H_2\}. \tag{5.75}$$

We now proceed to derive approximate expressions for the probabilities of miss and false alarm. We assume that K is large enough so that asymptotic statistics for the eigenvalues can be used. We also let $L \gg 2$. Then as discussed in [30], to a good approximation $\hat{\lambda}_i/a(2)$ may be replaced by $\hat{\lambda}_i/\sigma_n^2$ in (5.74) and (5.75). Also if X is a zero-mean random variable with standard deviation much smaller than one, we may use the following approximation in our derivations:

$$\log(1 + x) \approx x - \frac{1}{2}x^2.$$

To obtain an expression for the *miss probability,* let $\eta \equiv \lambda_2 - \sigma_n^2$ and $\hat{\lambda}_2/\lambda_2 = 1 + \beta$. Then, β is asymptotically normal with mean zero and variance $1/K$ (see Section 5.3). Since the interesting detection performance region is near the threshold, we may assume the combination of the signal-to-noise ratio, source correlation, and angular separation between the sources to be such that $\eta \leq (L - 1)\sigma_n^2$ (see Section 5.4 for the dependence of η on the array size, source correlation, and SNR). Then

$$\log Q_M \simeq \{\frac{\eta}{\sigma_n^2}[1 - \frac{\eta}{2\sigma_n^2(L-1)}] - \log(\frac{\lambda_2}{\sigma_n^2})\} + [\frac{\lambda_2}{\sigma_n^2} - \frac{\lambda_2\eta}{(L-1)\sigma_n^4} - 1]\beta$$
$$+ \frac{1}{2}[1 - \frac{\lambda_2{}^2}{\sigma_n^4(L-1)}]\beta^2. \tag{5.76}$$

Let β_1 and β_2, $\beta_1 \geq \beta_2$, be the roots of the equation

$$\log Q_M - \frac{p(2, K) - p(1, K)}{K} = 0$$

that under the approximation in (5.76) is quadratic in β. If the coefficient of β^2 is positive, we have

$$P_M(2) \simeq P[\beta_2 < \beta < \beta_1] = \mathrm{Erf}(\sqrt{K}\,\beta_1) - \mathrm{Erf}(\sqrt{K}\,\beta_2), \tag{5.77}$$

where

$$\mathrm{Erf}(x) \triangleq \frac{1}{\sqrt{2\pi}} \int_{-\infty}^{x} e^{-\frac{u^2}{2}}\, du.$$

If the coefficient of β^2 is negative, then

$$P_M(2) \simeq \mathrm{Erf}(\sqrt{K}\,\beta_2) - \mathrm{Erf}(\sqrt{K}\,\beta_1) + 1. \tag{5.78}$$

Determination of an asymptotic expression for the *false alarm probability* requires the asymptotic probability density functions (PDFs) for $\hat{\lambda}_{d+1}, \ldots, \hat{\lambda}_L$. For the case of $d = 2$ the PDF of $\hat{\lambda}_3$ is needed. We use the empirical results given in Section 5.3 to characterize $\hat{\lambda}_3$.

Let $\bar{\lambda}_3$ be the expected value of $\hat{\lambda}_3$, $\tilde{\lambda}_3 = \hat{\lambda}_3 - \bar{\lambda}_3$, $\bar{\lambda}_3/\sigma_n^2 = \alpha > 1$, and $\hat{\lambda}_3/\sigma_n^2 = \gamma$. The probability density function of $\hat{\lambda}_3$ may be well approximated by a Gaussian density with mean $\{1 + [b(L-2, K)/\sqrt{K}\}\sigma_n^2$ and variance $[c(L-2, K)/K]\sigma_n^4$. Then $\alpha = 1 + [b(M-2, K)/\sqrt{K}]$ and γ is a random variable with mean zero and variance $c(M-2, K)/K$. Since with a high probability $\hat{\lambda}_3 - \sigma_n^2 \ll (L-1)\sigma_n^2$, $\log Q_F$ in (5.75) can be approximated by

$$\log Q_F \simeq [1 - \alpha + \log \alpha - \frac{(1-\alpha)^2}{2(L-1)}] - [1 - \frac{1}{\alpha} + \frac{1-\alpha}{L-3}]\gamma - \frac{1}{2\alpha^2}\gamma^2. \qquad (5.79)$$

Let γ_1 and γ_2, $\gamma_1 \geq \gamma_2$, be the roots of the equation

$$\log Q_F + \frac{p(3, K) - p(2, K)}{K} = 0,$$

where $\log Q_F$ is given in (5.79). Then

$$P_F(2) \simeq P[\gamma \leq \gamma_2 \text{ or } \gamma \geq \gamma_1]. \qquad (5.80)$$

Under the normality assumption for γ, $P_F^{(2)}$ is given by

$$P_F(2) \simeq \text{Erf}(\sqrt{\frac{K}{c(L-2, K)}}\gamma_2) - \text{Erf}(\sqrt{\frac{K}{c(L-2, K)}}\gamma_1) + 1. \qquad (5.81)$$

From the results given here as well as those in [30], we may draw the following general conclusions:

1. Given the signal, noise, and array parameters, the MDL penalty function leads to a larger miss probability but a smaller false alarm rate than the AIC penalty function, and neither of the two penalties is "the optimal" in any sense.
2. The false alarm probabilities are very weakly dependent on the signal, noise, and array parameters and are usually small. Thus, the threshold behavior of the detection probability is almost entirely determined by the miss probability threshold.
3. From (5.76) and (5.77) we can see that the probability of underestimating the number of sources given H_2, $P_M(2)$, not only depends on the angular separation, source correlation, and SNRs but also on the array geometry, since all these parameters affect the second largest eigenvalue, λ_2.

In the following we present examples of detection performance for a linear array with uniformly spaced elements and two closely spaced sources. For simplicity we assume that all sensors are omnidirectional with unity gain. Assuming, as observed experimentally, convexity of $\Lambda(d, P, K)$ as a function of d with a high probability, we define the ***probability of detection error,*** given H_2, by

$$P_E(2) = P_M(2) + P_F(2). \tag{5.82}$$

Figures 5.9–5.12 show $P_E(2)$ as function of signal-to-noise ratio, the number of snapshots, angle separation, and source correlation for a linear array with $L = 16$ sensors with half-wavelength element spacing. The two sources are assumed to have the same signal-to-noise ratio and $\theta_1 = -\theta_2$. Figure 5.9 gives a comparison of the theoretical and simulation results on $P_E(2)$ for two uncorrelated sources. We suspect the 1- or 2-dB difference between the onset of threshold for the theoretical and experimental results to be due to the asymptotic statistics that were used. For example, it is known that $a(2)$ is a biased estimator of σ_n^2, whereas we assumed the bias to be asymptotically (large L and K) zero. We have experimentally found the sample average of $a(2)$ for this example to be about $0.95\sigma_n^2$. If this value is used in the derivations the theoretical curves approach the experimental ones by nearly 0.5 dB. The same type of argument may be made with respect to the asymptotic versus finite sample model that is used for $\hat{\lambda}_2$. Figures 5.10–5.12 show the theoretical results for $P_E(2)$. It is clear from these results that the AIC penalty function leads to better detection performance than the MDL penalty function. As we pointed out earlier in this section, the MDL penalty function has a better performance in terms of the probability of overestimating the number of sources given H_2 than the AIC penalty function. The dominant term of $P_E(2)$ in this example is the probability of underestimating the number of the sources given H_2 as one can expect in the situations of

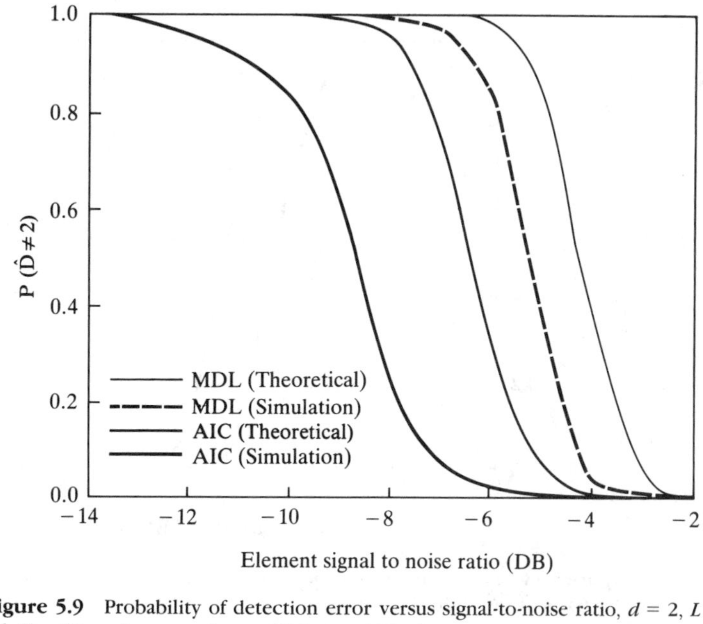

Figure 5.9 Probability of detection error versus signal-to-noise ratio, $d = 2$, $L = 16$, $K = 64$, angle separation = 0.5 beamwidth, $\psi = 0$.

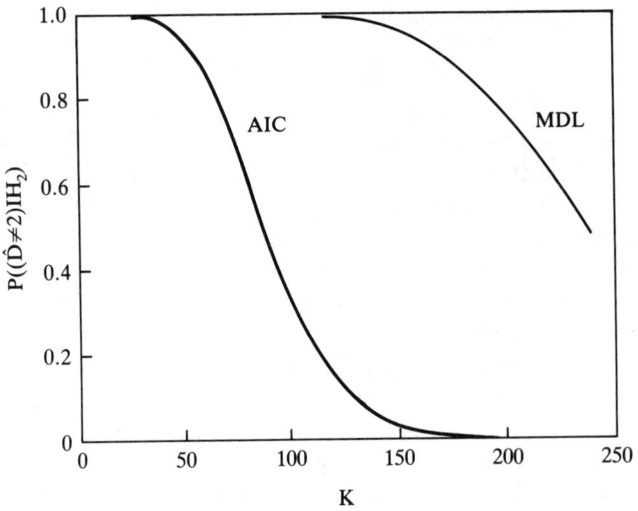

Figure 5.10 Probability of detection error versus number of snapshots, $d = 2$, $L = 16$, SNR $= -6.0$ dB, angle separation $= 0.5$ beamwidth, $\psi = 0.5$.

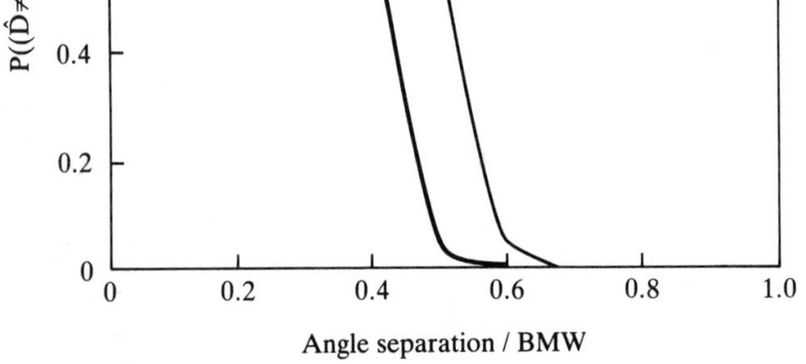

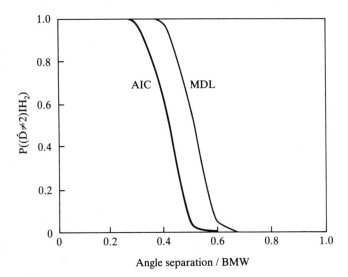

Figure 5.11 Probability of detection error versus angle separation, $d = 2$, $L = 16$, $K = 64$, SNR $= -3$ dB, $\psi = 0.5$.

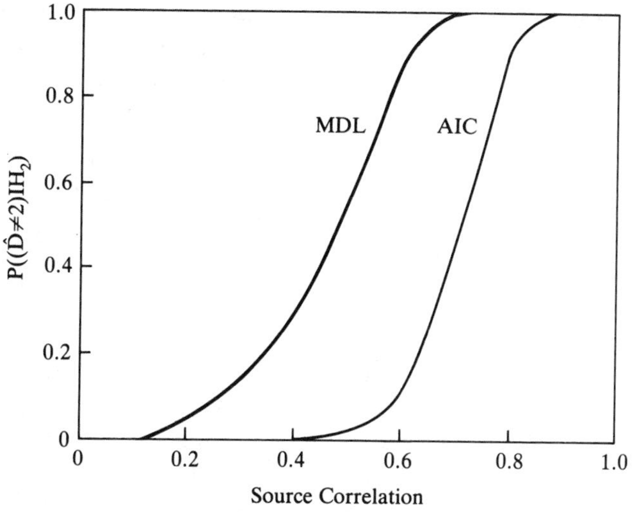

Figure 5.12 Probability of detection error versus source correlation coefficient, $d = 2$, $L = 16$, $K = 64$, SNR $= -3$ dB, angle separation $= 0.5$ beamwidth.

lower SNR, shorter observation time, smaller angular separation, and/or higher source correlation. In other examples with larger number of sources, we have observed sizable probability of overestimation, usually by one, by AIC. MDL, however, continues to be relatively stable with essentially zero probability of false alarm.

5.6. PROPERTIES OF SIGNAL-SUBSPACE ESTIMATES

Having estimated the number of sources, the next step of signal-subspace processing is often to search in the estimated signal subspace, span of $\hat{\mathbf{E}}_s$, for estimates of the source directions, θ_i, $i = 1, 2, \ldots, \hat{d}$. Since for finite observation time the estimated signal subspace is never equal to the true signal subspace even if $\hat{d} = d$, and since regardless of the way we perform the search, the quality of span of $\hat{\mathbf{E}}_s$ is always the dominant factor in the performance of the estimator, it is important to evaluate it directly and to relate it to the signal, noise, and array parameters. In this section we address this issue instead of conducting performance analysis with a particular choice of the subspace search method. We introduce a scalar measure to evaluate the quality of span of $\hat{\mathbf{E}}_s$. This measure is based on the concept of angles between subspaces to describe their closeness. First, we give the following definition of angles between subspaces.

Definition. Let $\hat{E}_s$ and E_s be subspaces in the L-dimensional complex vector space with the same dimension d. The principal angles $\alpha_1, \alpha_2, \ldots, \alpha_d \varepsilon [0, \pi/2]$ between $\hat{\mathbf{E}}_s$ and $\mathbf{E}_s$ are defined recursively as follows. For $\|\mathbf{u}\| = \|\mathbf{v}\| = 1$, we define

$$\cos(\alpha_1) \triangleq \min_{\mathbf{u}\varepsilon\hat{E}_s} \min_{\mathbf{v}\varepsilon E_s} \mathbf{u}^\dagger \mathbf{v} = \mathbf{u}_1{}^\dagger \mathbf{v}_1 \tag{5.83}$$

$$\cos(\alpha) \triangleq \min_{\mathbf{u}\varepsilon\hat{E}_s} \min_{\mathbf{v}\varepsilon E_s} \mathbf{u}^\dagger \mathbf{v} = \mathbf{u}_i{}^\dagger \mathbf{v}_i, \qquad i = 2, \ldots, d \tag{5.84}$$

with

$$\mathbf{u}^\dagger \mathbf{u}_j = 0, \qquad j = 1, \ldots, i-1, \tag{5.85}$$

and

$$\mathbf{v}^\dagger \mathbf{v}_j = 0, \qquad j = 1, \ldots, i-1. \tag{5.86}$$

Note that $\alpha_1, \alpha_2, \ldots, \alpha_d$ so defined have the property that

$$\pi/2 \geq \alpha_1 \geq \alpha_2 \geq \cdots \geq \alpha_d \geq 0. \tag{5.87}$$

For our purpose, α_1 is the most important angle, since it represents the largest deviation of $\hat{\mathbf{E}}_s$ from $\mathbf{E}_s$. If $\hat{\mathbf{E}}_s = \mathbf{E}_s$ then $\alpha_1 = 0$ and so are the rest. Thus we will call α_1 *the angle between* $\hat{E}_s$ *and* E_s, and denote it simply as α for convenience.

Since $\hat{\mathbf{E}}_s{}^\dagger \hat{\mathbf{E}}_s = \mathbf{I}$ and $\mathbf{E}_s{}^\dagger \mathbf{E}_s = \mathbf{I}$, we can prove by the definition of the *singular value* (*SV*), [31]:

$$\cos\alpha = \text{the smallest SV of } \hat{\mathbf{E}}_s{}^\dagger \mathbf{E}_s. \tag{5.88}$$

This provides a convenient way to calculate the angle between $\hat{\mathbf{E}}_s$ and $\mathbf{E}_s$. As $\hat{\mathbf{E}}_s$ is random, we may use the statistical mean-squared difference between $\cos\alpha$ and its perfect value of 1.0, given by

$$\text{MSE} \triangleq E\{(1 - \cos\alpha)^2\}, \tag{5.89}$$

to evaluate the quality of the estimated signal subspace.

The MSE as defined in (5.89) can be used to relate the quality of the estimated signal subspace to the signal, noise, and array parameters. The advantage of using this scalar MSE over the conventional way of using the $dL \times L$ covariance matrices of the estimated eigenvectors lies in its simplicity and clarity. It condenses the information in these covariance matrices and possesses an easily understandable geometrical character. Another nice feature of this scalar measure is that it is not limited to specific number of sources, array geometry, or any other parameters of source signals and noise.

In the following we derive the approximate closed-form expression of the MSE for the case of two sources and spatially white noise.

The singular values of a matrix $\mathbf{Q}$ are equal to the square roots of the eigenvalues of $\mathbf{Q}^\dagger\mathbf{Q}$. Since we need the smallest singular value of $\hat{\mathbf{E}}_s{}^\dagger\mathbf{E}_s$, we need to calculate the smallest eigenvalue of the 2×2 matrix $\hat{\mathbf{E}}_s{}^\dagger \mathbf{E}_s \mathbf{E}_s{}^\dagger \hat{\mathbf{E}}_s$. We have

$$\hat{\mathbf{E}}_s{}^\dagger \mathbf{E}_s \mathbf{E}_s{}^\dagger \hat{\mathbf{E}}_s = \begin{bmatrix} \hat{\mathbf{e}}_1{}^\dagger \mathbf{e}_1 \mathbf{e}_1{}^\dagger \hat{\mathbf{e}}_1 + \hat{\mathbf{e}}_1{}^\dagger \mathbf{e}_2 \mathbf{e}_2{}^\dagger \hat{\mathbf{e}}_1 & \hat{\mathbf{e}}_1{}^\dagger \mathbf{e}_1 \mathbf{e}_2{}^\dagger \hat{\mathbf{e}}_1 + \hat{\mathbf{e}}_1{}^\dagger \mathbf{e}_2 \mathbf{e}_2{}^\dagger \hat{\mathbf{e}}_2 \\ \hat{\mathbf{e}}_2{}^\dagger \mathbf{e}_1 \mathbf{e}_1{}^\dagger \hat{\mathbf{e}}_1 + \hat{\mathbf{e}}_2{}^\dagger \mathbf{e}_2 \mathbf{e}_2{}^\dagger \hat{\mathbf{e}}_1 & \hat{\mathbf{e}}_2{}^\dagger \mathbf{e}_1 \mathbf{e}_1{}^\dagger \hat{\mathbf{e}}_2 + \hat{\mathbf{e}}_2{}^\dagger \mathbf{e}_2 \mathbf{e}_2{}^\dagger \hat{\mathbf{e}}_2 \end{bmatrix}. \tag{5.90}$$

Since $|\hat{\mathbf{e}}_1{}^\dagger \mathbf{e}_1| \gg |\hat{\mathbf{e}}_1{}^\dagger \mathbf{e}_2| \simeq 0$ and $|\hat{\mathbf{e}}_2{}^\dagger \mathbf{e}_2| \gg |\hat{\mathbf{e}}_2{}^\dagger \mathbf{e}_1| \simeq 0$ (see Section 5.4), we have

$$\hat{\mathbf{E}}_s{}^\dagger \mathbf{E}_s \mathbf{E}_s{}^\dagger \hat{\mathbf{E}}_s \simeq \begin{bmatrix} |\hat{\mathbf{e}}_1{}^\dagger \mathbf{e}_1|^2 & 0 \\ 0 & |\hat{\mathbf{e}}_2{}^\dagger \mathbf{e}_2|^2 \end{bmatrix}. \tag{5.91}$$

It is known that the perturbation of the eigenvector corresponding to the smaller eigenvalue is larger than that of the eigenvector of a larger eigenvalue [31, p. 204]; that is,

$$|\hat{\mathbf{e}}_2{}^\dagger \mathbf{e}_2| < |\hat{\mathbf{e}}_1{}^\dagger \mathbf{e}_1|. \tag{5.92}$$

Therefore, we have

$$\cos \alpha \cong |\hat{\mathbf{e}}_2{}^\dagger \mathbf{e}_2| = |\mathbf{e}_2{}^\dagger \hat{\mathbf{e}}_2|. \tag{5.93}$$

Let $\tilde{\mathbf{e}}_2$ be an unnormalized version of the second eigenvector, so

$$\hat{\mathbf{e}}_2 = \tilde{\mathbf{e}}_2 / \|\tilde{\mathbf{e}}_2\|.$$

We can write

$$\hat{\mathbf{e}}_2{}^\dagger \hat{\mathbf{e}}_2 = \mathbf{e}_2{}^\dagger \tilde{\mathbf{e}}_2 / \|\tilde{\mathbf{e}}_2\|. \tag{5.94}$$

As discussed in Section 5.3, the $o(K^{-1})$ asymptotic covariance matrix of $\tilde{\mathbf{e}}_2$ is the same as that of $\hat{\mathbf{e}}_2$, and the mean vector of $\tilde{\mathbf{e}}_2$ is determined by (5.27). It is easy to see from (5.27) and (5.38) that for large K, $\mathbf{e}_2{}^\dagger \tilde{\mathbf{e}}_2$ is a random variable with mean equal to 1.0 and variance equal to 0, that is, a constant. Therefore, $\mathbf{e}_2{}^\dagger \hat{\mathbf{e}}_2 = \mathbf{e}_2{}^\dagger \tilde{\mathbf{e}}_2 / \|\tilde{\mathbf{e}}_2\|$ asymptotically has the same distribution as $\|\tilde{\mathbf{e}}_2\|^{-1}$.

It is easy to see from (5.38) that $\text{cov}\{\tilde{\mathbf{e}}_2, \tilde{\mathbf{e}}_2\}$ has a rank of $L - 1$ with two nonzero eigenvalues:

$$\nu_1 = \frac{\lambda_1 \lambda_2}{K(\lambda_1 - \lambda_2)^2} \tag{5.95}$$

and

$$\nu_2 = \frac{\sigma_n^2 \lambda_2}{K(\lambda_2 - \sigma_n^2)^2} \tag{5.96}$$

with ν_2 having a multiplicity of $L - 2$. Furthermore the eigenspace corresponding to ν_1 has a dimension of one and the eigenspace corresponding to ν_2 has a dimension of $L - 2$. Note that $E\{\tilde{\mathbf{e}}_2\}$ is not in the range space of $\text{cov}\{\tilde{\mathbf{e}}_2, \tilde{\mathbf{e}}_2\}$. Thus it follows from [32, p. 188], that $\tilde{\mathbf{e}}_2{}^\dagger \tilde{\mathbf{e}}_2$ has the same distribution as

$$1 + \nu_1 \xi_1 + \nu_2 \xi_2, \tag{5.97}$$

where ξ_1 and ξ_2 are independent chi-squared random variables with their degrees of freedom being 2 and $2(L-2)$, respectively.

Therefore, we finally conclude that the cosine of the angle between span $\mathbf{E}_s$ and span $\hat{\mathbf{E}}_s$, for the situation of two sources in the spatially white noise, has asymptotically the same distribution as $\sqrt{[1+\nu_1\xi_1+\nu_2\xi_2]}$. By definition of the MSE, we now have for the above situation

$$\begin{aligned} \text{MSE} &= E\{(1-\cos\alpha)^2\} \\ &\cong E\{[1-(1+\nu_1\xi_1+\nu_2\xi_2)^{-1/2}]^2\}. \end{aligned} \tag{5.98}$$

For large number of snapshots, with respect to the number of sensors, that is, $K/L \gg 1$, the random fluctuation terms in (5.98) are much less than 1.0; that is, $\cos\alpha$ is not far away from 1.0. Thus, we may use

$$(1+x)^{-1/2} \cong 1 - \frac{x}{2}$$

to further simplify (5.98), resulting in

$$\text{MSE} \cong E\left\{\left[\frac{1}{2}(\nu_1\xi_2+\nu_2\xi_2)\right]^2\right\} \tag{5.99}$$

If the number of sensors, L, is much larger than 2, $\nu_2\xi_2$ has a variance much larger than that of $\nu_1\xi_1$. Therefore, it is reasonable to drop the $\nu_1\xi_2$ term, resulting in

$$\text{MSE} \cong E\left\{(\frac{1}{2}\nu_2\xi_2)^2\right\}. \tag{5.100}$$

Since the second-order moment of the chi-squared random variable is $4(L-2) + [2(L-2)]^2 \simeq [2(L-2)]^2$ for large L, we obtain

$$\text{MSE} \cong (L-2)^2\nu_2^2$$

or

$$\text{RMSE} \triangleq \sqrt{\text{MSE}} = \frac{(L-2)\sigma_n^2\lambda_2}{K(\lambda_2-\sigma_n^2)^2}. \tag{5.101}$$

In summary this formula can be used to evaluate the quality of the estimated signal subspace under the following conditions:

1. Two sources in the spatially white noise.
2. $L \gg 2$; that is, the array consists of many sensors.
3. $K/L \gg 1$.

Using the expression of λ_2 for a given array geometry we can now immediately relate the quality of the estimated signal subspace to the signal, noise, and array parameters.

For example, consider a linear uniform array of $L = 16$ omnidirectional sensors with unity gain and assume that the two sources have the same SNR. Figures 5.13–5.15 show the RMSE as functions of signal-to-noise ratio, angle separation, and source correlation, respectively. Threshold effects can be seen from these plots; that is, the qualities of the estimated signal subspace degrade very rapidly when the signal-to-noise or angle separation are lowered below certain values, or the source correlation rises above a critical point. Improving the quality of the signal subspace clearly requires an increase in the observation time. In the threshold region, this required increase can be quite severe.

The following analytical expressions may serve to make the dependence of the RMSE on signal and array parameters more clear. Assume that the two sources have the same power and are uncorrelated. Using (5.101) and (5.52), we have

$$\text{RMSE} \cong \frac{(L-2)[1+L(\text{SNR})(1-|\phi|)]}{L^2(\text{SNR})^2(K)(1-|\phi|)^2}. \tag{5.102}$$

For a linear array with uniform spacing and closely spaced sources, the term $1 - |\phi|$ can be approximated by [25, p. 54]

$$1-|\phi| \cong \frac{1}{24}(2\pi D_e/\lambda)^2(\sin\theta_1 - \sin\theta_2)^2(L^2-1). \tag{5.103}$$

For the high-SNR region, that is, $L(\text{SNR})(1-|\phi|) \gg 1$, we obtain

$$\begin{aligned}\text{RMSE} &\cong \frac{24}{L^2(\text{SNR})(K)(2\pi D_e/\lambda)^2(\sin\theta_1-\sin\theta_2)^2}\\ &\cong \frac{0.152}{(\text{SNR})(K)\{\Delta\theta/[\lambda/(LD_e)]\}^2\cos^2\bar{\theta}},\end{aligned} \tag{5.104}$$

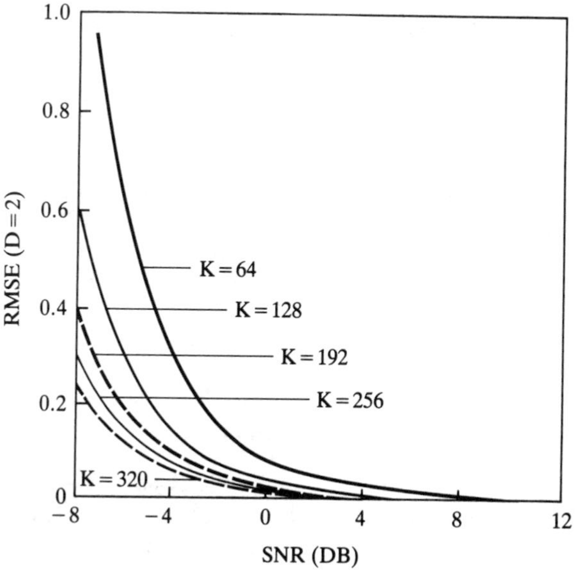

Figure 5.13 Quality of estimated signal-subspace versus signal-to-noise ratio, $d = 2$, $L = 16$, $\psi = 0.5$, angle separation = 0.5 beamwidth.

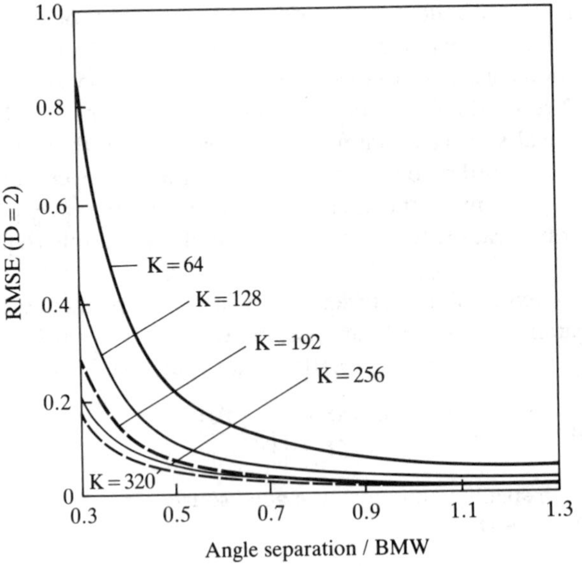

Figure 5.14 Quality of estimated signal subspace versus angle separation, $d = 2$, $L = 16$, SNR = -10 dB, $\psi = 0.5$.

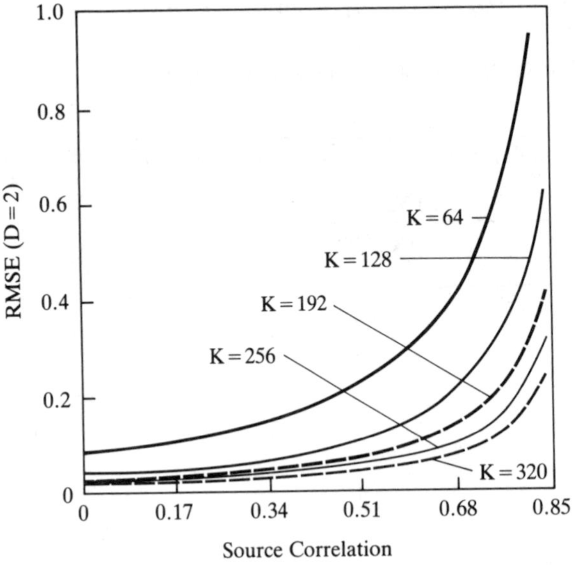

Figure 5.15 Quality of estimated signal subspace versus source correlation coefficient, $d = 2$, $L = 16$, SNR = -10 dB, angle separation = 0.5 beamwidth.

where $\Delta\theta = \theta_1 - \theta_2$ and $\bar{\theta} \triangleq (\theta_1 + \theta_2)/2$. On the other hand, at low SNR

$$\text{RMSE} \cong \frac{0.023}{(\text{SNR})^2(K)\{\Delta\theta/[\lambda/(LD_e)]\}^4\cos^4\bar{\theta}}. \tag{5.105}$$

Comparing (5.104) with (5.105) we can see that at low SNR and/or small angle separation, K has to be increased much more to combat a decrease in SNR and/or angle separation than at high-SNR and large-angle separation—an indication of the difficulty one would have for high-resolution direction finding at low signal-to-noise ratios.

It is also interesting to note that the estimation performance becomes worse when the two closely spaced sources are in the proximity of the end fire of the array. Again at low SNR, this effect is more serious.

5.7. RESOLUTION THRESHOLD FOR MUSIC

As mentioned in Section 5.2, a complete analytical evaluation of the statistical properties of nonlinear spectral estimators is, at best, a difficult task. An earlier attempt at the statistical characterization of narrow-band signal subspace algorithms is given in [33]. A feature that is often of great interest is the threshold SNR beyond which the estimator has a high probability of resolving sources that are closely spaced with respect to the array beamwidth. In this section we use a plausibility argument to obtain an asymptotic expression for the resolution threshold of the MUSIC estimator (also see [34]). Our approach is based on the average deviation of the estimated null spectrum of MUSIC from its exact value (one that is calculated based on the eigenvectors of $\mathbf{R}$), in the neighborhood of the source directions. A similar analysis is carried out for min-norm in the next section.

5.7.1 Mean of $\hat{D}(\omega)$

We begin by evaluating $E\{\hat{D}(\omega)\}$ and $\text{var}\{D(\omega)\}$ to the order K^{-1}. The estimated null spectrum for this method is given by

$$\hat{D}(\omega) = 1 - \mathbf{V}^{\dagger}(\omega)\Big(\sum_{i=1}^{d} \hat{\mathbf{e}}_i \hat{\mathbf{e}}_i^{\dagger}\Big)\mathbf{V}(\omega). \tag{5.106}$$

The expected value of $\hat{D}(\omega)$, using the definition of $\hat{\mathbf{e}}$ in Section 5.3, can be expressed as

$$\begin{aligned} E\{\hat{D}(\omega)\} = 1 - \mathbf{V}^{\dagger}(\omega)\Big(\sum_{i=1}^{d} \mathbf{e}_i \mathbf{e}_i^{\dagger}\Big)\mathbf{V}(\omega) \\ - \mathbf{V}^{\dagger}(\omega)\, E\Big\{\Big[\sum_{i=1}^{d} \boldsymbol{\eta}_i \boldsymbol{\eta}_i^{\dagger}\Big]\mathbf{V}(\omega) - 2\text{Re}\Big[\mathbf{V}^{\dagger}(\omega)\Big(\sum_{i=1}^{d} \mathbf{e}_i \boldsymbol{\eta}_i^{\dagger}\Big)\mathbf{V}(\omega)\Big\}. \end{aligned} \tag{5.107}$$

Substituting from (5.40) and (5.41) for the expectations in (5.107) gives

$$E\{\hat{D}(\omega)\} \simeq D(\omega) - \mathbf{V}^{\dagger}(\omega)\left[\sum_{i=1}^{d} \sum_{\substack{j=1 \\ j\neq i}}^{L} \frac{\lambda_i \lambda_j}{K(\lambda_i - \lambda_j)^2} (\mathbf{e}_j \mathbf{e}_j^{\dagger} - \mathbf{e}_i \mathbf{e}_i^{\dagger})\right]\mathbf{V}(\omega), \tag{5.108}$$

where $\simeq$ is used to indicate retention of order K^{-1} terms only.

Since our primary objective is the investigation of the resolving capability of this method, we may concentrate on the statistics of $\hat{D}(\omega)$ in the neighborhood of ω_i, $i = 1, \ldots, d$. The results for $M = 1$ and $M = 2$ follow immediately, noting that $\mathbf{V}(\omega_i) \in$ column span of $\mathbf{E}_s$ and therefore $\mathbf{V}(\omega_i)$ is orthogonal to $\mathbf{e}_i$, $i = d + 1, \ldots, L$. Since $D(\omega_i) = 0$, $i = 1, \ldots, d$, it follows that for $d = 1$,

$$E\{\hat{D}(\omega_1)\} \simeq \frac{\lambda_1 \sigma_n^2 (L-1)}{K\lambda_1'^2}. \tag{5.109}$$

Similarly for $M = 2$,

$$E\{\hat{D}(\omega_k)\} \simeq \sigma_n^2 \mathbf{V}^\dagger(\omega_k)(L-2)\left[\frac{\lambda_1}{\lambda_1'^2}\mathbf{e}_1^\dagger \mathbf{e}_1 + \frac{\lambda_2}{\lambda_2'^2}\mathbf{e}_2^\dagger \mathbf{e}_2\right]\mathbf{V}(\omega_k)/K. \tag{5.110}$$

It is conjectured that $E\{\hat{D}(\omega_k)\}$ is an indicator of the resolving capability of the estimator. That is, as $E\{\hat{D}(\omega_k)\}$ approaches zero, ideal resolution is achieved. Of course, true resolution does not depend only on the absolute value of $\hat{D}(\omega_k)$, but on its value relative to $\hat{D}(\omega)$, $\omega \neq \omega_k$. Furthermore, for $E\{\hat{D}(\omega_k)\}$ to be an indicator of the resolving ability of the method, $\text{var}\{\hat{D}(\omega)\}$ must be relatively small. An order K^{-1} approximation for this statistic of $\hat{D}(\omega)$ is, therefore, derived next.

5.7.2 Variance of $\hat{D}(\omega)$

To derive an expression for $\text{var}\{\hat{D}(\omega)\}$, it is convenient to express $D(\omega)$ as

$$D(\omega) = 1 - D_c(\omega),$$

where $D_c(\omega)$ is the null-complement spectrum. The variance of $D(\omega)$ is then given by

$$\text{var}\{\hat{D}(\omega)\} = \text{var}\{\hat{D}_c(\omega)\}.$$

Therefore, we first derive an expression for $E\{\hat{D}_c^2(\omega)\}$. From the definition of $\hat{D}(\omega)$, it follows that

$$\hat{D}_c^2(\omega) = \sum_{i=1}^{d}\sum_{j=1}^{d} |\mathbf{V}^\dagger(\omega)\hat{\mathbf{e}}_i|^2| \, \mathbf{V}^\dagger(\omega)\hat{\mathbf{e}}_j|^2. \tag{5.111}$$

Substituting $\hat{\mathbf{e}}_i = \mathbf{e}_i + \boldsymbol{\eta}_i$ in (5.111) and defining $u_i \triangleq \mathbf{V}^\dagger(\omega)\mathbf{e}_i$ and $w_i \triangleq \mathbf{V}^\dagger(\omega)\boldsymbol{\eta}_i$, we obtain

$$\begin{aligned}\hat{D}_c^2(\omega) = \sum_{i=1}^{d}\sum_{j=1}^{d} [&|u_i|^2\,|u_j|^2 + 2|u_j|^2\,u_i w_i^* \\ &+ 2|u_j|^2 u_i^* w_i + 2|u_j|^2\,|w_i|^2 + u_i u_j w_i^* w_j^* \\ &+ 2u_j u_i^* w_j^* w_i + u_j^* u_i w_j w_i + 4\text{Re}(u_j w_j^* |w_i|^2) + |w_i|^2\,|w_j|^2].\end{aligned} \tag{5.112}$$

The order K^{-1} expected value of $\hat{D}_c^2(\omega)$ is then given by

$$E\{\hat{D}_c^2(\omega)\} \simeq \sum_{i=1}^{d}\sum_{j=1}^{d} \{|u_i|^2\,|u_j|^2 + 4c_i|u_i|^2\,|u_j|^2$$

$$+ 2\frac{\lambda_i}{K}\left[\sum_{\substack{k=1\\k\neq i}}^{L} \frac{\lambda_k}{(\lambda_i-\lambda_k)^2}\,|u_k|^2\right]|u_j|^2$$

$$- \frac{\lambda_i\lambda_j}{K(\lambda_i-\lambda_j)^2}\,|u_i|^2\,|u_j|^2\,(1-\delta_{ij})$$

$$+ 2\frac{\lambda_i}{K}\left[\sum_{\substack{k=1\\k\neq i}}^{L} \frac{\lambda_k}{(\lambda_i-\lambda_k)^2}\,|u_k|^2\right]|u_j|^2\,\delta_{ij}$$

$$- \frac{\lambda_i\lambda_j}{K(\lambda_i-\lambda_j)^2}\,|u_i|^2\,|u_j|^2\,(1-\delta_{ij})\}, \tag{5.113}$$

where c_i is given in (5.40).

At the ℓth true angle corresponding to ω_ℓ, $\sum_{i=1}^{d}|u_i|^2 = 1$ and $u_i = 0$, $i = d+1, \ldots, K$. Therefore,

$$E\{\hat{D}_c^2(\omega_\ell)\} \simeq 1 - 2\sum_{i=1}^{d}\frac{\lambda_i}{K}\left[\sum_{\substack{k=1\\k\neq i}}^{L}\frac{\lambda_k}{(\lambda_i-\lambda_k)^2}\,|u_i|^2\right]$$
$$+ 2\sum_{i=1}^{d}\frac{\lambda_i}{K}\left[\sum_{\substack{k=1\\k\neq i}}^{d}\frac{\lambda_k}{(\lambda_i-\lambda_k)^2}\,|u_k|^2\right]. \tag{5.114}$$

Variance of $\hat{D}(\omega_\ell)$ to an order K^{-1} approximation is then given by

$$\mathrm{var}\{\hat{D}(\omega_\ell)\} \simeq E\{\hat{D}_c^2(\omega_\ell)\} - 1 + 2E\{\hat{D}(\omega_\ell)\} = 0. \tag{5.115}$$

So, at the true angles, the variance of $\hat{D}(\omega_\ell)$ does not have order K^{-1} dependence, but varies as K^{-2}. This, of course, implies that the dependence of the standard deviation of $\hat{D}(\omega_\ell)$ is on the order K^{-1}, which is the same order of dependence as the bias of $\hat{D}(\omega_\ell)$.

Resolution Threshold for Two Equipowered Uncorrelated Sources

In this section we use the expressions for λ_i and $\mathbf{e}_i$ given in Section 5.4 to evaluate a resolution threshold for two uncorrelated sources with equal powers. The assumptions of uncorrelatedness between the sources and of equal powers are not necessary for this type of analysis. If the sources are correlated but not coherent, that is, $|\psi| \neq 1$ in (5.52) and (5.53), similar expressions as given in this section can be derived. First, it is useful to determine explicitly the dependence of $E\{\hat{D}(\omega_i)\}$ on the physical parameters of the problem such as signal and noise powers, number of array elements, and spatial frequency or angular separation of the sources.

For one signal in noise, $E\{\hat{D}(\omega_1)\}$ for $LP_1 \gg \sigma_n^2$ is given by

$$E[\hat{D}(\omega_1)] \simeq \frac{(L-1)\sigma_n^2}{KLP_1}. \tag{5.116}$$

Thus, the bias at ω_1 is directly proportional to the noise-to-signal ratio and inversely proportional to the number of snapshots, a behavior that is not unexpected.

Approximate expressions for $E\{\hat{D}(\omega_k)\}$ can be similarly obtained for two sources with equal powers and *array signal-to-noise ratio* (*ASNR*) much greater than one. Retaining the largest two terms in (5.110), we get

$$E\{\hat{D}(\omega_k)\} \simeq \frac{(L-2)}{K}\left[\frac{1}{(\text{ASNR})} + \frac{1}{(\text{ASNR})^2\Delta^2}\right], \tag{5.117}$$

where ASNR $\underline{\Delta}$ LP_1/σ_n^2 is the array signal-to-noise ratio and $\Delta = [L(\omega_1 - \omega_2)]/2\sqrt{3}$. For the large dynamic range situation, that is, $P_1 \gg P_2$, a similar expression is given in [20].

In the previous section we showed that the order of the standard deviation of $\hat{D}(\omega_k)$ is K^{-1}. Unfortunately, an expression for this order approximation of the standard deviation (order K^{-2} approximation for the variance) is very difficult to obtain. However, simulation results have shown that standard deviation $[\hat{D}(\omega_k)] \ll$ bias $[\hat{D}(\omega_k)]$ (see, for example, [21]). Thus, we expect that the resolution threshold is mainly determined by $E\{\hat{D}(\omega)\}$ and proceed to base the derivation of the threshold on this premise.

The foregoing expressions for $E\{\hat{D}(\omega_k)\}$ indicate the general behavior of the null-spectrum bias as a function of source and array parameters. To obtain a quantitative measure of the resolution threshold for two closely spaced equipowered sources, we now introduce a plausible, nonprobabilistic approach based on $E\{\hat{D}(\omega)\}$. We use the signal-to-noise ratio at which $E\{\hat{D}(\omega_1)\} = E\{\hat{D}(\omega_2)\} = E\{\hat{D}(\omega_m)\}$, where $\omega_m = (\omega_1 + \omega_2)/2$, as an approximation to this threshold. The reasons for this conjecture are as follows. Resolution is achieved when $\hat{D}(\omega_1)$ and $\hat{D}(\omega_2)$ are *both* less than $\hat{D}(\omega_m)$. When the preceding equality is true, the probability of resolution ranges from approximately 0.33, for the case of totally independent variations of $\hat{D}(\omega_1)$, $\hat{D}(\omega_2)$, and $\hat{D}(\omega_m)$, to nearly 0.5 for the situation when $\hat{D}(\omega_1)$ and $\hat{D}(\omega_2)$ are maximally correlated.

Figure 5.1 showed the resolution probability of the MUSIC algorithm for a typical array. It is clear that the SNR for probabilities of 0.33 and 0.5 are within about 1 or 2 dB of each other. Thus, the proposed method should give the approximate SNR for the 0.3 to 0.5 probability of resolution threshold region. Note also that equating the averages of $\hat{D}(\omega_1)$, $\hat{D}(\omega_2)$, and $\hat{D}(\omega_m)$ is justified due to the small standard deviation of $\hat{D}(\omega_k)$ relative to its mean. We again substitute the expressions from Section 5.4 in (5.110) to obtain $E\{\hat{D}(\omega_1)\}$ and $E\{\hat{D}(\omega_2)\}$, and (5.108) to find $E\{\hat{D}(\omega_m)\}$. Threshold occurs at the ASNR for which $E\{\hat{D}(\omega_{1(2)})\} = E\{\hat{D}(\omega_m)\}$ with the resolution probability increasing as $E\{\hat{D}(\omega_{1(2)})\}$ becomes increasingly less than $E\{\hat{D}(\omega_m)\}$. Let the threshold ASNR be denoted by ξ_T. Then, for $L \gg 1$, $\Delta \ll 1$ and ASNR $\gg 1$,

$$\xi_T \simeq \frac{1}{K}\{20(L-2)\,\Delta^{-4}\,[1 + \sqrt{1 + \frac{K}{5(L-2)}\,\Delta^2}]\}. \tag{5.118}$$

Several features of ξ_T can be observed from (5.118). First, for large L, ξ_T is approximately proportional to L. This means that we can define a threshold element SNR that does not change significantly when L changes. Second, for $K \ll 5L\Delta^{-2}$, ξ_T varies as $\Delta^{-4}K^{-1}$, whereas for $K \gg 5L\Delta^{-2}$, it varies as $\Delta^{-3}K^{-1/2}$.

Figures 5.16 and 5.17 show plots of ξ_T as a function of normalized separation between the spatial frequencies, Δ', defined by

$$\Delta' = \frac{L(\omega_1 - \omega_2)}{2\pi}.$$

These plots were obtained by equating $E\{\hat{D}(\omega_1)\}$ and $E\{\hat{D}(\omega_m)\}$ calculated exactly according to (5.108). The results, however, are essentially indistinguishable from ξ_T as given approximately by the closed-form expression in (5.118) and confirm the general behavior of ξ_T given. For the values of L, N, and Δ' used in the figures, ξ_T shows a very strong (Δ^{-2} to Δ^{-4}) dependence on angular separation for smaller Δ'. In the next section, similar calculations are performed for the min-norm algorithm, and the results are compared with those of MUSIC.

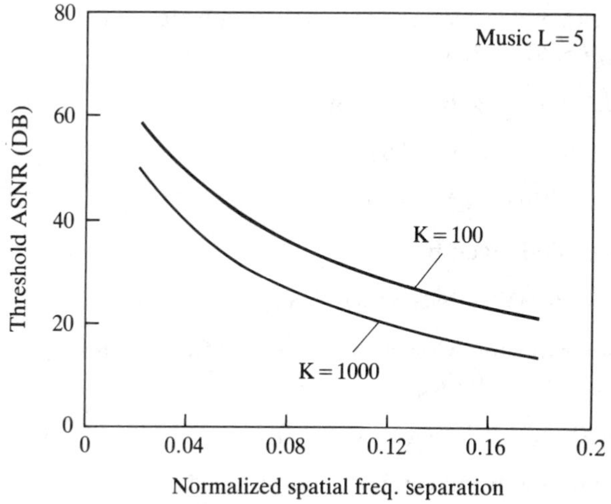

Figure 5.16 MUSIC threshold ASNR as a function of the normalized spatial frequency separation, $L = 5$.

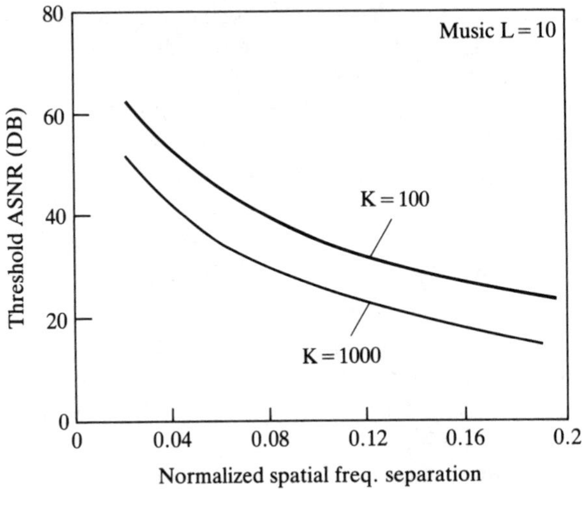

Figure 5.17 Same as Fig. 5.11, except $L = 10$.

5.8. RESOLUTION THRESHOLD FOR THE MIN-NORM ALGORITHM

In this section the asymptotic expected value of the null spectrum for the min-norm algorithm is derived. In general, the evaluation of the moments of the min-norm $\hat{D}(\omega)$ is considerably more complex than those of MUSIC. We shall, therefore, not give a derivation for var$\{\hat{D}(\omega)\}$ and conjecture that, as in the case of MUSIC, the standard deviation of $\hat{D}(\omega)$ is small compared to $E\{\hat{D}(\omega)\}$, for ω in the neighborhood of the true spatial frequencies. Therefore, in the following only $E\{\hat{D}(\omega)\}$ is evaluated followed by numerical computation of resolution thresholds and a comparison of the MUSIC and min-norm resolution thresholds.

We base the evaluation of $E\{\hat{D}(\omega)\}$ on a perturbation of the exact min-norm coefficient vector. The estimate of this vector is obtained by substituting estimates of signal-subspace quantities in (5.22). Then

$$\hat{\mathbf{h}}^T = [1, \ -\hat{\mathbf{g}}^\dagger \hat{\mathbf{E}}'^T/(1 - \hat{\mathbf{g}}^\dagger \hat{\mathbf{g}})]. \tag{5.119}$$

The null-spectrum estimate follows from (5.16) as

$$\hat{D}(\omega) = |\mathbf{V}^\dagger(\omega)\hat{\mathbf{h}}|^2 = \mathbf{V}^\dagger(\omega)\hat{\mathbf{h}}\hat{\mathbf{h}}^\dagger\mathbf{V}(\omega). \tag{5.120}$$

Let $\hat{\mathbf{h}} = \mathbf{h} + \boldsymbol{\alpha}$. $\hat{D}(\omega)$ is then given by

$$\hat{D}(\omega) = D(\omega) + 2\mathrm{Re}[\mathbf{V}^\dagger(\omega)\boldsymbol{\alpha}\mathbf{h}^\dagger\mathbf{V}(\omega)] + \mathbf{V}(\omega)^\dagger\boldsymbol{\alpha}\boldsymbol{\alpha}^\dagger\mathbf{V}(\omega). \tag{5.121}$$

It is of interest again to consider the behavior of $\hat{D}(\omega)$ at signal frequencies ω_k. Using the fact that $D(\omega_k) = \mathbf{h}^\dagger\mathbf{V}(\omega_k) = 0$, $E\{\hat{D}(\omega_k)\}$ reduces to

$$E\{\hat{D}(\omega_k)\} = \mathbf{V}^\dagger(\omega_k)\boldsymbol{\alpha}\boldsymbol{\alpha}^\dagger\mathbf{V}(\omega_k). \tag{5.122}$$

Thus, we have to find an expression for the correlation matrix of $\boldsymbol{\alpha}$. Denote the random perturbation in $\hat{\mathbf{E}}'$ and $\hat{\mathbf{g}}$ by $\boldsymbol{\epsilon}'$ and $\boldsymbol{\gamma}$, respectively. Then

$$\hat{\mathbf{E}}' = \mathbf{E}' + \boldsymbol{\epsilon}' \qquad \text{and} \qquad \hat{\mathbf{g}} = \mathbf{g} + \boldsymbol{\gamma}. \tag{5.123}$$

We now find an approximation to $\boldsymbol{\alpha}$ and then derive an expression for $\overline{\boldsymbol{\alpha}\boldsymbol{\alpha}^{\dagger}}$. Note that the first- and second-order statistics of $\boldsymbol{\epsilon}'$ and $\boldsymbol{\gamma}$ are those of the appropriate elements of the set $\{\boldsymbol{\eta}_i\}$, $i = 1, \ldots, d$. By definition we have

$$\boldsymbol{\alpha} = \begin{bmatrix} 0 \\ \dfrac{\mathbf{E}'\,\mathbf{g}^*}{1 - \mathbf{g}^{\dagger}\mathbf{g}} - \dfrac{\hat{\mathbf{E}}'\,\hat{\mathbf{g}}^*}{1 - \hat{\mathbf{g}}^{\dagger}\hat{\mathbf{g}}} \end{bmatrix} = \begin{bmatrix} 0 \\ \boldsymbol{\alpha}' \end{bmatrix}. \tag{5.124}$$

Therefore, $\hat{D}(\omega_k) = \mathbf{V}'^{\dagger}(\omega_k)\,\boldsymbol{\alpha}'\,\boldsymbol{\alpha}'^{\dagger}\,\mathbf{V}'(\omega_k)$. $\mathbf{V}'$ is a vector formed from the second to the Lth elements of $\mathbf{V}$. We now develop an approximate expression for $\boldsymbol{\alpha}'$ in terms of the lowest-order perturbations as follows

$$\frac{1}{1 - \hat{\mathbf{g}}^{\dagger}\hat{\mathbf{g}}} \simeq \frac{1 + \rho}{1 - \mathbf{g}^{\dagger}\mathbf{g}} \tag{5.125}$$

where

$$\rho = \frac{2\mathrm{Re}[\mathbf{g}^{\dagger}\boldsymbol{\gamma}]}{1 - \mathbf{g}^{\dagger}\mathbf{g}} \tag{5.126}$$

and

$$\boldsymbol{\alpha}' \simeq \frac{-1}{1 - \mathbf{g}^{\dagger}\mathbf{g}}\,[\rho\mathbf{E}'\mathbf{g}^* + \mathbf{E}'\boldsymbol{\gamma}^* + \boldsymbol{\epsilon}'\mathbf{g}^*]. \tag{5.127}$$

Consequently,

$$E[\boldsymbol{\alpha}'\,\alpha'^{\dagger}] \simeq \frac{1}{[1 - \mathbf{g}^{\dagger}\mathbf{g}]^2}\,[\overline{\rho^2}\mathbf{E}'\mathbf{g}^*\mathbf{g}^T\mathbf{E}'^{\dagger} + \mathbf{E}'\,\overline{\boldsymbol{\gamma}^*\boldsymbol{\gamma}^T}\mathbf{E}'^{\dagger} + \overline{\boldsymbol{\epsilon}'\mathbf{g}^*\mathbf{g}^T\boldsymbol{\epsilon}'^{\dagger}} + \mathbf{Q} + \mathbf{Q}^{\dagger}] \tag{5.128}$$

where

$$\mathbf{Q} = \mathbf{E}'\mathbf{g}^*\overline{\rho\boldsymbol{\gamma}^T}\mathbf{E}'^{\dagger} + \mathbf{E}'\mathbf{g}^*\mathbf{g}^T\overline{\boldsymbol{\epsilon}'^{\dagger}\rho} + \mathbf{E}'\overline{\boldsymbol{\gamma}^*\mathbf{g}^T\boldsymbol{\epsilon}'^{\dagger}}. \tag{5.129}$$

The quantities under the expectation operations are evaluated in the appendix to this chapter. In the following, these results are used to obtain $E\{\hat{D}(\omega_k)\}$ for $d = 1$ and $d = 2$.

For $d = 1$ we have $g = e_1(1) = 1/\sqrt{L}$ and

$$\frac{1}{(1 - 1/L)^2}\,\mathbf{V}'^{\dagger}(\omega_1)\mathbf{E}'\mathbf{E}'^{\dagger}\mathbf{V}'(\omega_1) = 1. \tag{5.130}$$

We now make a further assumption that $L \gg 1$. This assumption is not crucial but allows for a more manageable expression for $E\{\hat{D}(\omega_k)\}$. Substituting the various quantities from the appendix into (5.128) and noting the dominance of the second term in (5.128) results in

$$E\{\hat{D}(\omega)\} \simeq \mathbf{V}'^{\dagger}(\omega)\ \mathbf{E}'\ \overline{\boldsymbol{\gamma}^*\boldsymbol{\gamma}^T}\mathbf{E}'^{\dagger}\ \mathbf{V}'\ (\omega) \cdot \frac{1}{(1-\frac{1}{L})^2}$$

$$= \frac{\sigma_n^2}{LP_1(1-\frac{1}{L})^2} \mid \mathbf{V}'^{\dagger}(\omega)\ \mathbf{E}'|^2 \sum_{k=2}^{L} |e_k(1)|^2. \quad (5.131)$$

The sum in (5.131) can be found using the orthonormality of $\{\mathbf{e}_k\}$ as

$$\sum_{k=2}^{L} |e_k(1)|^2 = 1 - \frac{1}{L}.$$

Then

$$E\{\hat{D}(\omega_1)\} \simeq \frac{\sigma_n^2}{LP_1K}(1-\frac{1}{L}). \quad (5.132)$$

For $d = 2$ we consider the case of closely spaced sources with equal powers. Again the second term in (5.128) dominates, resulting in

$$E\{\hat{D}(\omega_k)\} \simeq \frac{1}{1-(|e_1(1)|^2 + |e_2(1)|^2)} \{\Gamma_1|\mathbf{V}'^{\dagger}(\omega_k)\mathbf{F}_1|^2 + \Gamma_2|\mathbf{V}'^{\dagger}(\omega_k)\mathbf{F}_2|^2\} \quad (5.133)$$

where $\Gamma_i = [\overline{\boldsymbol{\gamma}^*\boldsymbol{\gamma}^T}]_{ii}$ and $\mathbf{F}_i$ is defined in the appendix as the vector of the second to Lth elements of $\mathbf{e}_i$. Following the same procedure as for $d = 1$, we obtain

$$K\Gamma_i \simeq \frac{\lambda_i\sigma_n^2(1-|e_i(1)|^2)}{\lambda'^2_i} + \lambda_i\left[\frac{\lambda_{3-i}}{(\lambda_1-\lambda_2)^2} - \frac{\sigma_n^2}{\lambda'^2_i}\right]|e_{3-i}(1)|^2.$$

Substituting the values from the appendix, we obtain

$$E\{\hat{D}(\omega_k)\} \simeq \frac{1}{(1-\frac{4}{L})K}\left[\frac{1}{(\mathrm{ASNR})} + \frac{1}{(\mathrm{ASNR})^2\Delta^2}\right]. \quad (5.135)$$

These results on $E\{\hat{D}(\omega_k)\}$ are somewhat optimistic for min-norm. There are other, smaller terms in (5.128) that contribute to the null-spectrum bias. Nevertheless, the dominant makeup of this quantity does indicate a smaller bias (by nearly a factor of L) as compared with MUSIC. This might be expected to carry over to the relative resolution thresholds for the two methods, making the min-norm threshold lower than that of MUSIC. This difference in the lower resolution threshold of min-norm relative to that of MUSIC resolution thresholds has been verified in numerous Monte Carlo simulations. In the following we give some comparisons of the resolution ASNRs for min-norm and MUSIC based on the signal-to-noise ratios for which $E\{\hat{D}(\omega_{1(2)})\} = E\{\hat{D}(\omega_m)\}$ for each method.

We begin by observing the behavior of $E\{\hat{D}(\omega_1)\}$ and $E\{\hat{D}(\omega_m)\}$ as functions of various parameters of two uncorrelated sources with equal powers, array size, and number of snapshots for both min-norm and MUSIC. Figures 5.18–5.21 show typical comparative plots of $E\{\hat{D}(\omega)\}$ at $\omega = \omega_1$ and $\omega = \omega_m$ for two equipowered, uncor-

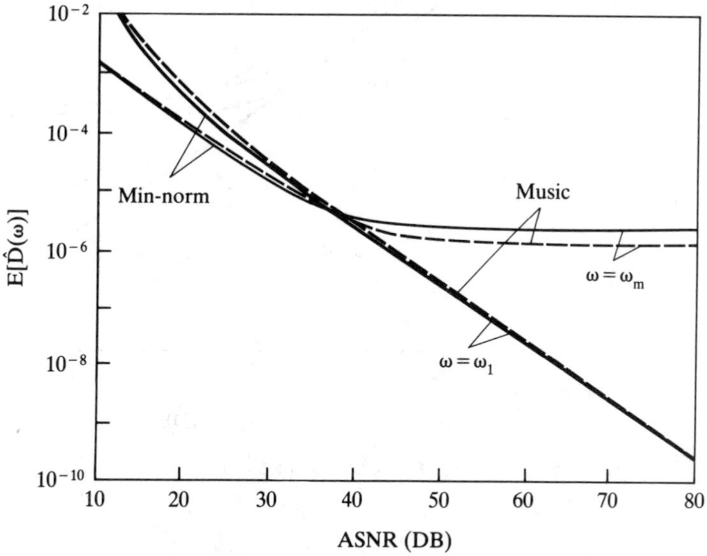

Figure 5.18 $E[\hat{D}(\omega_1)]$ and $E[\hat{D}(\omega_m)]$ for MUSIC and min-norm, $L = 5$, $K = 100$, $\Delta' = 0.06$.

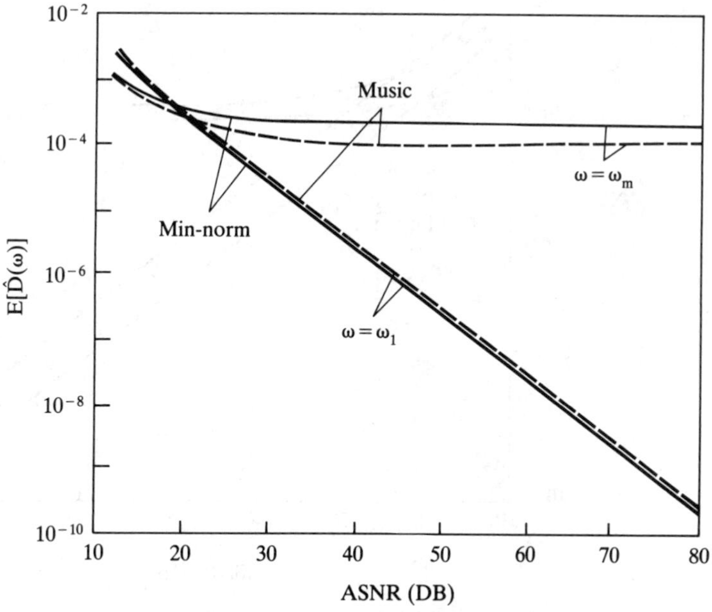

Figure 5.19 $E[\hat{D}(\omega_1)]$ and $E[\hat{D}(\omega_m)]$ for MUSIC and min-norm, $L = 5$, $K = 100$, $\Delta' = 0.18$.

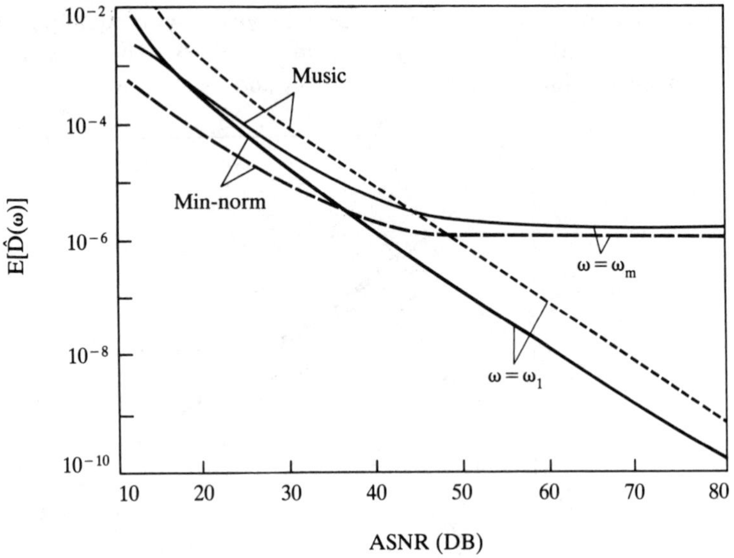

Figure 5.20 Same as Fig. 5.18, except $L = 10$.

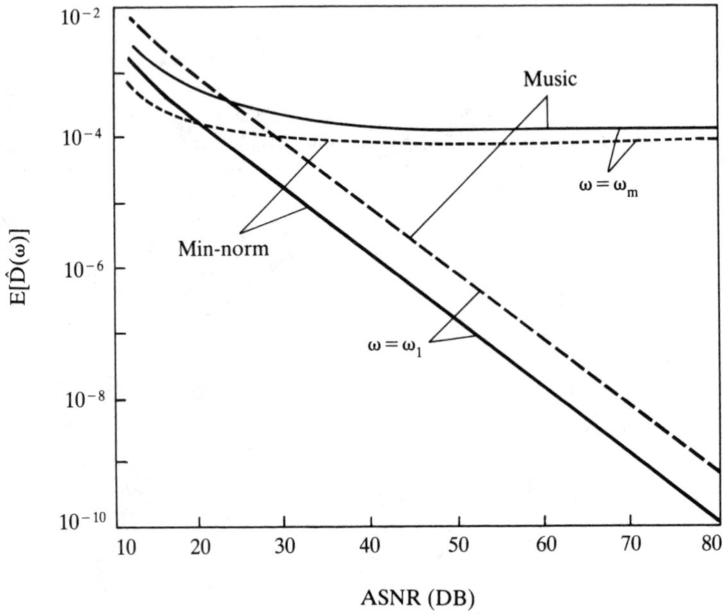

Figure 5.21 Same as Fig. 5.19, except $L = 10$.

related sources as functions of the ASNR. The MUSIC plots are again based on (5.108), and the min-norm plots were computed according to the expected value of (5.121) for $\omega = \omega_m$ and from (5.122) for $\omega = \omega_1$. These figures show that the general behavior of $E\{\hat{D}(\omega_1)\}$ for MUSIC and min-norm algorithms is as predicted by the approximate formulas given by (5.117) and (5.135), respectively. It is also obvious that $E\{\hat{D}(\omega)\}$ at both ω_1 and ω_m have similar trends for both MUSIC and min-norm, attesting to the fact that the fundamental limitations of all signal-subspace algorithms stem from the estimation errors in $\hat{\mathbf{E}}_s$ or $\hat{\mathbf{E}}_N$.

Using the cross-over ASNRs for $E\{\hat{D}(\omega_1)\}$ and $E\{\hat{D}(\omega_m)\}$ from calculations such as those shown in Figs. 5.18–5.21 the resolution threshold ASNR, ξ_T, for min-norm is also determined. Figures 5.22 and 5.23 show the dependence of ξ_T on Δ', K, and L.

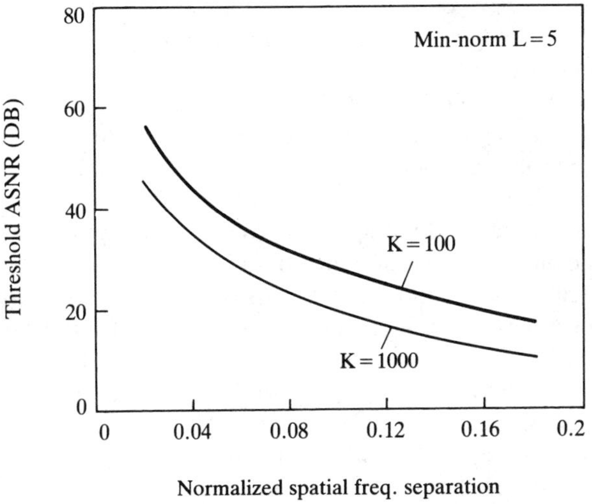

Figure 5.22 Min-norm threshold ASNR as a function of the normalized spatial frequency separation, $L = 5$.

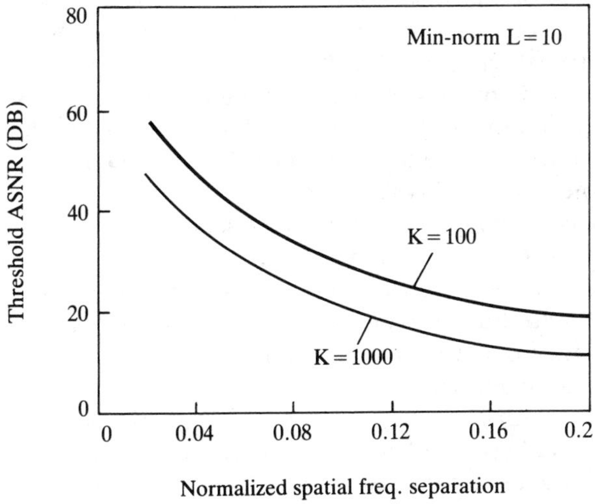

Figure 5.23 Same as Fig. 5.22, except $L = 10$.

Comparing these figures with Figs. 5.16 and 5.17, it becomes clear that for two equi-powered sources observed by a linear uniform array, the min-norm resolution threshold is lower than that for MUSIC. This difference appears to increase somewhat with an increase in L.

5.9. RECENT RESULTS

Since the original writing of this chapter a number of theoretical papers on the statistical performance of signal-subspace methods have appeared in the literature. In this section we give a brief review of some of these results that are closely related to the analyses presented here.

Several of the recent papers have dealt with the resolution thresholds and the accuracy of parameter estimates for signal-subspace methods that are applicable only to *linear uniform arrays.* Since the array manifold of a linear uniform array admits a polynomial representation, lower resolution thresholds can be obtained with the root versions of MUSIC and minimum-norm [35]. References [36] and [37] give asymptotic expressions for the variances of the estimates of these roots. It is shown in [36], for example, that the perturbations of the roots of MUSIC at higher signal-to-noise ratios are primarily in the radial direction, resulting in the disappearance of distinct spectral peaks and the onset of resolution threshold for spectral MUSIC. This idea was the basis of the definition of resolution threshold used in this chapter. Similar analysis of the angle estimation error has been reported [37] for the signal-subspace method ESPRIT [38].

When an array is linear and uniform, the sample correlation matrix of the array data can be spatially smoothed. Smoothing may be performed to combat coherence of sources and/or to improve accuracy of the parameter estimates. A generalization of the MUSIC resolution threshold results for the smoothed correlation matrix has been presented in [39]. It is shown, for example, that for two incoherent sources, forward-backward smoothing increases the accuracy of the estimated subspaces, thereby lowering the resolution threshold of MUSIC.

For the related problem of frequency estimation of uniformly sampled sinusoids in noise, a different approach to the determination of the threshold characteristics of signal-subspace methods that use singular value decomposition of a data matrix, is reported in [40]. In this text, threshold signal-to-noise ratio is defined in the traditional manner, as the point of rapid departure of the standard deviation of the estimated frequencies from the Cramér-Rao lower bound. The basic tool of this analysis is the probability of outliers in the estimation of the signal subspace.

Using the perturbation of the peak locations of MUSIC spectra, two papers have presented results on the behavior of the angle estimation errors. In [41] the relative efficiency of the angle estimate of one source in noise is derived.

Beamspace processing of signals is gaining attention, in particular for large arrays. For spectral MUSIC operating on the output of a matrix beamformer or in the presence of directional constraints, references [43] and [44] present resolution threshold expressions which are similar in form to the one presented in this chapter. These papers show that the resolution threshold signal-to-noise ratio is reduced due to the spatial filtering of out-of-sector noise. These and other unpublished results confirm, for example, that the rather severe Δ^{-4} or Δ^{-3} dependence of the threshold for methods such as MUSIC is applicable to many array configurations.

To obtain simple, closed-form expressions for the probabilities of error of AIC and MDL several assumptions were made in this chapter concerning the number of elements of the array and the probability distributions of the estimated noise-subspace eigenvalues. For example, it was assumed that the number of elements are much greater than the number of sources. In addition, probabilities of error for the case of at most two uncorrelated sources were obtained, since for these situations the expressions can be explicitly given in terms of source and array parameters via (5.52). It was argued that the approximations accounted for the differences in the onset of threshold of detection and for the somewhat optimistic prediction of the false alarm probability of AIC. A recent paper [45] presents more accurate results on the same subject by forgoing some of the approximations made here. The expressions of the probabilities of error are derived for any number of sources that is fewer than the number of elements. The evaluation of the false alarm probabilities, as indicated earlier in the chapter, then require multidimensional numerical integration of a rather complicated joint probability density function. The general conclusions of [45] on the relative merits of AIC and MDL are the same as those presented in this chapter. The expressions in [45] give virtually the same theoretical probability of error curve for MDL as shown in Fig. 5.9.

5.10. CONCLUSIONS

This chapter introduced some measures for characterizing the performances of signal-subspace methods in general (the MUSIC and minimum-norm methods, in particular) for estimating the number and angles of arrival of multiple narrow-band plane waves received by noisy elements of sensor arrays. Accuracy of order determination schemes that use criteria such as the AIC and the MDL was characterized theoretically in terms of the probabilities of erroneously estimating the number of sources. It was argued that the effectiveness of all signal-subspace methods are ultimately dependent on the accuracy of the estimated signal subspace. The quality of such an estimated subspace was quantified according to the mean-square deviation from the ideal value of one, of the cosine of an effective angle between the estimated and true subspaces. All these measures were explicitly related to physically significant parameters of the sources and the measurement system.

Nonlinear parameter estimators possess thresholds below which measures of performance degrade very rapidly. Usually, the measure of interest is the mean-square estimation error, which is not always theoretically obtainable in closed form for the angle estimation algorithms addressed here. Alternatively, the resolution performance of closely spaced sources was used to characterize the threshold behavior of two popular signal-subspace methods: MUSIC and min-norm. Asymptotic statistical perturbation analysis was used to obtain resolution threshold signal-to-noise ratios for both methods in terms of the angular proximity of the sources, array aperture, and number of snapshots. It was shown that based on the resolution criterion used in this chapter, min-norm has a lower resolution threshold for scenarios considered here. Both thresholds, however, were shown to be very strongly dependent on the angular separation of the sources [threshold SNR $\propto$ (angular separation)$^{-\alpha}$, $3 \leq \alpha \leq 4$].

It should be emphasized that the performance threshold of "spectral signal subspace methods" such as MUSIC is bais−, rather than variance-limited. Therefore, resolution threshold may, in fact, be the natural quantifier of the limiting performance of such methods.

ACKNOWLEDGMENTS

Our work on signal-subspace processing at the University of Minnesota has been supported in part by the National Science Foundation under Grants ECS-8105962 and ECS-8414316 and in part by the Office of Naval Research under Grant N00014-86-K-0410. The simulations and computations leading to Figs. 5.1, 5.2, 5.4–12, and 5.16–21 were carried out by H. Hung and J. Yang of the University of Minnesota.

APPENDIX

In this appendix we derive approximate expressions for the expectations in (5.133). Again, we only keep the first-order terms.

1. Let $[\mathbf{A}]_{ij}$ denote the ijth element of a matrix $\mathbf{A}$. The second-order moments of $\boldsymbol{\gamma}$ are given by

$$\overline{[\boldsymbol{\gamma}^*\boldsymbol{\gamma}^T]}_{ij} = \overline{[\boldsymbol{\gamma}\boldsymbol{\gamma}^{\dagger *}]}_{ij} \simeq \sum_{\substack{k=1\\k\neq i}}^{L} \frac{\lambda_i\lambda_k}{(\lambda_i - \lambda_k)^2 K} |e_k(1)|^2\, \delta_{ij} \tag{5.A.1}$$

and

$$\overline{[\boldsymbol{\gamma}\boldsymbol{\gamma}^T]}_{ij} = \overline{[\boldsymbol{\gamma}^*\boldsymbol{\gamma}^\dagger]}^*_{ij} \simeq \frac{-\lambda_i\lambda_j}{K(\lambda_i - \lambda_j)^2} e_i(1)e_j(1)(1 - \delta_{ij}). \tag{5.A.2}$$

2. $$\overline{\rho^2} = \frac{2}{(1 - \mathbf{g}^\dagger\mathbf{g})^2} [\mathbf{g}^\dagger\overline{\boldsymbol{\gamma}\boldsymbol{\gamma}^\dagger}\mathbf{g} + \mathrm{Re}(\mathbf{g}^T\overline{\boldsymbol{\gamma}^*\boldsymbol{\gamma}^\dagger}\mathbf{g})]. \tag{5.A.3}$$

Simplifying, we obtain

$$\overline{\rho^2} \simeq \frac{2}{(1 - \mathbf{g}^\dagger\mathbf{g})^2} \sum_{i=1}^{d} \sum_{k=M+1}^{L} \frac{\lambda_i\lambda_k}{K(\lambda_i - \lambda_k)^2} |e_k(1)|^2 |e_i(1)|^2. \tag{5.A.4}$$

3. $\boldsymbol{\epsilon}'\mathbf{g}^*\mathbf{g}^T\boldsymbol{\epsilon}'^\dagger$ is derived by noting that $[\boldsymbol{\epsilon}']_{ij} = \eta_j(i+1)$ and $g(i) = e_i(1)$

$$\overline{\boldsymbol{\epsilon}'\mathbf{g}^*\mathbf{g}^T\boldsymbol{\epsilon}'^\dagger} \simeq \sum_{i=1}^{d} |e_i(1)|^2 \sum_{j\neq i}^{L} \frac{\lambda_i\lambda_j}{K(\lambda_i - \lambda_j)^2}\mathbf{F}_j\mathbf{F}_j^\dagger, \tag{5.A.5}$$

where $\mathbf{F}_j^T = e_j(2), e_j(3), \ldots, e_j(L)$.

4. $$\overline{\rho\boldsymbol{\gamma}^T} = \frac{1}{1 - \mathbf{g}^\dagger\mathbf{g}}[\mathbf{g}^\dagger\overline{\boldsymbol{\gamma}\boldsymbol{\gamma}^T} + \overline{\boldsymbol{\gamma}^\dagger\mathbf{g}\boldsymbol{\gamma}^T}]. \tag{5.A.6}$$

From (5.A.2), we get

$$[\mathbf{g}^\dagger\overline{\boldsymbol{\gamma}\boldsymbol{\gamma}^T}]_j \simeq \frac{-\lambda_j e_j(1)}{K}\sum_{i\neq j}^{d} \frac{\lambda_i}{(\lambda_i - \lambda_j)^2}|e_i(1)|^2. \tag{5.A.7}$$

The second term follows from (5.A.1) as

$$[\overline{\boldsymbol{\gamma}^\dagger\mathbf{g}\boldsymbol{\gamma}^T}]_j \simeq \sum_{k\neq j}^{L} \frac{\lambda_j\lambda_k}{(\lambda_j - \lambda_k)^2 K} e_j(1)\,|e_k(1)|^2. \tag{5.A.8}$$

5. $$\overline{\rho\boldsymbol{\epsilon}'^\dagger} = \frac{1}{1 - \mathbf{g}^\dagger\mathbf{g}}[\mathbf{g}^\dagger\overline{\boldsymbol{\gamma}\boldsymbol{\epsilon}'^\dagger} + \overline{\boldsymbol{\gamma}^\dagger\mathbf{g}\,\boldsymbol{\epsilon}'^\dagger}]. \tag{5.A.9}$$

Again, we first consider the first term in the brackets:

$$\mathbf{g}^\dagger\overline{\boldsymbol{\gamma}\boldsymbol{\epsilon}'^\dagger} = \sum_{i=1}^{d} e_i^*(1)\overline{\eta_i(1)\boldsymbol{\epsilon}'^\dagger}. \tag{5.A.10}$$

Therefore,

$$[\mathbf{g}^\dagger\overline{\boldsymbol{\gamma}\,\boldsymbol{\epsilon}'^\dagger}]_{ij} \simeq \sum_{k\neq i}^{L} \frac{\lambda_i\lambda_k\,e_i^*(1)}{(\lambda_i - \lambda_k)^2\,K} e_k(1)\,e_k^*(j+1), \qquad j = 1, \ldots, L-1. \tag{5.A.11}$$

The second term is given by

$$\overline{\boldsymbol{\gamma}^\dagger\mathbf{g}\,\boldsymbol{\epsilon}'^\dagger} = \sum_{i=1}^{d} e_i(1)\overline{\eta_i^*(1)\boldsymbol{\epsilon}'^\dagger}, \tag{5.A.12}$$

resulting in the ijth element of the matrix as

$$[\overline{\boldsymbol{\gamma}^\dagger\mathbf{g}\,\boldsymbol{\epsilon}'^\dagger}]_{ij} \simeq -\sum_{k\neq i}^{d} \frac{\lambda_i\lambda_k}{(\lambda_i - \lambda_k)^2 K} e_k(1)e_k^*(j+1)e_i^*(1). \tag{5.A.13}$$

6. The final term of interest is $\overline{\boldsymbol{\gamma}^* \mathbf{g}^T \boldsymbol{\epsilon}'^\dagger}$. First, consider the jth element of $\mathbf{g}^T \boldsymbol{\epsilon}'^\dagger$:

$$[\mathbf{g}^T \boldsymbol{\epsilon}'^\dagger]_j = \sum_{k=1}^{d} e_k(1)\, \eta_k^*(j+1). \tag{5.A.14}$$

Then

$$\begin{aligned} \overline{[\boldsymbol{\gamma}^* \mathbf{g}^T \boldsymbol{\epsilon}'^\dagger]}_{ij} &= \sum_{k=1}^{d} e_k(1)\, \overline{\eta_i^*(1)\, \eta_k^*(j+1)} \\ &= -\sum_{k \neq i}^{d} \frac{\lambda_i \lambda_k}{K(\lambda_i - \lambda_k)^2} |e_k(1)|^2 e_i^*(j+1)\,. \end{aligned} \tag{5.A.15}$$

REFERENCES

1. R. O. Schmidt, "Multiple Emitter Location and Signal Parameter Estimation," *Proc. RADC Spectrum Estimation Workshop*, Rome Air Development Center, Rome, New York, 1979, pp. 243–258.
2. R. Kumaresan and D. W. Tufts, "Estimating the Angles of Arrival of Multiple Plane Waves," *IEEE Trans. Aerospace and Electronic Systems*, **19** (1983), 134–139.
3. H. L. Van Trees, *Detection, Estimation, and Modulation Theory,* Part I (New York: John Wiley, 1968).
4. D. R. Brillinger, *Time Series: Data Analysis and Theory*, expanded ed. (San Francisco: Holden-Day, 1981).
5. A. Papoulis, *Probability, Random Variables, and Stochastic Processes* (New York: McGraw-Hill, 1984).
6. H. Wang and M. Kaveh, "Coherent Signal-Subspace Processing for the Detection and Estimation of Angles of Arrival of Multiple Wideband Sources," *IEEE Trans. Acoustics, Speech, and Signal Processing,* **33** (1985), 823–831.
7. H. Wang and M. Kaveh, "Performance of Singal-Subspace Processing Methods," Part Two: "Coherent Wideband Systems," *IEEE Trans. Acoustics, Speech, and Signal Processing*, **35** 1583–1591 (November 1987).
8. K. Buckley and L. J. Griffiths, "Eigenstructure Broadband Source Location Estimation," *Proc. IEEE Int. Conf. on Acoustic, Speech, and Signal Processing,* Tokyo, Japan, 1986, pp. 1869–1872.
9. V. F. Pisarenko, "The Retrieval of Harmonics from a Covariance Function," *Geophys. Jour. Royal Ast. Soc.,* **33** (1973), 347–366.
10. S. S. Reddi, "Multiple Source Location—A Digital Approach," *IEEE Trans. Aerospace and Electronic Systems,* **15** (1979), 95–105.
11. W. S. Ligget, "Passive Sonar: Fitting Models to Multiple Time Series," in *Signal Processing*, ed. J. W. R. Griffiths (New York: Academic Press, 1973), pp. 327–345.
12. N. L. Owsley, "A Recent Trend in Adaptive Spatial Processing for Sensor Arrays," in *Signal Processing,* ed. J. W. R. Griffiths (New York: Academic Press, 1973), pp. 591–604.

13. D. H. Johnson, "The Application of Spectral Estimation Methods to Bearing Estimation Problems," *Proc. IEEE,* **70** (1982), 1018–1028.

14. G. Bienvenu, "Eigensystem Properties of the Sampled Space Correlation Matrix," *Proc. IEEE Int. Conf. on Acoustics, Speech, and Signal Processing,* Boston, 1983, pp. 332–335.

15. S. Haykin, "Radar Array Processing for Angle of Arrival Estimation," in *Array Signal Processing.* ed. S. Haykin (Englewood Cliffs, N.J.: Prentice-Hall, 1985), pp. 194–292.

16. B. Noble, *Applied Linear Algebra* (Englewood Cliffs, N.J.: Prentice-Hall, 1969).

17. T. W. Anderson, "Asymptotic Theory for Principal Component Analysis," *Ann. Math. Stat.*, **34** (1963), 122–148.

18. R. P. Gupta, "Asymptotic Theory for Principal Component Analysis in the Complex Case," *J. Indian Stat. Assoc.,* **3** (1965), 97–106.

19. J. H. Wilkinson, *The Algebraic Eigenvalue Problem* (New York: Oxford University Press, 1965).

20. M. Kaveh and A. J. Barabell, "The Statistical Performance of the MUSIC and the Minimum-Norm Algorithms in Resolving Plane Waves in Noise," *IEEE Trans. Acoustics, Speech, and Signal Processing,* **34** (1986), 331–341, correction of the paper in *IEEE Trans. Acoustics, Speech, and Signal Processing*, **34**, 633, 1986.

21. D. J. Jeffries and D. R. Farrier, "Asymptotic Results for Eigenvector Methods," *IEEE Proc.,* Part F, **132** (1985), 589–594.

22. P. R. Krishnaiah, "Some Recent Developments on Complex Multivariate Distributions," *J. Multivariate Anal.* (1976), 1–30.

23. C. G. Khatri, "Distribution of the Largest or the Smallest Characteristic Roots Under Null Hypothesis Concerning Complex Multivariate Normal Populations," *Ann. of Math. Stat.,* (1964), 1807–1810.

24. M. Kaveh, H. Wang, and H. Hung, "On the Theoretical Performance of a Class of Estimators of the Number of Narrowband Sources," *IEEE Trans. Acoustics, Speech, and Signal Processing,* **35** (September 1987).

25. J. E. Hudson, *Adaptive Array Principles* (Stevenage, U. K.: Peter Peregrinus, 1982).

26. D. N. Simkins, "Multichannel Angle of Arrival Estimation," Ph. D. dissertation, Stanford University, Stanford, California, 1980.

27. H. Akaike, "A New Look at Statistical Model Identification," *IEEE Trans. Auto. Control,* **19** (1974), 716–723.

28. J. Rissanen, "Modeling by Shortest Data Description," *Automatica*, **14** (1978), 465–471.

29. M. Wax and T. Kailath, "Detection of Signals by Information Theoretic Criteria," *IEEE Trans. Acoustics, Speech, and Signal Processing,* **33** (1985), 387–392.

30. H. Wang and M. Kaveh, "Performance of Signal-Subspace Processing Methods," Part I: "Narrowband Systems," *IEEE Trans. Acoustics, Speech, and Signal Processing,* **34** (1986), 1201–1209.

31. G. H. Golub and C. F. Van Loan, *Matrix Computation* (Baltimore, Md.: Johns Hopkins University Press, 1983).

32. C. R. Rao, *Linear Statistical Inference and Its Applciations* (New York: John Wiley, 1973).

33. J. F. Böhme, "On the Sensitivity of Orthogonal Beamforming," *Proc. IEEE Int. Conf. on Acoustics, Speech, and Signal Processing,* Paris, 1982, pp. 787–790.

34. K. C. Sharman and T. S. Durrani, "Resolving Power of Signal Subspace Methods for Finite Data Lengths," *Proc. IEEE Int. Conf. on Acoustics, Speech, and Signal Processing,* Tampa, Florida, 1985, pp. 1501–1504.

35. A. J. Barabell, "Improving the Resolution Performance of Eigenstructure-Based Direction-Finding Algorithms," *IEEE Int. Conf. on Acoustics, Speech, and Signal Processing*, Boston, 1983, pp. 336–339.

36. B. D. Rao and K. V. S. Hari, "Statistical Performance Analysis of the Minimum-Norm Method," *Proc. IEEE Int. Conf. on Acoustics, Speech, and Signal Processing,* Glasgow, Scotland, 1989, pp. 2760–2763.

37. B. D. Rao and K. V. S. Hari, "Performance Analysis of Subspace-Based Methods," *Proc. Fourth ASSP Workshop on Spectrum Estimation and Modeling,* Minneapolis, Minnesota, 1988, pp. 92–97.

38. R. Roy, A. Paulraj, and T. Kailath, "Estimation of Signal Parameters via Rotational Invariance Techniques-ESPRIT," *Proc. Asilomar Conf. on Circuits, Systems and Computing,* Monterey, California, 1985.

39. B. H. Kwon and S. U. Pillai, "Performance Analysis of Eigenvector-Based High-Resolution Estimators for Direction Finding in Correlated and Coherent Scenes," *Proc. Fourth ASSP Workshop on Spectrum Estimation and Modeling,* Minneapolis, Minnesota, 1988, pp. 103–107.

40. D. W. Tufts, A. C. Kot, and R. J. Vaccaro, "The Threshold Analysis of SVD-Based Algorithms," *Proc. IEEE Int. Conf. on Acoustics, Speech, and Signal Processing,* New York, 1988, pp. 2416–2419.

41. B. Porat and B. Friedlander, "Analysis of the Relative Efficiency of the MUSIC Algorithm," *IEEE Trans. Acoustics, Speech, and Signal Processing,* **36** (1988), 532–534.

42. P. Stoica and A. Nehorai, "MUSIC, Maximum-Likelihood, and Cramér-Rao Lower Bound," *IEEE Trans. Acoustics, Speech, and Signal Processing,* **37** (1989), 720–741.

43. H. B. Lee and M. S. Wengrowitz, "Improved High-Resolution Direction Finding Through Use of Homogeneous Constraints," *Proc. Fourth ASSP Workshop on Spectrum Estimation and Modeling,* Minneapolis, Minnesota, 1988, pp. 152–157.

44. X. L. Xu and K. M. Buckley, "Statistical Performance Comparison of MUSIC in Element-Space and Beam-Space," *Proc. IEEE Int. Conf. on Acoustics, Speech, and Signal Processing,* Glasgow, Scotland, 1989, pp. 2124–2127.

45. Q. T. Zhang, K. M. Wong, P. C. Yip, and J. P. Reilly, "Statistical Analysis of the Performance of Information Theoretic Criteria in the Detection of the Number of Signals in Array Processing," *IEEE Trans. Acoustics, Speech, and Signal Processing,* **37** (1989), pp. 1557–1567.

6

Focused Wide-Band Array Processing for Spatial Spectral Estimation

Jeffrey Krolik

6.1 INTRODUCTION

This chapter addresses the problem of detecting and localizing multiple wide-band acoustic sources given the outputs of a passive sensor array. Since far-field sources generate plane-wave components in the acoustic field, a critical step in source detection and localization is the discrimination of multiple plane-wave signals from a background of diffuse ambient noise. This can be achieved by mapping the field directionality, also known as the spatial spectrum, and associating local power maxima with the directions of the far-field sources. Although the conventional delay-and-sum (DS) beamformer is commonly used to map field directionality, physical limitations on array aperture size constrain its ability to resolve closely spaced arrivals. As a result, "high-resolution" spatial spectral estimation methods have been the focus of considerable study (cf. [1] and the references contained therein). In most cases, these methods have been developed for narrow-band signals and require the estimation of the cross-spectral density matrix (CSDM) of array outputs. Extension of these techniques to the wide-band problem has then been achieved by estimating a series of cross-spectral density matrices over the receiver band and then combining the results of narrow-band spatial spectral estimation at each frequency [2]. A serious limitation of this approach, however, has been the requirement for long observation times to obtain statistically stable measurements of the necessary frequency-dependent cross-spectral density matrices [3]. In practice, because changing source locations and nonstationary field properties limit the available observation time, the use of narrow-band high-resolution techniques and their wide-band extensions is often precluded.

In this chapter, focused wide-band array processing techniques are studied as methods that can more effectively exploit larger signal bandwidths to reduce significantly the observation time required to achieve high-resolution performance. The objective of wideband focusing is to preprocess the sensor outputs so that broadband sources can be represented by rank-one models. This approach is motivated by the fact that in limited observation time settings, the parameters of low-rank models can be estimated with greater statistical stability. In contrast to the narrow-band case, however, broad-band sources cannot be represented by rank-one models without preprocessing the array outputs [4]. The idea of wide-band focusing was first proposed as a means of lowering the threshold signal-to-noise ratio achievable by eigenvector-based spatial spectral estimation methods [5]. The resulting wide-band techniques are called coherent signal-subspace (CSS) methods because focusing matrices are used to align the signal subspaces associated with narrow-band cross-spectral density matrices in the receiver bandwidth [5–13]. After signal-subspace alignment, the estimated narrow-band CSDMs are averaged to obtain a focused covariance matrix wherein each source ideally contributes a rank-one component. As shown in [7], the resulting focused covariance matrix can be estimated with an accuracy that reflects the full time-bandwidth product of the sources. Although CSS methods offer more statistically stable performance, particularly in "multigroup" scenarios where the sources are clustered around several widely separated directions, formation of the necessary focusing matrices has until recently required preliminary estimates of the source directions. In the multigroup case, preliminary source location estimates have been used to prewhiten the field spatially, thereby reducing it to the single-group situation [5], or to form rotational signal-subspace focusing matrices [9]. Aside from the computational burden of forming preliminary location estimates, these focusing methods are also vulnerable to source location bias resulting from uncertainty in the temporal spectra of the signals [14].

In addition to coherent signal-subspace methods, several other spatial spectral estimation methods have been proposed for use with larger bandwidth sources [16–18]. A common feature of these methods is their use of low-rank models for representing broad-band sources received at the sensor array. For example, in broad-band signal-subspace spatial spectral estimation (BASS–ALE) [15], low-rank signal representations are used to approximate the space-time observations of broad-band sources. In BASS–ALE methods, source location bias resulting from an uncertain signal spectrum can be controlled by increasing the number of location vectors used to span the low-rank signal subspace. The approaches examined in this chapter avoid the need for preliminary estimates of the source locations and are not biased as a result of uncertainty in the source spectra.

In this chapter, focusing techniques are examined that ensure that broad-band sources can always be characterized by rank-one models, regardless of their temporal spectra or location. In the first approach, the covariance matrix of sensor outputs is formed after steering delays are inserted to form a conventional DS beamformer beam. The resulting space-time statistic, called the steered covariance matrix (STCM), effectively focuses wide-band arrivals from its steering direction. In other words, a wide-band source from the steering direction of the STCM contributes a rank-one component to the matrix, regardless of its temporal spectrum. Thus using a different STCM for each direction of interest, every source in the field can be handled by a rank-one model. As shown in this chapter, the advantage of using the STCM in broad-band settings is that it can be estimated with much greater statistical stability than individual CSDMs. In Section 6.4.1, the stability of the STCM is used to facilitate a significant reduction in the threshold observation time required to perform minimum variance (or equivalently maximum likelihood [19]) spatial spectral estimation [20, 21]. Further, a simulation study and experimental results with actual towed-array data are presented that suggest the steered minimum-variance (STMV) method can facilitate statistically stable high-resolution mapping of field directionality using acceptably short observation times.

Although high-resolution steered covariance methods have lower threshold observation times than their CSDM-based counterparts, their implementation is hampered by excessive computational requirements. The computational burden of steered covariance methods arises from the need to compute a different broad-band covariance matrix for each steering direction of interest. One possible means of reducing this computational load consists of computing steered covariance matrices only for a few primary steering directions and then using secondary steering vectors to form spatial spectral estimates. This technique has been called the doubly steered coherent signal-subspace method [13]. Although the doubly steered approach yields performance comparable to other coherent signal-subspace methods, it also suffers from source location bias depending on source direction and spectral content [14].

Clearly, a more computationally efficient approach to wide-band focusing could be achieved by using a transformation that focuses arrivals accurately from all directions simultaneously. In this case, the entire spatial spectrum could be estimated using a single focused wide-band covariance matrix. Indeed, this is the original objective of coherent signal-subspace methods. Nevertheless, it is only recently that focusing methods have been proposed which can reduce each wide-band source in multigroup scenarios to a rank-one representation without preliminary estimates of the group locations. These new methods are based on adjusting the spatial sampling rate or "spatially resampling" the array outputs as a function of temporal frequency so that broad-band sources are aligned in the spatial frequency domain [10, 11, 12]. The advantages of spatial resampling are that it can facilitate statistically stable broad-band spatial spectral estimation with minimal source location bias and relatively modest computational requirements. A limitation of present spatial resampling methods is that they apply only to the reception of plane-wave sources with uniformly spaced linear arrays. In this chapter, the design and performance of linear shift-variant filters for focused wide-band spatial spectral estimation via spatial resampling is examined. The application of spatial resampling to minimum-variance spatial spectral estimation is also considered.

The remainder of this chapter is divided into three principal sections. In the next section, the concept and motivation for focusing wide-band sensor array outputs are presented. In Section 6.3, estimation of focused wide-band covariance matrices by presteering and spatially resampling is considered. And finally in Section 6.4, the performance of presteered and spatially resampled minimum-variance spatial spectral estimation is examined.

6.2 FOCUSED WIDE-BAND ARRAY PROCESSING CONCEPTS

In this section, focused wide-band processing is motivated as a means of forming frequency-independent wide-band covariance matrices which can be estimated with a statistical stability which reflects the full observation time-bandwidth product of wide-band sources. In this section, steered covariance and spatial resampling methods are considered as approaches for achieving wide-band focusing objectives.

6.2.1 Model Formulation and the Wide-Band Problem

The concept of wide-band focusing can be motivated by considering an array of wide-band sensors which measure a field generated by P broad-band point sources in the presence of diffuse ambient noise. It is assumed that the sources and background noises are statistically independent. Denote the complex analytic representation of the continuous-time output of a sensor at location x_m by $y(m, t)$, for $m = -K, \ldots, K$. The total number of sensors is $M = 2K + 1$. Under the assumed model, $y(m, t)$, observed over a time interval of T seconds, can be expressed as

$$y(m, t) = \sum_{i=1}^{P} s_i[t - \tau_m(\theta_i)] + v(m, t) \tag{6.1}$$

where $s_i(t)$, $i = 1, \ldots, P$ are stationary, zero-mean, random processes corresponding to each received source signal and $v(m, t)$ is the spatially and temporally stationary, zero-mean, background noise process at the mth sensor. For a source at location θ_i, $\tau_m(\theta_i)$ is the signal propagation delay to the mth sensor, as measured relative to the array coordinate origin. The locations, θ_i, $i = 1, \ldots, P$, of the P sources are the parameters to be estimated from a finite-time observation of the sensor outputs. Except under multipath propagation conditions, the sources are assumed to be mutually uncorrelated. In the multipath case, the correlation between the ith and nth multipath source signals is modeled by

$$s_i(t) = a_{in} s_n(t - \xi_{in}) \tag{6.2}$$

where a_{in} is the relative attenuation and ξ_{in} is the relative multipath delay between the ith and nth arrivals.

For purposes of understanding the wide-band array processing problem, it is useful to consider a frequency-domain representation of the sensor outputs. In particular, the time-domain vector of outputs, $\mathbf{y}(t) = [y(-K, t), \ldots, y(K, t)]^T$, can be represented over the time interval from $-T/2$ to $T/2$, by the frequency-domain vectors, $\mathbf{Y}(\omega_k) = [Y(-K, \omega_k), \ldots, Y(K, \omega_k)]^T$, where superscript T denotes transpose. The elements, $Y(m, \omega_k)$, of $\mathbf{Y}(\omega_k)$ correspond to the Fourier series coefficients of $y(m, t)$ at frequency $\omega_k = 2\pi k/T$. For large T, the frequency-domain vectors, $\mathbf{Y}(\omega_k)$ and $\mathbf{Y}(\omega_r)$, are uncorrelated for $k \neq r$. Using this result and assuming the sensor outputs are approximately bandlimited to $\omega_l \leq \omega_k \leq \omega_h$, the second-order statistics of the field are completely specified by the set of narrow-band cross-spectral density matrices, $R(\omega_k) = E\{\mathbf{Y}(\omega_k)\mathbf{Y}(\omega_k)^+\}$, for $k = l, \ldots, h$, where $E\{\cdot\}$ denotes expectation and superscript $+$ indicates conjugate transpose. For the model of (6.1), $R(\omega_k)$ can be expressed as [21]

$$R(\omega_k) = A(\omega_k, \boldsymbol{\theta}) P_s(\omega_k) A(\omega_k, \boldsymbol{\theta})^+ + R_v(\omega_k) \tag{6.3}$$

where $A(\omega_k, \boldsymbol{\theta}) = [\mathbf{a}(\omega_k, \theta_1), \mathbf{a}(\omega_k, \theta_2), \ldots, \mathbf{a}(\omega_k, \theta_P)]$ is the $M \times P$ source direction matrix, $\mathbf{a}(\omega_k, \theta_i) = [e^{-j\omega_k \tau_{-K}(\theta_i)}, \ldots, e^{-j\omega_k \tau_K(\theta_i)}]^T$ is the direction vector of the ith source, $P_s(\omega_k)$ is a $P \times P$ source spectral density matrix, $\boldsymbol{\theta} = [\theta_1, \ldots, \theta_P]$, and $R_v(\omega_k)$ represents the $M \times M$ noise covariance matrix.

The underlying problem in wide-band spatial spectral estimation stems from the structure of $R(\omega_k)$ given in (6.3). In particular, although the source location vector, $\boldsymbol{\theta}$, is frequency independent, this parameter vector is embedded in source direction matrices, $A(\omega_k, \boldsymbol{\theta})$, $k = l, \ldots, h$, which are clearly frequency dependent. For wide-band fields, this means that the same source contributes a *different* rank-one component to $R(\omega_k)$ at each temporal frequency. Hence without preprocessing, the frequency-domain vectors, $\mathbf{Y}(\omega_k)$ for $k = l, \ldots, h$, cannot be used to form a single wide-band covariance matrix wherein each source has a rank-one representation. Thus wide-band spatial spectral estimation has commonly consisted of first forming narrow-band spatial spectral estimates at each frequency and then combining the results to obtain so-called "incoherent" wide-band methods [2, 5]. The drawback of this approach is that the threshold observation time for incoherent methods is dictated by estimation of individual narrow-band cross-spectral density matrices at each frequency in the receiver band. The aim of wide-band focusing is to preprocess $\mathbf{Y}(\omega_k)$ so that components of interest in the transformed sensor output vectors have the same rank-one description at all temporal frequencies. The narrow-band covariance matrices of the preprocessed outputs may then be summed across frequency to obtain a focused wide-band covariance matrix wherein components of interest can be estimated with a statistical stability which reflects their full observation time-bandwidth product.

6.2.2 Steered Covariance Methods

In steered covariance methods, the sensor outputs are preprocessed to ensure that arrivals from a particular direction have the same rank-one representation at all temporal frequencies. Letting θ denote the direction of interest, the focusing property for arrivals from direction θ can be achieved by forming $T_s(\theta, \omega_k)\mathbf{Y}(\omega_k)$, where $T_s(\theta, \omega_k)$ is defined by

$$T_s(\theta, \omega_k) = \begin{bmatrix} e^{j\omega_k \tau_{-K}(\theta)} & & \cdot & 0 \\ 0 & e^{j\omega_k \tau_{-K+1}(\theta)} & & \cdot \\ \cdot & & & \\ 0 \cdot \cdot & & \cdot & e^{j\omega_k \tau_K(\theta)} \end{bmatrix} \tag{6.4}$$

The matrix $T_s(\theta, \omega_k)$ is called the steering matrix since $T_s(\theta, \omega_k)\mathbf{Y}(\omega_k)$ can be used to construct a vector of steered time-domain sensor outputs,

$$\mathbf{y}_s(t, \theta) = \sum_{k=l}^{h} T_s(\theta, \omega_k)\mathbf{Y}(\omega_k)e^{j\omega_k t} \tag{6.5}$$

where substituting (6.4) into (6.5), $\mathbf{y}_s(t, \theta)$ is given by

$$\mathbf{y}_s(t, \theta) = \{y[-K, t + \tau_{-K}(\theta)], y[-K+1, t + \tau_{-K+1}(\theta)], \ldots, y[K, t + \tau_K(\theta)]\}^T \tag{6.6}$$

Note that the steering delays in (6.6) are precisely those required to form a delay-and-sum beamformer beam in steering direction, θ.

The fact that a source in direction θ has the same rank-one representation in the steered sensor outputs at all frequencies can be clearly seen by considering $E\{T_s(\theta, \omega_k)\mathbf{Y}(\omega_k)\mathbf{Y}(\omega_k)^+ T_s(\theta, \omega_k)^+\} = T_s(\theta_p, \omega_k)R(\omega_k)T_s(\theta_p, \omega_k)^+$ using $R(\omega_k)$ of (6.3) when θ equals a particular source direction, θ_p. In general, summing presteered narrow-band covariance matrices over frequency results in a focused wide-band covariance matrix given by

$$R_s(\theta) = \sum_{k=l}^{h} T_s(\omega_k, \theta)R(\omega_k)T_s(\omega_k, \theta)^+ \tag{6.7}$$

From (6.5), $R_s(\theta)$ is equal to $E\{\mathbf{y}_s(t, \theta)\mathbf{y}_s(t, \theta)^+\}$, the covariance of time-domain sensor outputs when delays have been inserted to form a conventional delay-and-sum beamformer beam. Thus $R_s(\theta)$ is referred to as the steered covariance matrix corresponding to direction θ.

The structure of $R_s(\theta)$ under the model of (6.1) can be used to motivate a variety of wide-band spatial spectral estimation methods. For example, consider a horizontal line array with sensors equally spaced a distance, d, apart at positions, $x_m = md$, $m = -K, \ldots, K$. In this case, θ is the bearing of the source relative to array broadside, $\tau_m(\theta) = m \cdot \tau(\theta)$, where $\tau(\theta) = d/c \cdot \sin(\theta)$, and c is the propagation speed. Using the model of (6.1), the jkth element of $R_s(\theta)$ is a function of $m = j - k$

given by

$$\{R_s(\theta)\}_{jk} = \sum_{i=1}^{P} \rho_i\{m \cdot [\tau(\theta) - \tau(\theta_i)]\} + \eta[m, m \cdot \tau(\theta)] \tag{6.8}$$

where $\rho_i(\tau) = E\{s_i(t + \tau)s_i(t)^*\}$ is the autocorrelation function of the ith source and $\eta(m, \tau) = E\{v(0, t + \tau)v(m, t)^*\}$ is the cross-correlation function of the noise received at the array coordinate origin and the mth sensor. Superscript * denotes complex conjugation. Note that (6.8) is valid under the condition that either the sources are uncorrelated or the relative multipath delays, ξ_{in}, are much longer than the correlation time of the source signals plus the propagation delay across the array. This insensitivity of the STCM to multipath correlation implies that STCM-based techniques can be used to resolve coherent multipath sources without modification [22]. Further, when the steering direction, θ, is equal to θ_p, then (6.8) becomes

$$\{R_s(\theta_p)\}_{jk} = \sum_{\substack{i=1 \\ i \neq p}}^{P} \rho_i\{m \cdot [\tau(\theta_p) - \tau(\theta_i)]\} + \sigma_p^2 + \eta[m, m \cdot \tau(\theta_s)] \tag{6.9}$$

where $\sigma_p^2 = \rho_p(0)$ is the power of the pth source. Thus for a source aligned with the steering direction of the array, the STCM contains a constant, rank-one component equal to the source power regardless of its spectral signature. Since for broadband sources each $\rho_i(\tau)$ will be a slowly decaying function of τ, the off-steering direction terms in the summation of (6.9) will tend to zero with increasing m. These signals can be interpreted as partially coherent components with decorrelation lengths which decrease as their separation, $|\tau(\theta_p) - \tau(\theta_i)|$, from the steering direction increases. The last term in (6.9) contributes a partially coherent component to $R_s(\theta)$ due to the noise field. For diffuse ambient noise, this term decays with increasing m regardless of steering direction, θ. The rate at which this term decays depends on the spatial and temporal correlation of the diffuse ambient noise field. For example, in spatially white noise, $\eta[m, m\ \tau(\theta)] = \sigma_v^2\delta_K(m)$, where σ_v^2 is the noise power and $\delta_K(m)$ is the Kronecker delta. In this case, the noise contributes to the STCM only on the main diagonal regardless of its autopower spectral density. From the foregoing discussion, the power level of a point source in the steering direction, θ, can be estimated by measuring the level of the constant, or "dc," term in (6.9). Performing this estimation for a closely spaced set of steering directions yields a broad-band spatial power spectral estimate. Note that delay-and-sum beamforming corresponds to making a conventional Blackman-Tukey estimate of this dc component. In Section 6.4.1, minimum-variance spatial spectral estimation is examined as a method of estimating the dc component of (6.9) for each steering direction, θ.

It is worth noting that the STCM expressed by (6.7) with the steering matrix of (6.4) is of the same form as the focused covariance matrix used in the coherent signal-subspace method proposed by Wang and Kaveh [5] for the case where all sources in the field are in a single-group, unresolved by a conventional DS beamformer. In [5], $R_s(\theta)$ is steered using a preliminary estimate of the group location so that a single $R_s(\theta)$ can be used to *approximately* focus all sources within the group. In the STCM-based methods discussed here, a different $R_s(\theta)$ is computed for each steering direction, θ, of interest without using preliminary estimates of the source locations. The need to compute $R_s(\theta)$ as a function of θ makes STCM-based methods more computationally intensive than the approach taken in [5]. However, it also avoids source location bias which can result when a single steered covariance matrix is used to estimate the spatial spectrum in directions not corresponding to the precise steering direction, θ [14].

6.2.3 Spatial Resampling Methods

The need to compute steered covariance matrices in all directions of interest is a result of the fact that, strictly speaking, each STCM only focuses wide-band arrivals in a single steering direction, θ. Clearly, a transformation that could accurately focus arrivals from all directions simultaneously would result in a more computationally efficient approach. Ideally, this focusing operation would transform each narrow-band covariance matrix so that every source in the field has a frequency-independent rank-one characterization, regardless of its location. In this case, a single focused wide-band covariance matrix could be used to estimate the spatial spectrum in all directions, thereby offering a significant computational advantage over steered covariance methods. For equispaced linear arrays, an approach which achieves this focusing objective without using preliminary estimates of the source directions is spatial resampling [10, 11, 12].

The spatial resampling concept is motivated by treating the outputs of a discrete M sensor array as the result of spatially sampling a continuous linear array. For a horizontal line array with sensors spaced a distance, d, apart at positions $x_m = md$, $m = -K, \ldots, K$, the mth element of the frequency-domain sensor output vector, $\mathbf{Y}(\omega_k)$, can be expressed as

$$Y(m, \omega_k) = \sum_{i=1}^{P} S_i(\omega_k)e^{j\omega_k md \sin(\theta_i)/c} + V(m, \omega_k) \tag{6.10}$$

where $S_i(\omega_k)$ represents the ith source component at frequency ω_k and $V(m, \omega_k)$ denotes the noise output of the mth sensor at frequency ω_k. As in (6.8), θ_i, $i = 1, \ldots, P$, are the bearings of the sources relative to array broadside. Note that for an equispaced horizontal line array, the narrow-band CSDM at frequency, ω_k, is given by $R(\omega_k)$ in (6.3) where the ith column of the source direction matrix, $A(\omega_k, \boldsymbol{\theta})$, is $\mathbf{a}(\omega_k, \theta_i) = [e^{j\omega_k Kd \sin(\theta_i)/c}, \ldots, e^{-j\omega_k Kd \sin(\theta_i)/c}]^T$. Letting $Y_f(x, \omega_k)$ denote the field incident upon a *continuous* array positioned along the x-axis at frequency, ω_k, $Y(m, \omega_k)$ can be viewed as the result of spatially sampling

$$Y_f(x, \omega_k) = \sum_{i=1}^{P} S_i(\omega_k)e^{j\omega_k x \sin(\theta_i)/c} + V_f(x, \omega_k) \tag{6.11}$$

at positions $x = md$ and $m = -K, \ldots, K$, where $V_f(x, \omega_k)$ represents the noise component of the field such that $V(m, \omega_k) = V_f(md, \omega_k)$.

The objective of simultaneously focusing all sources in the field can be achieved by transforming the sensor outputs, $Y(m, \omega_k)$, so that the resulting source direction matrix is constant for all frequencies within their common bandwidth. Let $\tilde{\mathbf{Y}}(\omega_k)$ and $\tilde{A}(\omega_k, \boldsymbol{\theta})$ denote the focused sensor output vector and focused source direction matrix, respectively, and let ω_o denote the focusing frequency. Then the required wide-band focusing objective corresponds to finding frequency-dependent focusing matrices $T_r(\omega_k)$ such that

$$\tilde{A}(\omega_k, \boldsymbol{\theta}) \equiv T_r(\omega_k)A(\omega_k, \boldsymbol{\theta}) = A(\omega_o, \boldsymbol{\theta}) \tag{6.12}$$

so that each broad-band source in the focused covariance matrix, $\tilde{R}(\omega_o)$, given by

$$\tilde{R}(\omega_o) = \sum_{k=l}^{h} T_r(\omega_k)R(\omega_k)T_r(\omega_k)^+ \tag{6.13}$$

can be characterized by a rank-one covariance matrix. To obtain $\tilde{A}(\omega_k, \boldsymbol{\theta}) = A(\omega_o, \boldsymbol{\theta})$, for all ω_k within the common source bandwidth, the spatial resampling approach is simply to adjust the spatial sampling interval, d, as a function of temporal frequency, ω_k. Denoting the frequency-dependent spatial sampling interval by $d(\omega_k)$, the result of ideal wide-band focusing is to resample the incident field at $x = nd(\omega_k) = nd\omega_o/\omega_k$, where d is the physical sensor separation of the array. Substituting $d(\omega_k)$ into (6.11) gives an expression for the ideal focused array outputs, $\tilde{Y}(n, \omega_k)$:

$$\tilde{Y}(n, \omega_k) = \sum_{i=1}^{P} S_i(\omega_k)e^{j\omega_o\alpha_i nd} + \tilde{V}(n, \omega_k), \qquad -\tilde{K} \le n \le \tilde{K} \tag{6.14}$$

where $\tilde{M} = 2\tilde{K} + 1$ is the length of the resampled array and $\tilde{V}(n, \omega_k)$ represents the noise component of the resampled sensor output. When $\tilde{K} = K$, clearly $\tilde{Y}(n, \omega_k)$ of (6.14) implies $\tilde{A}(\omega, \boldsymbol{\theta}) = A(\omega_o, \boldsymbol{\theta})$ for all ω_k as required. To avoid spatial aliasing, $d(\omega_k)$, must be chosen such that $\omega_k/c \le \pi/d(\omega_k)$, which implies $\omega_o \le \pi c/d$. For a physical sensor spacing of a half-wavelength at the highest source frequency, ω_h, this condition implies the array must be focused to a frequency $\omega_o \le \omega_h$.

The digital transformation required to achieve perfect spatial resampling can be obtained by examining the focusing operation in the spatial frequency domain. Let $Y_f(j\Psi, \omega_k)$ denote the spatial Fourier transform of the incident field $Y_f(x, \omega_k)$ in (6.11) and consider the effect of sampling the field component corresponding to plane waves propagating at speed c. In this case, $Y_f(x, \omega_k)$ is approximately spatially bandlimited with $Y_f(j\Psi, \omega_k)$ vanishing outside the region, $-\omega_k/c \le \Psi \le \omega_k/c$. Sampling the field at intervals of d meters, the Fourier transform of $Y(m, \omega_k)$, denoted by $Y(e^{j\psi}, \omega_k)$, is given by the familiar relationship [23]:

$$Y(e^{j\psi}, \omega_k) = \frac{1}{d}\sum_{r=-\infty}^{\infty} Y_f\left(j\frac{\psi}{d} + j\frac{2\pi r}{d}\right) \tag{6.15}$$

as the number of sensors, $M \to \infty$. Similarly, the Fourier transform of the resampled field, $\tilde{Y}(n, \omega_k)$, as $\tilde{M} \to \infty$, can be expressed as

$$\tilde{Y}(e^{j\tilde{\psi}}, \omega_k) = \frac{\omega_k}{\omega_o d} \sum_{r=-\infty}^{\infty} Y_f\left(j\frac{\tilde{\psi}}{d}\frac{\omega_k}{\omega_o} + j\frac{2\pi r}{d}\frac{\omega_k}{\omega_o}\right) \tag{6.16}$$

Comparison of (6.15) and (6.16) for $Y_f(j\Psi, \omega_k)$ bandlimited to $\Psi \le |\omega_k/c|$ and $d \le \pi c/\omega_k$ reveals that $\tilde{Y}(e^{j\tilde{\psi}}, \omega_k)$ can be obtained from $Y(e^{j\psi}, \omega_k)$ by the relation,

$$\tilde{Y}(e^{j\tilde{\psi}}, \omega_k) = \begin{cases} \dfrac{\omega_k}{\omega_o} Y(e^{j\tilde{\psi}\frac{\omega_k}{\omega_o}}, \omega_k), & \text{for } |\tilde{\psi}| \le \min(\pi, \pi \cdot \dfrac{\omega_o}{\omega_k}) \\ 0 \quad , & \text{for } \min(\pi, \pi \cdot \dfrac{\omega_o}{\omega_k}) < |\tilde{\psi}| \le \pi \end{cases} \tag{6.17}$$

where min(x, y) denotes the lesser of x or y, and the focusing frequency satisfies $\omega_o \le \pi c/d$. The relation of (6.17) indicates that ideal spatial resampling rescales the spatial frequency axis, ψ, at each temporal frequency, ω_k, by the factor ω_o/ω_k so that a plane-wave signal at $\psi_i = \omega_k\ \alpha_i d$ is mapped to $\tilde{\psi}_i = \omega_k\ \alpha_i d \cdot \omega_o/\omega_k = \omega_o\ \alpha_i d$. This frequency mapping operation is the key to simultaneously focusing all wide-band sources in the field since it ensures that the spatial frequency of each broadband source is the same over its entire temporal bandwidth. This permits wideband sources in the spatially resampled sensor outputs to be treated with rank-one models. Clearly, no linear space-*invariant* transformation exists that will effect such a frequency rescaling. In the class of linear space-*variant* transformations, however, such frequency-mapping filters have been studied in the context of multirate signal processing [23, 24]. In Section 6.3, the design of realizable resampling filters and the properties of focused wide-band covariance matrices which result from spatial resampling are examined.

6.3 ESTIMATION OF FOCUSED WIDE-BAND COVARIANCE MATRICES

In this section, the finite-time estimation of focused covariance matrices obtained by presteering and spatially resampling wide-band sensor outputs is examined. Practical methods of estimating $R_s(\theta)$ and $\tilde{R}(\omega_o)$ can be obtained by substituting a finite-time estimate, $\hat{R}(\omega_k)$, of each narrow-band cross-spectral density matrix, $R(\omega_k)$, in place of its true value in (6.7) and (6.13), respectively. A common method of forming $\hat{R}(\omega_k)$ from discrete-time sensor outputs is to divide the T second observation into N nonoverlapping segments of ΔT seconds each, and then apply the discrete Fourier transform (DFT) to obtain uncorrelated frequency domain vectors, $\mathbf{Y}_n(\omega_k)$, for each segment $n = 1, \ldots, N$. The cross-spectral density matrix at frequency, ω_k, is then estimated by taking:

$$\hat{R}(\omega_k) = \frac{1}{N} \sum_{n=1}^{N} \mathbf{Y}_n(\omega_k)\mathbf{Y}_n(\omega_k)^+ \tag{6.18}$$

Substituting $\hat{R}(\omega_k)$ in place of its true value, $R(\omega_k)$, in (6.7) and (6.13) gives the estimates:

$$\hat{R}_s(\theta) = \sum_{k=l}^{h} T_s(\omega_k, \theta)\hat{R}(\omega_k)T_s(\omega_k, \theta)^+ \tag{6.19}$$

and

$$\tilde{R}(\omega_o) = \sum_{k=l}^{h} T_r(\omega_k)\hat{R}(\omega_k)T_r(\omega_k)^+ \tag{6.20}$$

Note that efficient computation of (6.19) and (6.20) can be achieved by using the fast Fourier transform (FFT) to obtain $\mathbf{Y}_n(\omega_k)$ from discrete-time sensor outputs.

6.3.1 Properties of the Estimated Steered Covariance Matrix

An attractive feature of the estimated steered covariance matrix of (6.19) is that its statistics can be expressed in terms of the Wishart characteristic function. Under the assumption that the sensor outputs are jointly Gaussian random processes, $Y_1(\omega_k)$, $Y_2(\omega_k), \ldots, Y_N(\omega_k)$ are approximately independent, identically distributed, M variate complex Gaussian random vectors [25]. In this case, the statistics of the matrix $\hat{R}(\omega_k)$ with elements $\{\kappa_{ij}\}$ is defined by the characteristic function of the $M(M+1)/2$ complex variates, $\kappa_{11}, \kappa_{22}, \ldots, \kappa_{MM}, \kappa_{12}, \ldots, \kappa_{M-1,M}$, given by [26]:

$$\phi_k(\mathbf{t}) = \det[I - jR(\omega_k)\Gamma]^{-N} \tag{6.21}$$

where the ijth element of Γ is $\Gamma_{ij} = (1 + \delta_{ij})t_{ij}$, $\mathbf{t} = [t_{11}, t_{12}, \ldots, t_{1M}, t_{22}, t_{23}, \ldots, t_{MM}]$, $t_{ij} = t_{ji}^*$ for $i > j$, and δ_{ij} is the Kronecker delta. In the case where $N \geq M$, the matrix $\hat{R}(\omega_k)$ has the complex Wishart probability density, denoted $CW[N, M; R(\omega_k)]$. When $N < M$, the density of $\hat{R}(\omega_k)$ does not exist although the characteristic function of (6.21) remains valid. The latter situation is of particular importance in this discussion since the use of only a few snapshots of data (e.g., $N = 1$) will be considered.

The characteristic function of $\hat{R}_s(\theta)$ follows directly from (6.19) and (6.21). To obtain the statistics of each of the terms of (6.19), note that the characteristic function of $T_s(\omega_k, \theta)\hat{R}(\omega_k)T_s(\omega_k, \theta)^+$ is given by $\det[I - jT_s(\omega_k, \theta)R(\omega_k)T_s(\omega_k, \theta)^+\Gamma]^{-N}$ for any nonsingular $M \times M$ matrix $T_s(\omega_k, \theta)$ [26]. Hence, since the $\hat{R}(\omega_k)$ are independent by virtue of the uncorrelatedness of the complex Gaussian $Y(\omega_k)$, $k = l, \ldots, h$, the characteristic function of $\hat{R}_s(\theta)$, denoted by $\phi(t)$, is given by the product of characteristic functions:

$$\phi(\mathbf{t}) = \prod_{k=l}^{h} \det[I - jT_s(\omega_k, \theta)R(\omega_k)T_s(\omega_k, \theta)^+\Gamma]^{-N} \tag{6.22}$$

which is the required result.

The advantage of using the STCM for broad-band source location lies in the fact that it can be estimated with much greater statistical stability than can individual cross-spectral density matrices. To demonstrate this, consider the form of (6.22) under the model of (6.1) when the sources are uncorrelated and the noise is spatially white. In this case, the CSDM is given by [21],

$$R(\omega_k) = \sum_{p=1}^{P} \sigma^2_{p,k}\mathbf{a}(\omega_k, \theta_p)\mathbf{a}(\omega_k, \theta_p)^+ + \sigma^2_{v,k}I \tag{6.23}$$

where $\sigma^2_{v,k}$ and $\sigma^2_{p,k}$ are, respectively, the noise and signal powers at frequency ω_k, and I is an $M \times M$ identity matrix. Substituting (6.23) into (6.22) and noting that $T_s(\omega_k, \theta)\mathbf{a}(\omega_k, \theta_p) = \mathbf{a}(\omega_k, \theta_p - \theta)$ and $T_s(\omega_k, \theta)T_s(\omega_k, \theta)^+ = I$, (6.22) can be expressed as:

$$\phi(\mathbf{t}) = \prod_{k=l}^{h} \det\left\{I - j\left[\sum_{p=1}^{P} \sigma^2_{p,k}\mathbf{a}(\omega_k, \theta_p - \theta)\mathbf{a}(\omega_k, \theta_p - \theta)^+ + \sigma^2_{v,k}I\right]\Gamma\right\}^{-N} \tag{6.24}$$

Although in theory, the probability density function of $\hat{R}_s(\theta)$ can be obtained by taking the $M(M+1)/2$-dimensional inverse Fourier transform of (6.24), this is a clearly formidable task. In the interest of finding tractable expressions for the probability distribution of $\hat{R}_s(\theta)$, consider the situations in which the inner bracketed terms in (6.24) are approximately constant for all $k = l, \ldots, h$. In these cases, the probability distribution associated with the product of (6.24) can be determined by inspection. As a first scenario, suppose all P sources have flat autospectra (i.e., $\sigma^2_{p,k} = \sigma^2_{p,h}, k = l, \ldots, h$) with directions, $\theta_p, p = 1, \ldots, P$ in the neighborhood of the steering direction, θ, such that:

$$\mathbf{a}(\omega_k, \theta_p - \theta) \cong \mathbf{a}(\omega_a, \theta_p - \theta), \qquad \text{for } k = l, \ldots, h \tag{6.25}$$

where $\omega_a = (\omega_h + \omega_l)/2$. This is the single group case of [5]. Further, assuming that the noise autospectrum is flat over this interval (i.e., $\sigma^2_{v,k} = \sigma^2_{v,h}$, for $k = l, \ldots, h$), then (6.24) can be simplified to:

$$\phi(\mathbf{t}) \cong \det\left\{I - j\left[\sum_{p=1}^{P} \sigma^2_{p,h}\mathbf{a}(\omega_k, \theta_p - \theta)\mathbf{a}(\omega_k, \theta_p - \theta)^+ + \sigma^2_{v,h}I\right]\Gamma\right\}^{-NB} \tag{6.26}$$

where $B = h - l + 1$. By comparison of (6.26) and (6.21), $\phi(\mathbf{t})$ can be recognized as the Wishart characteristic function with NB degrees of freedom. Thus, provided that $NB \geq M$, the probability distribution of $\hat{R}_s(\theta)$ is approximately the complex Wishart density, $CW[NB, M; R_s(\theta)]$. As a second scenario, note that in the low signal-to-noise ratio case with flat noise autospectrum (i.e., $\sigma^2_{p,h} \ll \sigma^2_{v,h}$, for all $p = 1, \ldots, P$), $\hat{R}_s(\theta)$ is again approximately Wishart distributed with density $CW[NB, M; R_s(\theta)]$ by a similar argument. Note that $NB \cong (T/\Delta T) \cdot \Delta T(\omega_h - \omega_l) = T(\omega_h - \omega_l)$ represents the full time-bandwidth product of the source signals. Clearly, when $B \gg 1$, the estimation of $\hat{R}_s(\theta)$ can be achieved with much greater statistical stability than the individual $\hat{R}(\omega_k)$. This is the key advantage of using methods based on the steered covariance matrix versus individual cross-spectral density matrices for broad-band spatial power spectral estimation.

Although the preceding discussion is strictly valid only for the single group or weak signal cases, the simulation results of Section 6.4.3 indicate that $\hat{R}_s(\theta)$ remains nonsingular for $NB > M$, even when the sources are strong and far apart with θ not in their neighborhood. This property of the STCM is what permits the steered covariance methods to achieve a significant reduction in threshold observation time. In terms of source location estimation error, the statistical stability of the steered covariance matrix method could be expected to translate into decreased location estimate variance. As is common in parameter estimation problems, however, this decrease in variance is accompanied by an increase in location estimate bias above the threshold observation time. This trade-off is studied in Section 6.4.3 by a computer simulation of steered minimum variance spatial spectral estimation.

6.3.2 Ideal Focusing Transformations for Spatial Resampling

While focusing wide-band arrivals from a single direction can be precisely realized using the steering matrix of (6.4), the transformation necessary to focus all arrivals simultaneously can only be approximated using a finite-length array. To achieve perfect spatial resampling corresponding to the spatial frequency mapping operation implied by (6.17), let $h_k(n, m)$ denote the impulse response of the desired linear space-variant filter to be applied at temporal frequency, ω_k. Using $h_k(n, m)$, a general linear transformation can be expressed as:

$$\tilde{Y}(n, \omega_k) = \sum_{m=-\infty}^{\infty} h_k(n, m) Y(m, \omega_k) \tag{6.27}$$

and defining $H_k(n, e^{j\psi}) = \sum_{m=-\infty}^{\infty} h_k(n, m) e^{+j\psi m}$, (6.27) can also be expressed as:

$$\tilde{Y}(n, \omega_k) = \frac{1}{2\pi} \int_{-\pi}^{\pi} Y(e^{j\psi}, \omega_k) H_k(n, e^{j\psi}) d\psi \tag{6.28}$$

Taking the Fourier transform of $Y(n, \omega_k)$ in (6.28) gives:

$$\tilde{Y}(e^{j\tilde{\psi}}, \omega) = \frac{1}{2\pi} \int_{-\pi}^{\pi} Y(e^{j\psi}, \omega_k) H_k(e^{j\tilde{\psi}}, e^{j\psi}) d\psi \tag{6.29}$$

where

$$H_k(e^{j\tilde{\psi}}, e^{j\psi}) = \sum_{n=-\infty}^{\infty} H_k(n, e^{j\psi}) e^{-j\tilde{\psi} n} \tag{6.30}$$

is the bifrequency response of the filter. The focusing requirement of (6.16) implies that the bifrequency response of an ideal resampling filter, denoted $H_k^o(e^{j\tilde{\psi}}, e^{j\psi})$, should have the following form when $\omega_o \leq \omega_k$,

$$H_k^o(e^{j\tilde{\psi}}, e^{j\psi}) = \begin{cases} \delta_p\left(\tilde{\psi} - \psi\dfrac{\omega_o}{\omega_k}\right), & \text{for } |\tilde{\psi}| \leq \dfrac{\omega_o}{\omega_k}\pi \\ 0 \quad , & \text{for } \dfrac{\omega_o}{\omega_k}\pi < |\tilde{\psi}| \leq \pi \end{cases} \tag{6.31a}$$

and when $\omega_o > \omega_k$,

$$H_k^o(e^{j\tilde{\psi}}, e^{j\psi}) = \begin{cases} \delta_p\left(\tilde{\psi} - \psi\dfrac{\omega_o}{\omega_k}\right), & \text{for } |\psi| \leq \dfrac{\omega_k}{\omega_o}\pi \\ 0 \quad , & \text{for } \dfrac{\omega_k}{\omega_o}\pi < |\psi| \leq \pi \end{cases} \tag{6.31b}$$

where $\delta_p(\psi) = 2\pi \sum_{k=-\infty}^{\infty} \delta(\psi + 2\pi k)$ and $\delta(\psi)$ is the Dirac impulse function. This specification of $H_k^o(e^{j\tilde{\psi}}, e^{j\psi})$ can be verified by substituting (6.31) into (6.29) and comparing the result with (6.16). The spatial frequency mappings described by (6.31) are illustrated in Fig. 6.1. When $\omega_o < \omega_k$ as shown in Fig. 6.1(a), the transformation of (6.31a) corresponds to interpolation. Note that the filter rejects imaged versions of $Y(e^{j\psi}, \omega_k)$, thereby ensuring that the mapping is one to one. Similarly, when $\omega_o > \omega_k$ as illustrated in fig. 6.1(b), (6.31b) corresponds to decimation. In this case, aliased versions of $Y(e^{j\psi}, \omega_k)$ are rejected to give a one-to-one mapping. The impulse response of the ideal resampling filter can be obtained by taking the inverse Fourier transform of (6.31) with respect to ψ and $\tilde{\psi}$ to obtain

$$h_k^o(n, m) = \frac{1}{\pi} \frac{\sin\left[\Psi_k\left(\dfrac{\omega_o}{\omega_k} \cdot n - m\right)\right]}{\left[\dfrac{\omega_o}{\omega_k} \cdot n - m\right]} \tag{6.32}$$

where $\Psi_k = \min[\pi, (\omega_k/\omega_o)\pi]$. From (6.32), the impulse response of the ideal resampling filter is of infinite length in the index, m. This implies that the filter is strictly nonrealizable with a finite-length array. Despite this fact, however, (6.32) is useful as a target response for approximation by realizable finite-length resampling filters as discussed in Section 6.3.3.

In addition to its use in filter design, the ideal response of (6.31) is also valuable in understanding the effect of spatial resampling on the diffuse ambient noise component, $V(m, \omega_k)$. For the general linear shift-variant filtering operation of (6.27), the noise component of output sequence at frequency, ω_k, is given by:

$$\tilde{V}(n, \omega_k) = \sum_{m=-\infty}^{\infty} h_k(n, m) V(m, \omega_k) \tag{6.33}$$

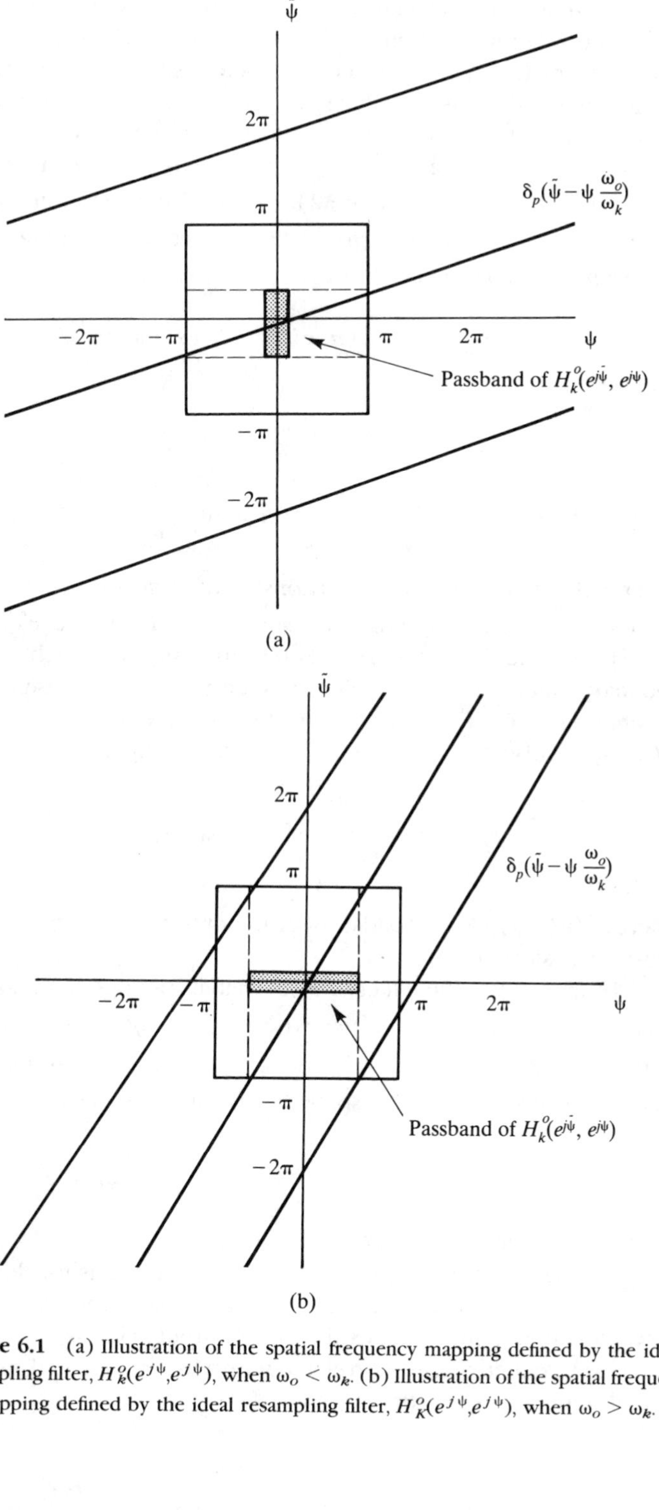

Figure 6.1 (a) Illustration of the spatial frequency mapping defined by the ideal resampling filter, $H_k^o(e^{j\bar{\psi}}, e^{j\psi})$, when $\omega_o < \omega_k$. (b) Illustration of the spatial frequency mapping defined by the ideal resampling filter, $H_K^o(e^{j\bar{\psi}}, e^{j\psi})$, when $\omega_o > \omega_k$.

where $V(m, \omega_k)$ is a zero-mean stationary noise process defined in (6.10). Denote the autocorrelation function of $V(m, \omega_k)$ by $\eta_v(m, \omega_k) = E\{V(q, \omega_k)V^*(q + m, \omega_k)\}$, where superscript * indicates complex conjugation. In general, linear shift-variant filtering of a wide-sense stationary input does not result in a wide-sense stationary output. This fact considerably complicates the analysis of $\tilde{V}(n, \omega_k)$ for general $h_k(n, m)$. However, in the case where a realizable $h_k(n, m)$ can be approximated by the ideal response, $h_k^o(n, m)$ of (6.32), the resulting resampled noise sequence does retain its wide-sense stationarity. This property can be obtained by examining the transfer function, $H_k^o(n, e^{j\psi})$ of $h_k^o(n, m)$, given by

$$H_k^o(n, e^{j\psi}) = G_k(e^{j\psi})e^{j\frac{\omega_o}{\omega_k}\psi n} \tag{6.34}$$

where

$$G_k(e^{j\psi}) = \begin{cases} 1, & \text{for } |\psi| \leq \min(\pi, \frac{\omega_k}{\omega_o}\pi) \\ 0, & \text{for } \min(\pi, \frac{\omega_k}{\omega_o}\pi) < |\psi| \leq \pi. \end{cases}$$

Observe that $H_k^o(n, e^{j\psi})$ of (6.34) consists of a linear shift-invariant filter, $G_k(e^{j\psi})$, in cascade with an odd-mapping linear shift-variant filter, to use the terminology of [24]. In [24], a simple derivation reveals that the output of such a filter with wide-sense stationary input is also wide-sense stationary. Using the results of [24] and the transfer function of (6.34), the autocorrelation function of the resampled noise sequence, $\tilde{\eta}_v(n, \omega_k) = E\{\tilde{V}(q, \omega_k)\tilde{V}^*(q + n, \omega_k)\}$, can be expressed as

$$\tilde{\eta}_v(n, \omega_k) = \frac{1}{2\pi}\int_{-\pi}^{\pi} N_v(e^{j\psi}, \omega_k)|G_k(e^{j\psi})|^2 \cdot e^{-j\frac{\omega_o}{\omega_k}\psi n} d\psi \tag{6.35}$$

where $N_v(e^{j\psi}, \omega_k)$ is the spatial spectrum of the noise outputs obtained by taking the Fourier transform of $\eta_v(m, \omega_k)$.

To investigate the effect of ideal spatial resampling on uncorrelated noise outputs, let $\eta_v(m, \omega_k) = \sigma_{v,k}^2 \delta_K(m)$, where $\sigma_{v,k}^2$ is the noise variance at frequency ω_k. In this case, $N_v(e^{j\psi}, \omega_k) = \sigma_{v,k}^2$ for $|\psi| \leq \pi$ and substituting into (6.35) gives the following expression for the spatially resampled noise autocorrelation function:

$$\tilde{\eta}_v(n, \omega_k) = \sigma_{v,\mathrm{k}}^2 \frac{\omega_k}{\omega_o} \frac{\sin(\psi_k' n)}{\pi n} \tag{6.36}$$

where $\psi_k' = \min[\pi, (\omega_o/\omega_k)\pi]$.

Observe from (6.36) that when resampling calls for decimation of the sensor outputs (i.e., $\omega_o > \omega_k$), the resampled noise sequence is uncorrelated. However, when resampling interpolates the array outputs (i.e., $\omega_o < \omega_k$), the resulting noise process at frequency ω_k is correlated with a low-pass spatial spectrum bandlimited to spatial frequencies less than $\omega_o\pi/\omega_k$.

Although ideal spatial resampling requires an infinite-length array, the limiting characteristics of a large but finite-length array can be assessed by supposing the focused outputs $\tilde{Y}(n, \omega_k)$, $n = -\tilde{K}, \ldots, \tilde{K}$, have been obtained by ideal spatial resampling. In this case, substituting (6.3) and (6.12) into (6.13) gives an expression for the resulting focused wide-band covariance matrix,

$$\tilde{R}(\omega_o) = \tilde{A}(\omega_o, \boldsymbol{\theta})\left[\sum_{k=l}^{h} P_s(\omega_k)\right]\tilde{A}(\omega_o, \boldsymbol{\theta})^{\psi} + \sum_{k=l}^{h} \tilde{R}_V(\omega_k) \tag{6.37}$$

where the pqth element, $\{\tilde{R}_V(\omega_k)\}_{pq} = \tilde{\eta}_V(p - q, \omega_k)$ of (6.35). With spatial resampling, as with other coherent signal-subspace methods, broad-band spatial spectral estimation is achieved by using $\tilde{R}(\omega_o)$ as if it were a narrow-band covariance matrix at frequency ω_o. Note that the form of (6.37) indicates that as required, each broad-band source contributes a rank-one component to $\tilde{R}(\omega_o)$. Further, observe that frequency averaging source spectral density matrices to obtain $\tilde{P}_s = \sum_{k=l}^{h} P_s(\omega_k)$ effectively "decorrelates" correlated multipath arrivals provided that the relative multipath delays are longer than the inverse bandwidth of the sources, (that is, $2\pi/(\omega_h - \omega_l)$). The ability of spatial resampling methods to handle correlated sources is similar to that of the steered covariance methods discussed previously.

A unique feature of the spatial resampling approach is the form of the focused noise covariance matrix, $\tilde{R}_V(\omega_o) = \sum_{k=l}^{h} \tilde{R}_V(\omega_k)$. In particular, when the noise components of the array outputs are uncorrelated with variance $\sigma_{v,k}^2$ at each frequency, (6.36) implies

$$\{\tilde{R}_V(\omega_k)\}_{ij} = \sigma_{v,k}^2 \cdot \frac{\omega_k}{\omega_o} \cdot \frac{\sin[\psi_k'(i - j)]}{\pi(i - j)}.$$

Hence when $\omega_o > \omega_k$, the resulting focused noise covariance matrix is a multiple of the identity matrix, that is,

$$\tilde{R}_V(\omega_o) = \left[\sum_{k=l}^{h} \sigma_{vk}^2 \cdot \frac{\omega_k}{\omega_o}\right] I.$$

This property facilitates unbiased estimation of the source locations when using narrow-band spatial spectral estimation methods on $\tilde{R}(\omega_o)$. Further, the limiting form of $\tilde{R}_V(\omega_o)$ also validates the assumption that the largest P eigenvectors of $\tilde{R}(\omega_o)$ span the signal subspace defined by the columns of $\tilde{A}(\omega_o, \boldsymbol{\theta})$, a property critical when eigenvector methods for spatial spectral estimation such as the MUSIC technique are used on the focused covariance matrix [5, 10, 11, 12].

In the case where $\omega_o < \omega_k$, the limiting form of $\tilde{R}_V(\omega_o)$ with spatially uncorrelated noise is clearly not diagonal. From (6.36), the focused noise covariance matrix when the outputs at frequency ω_k are interpolated corresponds to a low-pass spatial spectrum with cutoff frequency, $\omega_o\pi/\omega_k$. For $\omega_o = \omega_l$, this implies that the sum of $\tilde{R}_V(\omega_k)$ gives a focused noise covariance matrix, $\tilde{R}_V(\omega_o)$, with a spatial spectrum which is flat from 0 to $\omega_o\pi/\omega_h$ and then decays gradually from $\omega_o\pi/\omega_h$ to π. Without taking steps to prewhiten the noise process, the focused noise correlation will result in biased bearing estimation for sources at spatial frequencies between $\omega_o\pi/\omega_h$ and $\omega_o d/c$. In theory, this bias could be alleviated by prewhitening the resampled sensor outputs using $\tilde{R}_V(\omega_o)^{-1}$. In contrast to the coherent signal-subspace methods of [5], this prewhitening operation would not require approximate knowledge of the source locations. As shown in the simulation examples of Section 6.4.3, however, the relatively slow decay of the focused broad-band spatial spectrum results in only modest bearing bias near the endfire directions. This fact, combined with the practical difficulties in accurately estimating $\tilde{R}_V(\omega_o)^{-1}$ when the physical noise covariance is unknown, suggests that spatial noise prewhitening may not be advisable. Further, experimental results using the MUSIC method with interpolated resampling points [10, 11, 12] indicate that although not exact, the largest P eigenvectors of $\tilde{R}(\omega_o)$ still quite accurately span the signal subspace defined by $\tilde{A}(\omega_o, \boldsymbol{\theta})$ despite the correlation of $\tilde{R}_v(\omega_o)$. This property can be expected to hold as long as the spatial spectrum corresponding to $\tilde{R}_V(\omega_o)$ is flat in the neighborhood of $\omega_o\ \alpha_i d$, $i = 1, \ldots, P$. It is important to note that the foregoing discussion describes only the limiting behavior of wide-band focusing under the assumption of ideal spatial resampling. In the discussion that follows, the design and characteristics of realizable resampling filters are examined using a finite-length sensor array.

6.3.3 Realizable Focusing Filters for Spatial Resampling

In preliminary work on spatial resampling [10], realizable resampling filters were obtained using designs taken directly from the multirate signal processing literature (see [23] and references contained therein). While the resulting filters were useful in demonstrating the spatial resampling concept, being tailored to a different application, they typically imposed constraints unnecessary in the wide-band focusing problem. For example, in common multirate signal processing systems, sampling rate conversions are generally restricted to a rational factor, and are designed assuming an infinite length input sequence. In the sensor array processing application, however, because sensors are at a premium, the problem of resampling *finite* length sequences is particularly important.

In this treatment, the linear shift-variant filter which minimizes the normalized maximum absolute resampling error at each resampled sensor output is derived. This minimax approach was originally proposed in [27] for the design of digital interpolators. In Fig. 6.2, the framework used for defining the spatial resampling error criterion is illustrated. The objective of the design procedure is to obtain a realizable filter, $h_k(n, m)$, nonzero only for $m \leq |K|$, which achieves minimum resampling error. The spatial resampling error at frequency ω_k is defined by subtracting the output of the realizable filter with the infinite-length ideal filter, $h_k^o(n, m)$, given in (6.32). The resulting error sequence is defined by $\tilde{E}(n, \omega_k) = \tilde{Y}_o(n,$

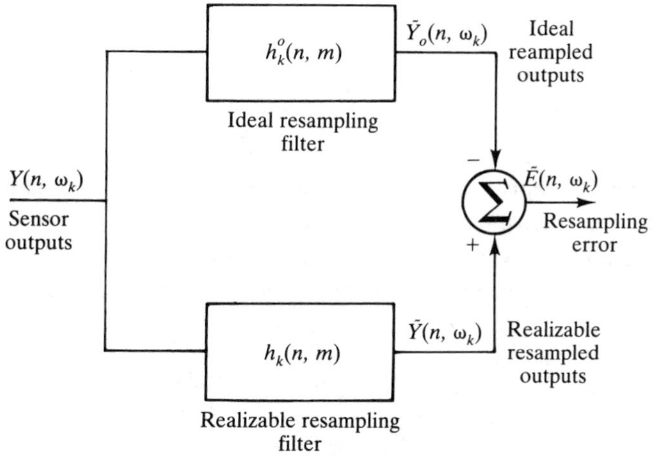

Figure 6.2 The framework used to define resampling error in the design of realizable linear shift-variant filters for spatial resampling.

$\omega_k) - \tilde{Y}(n, \omega_k)$, where $\tilde{Y}(n, \omega_k)$ is the realizable filter output and $\tilde{Y}_o(n, \omega_k)$ is the ideal filter output. The error criterion to be minimized is the normalized maximum absolute resampling error, $J_k(n)$, defined by

$$J_k(n) = \frac{\max |\tilde{Y}_o(n, \omega_k) - \tilde{Y}(n, \omega_k)|}{\|Y(\cdot, \omega_k)\|} \tag{6.38}$$

where the norm $\|Y(\cdot, \omega_k)\| = \left[\sum_{m=-\infty}^{\infty} |Y(m, \omega_k)|^2 \right]^{1/2}$ is assumed to be finite. The minimax design problem is to select $h_k(n, m)$ to minimize $J_k(n)$ for each $n \leq |\tilde{K}|$. The minimax solution follows immediately after noting that $\tilde{Y}_o(n, \omega_k)$ and $\tilde{Y}(n, \omega_k)$ can be expressed as $\langle Y(\cdot, \omega_k), h_k^o(n, \cdot)\rangle$ and $\langle Y(\cdot, \omega_k), h_k(n, \cdot)\rangle$, respectively, where $\langle x(\cdot), y(\cdot)\rangle = \sum_{m=-\infty}^{\infty} x(m)y(m)^*$ denotes the usual inner product operation. Substituting these inner product expressions into (6.38) gives:

$$J_k(n) = \frac{\max |\langle Y(\cdot, \omega_k), h_k(n, \cdot) - h_k^o(n, \cdot)\rangle|}{\|Y(\cdot, \omega_k)\|} \tag{6.39}$$

Applying the Cauchy-Schwarz inequality, (6.39) can be simplified to obtain:

$$J_k(n) = \|h_k(n, \cdot) - h_k^o(n, \cdot)\| \tag{6.40}$$

Minimization of $J_k(n)$, or equivalently $[J_k(n)]^2$, can be performed by applying Parseval's theorem to (6.40) and substituting the transfer function, $H_k^o(n, e^{j\psi})$, of

(6.34) to obtain:

$$[J_k(n)]^2 = \frac{1}{2\pi}\int_{-\pi}^{\pi}\left|\sum_{m=-K}^{K} h_k(n, m)e^{j\psi m} - G_k(e^{j\psi})e^{j\frac{\omega_o}{\omega_k}\psi n}\right|^2 d\psi \tag{6.41}$$

Since (6.41) represents the norm of the error in approximating $G_k(e^{j\psi})e^{j\frac{\omega_o}{\omega_k}\psi n}$ by a linear combination of the functions $\zeta_m(\psi) = e^{j\psi m}$, $m = -K, \ldots, K$, $[J_k(n)]^2$ can be minimized by making the error orthogonal to $\zeta_m(\psi)$. Thus the othogonality principle implies the optimum $h_k(n, m)$ satisfies the set of equations.

$$\frac{1}{2\pi}\int_{-\pi}^{\pi}\left[\sum_{m=-K}^{K} h_k(n, m)e^{j\psi m} - G_k(e^{j\psi})e^{j\frac{\omega_o}{\omega_k}\psi n}\right]e^{-j\psi r}d\psi = 0 \tag{6.42}$$

for $r = -K, \ldots, K$. Solving (6.42) leads easily to the optimum finite-length resampling filter, denoted $h_k^{\ddagger}(n, m)$, given by:

$$h_k^{\ddagger}(n, m) = \begin{cases} \dfrac{\frac{1}{\pi}\sin\left[\psi_k\left(\frac{\omega_o}{\omega_k}\cdot n - m\right)\right]}{\left[\frac{\omega_o}{\omega_k}\cdot n - m\right]}, & \text{for } -K \le m \le K \\ 0, \text{ otherwise} \end{cases} \tag{6.43}$$

where $\psi_k = \min\,[\pi, (\omega_k/\omega_o)\pi]$. Inspection of (6.43) indicates that the optimum realizable resampling filter is just a rectangularly windowed version of the ideal infinite-length filter given by (6.32).

The optimum minimax resampling error, denoted $J_k^{\ddagger}(n)$, can be obtained by substituting the optimum filter, $h_k^{\ddagger}(n, m)$, of (6.43) into (6.41), which yields:

$$J_k^{\ddagger}(n) = \left(\frac{\psi_k}{\pi} - \sum_{m=-K}^{K}\left\{\frac{1}{\pi}\,\frac{\sin\left[\psi_k\left(\frac{\omega_o}{\omega_k}\cdot n - m\right)\right]}{\left[\frac{\omega_o}{\omega_k}\cdot n - m\right]}\right\}^2\right)^{1/2} \tag{6.44}$$

Note that the filters of (6.43) also minimize the maximum total normalized resampling error,

$$J(n) = \max\left[\sum_{k=l}^{h}\frac{|\tilde{E}(n, \omega_k)|}{\|Y(\cdot, \omega_k)\|}\right].$$

Thus from (6.44), the total minimax error is given by $J^{\ddagger}(n) = \sum_{k=l}^{h} J_k^{\ddagger}(n)$. The total minimax error, $J^{\ddagger}(n)$, is useful in selecting the length, $\tilde{M} = 2\tilde{K} + 1$, of the resampled sensor array. In Fig. 6.3, $J^{\ddagger}(n)$ is plotted for three different choices of ω_o (that is, $\omega_o = 0.5\pi$, 0.85π, and π) when the array has $M = 17$ sensors and 31 fre-

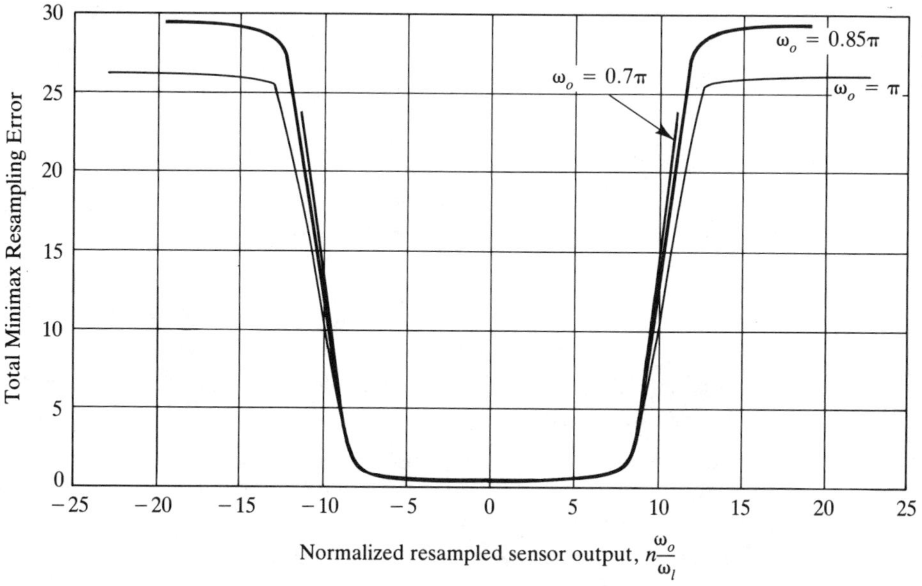

Figure 6.3 The total minimax resampling error obtained using different choices of focusing frequency, ω_o, with a 17 sensor array and receiver band from $\omega_l = 0.7\pi$ to $\omega_h = \pi$.

quency bins evenly spaced from $\omega_l = 0.7\pi$ to $\omega_h = \pi$, where $\omega = \pi$ is the half-wavelength frequency of the array. Note that the total minimax error is similar for all choices of ω_o and increases rapidly when $|n\omega_o/\omega_l| > 8$. Further examination of (6.44) suggests that the largest $\tilde{K}$ which avoids large resampling errors near the ends of the array is given by $\tilde{K} = \lfloor K\omega_l/\omega_o \rfloor$, where $\lfloor x \rfloor$ denotes the largest integer less than or equal to x. This implies a choice of $\tilde{M} = 2\tilde{K} + 1$ that is consistent with the example shown in Fig. 6.3. Intuitively, this choice of $\tilde{K}$ ensures that sensors on both sides of a resampled output are employed in generating the end points of the resampled array.

Since the objective of spatial resampling is to focus plane-wave arrivals, it is useful to consider the accuracy of the minimax resampling filter when faced with this type of input. To examine the response of (6.43) to a unit-amplitude plane-wave input sequence, $Y(m, \omega_k) = e^{j\omega_k \alpha_i dm}$ for $|m| \leq K$, consider the Fourier transform of its output, $\tilde{Y}(e^{j\psi}, \omega_k)$. Since the filter of (6.43) is nonzero only for $|m| \leq K$, its response to the plane-wave input sequence can be obtained simply by substituting $Y(e^{j\psi}, \omega_k) = \delta_p(\psi - \omega_k \alpha_i d)$ into (6.28) to obtain:

$$\tilde{Y}(e^{j\psi}, \omega_k) = H_k^{\ddagger}(e^{j\psi}, e^{j\omega_k \alpha_i d}) \tag{6.45}$$

where $H_k^{\ddagger}(e^{j\tilde{\psi}}, e^{j\omega_k \alpha_i d})$ is the bifrequency response of (6.43) evaluated at the spatial frequency of the input arrival. Straightforward evaluation of (6.45) using the bifrequency response definition of (6.30) and the filter of (6.43) yields:

$$\tilde{Y}(e^{j\psi}, \omega_k) = \frac{\omega_k}{\omega_o} G_k^{'}(e^{j\psi}) \frac{\sin\left[\frac{\omega_k}{\omega_o}(\psi - \omega_o\alpha_i d)\frac{M}{2}\right]}{\sin\left[\frac{\omega_k}{\omega_o}(\psi - \omega_o\alpha_i d)\frac{1}{2}\right]}, \qquad -\pi \leq \psi \leq \pi \tag{6.46}$$

where

$$G_k^{'}(e^{j\tilde{\psi}}) = \begin{cases} 1, & \text{for } |\psi| \leq \min\left(\pi, \frac{\omega_o}{\omega_k}\pi\right) \\ 0, & \text{for } \min(\pi, \frac{\omega_o}{\omega_k}\pi) < |\psi| \leq \pi \end{cases}$$

and (6.46) is periodically extended outside $|\psi| \leq \pi$. From (6.46), note that as the filter length, $M \rightarrow \infty$, $\tilde{Y}(e^{j\psi}, \omega_k)$ goes to $(\omega_k/\omega_o) \cdot G_k^{'}(e^{j\psi})\delta_p(\psi - \omega_o\alpha_i d)$, which is just the ideal response of (6.17) with the above plane-wave input sequence. For finite M, the ideal resampled sequence would be the windowed plane-wave sequence, $\tilde{Y}(n, \omega_k) = e^{j\omega_o\alpha_i dn}$, for $n \leq |\tilde{K}|$ and zero otherwise, where $\tilde{K} \cong K\omega_k/\omega_o$, from the discussion surrounding (6.44). The Fourier transform of this ideal finite-length resampled sequence is thus

$$\tilde{Y}_o(e^{j\psi}, \omega_k) \cong \frac{\omega_k}{\omega_o} \frac{\sin\left[\frac{\omega_k}{\omega_o}(\psi - \omega_o\alpha_i d)\frac{M}{2}\right]}{\sin\left[(\psi - \omega_o\alpha_i d)\frac{1}{2}\right]} \tag{6.47}$$

where the approximation arises from the fact that the ideal sequence is necessarily of odd integer length while $M\omega_k/\omega_o$ of (6.47) is generally not. Comparing (6.46) to (6.47) reveals that the dominant error in (6.46) arises from the truncation imposed by $G_k^{'}(e^{j\psi})$ when $\omega_o < \omega_k$, or alternatively the truncation effect caused by periodic extension when $\omega_o > \omega_k$. This error is greatest when the mainlobe of the sinc-type function in (6.46) is truncated, a situation which can occur for plane-waves close to the endfire direction at temporal frequencies near the Nyquist frequency of the array. Note that in practice, $\tilde{Y}(n, \omega_k)$ is rectangularly windowed in the region $n \leq K\omega_l/\omega_o$, which has the effect of smoothing the discontinuity in $\tilde{Y}(e^{j\psi}, \omega_k)$. Nevertheless the resampling error is still greatest near endfire as the simulation results of Section 6.4.3 demonstrate.

6.4 FOCUSED WIDE-BAND MINIMUM-VARIANCE SPATIAL SPECTRAL ESTIMATION

In this section, wideband focusing with the well-known minimum-variance method of spatial spectral estimation [19, 20, 21] is considered. In general, the minimum-variance (MV) spatial spectral estimate in direction θ can be thought of as the minimum beamformer output power that can be achieved subject to the constraint of unity gain in the θ direction. For an array at frequency ω_k, the narrow-band minimum-variance spatial spectral estimate, $Z_k(\theta)$, is given by

$$Z_k(\theta) = [\mathbf{a}(\omega_k, \theta)^+ R(\omega_k)^{-1} \mathbf{a}(\omega_k, \theta)]^{-1} \tag{6.48}$$

where $R(\omega_k)$ is the $M \times M$ cross-spectral density matrix and $\mathbf{a}(\omega_k, \theta)$ is defined by (6.3). A finite-time estimate, $\hat{Z}_k(\theta)$ of $Z_k(\theta)$, can be obtained by substituting the cross-spectral density matrix estimate, $\hat{R}(\omega_k)$ of (6.18), in place of $R(\omega_k)$ in (6.48). Note that a complete mapping of field directionality is formed by computing $\hat{Z}_k(\theta)$ for a set of directions that span the locations of interest.

For comparison with the focused wide-band MV methods discussed shortly, it is useful to let $\hat{Z}_{icmv}(\theta)$ be defined as the incoherent broad-band MV estimator obtained by summing narrow-band MV results over the band of interest. Substituting (6.18) into (6.48) and summing over the band from ω_l to ω_h, $\hat{Z}_{icmv}(\theta)$ can be expressed as

$$\hat{Z}_{icmv}(\theta) = \sum_{k=l}^{h} [\mathbf{a}(\omega_k, \theta)^+ \hat{R}(\omega_k)^{-1} \mathbf{a}(\omega_k, \theta)]^{-1} \tag{6.49}$$

In the remainder of this section, the incoherent MV method of (6.49) is used as a basis for comparing the performance of both presteering and spatial resampling approaches to focused wideband MV spatial spectral estimation.

6.4.1 The Steered Minimum-Variance Method

Following the discussion surrounding (6.9), the steered minimum-variance method consists of estimating the dc component of the STCM steered in each direction θ by means of a minimum-variance approach. For estimation of the spatial spectrum in direction θ, this technique has the effect of minimizing the contribution of power from sources in other directions. Applying the solution of (6.48) to estimating the dc component of $R_s(\theta)$, the resulting spatial spectral estimate, called here the STMV method, is given by:

$$Z_{stmv}(\theta) = [\mathbf{1}^+ R_s(\theta)^{-1} \mathbf{1}]^{-1} \tag{6.50}$$

where $\mathbf{1}$ is an $M \times 1$ vector of ones. Again, a finite-time estimate, $\hat{Z}_{stmv}(\theta)$, of $Z_{stmv}(\theta)$ can be obtained by substituting the estimate $\hat{R}_s(\theta)$ given in (6.19) in place of $R_s(\theta)$ in (6.50). A complete broad-band spatial power spectral estimate is then formed by computing $\hat{Z}_{stmv}(\theta)$ for a set of steering angles which span the locations of interest.

Comparison of the STMV method with the incoherent MV approach of (6.49) is made possible by expressing $R_s(\theta)$ as a sum of cross-spectral density matrices. Substituting (6.7) into (6.50) gives:

$$Z_{stmv}(\theta) = \frac{1}{\mathbf{1}^+ \left[\sum_{k=l}^{h} T_s(\omega_k, \theta) R(\omega_k) T_s(\omega_k, \theta)^+ \right]^{-1} \mathbf{1}} \tag{6.51}$$

It is straightforward to show that when $h = l$, (6.51) is precisely the narrow-band MV spatial spectral estimate of (6.48) [18]. Thus in the narrow-band case, the STMV reduces to the conventional narrow-band MV method. For broad-band sources, however, comparison of (6.51) and (6.49) reveals the essential difference between the focused and incoherent MV methods. Specifically, in (6.51), cross-spectral density matrices are averaged *prior* to matrix inversion, while in (6.49) the matrix inversion is applied to individual narrow-band CSDMs prior to averaging. While asymptotically the STMV method is strictly suboptimal, when only a limited number of data snapshots are available, the averaging in (6.51) provides a more statistically stable matrix to invert thus facilitating more accurate spatial spectral estimation.

The statistics of the STMV spatial spectral estimate can be approximated by using the results of Section 6.3.1. Although the characteristic function of $\hat{R}_s(\theta)$ given by (6.24) could formally be used to obtain the distribution of $\hat{Z}_{stmv}(\theta)$ in the general case, the nonlinear nature of the STMV estimate makes this result difficult to obtain in closed form. However in the single-group case described by (6.25), where all P sources are in the neighborhood of θ, previously derived results for the narrow-band MV estimate can be easily adapted to determine the statistics of $\hat{Z}_{stmv}(\theta)$. In particular, it is shown in [28] that the statistic $\hat{Z}_k(\theta) = [\mathbf{a}(\omega_k, \theta)^+\hat{R}(\omega_k)^{-1}\mathbf{a}(\omega_k, \theta)]^{-1}$ is a multiple of a chi-squared variable with $2(N - M + 1)$ degrees of freedom. From [28], the mean and variance of $\hat{Z}_k(\theta)$ are given by:

$$
\begin{aligned}
E[\hat{Z}_k(\theta)] &= \frac{N - M + 1}{N}\,[\mathbf{a}(\omega_k, \theta)^+R(\omega_k)^{-1}\mathbf{a}(\omega_k, \theta)]^{-1} \\
\mathrm{Var}[\hat{Z}_k(\theta)] &= \frac{1}{N - M + 1}\,\{E[\hat{Z}_k(\theta)]\}^2
\end{aligned}
\tag{6.52}
$$

under the condition that $\hat{R}(\omega_k)$ is $CW[N, M; R(\omega_k)]$ distributed. Now observe that $\hat{Z}_{stmv}(\theta) = [\mathbf{1}^+\hat{R}_s(\theta)^{-1}\mathbf{1}]^{-1}$ is of the same form as $\hat{Z}_k(\theta)$ under the condition that $\hat{R}_s(\theta)$ is Wishart distributed. From (6.26) it was demonstrated that in the single-group case, $\hat{R}_s(\theta)$ is approximately $CW[NB, M; R_s(\theta)]$ when $NB \geq M$. Thus by direct correspondence with (6.52), $\hat{Z}_{stmv}(\theta)$ is a multiple of a chi-squared variable with $2(NB - M + 1)$ degrees of freedom having mean and variance given by:

$$
\begin{aligned}
E[\hat{Z}_{stmv}(\theta)] &= \frac{NB - M + 1}{NB}\,[\mathbf{1}^+R_s(\theta)^{-1}\mathbf{1}]^{-1} \\
\mathrm{Var}[\hat{Z}_{stmv}(\theta)] &= \frac{1}{NB - M + 1}\,\{\mathrm{E}[\hat{Z}_{stmv}(\theta)]\}^2
\end{aligned}
\tag{6.53}
$$

To compare the statistics of the STMV with the incoherent MV method, note that the $\hat{Z}_{icmv}(\theta)$ of (6.49) consists of forming a sum of $\hat{Z}_k(\theta)$ for $k = l, \ldots, h$. Since from [28], $N\hat{Z}_k(\theta)/[\mathbf{a}(\omega_k, \theta)^+R(\omega_k)^{-1}\mathbf{a}(\omega_k, \theta)]^{-1}$ is $CW[N - M + 1, 1; 1]$ distributed and the $\hat{Z}_k(\theta)$ are independent, the sum given by:

$$
\hat{Z}_n(\theta) = N\sum_{k=l}^{h}\frac{\hat{Z}_k(\theta)}{[\mathbf{a}(\omega_k, \theta)^+R(\omega_k)\mathbf{a}(\omega_k, \theta)]^{-1}}
\tag{6.54}
$$

is $CW[B(N - M + 1), 1; 1]$ distributed, where a standard result for the distribution of a sum of independent, identically distributed Wishart variables has been employed [26]. Using (6.23) and (6.25), it is easy to show that the denominator, $[\mathbf{a}(\omega_k, \theta)^+ R(\omega_k)^{-1}\mathbf{a}(\omega_k, \theta)]^{-1}$ is nearly independent of ω_k for $k = l, \ldots, h$ in the situation where the P sources are in the neighborhood of θ. In this case, $\hat{Z}_{icmv}(\theta) \cong \beta(\theta)\hat{Z}_n(\theta)$ for some function $\beta(\theta)$, which is independent of frequency. Thus $\hat{Z}_{icmv}(\theta)$ is approximately a multiple of a chi-squared variable with $2B(N - M + 1)$ degrees of freedom having mean and variance given by:

$$E[\hat{Z}_{icmv}(\theta)] = \frac{(N - M + 1)}{N} \sum_{k=l}^{h} [\mathbf{a}(\omega_k, \theta)^+ R(\omega_k)^{-1}\mathbf{a}(\omega_k, \theta)]^{-1}$$

$$\text{Var}[\hat{Z}_{icmv}(\theta)] = \frac{1}{B(N - M + 1)} \{E[\hat{Z}_{icmv}(\theta)]\}^2 \tag{6.55}$$

The increased statistical stability offered by the STMV versus the incoherent MV method can be appreciated by comparing (6.53) and (6.55). For comparison purposes, a useful measure of estimator stability is the normalized variance, $\sigma_Z^2 \equiv \text{Var}[\hat{Z}(\theta)]/\{E[\hat{Z}(\theta)]\}^2$. From (6.53), the STMV method has a normalized variance, $\sigma_{stmv}^2 = 1/(NB - M + 1)$, versus the incoherent MV normalized variance, $\sigma_{icmv}^2 = 1/[B(N - M + 1)]$. Forming the ratio of normalized variances gives:

$$r_\sigma = \frac{\sigma^2{}_{stmv}}{\sigma^2{}_{icmv}} = 1 - \frac{(B - 1)(M - 1)}{NB - M + 1} \tag{6.56}$$

Clearly, $r_\sigma \leq 1$ provided $N \geq M$ with equality only in the limiting cases of B or M equal to 1. This demonstrates the improved stability offered by the STMV estimate. Note in particular the case where $N = M$ and $BN \gg 1$, corresponding to a relatively short time, broad-band source scenario. In this case, $r_\sigma \cong 1/M$ and the variance improvement of the STMV method over the incoherent MV can be large. The situation where $N < M$ leads to an even more striking difference between these two methods. In this case, $\hat{R}(\omega_k)$ is no longer Wishart distributed, and hence (6.52) and (6.55) are no longer valid. In fact, for $N < M$, $\hat{R}(\omega_k)$ will generally be singular, which implies that $\hat{Z}_{icmv}(\theta)$ cannot be computed. The steered covariance matrix estimate, $\hat{R}_s(\theta)$, however, is still approximately Wishart distributed as long as $NB \geq M$, which implies that $\hat{Z}_{stmv}(\theta)$ is still well defined with statistics given by (6.53). For broad-band signals, this result suggests that for $B \geq M$, the STMV method can yield spatial spectral estimates in the neighborhood of the sources with as little as one snapshot of data.

In summary, the steps used to perform the STMV method are as follows:

1. Form estimated cross-spectral density matrices, $\hat{R}(\omega_k)$, over the frequency band of interest (6.18).
2. Compute estimated steered covariance matrices, $\hat{R}_s(\theta)$, for each steering direction, θ, of interest (6.19).

3. Compute $\hat{R}_s(\theta)^{-1}$ and form $\hat{Z}_{stmv}(\theta) = [\mathbf{1}^+\hat{R}_s(\theta)^{-1}\mathbf{1}]^{-1}$ for each steering direction θ to obtain a broad-band spatial power spectral estimate.

Note that estimation of the source locations is achieved by determining the peak positions of the spatial power spectral estimate $\hat{Z}_{stmv}(\theta)$.

6.4.2 The Spatially Resampled Minimum-Variance Method

The broad-band spatially resampled minimum-variance (SRMV) method is obtained by applying the narrow-band minimum variance method of (6.48) to a finite-time estimate of the focused covariance matrix of (6.20). Note that the linear shift-variant transformation of (6.27) with a finite-length resampling filter, $h_k(n, m)$, is equivalent to performing $\tilde{Y}(\omega_k) = T_r(\omega_k)Y(\omega_k)$ where the pqth element of $T_r(\omega_k)$ is given by $\{T_r(\omega_k)\}_{pq} = h_k(p - \tilde{K} - 1, q - K - 1)$ for $p = 1, \ldots, \tilde{M}$, and $q = 1, \ldots, M$. Thus using the minimax resampling filter given by (6.43), the optimal $\tilde{M} \times M$ focusing matrix, $T_r^{\ddagger}(\omega_k)$, has elements given by

$$\{T_r^{\ddagger}(\omega_k)\}_{pq} = \frac{1}{\pi}\frac{\sin\left\{\Psi_k\left[\frac{\omega_o}{\omega_k}\cdot(p - \tilde{K} - 1) - (q - K - 1)\right]\right\}}{\left[\frac{\omega_o}{\omega_k}\cdot(p - \tilde{K} - 1) - (q - K - 1)\right]} \tag{6.57}$$

Substituting the focusing matrix of (6.57) and the estimate, $\hat{R}(\omega_k)$, of (6.18) into (6.20) gives the N snapshot finite-time estimate,

$$\tilde{R}_N(\omega_o) = \sum_{k=l}^{h} T_r^{\ddagger}(\omega_k)\hat{R}(\omega_k)T_r^{\ddagger}(\omega_k)^+ \tag{6.58}$$

of the focused covariance matrix, $\tilde{R}(\omega_o)$. Using the focused covariance matrix estimate of (6.58) and performing minimum-variance spatial spectral estimation at frequency ω_o yields the broad-band spatially resampled minimum-variance estimator, $\hat{Z}_{srmv}(\theta)$, in the direction θ,

$$\hat{Z}_{srmv}(\theta) = \frac{1}{\tilde{\mathbf{a}}(\omega_o, \theta)^+\left[\sum_{k=l}^{h} T_r^{\ddagger}(\omega_k)\hat{R}(\omega_k)T_r^{\ddagger}(\omega_k)^+\right]^{-1}\tilde{\mathbf{a}}(\omega_o, \theta)} \tag{6.59}$$

where the direction vector, $\mathbf{a}(\omega_o, \theta)$, is defined after (6.10).

Observe that the SRMV estimate of (6.59) resembles the STMV estimate of (6.51) in that it involves summing transformed cross-spectral density matrices across the receiver band prior to matrix inversion. However, in contrast to the STMV, in the SRMV method the focused wide-band covariance matrix, $\tilde{R}(\omega_o)$, represented by the summation in (6.59) does not depend on the look direction, θ. This is to be expected since $\tilde{R}(\omega_o)$ was designed to focus all arrivals simultaneously. The need to form and invert only one focused covariance matrix for all look directions, θ, makes the SRMV method considerably more computationally efficient than the STMV technique.

The finite-time performance of the SRMV method can be analyzed in a manner similar to that of the STMV technique given in Section 6.4.1. In fact, the results for restricted cases suggest that the statistical stability of the focused covariance matrix estimate, $\tilde{R}_N(\omega_o)$, of (6.58) is similar to that of the STCM estimate of (6.19). The discussion here treats the case where the diffuse noise is spatially uncorrelated, that is, $\eta_v(m, \omega_k) = \sigma_{v,k}^2\delta_K(m)$, and spatial resampling corresponds to decimation of the sensor outputs (that is, $\omega_o = \omega_h \geq \omega_k$, for all k). In this situation, each term in the $\tilde{M} \times \tilde{M}$ focused covariance matrix estimate of (6.58) has the expected value $E\{T_r^{\ddagger}(\omega_k)\hat{R}(\omega_k)T_r^{\ddagger}(\omega_k)^+\} = T_r^{\ddagger}(\omega_k)R(\omega_k)T_r^{\ddagger}(\omega_k)^+$, which can be expressed as

$$T_r^{\ddagger}(\omega_k)R(\omega_k)T_r^{\ddagger}(\omega_k)^+ \cong \tilde{A}(\omega_o, \theta)P_s(\omega_k)\tilde{A}(\omega_o, \boldsymbol{\theta})^+ + \left[\frac{\omega_k}{\omega_o}\right]\sigma_{v,k}^2 I \quad (6.60)$$

where the approximation of ideal resampling and the asymptotic result of (6.36) for the resampled noise covariance has been used. Note that for $\tilde{M} = M\omega_l/\omega_o$, the $\tilde{M} \times M$ focusing matrix $T_r^{\ddagger}(\omega_k)$ for $\omega_k \leq \omega_o$ has rank $\tilde{M} \leq M$. Thus for the cross-spectral density matrix estimate, $\hat{R}(\omega_k)$ of (6.18), with density $CW[N, M; R(\omega_k)]$, $T_r^{\ddagger}(\omega_k)\hat{R}(\omega_k)T_r^{\ddagger}(\omega_k)^+$ is also Wishart distributed with density, $CW[N, M; T_r^{\ddagger}(\omega_k) R(\omega_k)T_r^{\ddagger}(\omega_k)^+]$[26]. The focused covariance matrix estimate, $\tilde{R}_N(\omega_o)$ of (6.58), is thus the sum of $B = h - l + 1$ independent Wishart distributed random matrices.

To obtain expressions for the statistics of the SRMV method similar to those of the STMV method, consider a scenario where the temporal power spectrum of each source is flat over the receiver band, that is, $P_s(\omega_k) = P_s(\omega_o)$, for $k = l, \ldots, h$, and the power spectrum of the noise is such that $\sigma_{v,k}^2 = (\omega_o/\omega_k)\sigma_{v,h}^2$, for $k = l, \ldots, h$. Substituting into (6.60) gives that $T_r^{\ddagger}(\omega_p)R(\omega_p)T_r^{\ddagger}(\omega_p)^+ = T_r^{\ddagger}(\omega_q) R(\omega_q)T_r^{\ddagger}(\omega_q)^+$ for all p and q and therefore in this case, $\tilde{R}_N(\omega_o)$ is the sum of independent, *identically* distributed Wishart random matrices. From [26], $\tilde{R}_N(\omega_o)$ is thus Wishart distributed with density $CW[BN, \tilde{M}; \tilde{R}_\infty(\omega_o)]$, for $BN \geq \tilde{M}$, where $\tilde{R}_\infty(\omega_o) = \sum_{k=l}^{h} T_r^{\ddagger}(\omega_k)R(\omega_k)T_r^{\ddagger}(\omega_k)^+$. In this case, observe that $\hat{Z}_{\mathrm{srmv}}(\theta)$ of (6.59) is of the same form as $\hat{Z}_k(\theta)$ of (6.48) with Wishart distributed $\tilde{R}_N(\omega_o)$ replacing $R(\omega_k)$. Thus by direct correspondence with (6.52), $\hat{Z}_{srmv}(\theta)$ is a multiple of a chi-squared variable with $2(BN - \tilde{M} + 1)$ degrees of freedom having mean and variance given by

$$E\{\hat{Z}_{srmv}(\theta)\} = \frac{BN - \tilde{M} + 1}{BN}[\tilde{\mathbf{a}}(\omega_o, \theta)^+\tilde{R}_\infty(\omega_o)^{-1}\tilde{\mathbf{a}}(\omega_o, \theta)]^{-1}$$
$$\mathrm{Var}\{\hat{Z}_{srmv}(\theta)\} = \frac{1}{BN - \tilde{M} + 1}\{E[\hat{Z}_{srmv}(\theta)]\}^2 \quad (6.61)$$

Observe that under these conditions, the SRMV estimate for an arbitrary distribution of sources exhibits a statistical stability similar to the STMV method in a single group scenario. Particularly note that for $BN \gg \tilde{M}$, the number of statistical degrees of freedom available to the SRMV estimator reflects the full time-bandwidth product of the sources. This is the key factor which explains the increased statistical stability of focused wide-band spatial spectral estimation techniques. Again, even when $N < \tilde{M}$ and $\hat{Z}_{icmv}(\theta)$ cannot be computed, (6.61) implies that the SRMV estimate is still well defined as long as $BN \geq \tilde{M}$. For broad-band signals, this suggests that for $B \geq \tilde{M}$, the SRMV method can yield spatial spectral estimates, regardless of the source locations, with as little as one snapshot of date.

While the broad-band SRMV estimator clearly offers a lower threshold observation time that the incoherent MV method, it is important to realize the trade-off between resolution and statistical stability inherent in the approach. For example, consider the decimated resampled array in the previous scenario, where the resampled array is of length $\tilde{M} = M\omega_l/\omega_b < M$ sensors operating at a focusing frequency $\omega_o = \omega_b$. In this case, spatial resampling reduces the effective aperture of the array by a factor of ω_l/ω_b at the highest operating frequency. This loss of array aperture translates to a reduction in the asymptotic resolving power (ie., for large observation time) which is generally greater than that of the incoherent minimum variance method. As is common in spectral estimation methods, the SRMV estimator exhibits greater statistical stability at the price of lower resolution than incoherent methods. At observation times less than the threshold for incoherent methods, this trade-off is clearly desirable since it permits a reduction in the threshold observation time required. For longer observation times, the relative performance of spatially resampled and incoherent methods will in general depend upon the scenario under consideration. In a limited way, this trade-off is studied further in the simulation experiments presented in the Section 6.4.3.

To summarize, the steps used to perform spatially resampled minimum-variance spatial spectral estimation are as follows:

1. Form estimated cross-spectral density matrices, $\hat{R}(\omega_k)$ of (6.18), over the frequency band of interest.
2. Compute the focused broad-band covariance matrix, $\tilde{R}_N(\omega_o)$, of (6.58) using the optimal minimax focusing matrices, $T_{\Gamma}^{\ddagger}(\omega_k)$, given in (6.57).
3. Compute $\tilde{R}_N(\omega_o)^{-1}$ and form the SRMV spatial power spectral estimate, $\hat{Z}_{srmv}(\theta) = [\tilde{\mathbf{a}}(\omega_o, \theta)^+ \tilde{R}_N(\omega_o)^{-1} \tilde{\mathbf{a}}(\omega_o, \theta)]^{-1}$ of (6.59), for all directions of interest.

6.4.3 Simulation Results

The performance of the steered and spatially resampled minimum-variance methods of wide-band spatial spectral estimation is assessed here in terms of bearing estimate mean-squared error as measured in a series of simulation experiments. A uniformly spaced linear array of $M = 17$ omnidirectional sensors is considered with a half-wavelength interelement spacing at the normalized frequency of π radians. The Rayleigh limit of angular resolution for this array is approximately $2/(M - 1) \cong 0.133$ radians $\cong 7.62$ degrees. In realizing the model of (6.1), the sources are modeled as temporally stationary and mutually uncorrelated zero-mean Gaussian processes with bandpass autospectra that define the pqth element of the source spectral density matrix, $\{P_s(\omega_k)\}_{pq}$, by

$$\{P_s(\omega_k)\}_{pq} = \begin{bmatrix} \sigma_{p,b}^2 \delta_K(p-q), & \text{for } \omega_l \leq \omega_k \leq \omega_h \\ 0 \quad , & \text{otherwise} \end{bmatrix} \tag{6.62}$$

The noise outputs, $V(m, \omega_k)$, $m = -K, \ldots, K$, are modeled as stationary and mutually uncorrelated zero-mean Gaussian bandpass processes independent from the signals with autospectra such that the pqth element of the noise covariance matrix, $\{R_v(\omega_k)\}_{pq}$, is given by

$$\{R_v(\omega_k)\}_{pq} = \begin{bmatrix} \sigma_{v,b}^2 \delta_K(p-q), & \text{for } \omega_l \leq \omega_k \leq \omega_h \\ 0 \quad , & \text{otherwise} \end{bmatrix} \tag{6.63}$$

defined over the same frequency band as the source signals. The signal-to-noise ratio (SNR) of the pth source, denoted SNR_p, is defined from the foregoing by $\text{SNR}_p = \sigma_{p,b}^2/\sigma_{v,b}^2$. Estimated cross-spectral density matrices, $\hat{R}(\omega_k)$, were formed from the sensor outputs via (6.18). The number of segments or snapshots, N, employed to estimate each $\hat{R}(\omega_k)$, $k = l, \ldots, h$ is varied in this simulation study to examine the performance of the STMV and SRMV methods as a function of observation time.

To demonstrate the relative infinite-time performance of spatial resampling, a field consisting of five uncorrelated sources in a multigroup scenario was simulated. The sources and noise each had identical bandpass power spectra which were flat from $\omega_l = 0.7\pi$ to $\omega_h = \pi$. A total of $B = h - l + 1 = 31$ narrow-band cross-spectral density matrices, $R(\omega_k)$, equally spaced from ω_l to ω_h were used. For the spatially resampled array, a focusing frequency $\omega_o = \omega_l = 0.7\pi$ was employed corresponding to interpolation of the array outputs at each frequency. From the discussion following (6.44), a resampled array length of $\tilde{M} = M\omega_O/\omega_l = 17$ sensors was used. The source directions and signal-to-noise ratios are given by

Source number	1	2	3	4	5
$\text{Sin}(\theta_p)$	-0.91	-0.80	-0.11	0.0	0.79
SNR (dB)	15	15	9	9	-3

In Fig. 6.4, field directionality maps from five broad-band spatial spectral estimation methods are shown, where true cross-spectral density matrices, $R(\omega_k)$, have been used in place of estimates, $\hat{R}(\omega_k)$, to illustrate the infinite observation time characteristics of the methods. For clarity, an offset of 5 dB has been inserted between each plot in Fig. 6.4. The delay-and-sum beamformer trace in Fig. 6.4 corresponds to the expected beam power output from a broad-band Hamming-shaded DS beamformer in each of 201 steering directions equally spaced in the sine of bearing from $\sin(\theta) = -1$ to $\sin(\theta) = 1$. Because the angular separation between sources 1 and 2 and

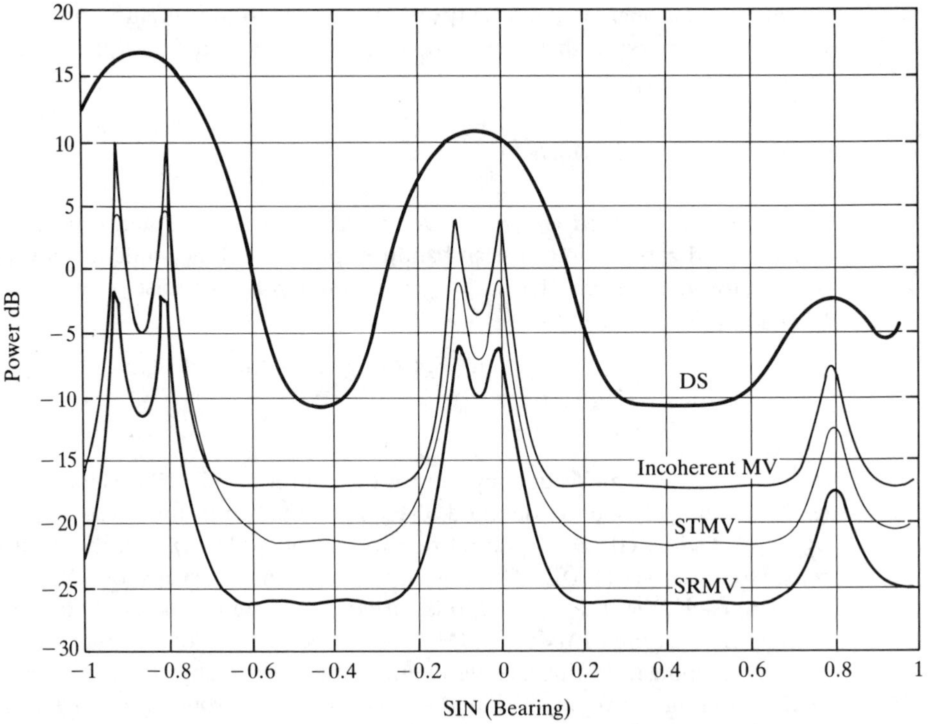

Figure 6.4 Broad-band spatial power spectral estimates for the Hamming-shaded DS, incoherent MV, STMV, and SRMV methods in an uncorrelated five source scenario under infinite observation time conditions. Each plot is offset by 5 dB.

between sources 3 and 4 is less than the Rayleigh limit, the broad-band Hamming-shaded DS beamformer clearly cannot resolve them. Note that although higher resolution can be obtained using a rectangularly shaded DS beamformer, its large sidelobe levels can create serious ambiguities in the field directionality map. The incoherent, steered, and spatially resampled minimum-variance spatial spectra in Fig. 6.4 were obtained using (6.49), (6.51), and (6.59), respectively, with the true $R(\omega_k)$ replacing the estimates, $\hat{R}(\omega_k)$. Note that all the broad-band minimum-variance methods resolve the five uncorrelated sources. Careful inspection of the SRMV peaks for sources 1 and 2, however, reveals that they are both slightly biased toward endfire. This phenomenon is consistent with the spatial coloration of the noise toward endfire as discussed in Section 6.3.2. Further note that the sharpness of the peaks suggests that the incoherent MV method, followed by the STMV method, appear to resolve the sources slightly better than does the SRMV method. Although the sharpness of spectral peaks must be used with caution in discussions concerning resolution, this infinite-time behavior qualitatively supports the notion that the incoherent MV method exhibits better asymptotic performance than do either the STMV or

SRMV techniques. Clearly, in large (in this example, infinite) observation time settings, the increased statistical stability of the STMV and SRMV methods is of little consequence. The advantages of the latter techniques only become apparent in shorter observation time situations.

To illustrate the finite observation time characteristics of the delay-and-sum, incoherent MV, steered MV, and spatially resampled MV methods, the previous experiment was repeated using estimated $\hat{R}(\omega_k)$ obtained with $N = M = 17$ snapshots of data at each frequency ω_k, $k = l, \ldots, h$. All other experimental parameters of the preceding scenario were maintained except that sources 1 and 2 were modified to simulate perfectly correlated multipath arrivals with a propagation delay much larger than the signal correlation time. In Fig. 6.5, spatial spectral estimates obtained in ten independent trials are shown for the Hamming-shaded delay-and-sum beamformer, the incoherent MV method, and the steered MV technique. Each family of plots is offset by 8 dB for clarity. Observe that the incoherent MV method exhibits considerably more variability from trial to trial than the DS result. Further, the well-

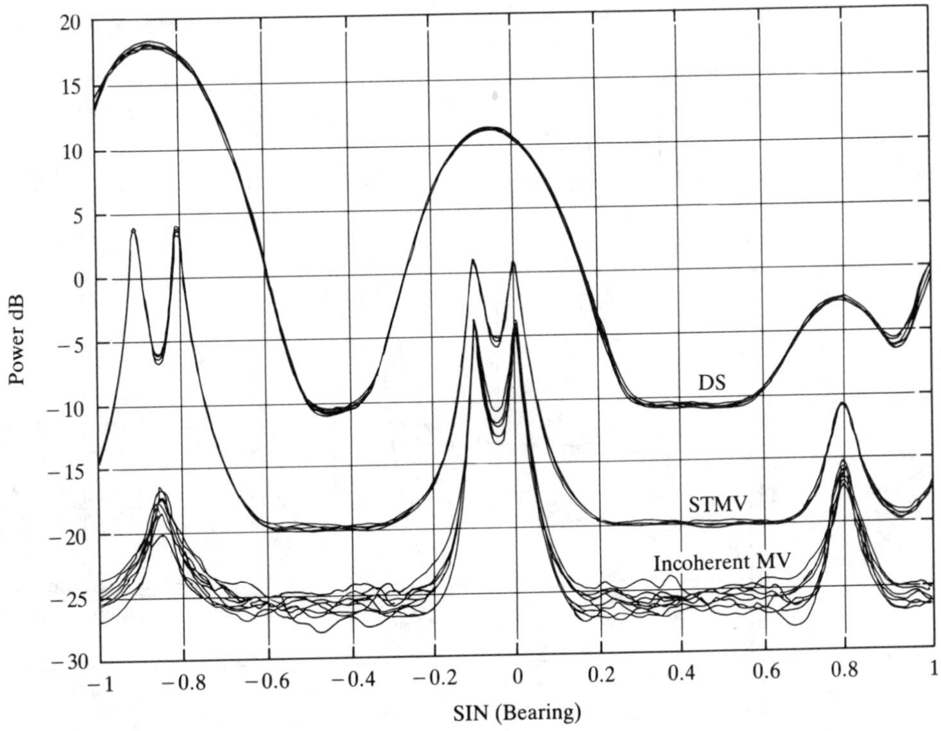

Figure 6.5 Broad-band spatial power spectral estimates for the Hamming-shaded DS, STMV, and incoherent MV methods in a correlated five source scenario under finite observation time conditions. Ten independent trials per method are shown with each set of traces offset by 8 dB.

known multipath cancellation behavior of the narrow-band MV method [21] results in an inability of the incoherent MV estimator to resolve the two correlated sources. In comparison, the STMV method resolves all five sources with a trial-to-trial variability that is considerably smaller than the incoherent MV technique. This behavior is consistent with the fact that the STMV spatial spectral estimator has greater statistical stability than does the incoherent MV method. The results of ten independent trials with the spatially resampled minimum variance method are shown in Fig. 6.6. Note that the SRMV method resolves all five sources, including the perfectly correlated sources numbers 1 and 2. Further, the trial-to-trial variability is comparable to the STMV results of Fig. 6.5. As a benchmark for evaluating the focusing capability of spatial resampling, the narrow-band MV spatial spectral estimator, $\hat{Z}_k(\theta)$ of (6.48) was computed using 527 (equal to the number of snapshots multiplied by the number of frequency components, i.e., BN) independent realizations of hypothetical vectors $\mathbf{Y}_n(\omega_o)$, $n = 1, \ldots, 527$ at the focusing frequency of $\omega_o = 0.7\pi$. Recall that an objective of spatial resampling was to achieve performance equivalent to this narrow-band benchmark. Note that each benchmark spatial spectral estimate was computed

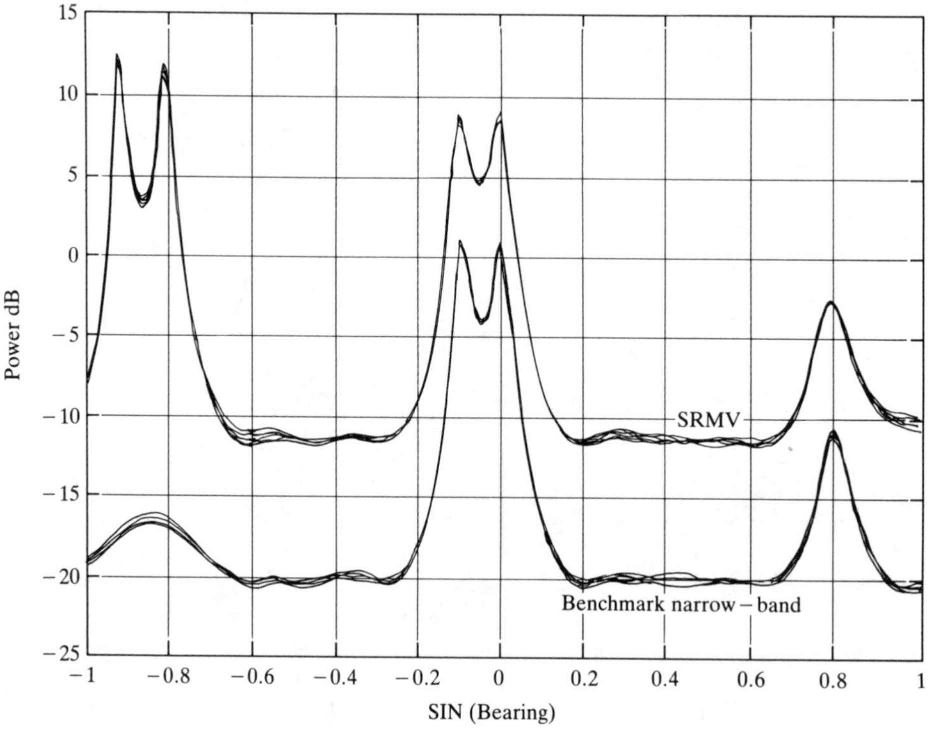

Figure 6.6 Broad-band spatial power spectral estimates for the SRMV and benchmark narrow-band methods in the same finite observation time scenario as Figure 6.5. Ten independent trials per method are shown with each set of traces offset by 8 dB.

using a single narrow-band covariance matrix, $\hat{R}(\omega_o)$, formed via (6.18) with 527 realizations of $\mathbf{Y}_n(\omega_o)$. Ten independent trials of the benchmark narrow-band MV benchmark are shown in the lower set of curves in Fig. 6.6. Note that as with the incoherent MV method, the narrow-band benchmark is unable to resolve the two correlated sources. Other than this difference, the SRMV method exhibits trial-to-trial variability and resolution performance similar to the benchmark results.

A more quantitative comparison of the incoherent, steered, and spatially resampled minimum-variance methods is achieved by measuring their mean-squared bearing error versus observation time. In the following experiments, the statistics of the spatial spectral estimate peak in the neighborhood of the third source at $\sin(\theta_3) = -0.11$ were estimated based on 100 independent trials at each observation time. The observed bias, standard deviation, and root-mean-squared error of bearing estimates achieved with the incoherent MV, STMV, SRMV, and benchmark narrow-band methods are summarized in Tables 6.1 and 6.2. Note that estimator bias was measured by subtracting the true source bearing from the average estimated bearing. The standard deviation results are plotted versus observation time in Fig. 6.7. At observation times corresponding to less than $N = 17$ snapshots, the incoherent MV

TABLE 6.1 Bearing Estimation Performance of the STMV and Incoherent MV Methods Versus Observation Time

Observation Time (snapshots)	Incoherent MV Method (bias	std. dev. (degrees)	rms)	STMV Method (bias	std. dev. (degrees)	rms)
1		Singular		0.100	0.158	0.187
2		Singular		0.109	0.097	0.139
4		Singular		0.118	0.064	0.134
8		Singular		0.116	0.037	0.122
17	0.015	0.0866	0.088	0.116	0.021	0.118
24	0.021	0.0263	0.034	0.114	0.018	0.115
32	0.019	0.0168	0.025	0.121	0.017	0.122

TABLE 6.2 Bearing Estimation Performance of the SRMV and Benchmark Narrow-Band Methods Versus Observation Time

Observation Time (snapshots)	SRMV Method (bias	std. dev. (degrees)	rms)	Benchmark MV Method (bias	std. dev. (degrees)	rms)
1	0.126	0.189	0.227	0.076	0.164	0.182
2	0.138	0.132	0.191	0.083	0.085	0.118
4	0.157	0.082	0.177	0.098	0.058	0.113
8	0.143	0.057	0.154	0.095	0.044	0.105
17	0.144	0.034	0.148	0.098	0.027	0.102
24	0.143	0.028	0.144	0.098	0.018	0.100
32	0.142	0.020	0.143	0.098	0.016	0.099

method could not be computed due to the singularity of the cross-spectral density matrix estimates, $\hat{R}(\omega_k)$. The broad-band focused covariance matrix estimates corresponding to the steered and spatially resampled minimum variance methods, however, remained nonsingular for observation times as small as one snapshot. The solid line in Fig. 6.7 is the Cramér-Rao lower bound (CRLB) for estimation of θ_3 in this scenario (cf. [25]). It should be noted that the CRLB used here assumes the receiver statistics are known a priori with the exception of θ_i, $i = 1, \ldots 5$, and further applies only to unbiased estimators. Since neither of these conditions apply to the estimators under consideration, the CRLB in the present study should be considered simply as a useful performance benchmark rather than as a strict lower bound. Observe that the STMV, SRMV, and benchmark narrow-band methods yielded bearing estimates with standard deviations that follow the CRLB even below the 17 snapshot threshold observation time of the incoherent MV method. Above the threshold observation time, the SRMV, STMV, and benchmark narrow-band methods exhibit lower bearing estimate standard deviation than do the incoherent MV estimates; however, this variance reduction is accompanied by an increase in bearing bias. These results suggest that the SRMV method achieves similar bearing estimation performance to the STMV

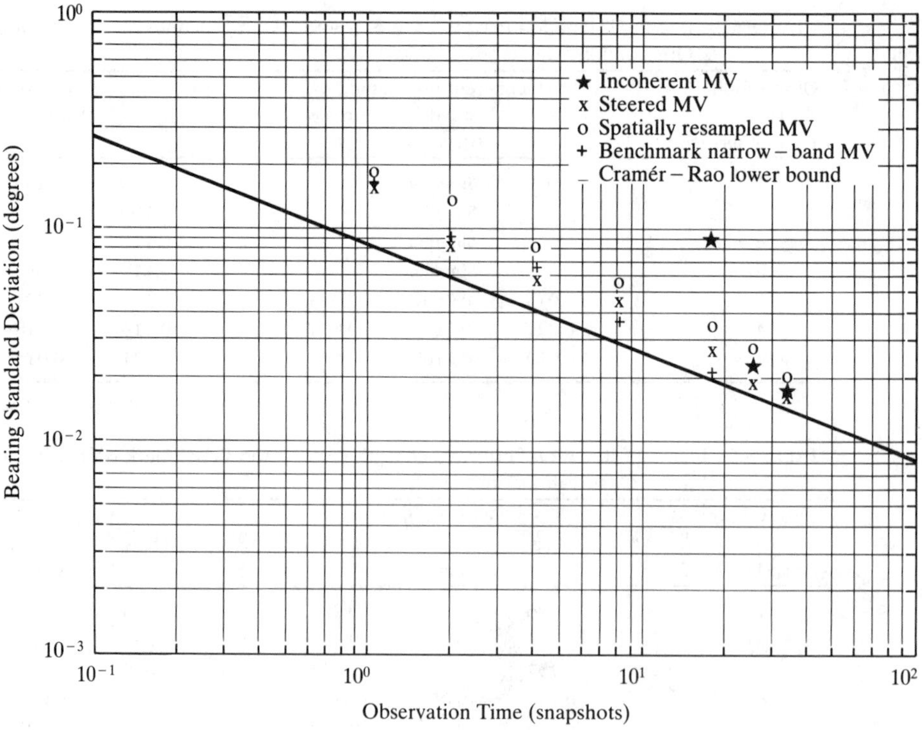

Figure 6.7 Bearing estimate standard deviation for incoherent MV, SRMV, STMV, and benchmark narrow-band methods versus the CRLB as a function of observation time as measured by the number of snapshots.

and benchmark narrow-band methods, achieving lower bearing estimate variance at the price of greater bearing bias. At observation times less than the threshold for the incoherent MV technique, this trade-off is highly desirable since it permits a considerable reduction in the observation time required for resolving the broad-band sources. At observation times greater than the threshold of the incoherent MV method, the overall bearing estimate mean-squared error may or may not be reduced by focused covariance methods depending on the specific scenario under consideration.

The simulation results of this section indicate that the SRMV and STMV methods offer similar broad-band bearing estimation performance in the scenario considered. As noted, however, the advantage of using the SRMV versus the STMV approach lies in its reduced computational complexity. To compute the SRMV estimate, $\hat{Z}_{srmv}(\theta)$, for L different values of θ (corresponding to L different look directions), an approximate calculation of the number floating point multiplies, $FPM(\hat{Z}_{srmv})$ gives

$$FPM(\hat{Z}_{srmv}) \cong BN(\tilde{M}M + \tilde{M}^2) + \tilde{M}^3 + L(\tilde{M}^2 + M) \tag{6.64}$$

In comparison, the number of floating point multiplies to compute the STMV spectral estimate, $\hat{Z}_{stmv}(\theta)$, in L different look directions is approximately

$$\mathrm{FPM}(\hat{Z}_{\mathrm{stmv}}) \cong L[BN(M^2 + M) + M^3] \tag{6.65}$$

In the simulation experiments considered, for example, $M = \tilde{M} = 17$, $B = 31$, $L = 201$, and for $N = 4$ snapshots substituting into (6.64) and (6.65) give $FPM(\hat{Z}_{stmv}) = 8.6 \times 10^6$ while $FPM(\hat{Z}_{srmv}) = 1.4 \times 10^5$, which implies the STMV method requires approximately 60 times more floating point multiplications than the SRMV method. In the case where $BN \gg M$, $M^2 \gg M$, and $\tilde{M} = M$, the ratio of $FPM(\hat{Z}_{srmv})$ to $FPM(\hat{Z}_{stmv})$ is approximately $(2BN + L)/LBN$, suggesting that the SRMV method is particularly efficient when the number of steering directions, L, is large.

In these simulations, the SRMV method has been implemented with the focusing frequency, ω_o, equal to the minimum source frequency, ω_l. The development of spatial resampling given in previous sections of this paper, however, only restricts $\omega_o \leq \pi c/d$. Choosing $\omega_o > \omega_l$ and therefore $\tilde{M} < M$ will result in slightly better statistical stability and less computational load due to the smaller dimension of the focused covariance matrix. To achieve these advantages, however, the focusing frequency ω_o should be chosen such that $K\omega_l/\omega_o$ is an integer, thereby ensuring that the effective resampled array aperture, as determined by $\tilde{K} = \lfloor K\omega_l/\omega_o \rfloor$, is maximized.

6.4.4 Results with Actual Towed-Array Data

To examine the performance of a focused wide-band approach in a practical setting, the STMV method was used to estimate the field directionality with actual towed-array data. The results reported in this section were obtained using towed-array data supplied by the Canadian Defence Research Establishment Atlantic located in Dartmouth, Nova Scotia. The data were provided in a heterodyned and decimated 16-bit complex integer format with a sampling rate of 47.619 Hz. The complex data represented the sensor outputs in a receiver bandwidth equal to the sampling rate of 47.619 Hz centered on a frequency of 296.875 Hz. The towed line array is made up of 32 hydrophones equispaced a distance of 3.0 meters apart for a half-wavelength frequency of 250 Hz. In these experiments, the first and last channels were not processed because of their large self-noise components. Data processing on the remaining sensor array consisted of first dividing the data into $\Delta T = 2.688$ sec (that is, 128 point) snapshots and performing discrete Fourier transforms on each channel to obtain frequency domain vectors, $\mathbf{Y}_n(k)$, $k = 1, \ldots, 128$. Cross-spectral density matrices at each frequency were then formed using (6.18) with an observation time of $T = 5.376$ sec equivalent to $N = 2$ snapshots. The total broad-band time-bandwidth product available per STCM estimate was therefore equal to 256 (that is, 5.376 sec $\times$ 47.619 Hz) compared to a CSDM time-DFT-binwidth product of 2 (that is, 5.376 sec $\times$ 0.372 Hz). Broad-band spatial spectral estimates were formed using 256 steering directions equispaced in sin-bearing from -80 to $+80$ degrees for each 5.376-sec frame of data. The complete 6-min data set permitted a total of 64 consecutive field directionality maps to be computed.

As a basis for comparison, the results of conventional Hamming-shaded delay-and-sum beamforming are shown in Fig. 6.8 The waterfall plot consists of 64 con-

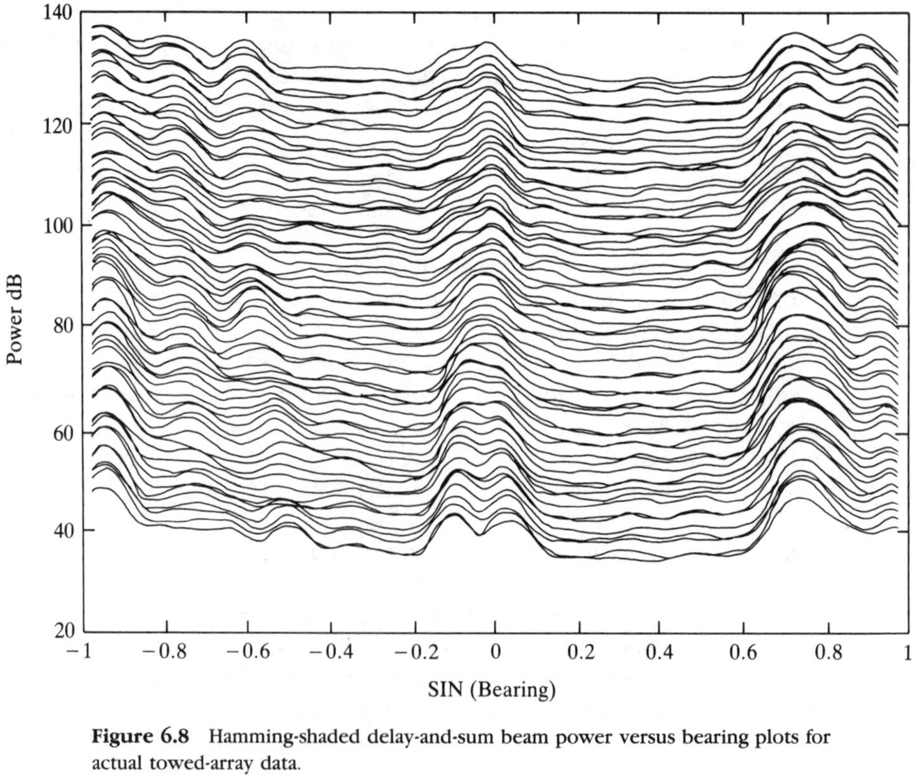

Figure 6.8 Hamming-shaded delay-and-sum beam power versus bearing plots for actual towed-array data.

secutive spatial spectral estimates each offset by 1.5 dB. Peaks in the spatial spectrum running up the plot correspond to broad-band plane-wave arrivals. Observe that the broad beam power peaks in Fig. 6.8 are consistent with the low resolution of the Hamming-shaded DS beamformer. Approximately 10 to 12 arrivals can be identified from the DS plots. In Figure 6.9, the result of applying the STMV method to the same data set is shown. Observe the increased spatial resolution apparent in the STMV plot without a significant decrease in statistical stability. The STMV method is stable because the available time-bandwidth product ($NB = 256$) is much larger than the number of sensors ($M = 30$). In contrast to the conventional DS result, about 20 to 22 arrivals can be resolved with the STMV method. Note that a CSDM-based MVDR method could not be computed with this short data length since the CSDM time-DFT-width product ($N = 2$) is less than the number of sensors ($M = 30$). Note that the ability of the STMV method to provide high-resolution spatial spectral estimates in such short observation time settings is a key feature of the technique.

While the STMV result indicates the effectiveness of the technique, its computation required the estimation and matrix inversion of 256 different steered covariance matrices for each horizontal trace in Fig. 6.9. This illustrates the heavy computational requirements of the STMV technique. In Figure 6.10, the waterfall plot of spatial spectral estimates obtained using the SRMV method is shown. The focusing

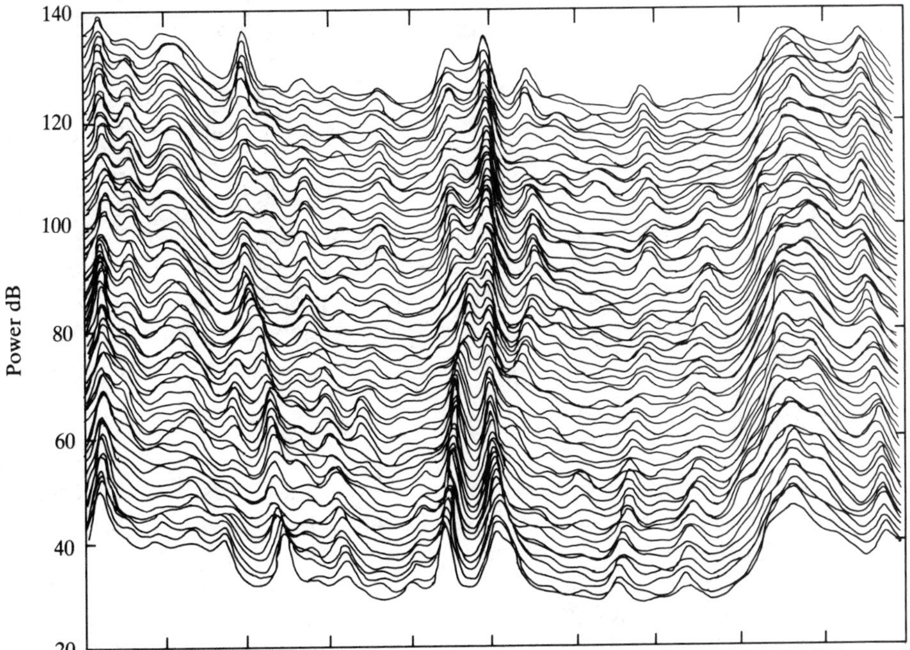

Figure 6.9 Broad-band steered minimum-variance spatial spectral estimates for actual towed-array data.

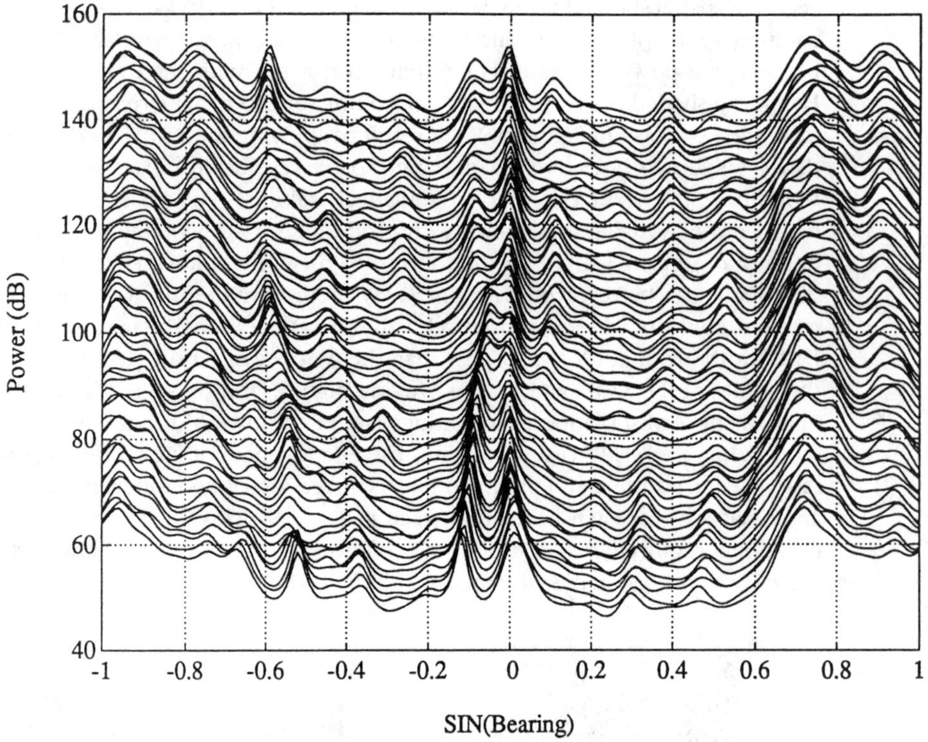

Figure 6.10 Broadband spatially resampled minimum variance spatial spectral estimates for actual towed-array data.

frequency employed corresponds to the center frequency of 296.875 Hz. Comparison of Figures 6.10 and 6.9 suggest that the SRMV and STMV methods provided similar resolution and statistical stability in this scenario. From (6.64) and (6.65), however, the SRMV method offers a considerably more efficient implementation.

6.5 CONCLUSION

This chapter has investigated focused wide-band processing as a means of reducing the threshold observation time required to perform high-resolution broad-band spatial spectral estimation. The aim of wide-band focusing is to preprocess array outputs so that broad-band sources can be represented by rank-one models. Low-rank representations are attractive because they lead to the formation of space-time statistics that can be estimated with an accuracy that reflects the full observation time-bandwidth product of wide-band sources. Thus high-resolution spatial spectral estimation methods based on wide-band focusing can be viewed as reaching their "asymptotic" performance at a lower threshold observation time than incoherent wide-band techniques. In this chapter, presteering and spatial resampling methods have been considered for wide-band focusing without preliminary estimates of the field directionality. In steered covariance methods, time-domain sensor outputs are simply steered in each direction of interest prior to forming wide-band covariance matrices. It has been shown that each steered covariance matrix effectively focuses wide-band arrivals from its particular steering direction. While the effective number of statistical degrees of freedom available to estimate each STCM can be much larger than for individual narrow-band cross-spectral density matrices, implementation of steered covariance methods is hampered by the computational burden of having to form a different STCM in each direction of interest. In spatial resampling methods, wide-band arrivals from all directions are focused simultaneously, thereby offering considerable computational savings relative to steered covariance methods. Spatial resampling consists of adjusting the spatial sampling rate of the array as a function of temporal frequency so that broad-band sources are aligned in the spatial frequency domain. The implementation of spatial resampling has been achieved here by designing realizable linear shift-variant filters which approximate the ideal spatial frequency-mapping transformations.

To demonstrate the advantages of wide-band focusing, the steered covariance and spatial resampling approaches have been used in conjunction with minimum-variance spatial spectral estimation to obtain the steered MV and spatially resampling MV methods, respectively. Analytical and simulations results have been presented which indicate that the statistical stability of the broad-band STMV and SRMV spatial spectral estimates reflects the full observation time-bandwidth product of wide-band sources. This increased statistical stability facilitates spatial spectral estimation at a significantly lower threshold observation time than the analogous incoherent wide-band MV approach. Finally, STMV and SRMV results with actual towed-array data suggest that focused wideband processing can facilitate practical high-resolution field directionality mapping using acceptably short observation times.

REFERENCES

1. D. H. Johnson, "The Application of Spectral Estimation Methods to Bearing Estimation Problems," *Proc. IEEE* **70**, no. 9 (September 1982), 1018–1028.
2. M. Wax, T. J. Shan, and T. Kailath, "Spatio-temporal Spectral Analysis by Eigenstructure Methods," *IEEE Trans. on Acoustics, Speech, and Signal Processing*, ASSP-32, no. 4 (August 1984), 817–827.
3. R. S. Walker, "Bearing Accuracy and Resolution Bounds of High-Resolution Beamformers," *Proc. IEEE*, ICASSP-85, Tampa, Florida, 1985, 1784–1787.
4. K. M. Buckley, "Spatial/Spectral Filtering with Linearly Constrained Minimum-Variance Beamformers," *IEEE Trans. on Acoustics, Speech, and Signal Processing*, ASSP-35, no. 3, (March 1987), 249–266.

5. H. Wang and M. Kaveh, "Coherent Signal-Subspace Processing for the Detection and Estimation of Angles of Arrival of Multiple Wideband Sources," *IEEE Trans. on Acoustics, Speech, and Signal Processing*, ASSP-33, (August 1985), 823–831.
6. J. Yang and M. Kaveh, "Wideband Adaptive Arrays Based on the Coherent Signal Subspace Transformation," *Proc. IEEE*, ICASSP-87, Dallas, (April 1987), 2011–2013.
7. H. Hung and M. Kaveh, "On the Statistical Sufficiency of the Coherently Averaged Covariance Matrix for Estimation of the Parameters of Wideband Sources," *Proc. IEEE*, ICASSP-87, Dallas, (April 1987), 33–36.
8. H. Wang, C. C. Li, and J. X. Zhu, "High-Resolution Direction Finding in the Presence of Multipath: A Frequency-Domain Smoothing Approach," *Proc. IEEE*, ICASSP-87, Dallas, (April 1987), 2276–2279.
9. H. Hung and M. Kaveh, "Focusing Matrices for Coherent Signal Subspace Processing," *IEEE Trans. on Acoustics, Speech, and Signal Processing, vol.* ASSP-36, (August 1988), 1272–1281.
10. J. Krolik and D. N. Swingler, "Focused Wideband Array Processing via Spatial Resampling," *IEEE Trans. on Acoustics, Speech, and Signal Processing*, ASSP-38, no. 2, (February 1990), 356–60.
11. G. Bienvenu, G. Vezzosi, L. Kopp, and F. Floring, "Coherent Wideband High Resolution Processing for a Linear Array," *Proc. IEEE*, ICASSP-89, Glasgow, (1989), 2799–2802.
12. H. Clergeot and O. Michel, "New Simple Implementation of the Coherent Signal Subspace Method for Wideband Direction of Arrival Estimation," *Proc. IEEE*, ICASSP-89, Glasgow, 1989, 2764–2767.
13. D. N. Swingler, R. S. Walker, and J. Krolik, "High-Resolution Broadband Beamforming Using a Doubly Steered Coherent Signal-Subspace Approach," *Proc. IEEE*, ICASSP-88, New York, (1988), 2658–2661.
14. D. N. Swingler and J. Krolik, "Source Location Bias in the Coherently Focused High-Resolution Broadband Beamformer," *IEEE Trans. on Acoustics, Speech, and Signal Processing*, ASSP-37, no. 1 (January 1989), 143–144.
15. K. M. Buckley and L. J. Griffiths, "Broadband Signal-Subspace Spatial-Spectrum (BASS–ALE) Estimation," *IEEE Trans. on Acoustics, Speech, and Signal Processing*, ASSP-36, no. 7 (July 1988), 953–964.
16. S. Nawab, F. Dowla, and R. Lacoss, "Direction Determination of Wideband Signals," *IEEE Trans. on Acoustics, Speech, and Signal Processing*, ASSP-33, no. 4 (Oct 1985), 1114–1122.
17. G. Su and M. Morf, "Signal Subspace Approach for Multiple Wideband Emitter Location," *IEEE Trans. on Acoustics, Speech, and Signal Processing*, ASSP-31, (Dec 1983), 1502–1522.
18. J. Krolik and D. N. Swingler, "Multiple Broadband Source Location Using Steered Covariance Matrices," *IEEE Trans. on Acoustics, Speech, and Signal Processing*, ASSP-37 (October 1989), 1481–1494.
19. J. Capon, "High-Resolution Frequency—Wavenumber Spectrum Analysis," *Proc. IEEE* **57** (August 1969), 1408–1418.
20. S. L. Marple, *Digital Spectral Analysis with Applications*, Prentice Hall Signal Processing Series, (Englewood Cliffs, N.J.; Prentice Hall, 1987).

21. N. Owsley, "Sonar Array Processing," S. Haykin, ed., in *Array Signal Processing* (Englewood Cliffs, N.J.; Prentice Hall, 1984).
22. H. Wang, C. C. Li, and J. X. Zhu, "High-Resolution Direction Finding in the Presence of Multipath: A Frequency-Domain Smoothing Approach," *Proc. IEEE*, ICASSP-87, Dallas, (1987), 2276–2279.
23. R. E. Crochiere, and L. R. Rabiner, *Multirate Digital Signal Processing* (Englewood Cliffs, N.J.; Prentice Hall, 1983).
24. B. Liu, and P. A. Franaszek, "A Class of Time-Varying Digital Filters," *IEEE Trans. on Circuit Theory*, CT-16, no. 4 (November 1969), 467–471.
25. W. J. Bangs, "Array Processing with Generalized Beamformers," Ph. D. dissertation, Yale University, University Microfilms, Ann Arbor, Mich., (1972).
26. Robb J. Muirhead, *Aspects of Multivariate Statistical Theory* (New York: John Wiley, 1982).
27. T. W. Parks and D. P. Kolba, "Interpolation Minimizing the Maximum Normalized Error for Band-Limited Signals," *IEEE Trans. on Acoustics, Speech, and Signal Processing*, ASSP-26, no. 4 (August 1978), 381–384.
28. J. Capon and N. R. Goodman, "Probability Distributions for Estimators of the Frequency-Wavenumber Spectrum," *Proc. IEEE*, **58** (October 1970), 1785–1786.

ACKNOWLEDGMENTS

The author would like to thank Dr. D. N. Swingler of Saint Mary's University, Halifax, N.S., for his important contributions to this work and Dr. R. S. Walker of the Defence Research Establishment Atlantic (DREA), Dartmouth, N.S., for many stimulating discussions concerning this research. This work was supported by DREA under contracts W7707-7-7934 and W7707-8-1160.

7 Statistical Efficiency Study of Direction Estimation Methods Part I: Analysis of MUSIC and Preliminary Study of MLM

Petre Stoica and Arye Nehorai

7.1. INTRODUCTION

Several important problems in the signal processing field, among them *direction finding* with narrow-band sensor arrays, *estimation of the parameters* of multiple superimposed exponential signals in noise, and *resolution of overlapping echoes* (see [1], [2], [9], [12] and the references there), can be reduced to estimating the parameters in the following model:

$$\boldsymbol{y}(t) = \boldsymbol{A}(\boldsymbol{\theta})\boldsymbol{x}(t) + \boldsymbol{e}(t), \qquad t = 1, 2, \ldots, N. \tag{7.1a}$$

In (7.1), $\boldsymbol{y}(t) \in C^{m \times 1}$ is the *noisy data vector*, $\boldsymbol{x}(t) \in C^{n \times 1}$ is the *vector of signal amplitudes*, $\boldsymbol{e}(t) \in C^{m \times 1}$ is an *additive noise vector*, and the matrix $\boldsymbol{A}(\boldsymbol{\theta}) \in C^{m \times n}$ has the following special structure

$$\boldsymbol{A}(\theta) = [\boldsymbol{a}(\omega_1)\ldots\boldsymbol{a}(\omega_n)], \tag{7.1b}$$

where $\{\omega_i\}$ are real parameters, $\boldsymbol{a}(\omega_i) \in C^{m \times 1}$ is a so-called *transfer vector* (between the ith signal and $\boldsymbol{y}(t)$), and $\boldsymbol{\theta} = [\omega_1, \ldots, \omega_n]^T$. In Section 7.2, we briefly discuss how the model (7.1) encompasses the data models used in some of the applications mentioned above, and introduce the basic assumptions on (7.1).

There are three main problems associated with fitting models of the form (7.1) to the data $\{y(1), \ldots, y(N)\}$:

1. *Estimation of the number of signals n.* Methods for estimating n are well documented in the literature (see, e.g., [3], [4], [21] and [22]) and will not be discussed here; see also Chapter 5. In this chapter we assume that the number of signals n is given.
2. *Estimation of the signal amplitudes* $\{\boldsymbol{x}(t)\}$. Once an estimate of $\boldsymbol{\theta}$ is available, the estimation of $\boldsymbol{x}(t)$ reduces (under reasonable conditions) to a simple least-squares fit. We will omit any explicit discussion on the problem of estimating $\{\boldsymbol{x}(t)\}$. However, estimates of $\{\boldsymbol{x}(t)\}$ will implicitly appear in the analysis that follows. Note that since it is required to estimate $\{\boldsymbol{x}(t)\}_{t=1}^{N}$ (and not their "average characteristics," such as their covariance matrix), we consider these variables to be *deterministic* (i.e., *fixed*). This assumption does not exclude the possibility that $x(1), \ldots, x(N)$ are samples from a random process. If so, then the distributional results derived in what follows should be interpreted as being conditioned on $\{\boldsymbol{x}(t)\}_{t=1}^{N}$.
3. *Estimation of the parameter vector* $\boldsymbol{\theta}$. Methods for accomplishing this task and their performance are the main topics to be dealt with in this chapter.

A class of methods for estimating $\boldsymbol{\theta}$ in (7.1), which has received significant attention is based on the *eigendecomposition of the sample covariance matrix* of $\boldsymbol{y}(t)$ [1]–[9], [15]–[22]. A representative member of this class is the *MUSIC (MUltiple SIgnal Characterization) algorithm* [1],[2]. There has been considerable interest recently in analyzing the statistical performance of the MUSIC. Some interesting and related studies of the resolvability of MUSIC have been reported in [7] and [8] (also see [21]). However, an expression for the covariance matrix, say $\boldsymbol{C}$, of the MUSIC estimate of $\boldsymbol{\theta}$ has not been derived in these papers. A preliminary analysis of the MUSIC performance expressed by $\boldsymbol{C}$ can be found in [5], but it appears to be incomplete. In Section 7.3, after a brief review of the MUSIC, we provide an explicit expression for $\boldsymbol{C}$ that holds for sufficiently large values of N. In this section we also discuss an improved MUSIC estimator introduced in [6], for which we present a performance analysis as well, and a computationally inexpensive modification that should improve its performance.

The expression of $\boldsymbol{C}$ derived in this paper can be used to compare the performance of the MUSIC with the performance achieved by other methods. In particular, comparison with the performance corresponding to the *Cramér-Rao lower bound (CRLB)* should be of interest. An expression for the CRLB on the covariance matrix of any unbiased estimator of the parameters $\boldsymbol{\theta}$ in the general model (7.1) does not appear to be available in the literature (only expressions for special cases corresponding to $N = 1$ and a specific "transfer vector" $\boldsymbol{a}(\omega)$ can be found, see, for example, [24]–[26]). In Section 7.4, we derive the CRLB on the covariance matrix under reasonable conditions. The behavior of that matrix when m, or N, or both increase is also studied.

The classical *maximum-likelihood* (*ML*) method can also be used, under appropriate assumptions, to estimate the parameter vector $\boldsymbol{\theta}$ in (7.1). The *ML estimator* (*MLE*) of $\boldsymbol{\theta}$ has been the topic of two interesting and related recent papers [10] and [12] (see also [11], [13], and [14] for studies of ML estimation of parameters in special cases of (7.1)). In Section 7.5, we briefly review the ML approach to estimation of $\boldsymbol{\theta}$, study the consistency properties of the MLE, and discuss its statistical (second-order) performance. The main issue of this section is the question of the asymptotic efficiency of the MLE. It is shown that the MLE is *not* statistically efficient if m is small, even if N is large, and that it can achieve the CRLB only if m is increased.

In Section 7.6, we investigate the relationship between the MUSIC and the ML estimators. Only some unsupported claims about this relationship can be found in the literature (see, e.g., [5] and [23]). We show that the MUSIC is a large sample (for $N >> 0$) realization of the MLE if and only if the signals are uncorrelated.

Finally, in Section 7.7, we present an analytic comparison between the MUSIC estimation error variance and the CRLB. For uncorrelated signals, the MUSIC variance is shown to approach the CRLB as m and N increase. For correlated signals, MUSIC is shown to be statistically inefficient. Also presented in this section are the results of a numerical study. Specifically, we perform a detailed comparison of the CRLB and MUSIC variance in the case of two complex sine waves, over the set of feasible angular frequencies and for several values of *SNR* (*signal-to-noise ratio*), m, and the coefficient of correlation between the two sine waves.

7.2. NOTATION, BASIC ASSUMPTIONS, AND SPECIAL CASES

We first list some notational conventions that will be used in this chapter:

A^T = the transpose of matrix $A \in C^{k \times p}$

A^* = the complex conjugate of A

$A^\dagger$ = the conjugate transpose of A

$\bar{A}$ = the real part of A

$\tilde{A}$ = the imaginary part of A

$\mathrm{tr}(A)$ = the trace of $A \in C^{k \times k}$

A_{ij} = the i, j element of A

$A \odot B$ = the Hadamard product of $A \in C^{k \times p}$ and $B \in C^{k \times p}$ $([A \odot B]_{ij} = A_{ij}B_{ij})$

$A \geq B$ = the difference matrix $A - B$ is positive semidefinite, with A and B being hermitian positive semidefinite matrices.

$\delta_{k,p}$ = the Kronecker delta (= 1 if $k = p$, and = 0 otherwise)

ω = a generic element of the vector $\boldsymbol{\theta}$; to avoid a complication of notation, the symbols ω and $\boldsymbol{\theta}$ are used to denote both the true and the unknown parameters.

$\boldsymbol{d}(\omega) = \mathrm{d}\boldsymbol{a}(\omega)/\mathrm{d}\omega$

E = the expectation operator; for deterministic signals, $\mathrm{E}\{\cdot\} = \lim_{N \to \infty} (1/N) \sum_{t=1}^{N} (\cdot)$.

Next, we introduce some basic assumptions on the model (7.1). The MUSIC and the ML methods are based on different sets of assumptions. However, some assumptions are common to both methods. The common assumptions are listed first:

A1: $m > n$, and the vectors $\boldsymbol{a}(\omega)$ corresponding to $(n + 1)$ different values of ω are linearly independent. (This is a weak assumption that guarantees the uniqueness of both the MUSIC and ML estimators.)

A2: $\mathrm{E}\{\boldsymbol{e}(t)\} = \mathbf{0}$, $\mathrm{E}\{\boldsymbol{e}(t)\boldsymbol{e}^{\dagger}(t)\} = \sigma^2\boldsymbol{I}$ and $\mathrm{E}\{\boldsymbol{e}(t)\boldsymbol{e}^{\mathrm{T}}(t)\} = \mathbf{0}$. (This is a more restrictive assumption that is essential for the MUSIC algorithm; for the ML method, relaxation of A2 is possible in principle, but would lead to considerable complications.)

The following additional assumption is needed for the MUSIC. AMU: The matrix

$$\boldsymbol{P} = \mathrm{E}\{\boldsymbol{x}(t)\boldsymbol{x}^{\dagger}(t)\} \tag{7.2}$$

is nonsingular (positive definite), and $N > m$:
The following one is needed for the MLE. AML:

$$\mathrm{E}\{\boldsymbol{e}(t)\boldsymbol{e}^{\dagger}(s)\} = \mathrm{E}\{\boldsymbol{e}(t)\boldsymbol{e}^{T}(s)\} = 0, \qquad \text{for } t \neq s,$$

and $e(t)$ is Gaussian distributed.

Assumption AML appears more restrictive than AMU (again AML could in principle be relaxed, but this would introduce significant complications). The distinction made between the assumptions used by MUSIC and MLE is important for realizing which one of these two estimators, if any, is usable in a certain situation (see the discussion that follows).

Next we describe briefly some applications of the general model (7.1). For other possible applications of (7.1) we refer to [9] and [12] and the references therein.

7.2.1 Direction Finding with Uniform Linear Sensor Arrays

The problem of determining the directions of n plane waves impinging on a linear uniform narrow-band array of m sensors, can be formulated as that of estimating the parameters $\boldsymbol{\theta}$ of the model (7.1), where $\boldsymbol{x}(t)$ is the vector of complex wave amplitudes, N is the number of "snapshots," and

$$a(\omega) = [1 \; e^{i\omega} \; \dots \; e^{i(m-1)\omega}]^T. \tag{7.3}$$

Note that in this case, $A(\theta)$ is a *Vandermonde matrix*, and therefore assumption A1 is satisfied. Assumption A2 and AML mean that the noise is spatially and temporally uncorrelated, and assumption AMU means that the plane waves are not "fully coherent" and the number of snapshots is greater than the number of sensors in the array. All these assumptions look reasonable and could be satisfied. Thus both the MUSIC and the MLE could be usable in this type of application.

7.2.2 Estimation of Complex Sine Wave Frequencies from Multiple-Experiment Data

Consider the following signal model

$$y_k(t) = \sum_{p=1}^{n} \gamma_p(t) e^{j\omega_p k} + e_k(t), \qquad \begin{matrix} k = 1, \dots, m \\ t = 1, \dots, N \end{matrix} \tag{7.4}$$

where m denotes the number of samples in an experiment, N is the number of experiments, $\{\gamma_p(t)\}$ and $\{\omega_p\}$ are the amplitudes and frequencies of the complex sine waves, and $\{e_k(t)\}$ is an additive noise. The model (7.4) can be written in the form (7.1) using the following definitions:

$$\begin{aligned} y(t) &= [y_1(t) \; \cdots \; y_m(t)]^T \\ x(t) &= [e^{j\omega_1}\gamma_1(t) \; \cdots \; e^{j\omega_n}\gamma_n(t)]^T \\ e(t) &= [e_1(t) \; \cdots \; e_m(t)]^T \\ a(\omega) &= [1 \; e^{j\omega} \; \cdots \; e^{j(m-1)\omega}]^T. \end{aligned} \tag{7.5}$$

The conditions A2 and AML mean that the noise within an experiment is white and that the noises of any two different experiments are uncorrelated, which is plausible. Note that in this case we may well have $m > N$, which implies that the MUSIC may not be usable.

7.2.3 Estimation of Complex Sine Wave Frequencies from Single-Experiment Data

The model for this application is given by (7.4) with t dropped:

$$y_k = \sum_{p=1}^{n} \gamma_p e^{j\omega_p k} + e_k, \qquad k = 1, \dots, m \tag{7.6}$$

which can be written in the form of (7.1) using the notation (7.5). Assumption A1 is satisfied if $m > n$, A2 means that the noise e_k is white and AML reduces to the requirement that e_k is Gaussian. Thus, the MLE may be usable. Since $N = 1$, assumption AMU is not satisfied for $n > 1$, and hence the MUSIC is not usable when there are at least two signals and the problem is stated as previously. To be able to use MUSIC, we must recast the model in a different form.

Let us denote the number of available samples by M (not m). Let m be some integer greater than n, and define

$$\begin{aligned} y(t) &= [y_t \cdots y_{t+m-1}]^T \\ a(\omega) &= [1\ e^{j\omega} \cdots e^{j(m-1)\omega}]^T \\ x(t) &= [\gamma_1 e^{j\omega_1 t} \cdots \gamma_n e^{j\omega_n t}]^T \qquad t = 1, \ldots, M - m + 1 \\ e(t) &= [e_t \cdots e_{t+m-1}]^T. \end{aligned}$$

Using the foregoing notation, we can write (7.6) in the form (7.1) with $N = M - m + 1$. In contrast to the multiple-experiment case, here $e(t)$ and $e(s)$ are correlated for $t \neq s$, and thus the MLE is not applicable. On the other hand, the model written in the foregoing form satisfies the MUSIC assumptions A1, A2, and AMU, provided $2n < 2m < M + 1$, which is readily achieved. Note that m can be chosen rather arbitrarily. This arbitrariness raises the question as to whether an optimal choice exists that minimizes the estimation errors of the MUSIC algorithm. This type of question will be considered in Section 7.7.

7.3. THE MUSIC ESTIMATOR

We begin by setting some additional notation (which will be used extensively in this paper) and a brief description of the MUSIC algorithm. Next, we establish the asymptotic distribution of the MUSIC estimator and derive an explicit expression for the covariance matrix of its estimation errors. Finally, we consider the improved MUSIC estimator introduced in [6] for which we also provide a performance analysis.

7.3.1 The MUSIC Algorithm

In this subsection, we assume that conditions A1, A2, and AMU hold. Under these assumptions, the covariance matrix of the observation vector $y(t)$ is given by

$$R \triangleq E\{y(t)y^\dagger(t)\} = A(\theta)P\,A^\dagger(\theta) + \sigma^2 I. \tag{7.7}$$

For notational convenience, we simply write A instead of $A(\theta)$ whenever there is no possibility of confusion. If $\hat{\theta}$ is an estimate of θ, then we also write $\hat{A}$ instead of $A(\hat{\theta})$.

Let $\lambda_1 \geq \lambda_2 \geq \cdots \geq \lambda_m$ denote the eigenvalues of R. Since rank $(APA^\dagger) = n$, it follows that

$$\lambda_i > \sigma^2, \text{ for } i = 1, \ldots, n, \quad \text{and} \quad \lambda_i = \sigma^2, \text{ for } i = n + 1, \ldots, m. \tag{7.8}$$

It will be assumed throughout this paper that $\{\lambda_i\}_{i=1}^n$ are distinct.

Denote the unit-norm eigenvectors associated with $\lambda_1, \ldots, \lambda_n$ by $s_1, \ldots, s_n$, and those corresponding to $\lambda_{n+1}, \ldots, \lambda_m$ by $g_1, \ldots, g_{m-n}$. Also define

$$\begin{aligned} S &= [s_1 \cdots s_n], \qquad (m \times n), \\ G &= [g_1 \cdots g_{m-n}], \qquad [m \times (m-n)]. \end{aligned} \tag{7.9}$$

Next, observe that

$$RG = APA^{\dagger}G + \sigma^2 G = \sigma^2 G,$$

which readily implies that

$$A^{\dagger} G = 0, \tag{7.10a}$$

or, equivalently

$$a^{\dagger}(\omega)GG^{\dagger}a(\omega) = 0, \qquad \text{for } \omega = \omega_1, \ldots, \omega_n. \tag{7.10b}$$

Since the normalized eigenvectors $\{s_i, g_j\}$ are orthonormal,

$$SS^{\dagger} + GG^{\dagger} = I, \tag{7.11}$$

it follows that (7.10) can also be written as

$$a^{\dagger}(\omega)[I - SS^{\dagger}]a(\omega) = 0, \qquad \text{for } \omega = \omega_1, \ldots, \omega_n. \tag{7.12}$$

It is not difficult to see that the true parameter values $\{\omega_1, \ldots, \omega_n\}$ are the only solutions of (7.10) or (7.12). The proof is by contradiction. Assume that there exists another solution that we denote by ω_{n+1}. The matrix $SS^{\dagger}$ in (7.12) is the orthogonal projection operator onto the subspace spanned by the columns of S. Thus, it would follow from (7.12) that the linearly independent vectors $\{a(\omega_i)\}_{i=1}^{n+1}$ (by assumption A2) belong to the column space of S. However, this is impossible since that space is of dimension equal to n.

The basic idea of the MUSIC algorithm is the exploitation of the property (7.10), or (7.12), of the true covariance matrix R. In practice R is unknown, but it can be consistently estimated from the available data. Let

$$\hat{R} = \frac{1}{N} \sum_{t=1}^{N} y(t)y^{\dagger}(t).$$

Similar to the eigendecomposition of R, let $\{\hat{s}_1, \ldots, \hat{s}_n, \hat{g}_1, \ldots, \hat{g}_{m-n}\}$ denote the unit-norm eigenvectors of $\hat{R}$, arranged in the descending order of the associated eigenvalues, and let $\hat{S}$ and $\hat{G}$ denote the matrices S and G made of $\{\hat{s}_i\}$ and, respectively, $\{\hat{g}_i\}$. Define

$$f(\omega) = a^{\dagger}(\omega)\hat{G}\hat{G}^{\dagger}a(\omega) \tag{7.13a}$$

$$= a^{\dagger}(\omega)[I - \hat{S}\hat{S}^{\dagger}]a(\omega). \tag{7.13b}$$

The MUSIC estimates of $\{\omega_i\}$ are obtained by picking the n values of ω for which $f(\omega)$ is minimized. Minimization of $f(\omega)$ is usually done by evaluating it at the points of a fine grid, using (7.13a) or (7.13b) [(7.13a) is preferred to (7.13b) if $n > m - n$, and vice versa].

There are several variants of the MUSIC algorithm described that are currently in use (see the excellent survey paper [6] and [15]–[17]). Several computationally efficient (adaptive or batch) implementations are also available [18]–[20], [30]. For the sake of conciseness, in this chapter we concentrate on the basic MUSIC algorithm and its improved version introduced in [6] (to be described later). Other variants of the MUSIC can be analysed similarly with respect to their statistical properties [37].

7.3.2 MUSIC Asymptotic Distribution

In this and the next subsection we assume that conditions A1, A2, AMU, and AML hold. As already explained, assumption AML is not necessary for the *application* of the MUSIC algorithm. However, it will be used in the *analysis* of the MUSIC estimator.

To establish the distribution of the MUSIC estimator, we need the following result on the statistics of the eigenvectors of the sample covariance matrix $\hat{\boldsymbol{R}}$.

Lemma 3.1. (a) The estimation errors $(\hat{s}_i - s_i)$ are asymptotically (for large N) jointly Gaussian distributed with zero means and covariance matrices given by

$$E\{(\hat{\boldsymbol{s}}_i - \boldsymbol{s}_i)(\hat{\boldsymbol{s}}_j - \boldsymbol{s}_j)^\dagger\} \tag{7.14a}$$

$$= \frac{\lambda_i}{N}\left[\sum_{\substack{k=1\\k\neq i}}^{n} \frac{\lambda_k}{(\lambda_k - \lambda_i)^2}\boldsymbol{s}_k\boldsymbol{s}_k^\dagger + \sum_{k=1}^{m-n}\frac{\sigma^2}{(\sigma^2-\lambda_i)^2}\boldsymbol{g}_k\boldsymbol{g}_k^\dagger\right]\delta_{i,j} \triangleq \boldsymbol{W}_i\delta_{i,j}$$

$$E\{(\hat{\boldsymbol{s}}_i - \boldsymbol{s}_i)(\hat{\boldsymbol{s}}_j - \boldsymbol{s}_j)^T\} = -\frac{\lambda_i\lambda_j}{N(\lambda_i - \lambda_j)^2}\boldsymbol{s}_j\boldsymbol{s}_i^T(1-\delta_{i,j}) \triangleq \boldsymbol{V}_{i,j} \tag{7.14b}$$

(b) The orthogonal projections of $\{\hat{\boldsymbol{g}}_i\}$ onto the column space of $\boldsymbol{S}$ are asymptotically (for large N) jointly Gaussian distributed with zero means and covariance matrices given by

$$E\{(\boldsymbol{S}\boldsymbol{S}^\dagger\hat{\boldsymbol{g}}_i)(\boldsymbol{S}\boldsymbol{S}^\dagger\hat{\boldsymbol{g}}_j)^\dagger\} = \frac{\sigma^2}{N}\left[\sum_{k=1}^{n}\frac{\lambda_k}{(\sigma^2-\lambda_k)^2}\boldsymbol{s}_k\boldsymbol{s}_k^\dagger\right]\delta_{i,j} \triangleq \frac{1}{N}\boldsymbol{U}\delta_{i,j} \tag{7.15}$$

$$E\{(\boldsymbol{S}\boldsymbol{S}^\dagger\hat{\boldsymbol{g}}_i)(\boldsymbol{S}\boldsymbol{S}^\dagger\hat{\boldsymbol{g}}_j)^T\} = 0, \qquad \text{for all } i, j. \tag{7.16}$$

Proof: The results (7.14) are standard, see [7], [8] and the references there. Result (7.15) has apparently been introduced in [5], [6] (based on results for the real case, presented in [31]) but a proof was not provided. Result (7.16) appears to be new. Proofs of (7.15) and (7.16) can be found in Appendix A.

We can now state and prove the result on the asymptotic distribution of the MUSIC estimator.

Theorem 3.1. The MUSIC estimation errors $\{\hat{\omega}_i - \omega_i\}$ are asymptotically (for large N) jointly Gaussian distributed with zero means and variances-covariances given by

$$E\{(\hat{\omega}_i - \omega_i)(\hat{\omega}_j - \omega_j)\} = \frac{1}{2N} \frac{\text{Re}\{\boldsymbol{d}^\dagger(\omega_j)\boldsymbol{G}\boldsymbol{G}^\dagger \boldsymbol{d}(\omega_i) \cdot \boldsymbol{a}^\dagger(\omega_i)\boldsymbol{U}\boldsymbol{a}(\omega_j)\}}{h(\omega_i)h(\omega_j)}, \tag{7.17a}$$

where $\boldsymbol{U}$ is defined in (7.15), $\boldsymbol{d}(\omega) = \text{d}\boldsymbol{a}(\omega)/\text{d}\omega$, and

$$h(\omega) = \boldsymbol{d}^\dagger(\omega)\boldsymbol{G}\boldsymbol{G}^\dagger \boldsymbol{d}(\omega). \tag{7.17b}$$

Proof: See Appendix B.

For the variance of the estimation error $(\hat{\omega}_i - \omega_i)$ we obtain from (7.17) the following expression

$$\begin{aligned} E\{(\hat{\omega}_i - \omega_i)^2\} &= \frac{1}{2N} \frac{\boldsymbol{a}^\dagger(\omega_i)\boldsymbol{U}\boldsymbol{a}(\omega_i)}{h(\omega_i)} \\ &= \frac{\sigma^2}{2N}\left[\sum_{k=1}^{n} \frac{\lambda_k}{(\sigma^2 - \lambda_k)^2} |\boldsymbol{a}^\dagger(\omega_i)\boldsymbol{s}_k|^2\right] \Big/ \left[\sum_{k=1}^{m-n} |\boldsymbol{d}^\dagger(\omega_i)\boldsymbol{g}_k|^2\right]. \end{aligned} \tag{7.18}$$

It is interesting to note that the MUSIC variance may take large values when some of the eigenvalues $\{\lambda_k\}_{k=1}^{n}$ are close to σ^2. This case corresponds to closely spaced signals (when the matrix $\boldsymbol{A}$ is almost rank deficient), to low signal-to-noise ratio, or to highly correlated signals (when the signal covariance matrix $\boldsymbol{P}$ is nearly singular). The variance may also be large when the vector $\boldsymbol{d}(\omega_i)$ is close to the column space of $\boldsymbol{A}$ (or $\boldsymbol{S}$) (and, therefore, quasi-orthogonal to $\{\boldsymbol{g}_k\}$). In such a case, the transfer vector $\boldsymbol{a}(\omega)$ is relatively insensitive to variations of ω around ω_i, which means that $f(\omega)$ has a flat minimum at $\omega = \omega_i$.

In Section 7.7, we derive an alternative formula to (7.18). Using that formula we will reinforce the earlier conclusions and will show that the MUSIC estimator variance has the tendency to decrease with increasing m (which is intuitively expected).

7.3.3 An Improved MUSIC Estimator and Its Asymptotic Distribution

An appealing (valid for large N) maximum-likelihood approach has been used recently in [6] to derive an improved MUSIC estimator. More exactly, the improved estimator is obtained by maximizing the ***likelihood*** of the vector

$$\epsilon_i \triangleq \boldsymbol{a}^\dagger(\omega)\hat{\boldsymbol{g}}_i, \qquad i = 1, \ldots, m - n. \tag{7.19}$$

Note that the MUSIC estimator minimizes $\Sigma_{i-1}^{m-n} |\epsilon_i|^2$. It is claimed in [6] that the estimator that maximizes the asymptotic (for $N >> 0$) likelihood of (7.19) is given by the minimizer of the following function:

$$\alpha(\omega) = f(\omega)/r(\omega). \tag{7.20a}$$

In (7.20a), $f(\omega)$ is the MUSIC function given by (7.13), and

$$r(\omega) = \boldsymbol{a}^{\dagger}(\omega)\hat{\boldsymbol{U}}\boldsymbol{a}(\omega), \tag{7.20b}$$

where $\hat{\boldsymbol{U}}$ is the matrix $\boldsymbol{U}$, (7.15), made of $\{\hat{\lambda}_k\}$ and $\{\hat{\boldsymbol{s}}_k\}$. As an aside, observe that $r(\omega)$ is related to the numerator in the variance formula (7.18). It is shown in the following that the estimates obtained by minimizing (7.20) for *any* (positive) function $r(\omega)$ have the same asymptotic (for large N) distribution as the MUSIC estimator.

Theorem 3.2. Assume that the function $r(\omega)$ satisfies the regularity condition $r(\omega_i) \neq 0$ for $i = 1, \ldots, n$, but is otherwise arbitrary. Then the estimates minimizing $f(\omega)$ and $\alpha(\omega)$ have the same asymptotic (for large N) distribution.

Proof: See Appendix C.

The result may explain the similarity in the performances of the MUSIC and the estimator minimizing (7.20), observed in some of the simulations in [6]. Note that since the result holds for a general function $r(\omega)$, improvement of the MUSIC performance should not be attempted by modifying the MUSIC function as in (7.20a) (at least, not for a "sufficiently large" N). In fact, we show in the following that an "exact" ML approach based on the "data" (7.19) does *not* lead to minimization of (7.20a)(7.20b).

In Appendix D, we show that the asymptotic (for large N) *negative log-likelihood function* of (7.19) is given by

$$-\ln L = \text{const} + (m - n)\ln[\boldsymbol{a}^{\dagger}(\omega)\boldsymbol{U}\boldsymbol{a}(\omega)] + \frac{Nf(\omega)}{\boldsymbol{a}^{\dagger}(\omega)\boldsymbol{U}\boldsymbol{a}(\omega)}. \tag{7.21}$$

Observe that (7.21) is O(1) for ω close to a true value, and O(N) otherwise. Thus, for ω close to ω_i for some i (which is the case of great interest), the dominant term of $-\ln L$ is affected if we neglect the second term in (7.21) as was implicitly done in [6]. On the other hand, the dominant term of (7.21) is not affected if $\boldsymbol{U}$ in the second and third terms of $-\ln L$ is replaced by a consistent estimate. We replace $\boldsymbol{U}$ by $\hat{\boldsymbol{U}}$ defined previously. Thus, we propose to determine the estimates of $\{\omega_i\}$ by minimizing the following function with respect to ω

$$\beta(\omega) = \frac{m - n}{N} \ln r(\omega) + \frac{f(\omega)}{r(\omega)}, \tag{7.22}$$

where $r(\omega)$ is defined in (7.20b).

Since the derivative of the first term in (7.22) with respect to ω is O($1/N$), it is not difficult to see that the asymptotic (for $N \gg 0$) distribution of the estimator that minimizes (7.22) is also identical to that of the MUSIC. However, in the finite sample case, use of (7.22) may lead to improved performance as compared to (7.20), since

(7.22) corresponds to a more exact (and computationally inexpensive) approximation of the likelihood function. Furthermore, when grid methods are used to locate approximately the minima of (7.22) (which is what is usually done), the first term in (7.22) may not be negligible, as explained. Numerical experience with the new eigenanalysis estimator introduced will be reported elsewhere.

7.4 THE CRAMÉR-RAO LOWER BOUND

In this section, we assume that conditions A1, A2, and AML hold. Under these conditions we derive the CRLB on the covariance matrix of any unbiased estimator of $\boldsymbol{\theta}$ and σ^2. In the following sections, we compare the performance of the MUSIC and the ML estimators (the MLE is discussed in Section 7.5) to the ultimate performance corresponding to the CRLB. The usefulness of the CRLB formula derived in the sequel is not, of course, limited to the performance studies reported in this chapter. It may also be used to establish the relative efficiency of other estimators for $\boldsymbol{\theta}$ and σ^2 proposed in the literature (see, e.g., [15]–[17], [20]).

The first result of this section is contained in the following theorem.

Theorem 4.1. Under the assumptions stated, the CRLB for $\boldsymbol{\theta}$ and σ^2 is given by

$$\mathrm{CRLB}(\boldsymbol{\theta}) = \frac{\sigma^2}{2}\left\{\sum_{t=1}^{N} \mathrm{Re}[\boldsymbol{X}^\dagger(t)\boldsymbol{D}^\dagger[\boldsymbol{I} - \boldsymbol{A}(\boldsymbol{A}^\dagger\boldsymbol{A})^{-1}\boldsymbol{A}^\dagger]\boldsymbol{D}\boldsymbol{X}(t)]\right\}^{-1} \tag{7.23}$$

and

$$\mathrm{var}_{\mathrm{CR}}(\sigma^2) = \frac{\sigma^4}{mN}, \tag{7.24}$$

where

$$\boldsymbol{X}(t) = \begin{bmatrix} x_1(t) & & O \\ & \ddots & \\ O & & x_n(t) \end{bmatrix}$$

$$\boldsymbol{D} = [\boldsymbol{d}(\omega_1) \ \cdots \ \boldsymbol{d}(\omega_n)],$$

(recall that $\boldsymbol{d}(\omega) = d\boldsymbol{a}(\omega)/d\omega$).

Proof: See Appendix E.

In the following, we drop the dependence of the CRLB on $\boldsymbol{\theta}$ for notational convenience. Instead, we stress the dependence of the CRLB on m and N. We would expect the CRLB to "decrease" when m or N increases. This intuitively expected result holds indeed, as shown in the next theorem.

Theorem 4.2. The CRLB covariance matrix (7.23) satisfies the following order relations

$$\mathrm{CRLB}(N) \geq \mathrm{CRLB}(N+1) \tag{7.25a}$$

$$\mathrm{CRLB}(m) \geq \mathrm{CRLB}(m+1). \tag{7.25b}$$

Proof: See Appendix F.

The foregoing results are valid for a general transfer vector $\boldsymbol{a}(\omega)$. In the following, we present some specialized results for transfer vectors of the form (7.3), which appear in several signal processing applications (see Section 7.2 for some examples).

7.4.1 CRLB for $n = 1$, $N = 1$, and $a(\omega)$ Given by (7.3)

In this case we have

$$\boldsymbol{A} = [1\ e^{j\omega}\ \cdots\ e^{j(m-1)\omega}]^T$$

$$\boldsymbol{D} = [0\ je^{j\omega}\ \cdots\ j(m-1)e^{j(m-1)\omega}]^T,$$

which gives

$$A^{\dagger}A = m$$

$$\boldsymbol{D}^{\dagger}\boldsymbol{D} = 1 + 2^2 + \cdots + (m-1)^2 = \frac{m(m-1)(2m-1)}{6} \tag{7.26}$$

$$\boldsymbol{D}^{\dagger}A = -j[1 + 2 + \cdots + (m-1)] = -j\frac{m(m-1)}{2}.$$

Inserting (7.26) into the expression (7.23) of the CRLB, we obtain

$$\mathrm{CRLB} = \frac{6\sigma^2}{|x|^2}\frac{1}{m(m^2-1)} = \frac{6}{\mathrm{SNR}}\frac{1}{m(m^2-1)} \simeq \frac{6}{m^3\mathrm{SNR}}, \tag{7.27}$$

which agrees with the result for this specialized case derived in [24] (see also [25]). In (7.27), $\mathrm{SNR} = |x|^2/\sigma^2$.

7.4.2 Asymptotic CRLB For $a(\omega)$ Given by (7.3)

According to Theorem 4.2, for increasing m or N, CRLB (m, N) forms a sequence of monotonically nonincreasing positive definite matrices. In particular, this implies that CRLB (m, N) has a well-defined limit when either m or N tends to infinity. In the following, we evaluate the limit (or asymptotic) matrices CRLB (m, ∞) and CRLB (∞, ∞). The formula derived for CRLB (m, ∞) holds for a general $\boldsymbol{a}(\omega)$. However, to obtain a formula for CRLB (∞, ∞) we need to specify $\boldsymbol{a}(\omega)$. The formula for CRLB (∞, ∞) provided in the following holds for $\boldsymbol{a}(\omega)$ given by (7.3).

Theorem 4.3. (a) For sufficiently large N, the CRLB is given by

$$\text{CRLB}\,(m, \infty) = \frac{\sigma^2}{2N}\{\text{Re}[\{\boldsymbol{D}^\dagger[\boldsymbol{I} - \boldsymbol{A}(\boldsymbol{A}^\dagger\boldsymbol{A})^{-1}\boldsymbol{A}^\dagger]\boldsymbol{D}\} \odot \boldsymbol{P}^T]\}^{-1}, \tag{7.28}$$

where $\boldsymbol{P}$ is defined in (7.2) and $\odot$ denotes the *Hadamard matrix product*.

(b) Let $\boldsymbol{a}(\omega)$ be given by (7.3). Then, for sufficiently large m and N, the CRLB is given by

$$\text{CRLB}\,(\infty, \infty) = \frac{6}{m^3 N}\begin{bmatrix} 1/\text{SNR}_1 & & O \\ & \ddots & \\ O & & 1/\text{SNR}_n \end{bmatrix}, \tag{7.29}$$

where $\text{SNR}_i = P_{ii}/\sigma^2$ is the signal-to-noise ratio for the ith signal.

Proof: See Appendix G.

The usefulness of the asymptotic CRLB formulas lies in the fact that they are (much) easier to evaluate than the exact finite-case formula (7.23), yet they may provide good approximations to the exact CRB for reasonably large values of m and N.

7.5. THE MAXIMUM-LIKELIHOOD ESTIMATOR

In this section, we assume that conditions A1, A2, and AML hold. The log-likelihood function of the observations $\{\boldsymbol{y}(t)\}_{t=1}^{n}$ is then given by

$$L = \text{const} - mN\ln\sigma^2 - \frac{1}{\sigma^2}\sum_{t=1}^{N}[\boldsymbol{y}(t) - \boldsymbol{A}\boldsymbol{x}(t)]^\dagger[\boldsymbol{y}(t) - \boldsymbol{A}\boldsymbol{x}(t)]; \tag{7.30}$$

see equation (7.E.1) in Appendix E. The likelihood (7.30) can be concentrated with respect to σ^2 and $\{\boldsymbol{x}(t)\}$. Some straightforward calculations show that the ML estimators of these parameters are given by (see, for example, equations (7.E.2a)–(7.E.2c) in Appendix E)

$$\hat{\boldsymbol{x}}(t) = [\boldsymbol{A}^\dagger\boldsymbol{A}]^{-1}[\boldsymbol{A}^\dagger\boldsymbol{y}(t)], \qquad t = 1, \ldots, N \tag{7.31}$$

$$\begin{aligned}\hat{\sigma}^2 &= \frac{1}{mN}\sum_{t=1}^{N}[\boldsymbol{y}(t) - \hat{\boldsymbol{A}}\hat{\boldsymbol{x}}(t)]^\dagger[\boldsymbol{y}(t) - \hat{\boldsymbol{A}}\hat{\boldsymbol{x}}(t)] \\ &= \frac{1}{m}\,\text{tr}[\boldsymbol{I} - \hat{\boldsymbol{A}}(\hat{\boldsymbol{A}}^\dagger\hat{\boldsymbol{A}})^{-1}\hat{\boldsymbol{A}}^\dagger]\hat{\boldsymbol{R}},\end{aligned} \tag{7.32}$$

where $\hat{\boldsymbol{A}}$ denotes the ML estimate of $\boldsymbol{A}$(i.e., $\hat{\boldsymbol{A}} = \boldsymbol{A}(\hat{\boldsymbol{\theta}})$, where $\hat{\boldsymbol{\theta}}$ is the MLE of $\boldsymbol{\theta}$). Inserting (7.31) and (7.32) (with $\hat{\boldsymbol{A}}$, which is not yet determined, replaced by $\boldsymbol{A}$) into (7.30), we obtain the *concentrated likelihood function*,

$$\text{const} - mN\ln F(\boldsymbol{\theta}),$$

where

$$F(\boldsymbol{\theta}) = \text{tr}[\boldsymbol{I} - \boldsymbol{A}(\boldsymbol{A}^{\dagger}\boldsymbol{A})^{-1}\boldsymbol{A}^{\dagger}]\hat{\boldsymbol{R}}. \tag{7.33}$$

Thus, the ML estimate of $\boldsymbol{\theta}$ is given by the minimizer of $F(\boldsymbol{\theta})$. Detailed discussions on the implementation of the MLE can be found in [10]–[12], [33]–[36], [42]. Here, we are interested in the properties of the MLE.

From the general theory of estimation, it is known that an MLE is asymptotically (as the number of data points tends to infinity) efficient, under some "regularity" conditions (see, e.g., [27]). In other words, its asymptotic covariance matrix attains the CRLB. However, what should "asymptotic" mean in our case where m or N or both could be large? Furthermore, does our estimation problem satisfy the "regularity" conditions? The essential regularity condition is that the MLE is consistent. In our case there are mN "data" from which $1 + n(N + 1)$ parameters are estimated. The ratio number of data to number of estimated parameters remains bounded if $m < \infty$, even if $N \to \infty$; this ratio increases without bound if and only if $m \to \infty$. This observation suggests that the CRLB cannot be achieved by increasing N; the essential requirement for attaining the CRLB should be to increase m. The preceding heuristical discussion is made more precise in the following (see also [37]).

7.5.1 The Case of Small m

We begin the analysis by studying the consistency properties of the MLE when $N \to \infty$ and $m < \infty$. Since $\hat{\boldsymbol{R}}$ tends to $\boldsymbol{R}$ as $N \to \infty$, it follows that the MLE of $\boldsymbol{\theta}$ tends to the minimizer of the following (asymptotic) criterion function

$$\begin{aligned}\text{tr}[\boldsymbol{I} - \hat{\mathbf{A}}(\hat{\mathbf{A}}^{\dagger}\hat{\mathbf{A}})^{-1}\hat{\mathbf{A}}^{\dagger}]\mathbf{R} &= \text{tr}[\boldsymbol{I} - \hat{\mathbf{A}}(\hat{\mathbf{A}}^{\dagger}\hat{\mathbf{A}})^{-1}\hat{\mathbf{A}}^{\dagger}][\boldsymbol{APA}^{\dagger} + \sigma^2\boldsymbol{I}] \\ &= \text{tr}[\boldsymbol{I} - \hat{\mathbf{A}}(\hat{\mathbf{A}}^{\dagger}\hat{\mathbf{A}})^{-1}\hat{\mathbf{A}}^{\dagger}]\boldsymbol{APA}^{\dagger} + \sigma^2(m - n) \geq \sigma^2(m - n).\end{aligned} \tag{7.34}$$

Clearly (7.34) is minimized by $\hat{\mathbf{A}} = \boldsymbol{A}$, which shows that the MLE of $\boldsymbol{\theta}$ is consistent when $N \to \infty$ (and $m < \infty$). The MLE of $[\boldsymbol{x}(t)]$ and σ^2, however, are inconsistent. This can be seen as follows. When $N \to \infty$, $\hat{\boldsymbol{x}}(t)$ and $\hat{\sigma}^2$ tend to the following limits

$$\hat{\boldsymbol{x}}(t) \to (\boldsymbol{A}^{\dagger}\boldsymbol{A})^{-1}\boldsymbol{A}^{\dagger}\boldsymbol{y}(t) = \boldsymbol{x}(t) + (\boldsymbol{A}^{\dagger}\boldsymbol{A})^{-1}\boldsymbol{A}^{\dagger}\boldsymbol{e}(t) \tag{7.35a}$$

$$\sigma^2 \to \frac{1}{m}\,\text{tr}[\boldsymbol{I} - \boldsymbol{A}(\boldsymbol{A}^{\dagger}\boldsymbol{A})^{-1}\boldsymbol{A}^{\dagger}]\boldsymbol{R} = \frac{m - n}{m}\,\sigma^2, \tag{7.35b}$$

which, for $m < \infty$, differ from the true values $\boldsymbol{x}(t)$ and σ^2. The inconsistency of $\boldsymbol{x}(t)$ and $\hat{\sigma}^2$ in the case of $m < \infty$ (anticipated in the previous discussion) implies that the MLE of $\boldsymbol{\theta}$ does not achieve the CRLB for large N if m is small. The following example illustrates this fact.

Consider the case of a single complex sine wave. The MLE of the sine wave frequency ω is given by the minimizer of (see (7.33))

$$F(\boldsymbol{\theta}) = \mathrm{tr}[\boldsymbol{I} - \boldsymbol{a}(\omega)[\boldsymbol{a}^{\dagger}(\omega)\boldsymbol{a}(\omega)]^{-1}\boldsymbol{a}^{\dagger}(\omega)]\hat{\boldsymbol{R}}, \tag{7.36a}$$

where $\boldsymbol{a}(\omega)$ is given by (7.3). Since $\boldsymbol{a}^{\dagger}(\omega)\boldsymbol{a}(\omega) \equiv m$, minimization of (7.36a) is equivalent to maximization of the following function (which, we note in passing, can be interpreted as an averaged periodogram),

$$g(\boldsymbol{\theta}) = \boldsymbol{a}^{\dagger}(\omega)\hat{\boldsymbol{R}}\boldsymbol{a}(\omega). \tag{7.36b}$$

We want to determine the asymptotic (for $N >> 0$) variance of the estimate $\hat{\omega}$ that maximizes (7.36b) and to show that this variance is strictly greater than the CRLB for $m < \infty$. After some calculations presented in Appendix H, we obtain

$$\mathrm{var}_{\mathrm{ML}}(\hat{\omega}) = \frac{6\sigma^2}{N}\frac{mP + \sigma^2}{m^2(m^2 - 1)P^2}. \tag{7.37}$$

The asymptotic (for large N) CRLB is given by (see (7.27) and (7.28))

$$\mathrm{var}_{\mathrm{CR}}(\hat{\omega}) = \frac{6\sigma^2}{N}\frac{1}{Pm(m^2 - 1)}.$$

Thus

$$\mathrm{var}_{\mathrm{ML}}(\hat{\omega})/\mathrm{var}_{\mathrm{CR}}(\hat{\omega}) = 1 + \frac{\sigma^2}{mP} = 1 + \frac{1}{m\mathrm{SNR}}, \tag{7.38}$$

which shows that *the MLE is inefficient for $m < \infty$, even though $N \to \infty$*. This example will be significantly generalized in Section 7.7.

7.5.2 The Case of Large *m*

We now assume that both N and m tend to infinity. Then the consistency of $\hat{\boldsymbol{\theta}}$ and $\hat{\sigma}^2$ follows from (7.34) and (7.35b). To establish the consistency of $\hat{\boldsymbol{x}}(t)$, we need to introduce the following additional assumption:

$$\boldsymbol{a}^{\dagger}(\omega)\boldsymbol{a}(\omega) \to \infty, \qquad \text{as } m \to \infty. \tag{7.39}$$

Observe that $\boldsymbol{a}(\omega)$ given by (7.3) satisfies (7.39). Under (7.39) the covariance matrix of the bias term in (7.35a),

$$\mathrm{E}\{(\boldsymbol{A}^{\dagger}\boldsymbol{A})^{-1}\boldsymbol{A}^{\dagger}\boldsymbol{e}(t)\boldsymbol{e}^{\dagger}(t)\boldsymbol{A}(\boldsymbol{A}^{\dagger}\boldsymbol{A})^{-1}\} = \sigma^2(\boldsymbol{A}^{\dagger}\boldsymbol{A})^{-1}$$

tends to zero as $m \to \infty$, which establishes the consistency of $\hat{\mathbf{x}}(t)$.

The condition (7.39) is not only sufficient but also necessary for the consistency of $\hat{\mathbf{x}}(t)$. In fact, without this condition the analysis of consistency would be meaningless. Indeed, the signals that do not satisfy (7.39) must be damped in some way (for example, exponentially damped). For such transient signals the behavior for large m is of no interest.

Once the consistency of the MLE has been established, its asymptotic efficiency essentially follows from the general theory of ML estimation [27]. Thus, *under* (7.39) *the MLE of* $\boldsymbol{\theta}$ *will achieve the CRLB for large* m *and* N. As a simple illustration of this property, observe from (7.38) that the asymptotic (for large N) ratio $\text{var}_{\text{ML}}(\hat{\omega})/\text{var}_{\text{CR}}(\hat{\omega})$ tends to one as m increases.

Remark. For uniformly spaced linear arrays or uniformly sampled undamped exponential signals the requirement on N to be large is not necessary. Indeed, the consistency and asymptotic (for large m) efficiency of the ML estimates of sine wave parameters have been established in [26] for the single-experiment case ($N = 1$).

Let us summarize the main results of this section. The MLE of $\boldsymbol{\theta}$ converges to the true values when N increases. However, if m is small, then the MLE does not achieve the CRLB even if N is increased without bound! For damped signal models there is no remedy to this situation and the CRLB cannot in general be attained. For undamped signal models (which satisfy (7.39)), the MLE achieves the CRLB as m becomes large; see also Section 7.7 and [37] for an analysis that reinforces the conclusions.

Remark. It is worth noting that the inefficiency of the ML estimator of $\boldsymbol{\theta}$ in the case of a small m is a direct consequence of the requirement to estimate the amplitude values $\{\mathbf{x}(1), \ldots, \mathbf{x}(N)\}$. If $[\mathbf{x}(t)\}_{t=1}^{N}$ can be assumed to be a sample from a Gaussian white process *and* if it is required to estimate the covariance matrix $\mathbf{P}$ only (a far less demanding requirement), then we may conjecture that in such a case the MLE of $\boldsymbol{\theta}$ will be statistically efficient for large $\mathbf{N}$ and any $m > n$.

The previous results provide some guidelines for choosing the values of m and N in a given application of the MLE (assuming that selection of m and N is at the disposal of the user, which, for some applications, such as array processing, where the number m of sensors is fixed, may not be the case). For damped signals we should proceed in the following rather obvious manner: m should be chosen according to some guess of the signal damping period and N should be increased as much as possible (under restrictions on computer and measurement time). For undamped signals we should select the values of m and N by a compromise between statistical efficiency and computational complexity. When m is increased, the MLE performance approaches the CRLB performance. Furthermore, the CRLB for undamped signal models is expected to decrease (much) faster with m than with N (for example, note from (7.29) that for complex sine waves the CRLB decreases as $1/m^3N$ as m and N increase). Thus, from the viewpoint of statistical efficiency the tendency should be to increase m rather than N. On the other hand, the computational burden associated with the MLE increases faster with m than with N (for example, evaluation of $F(\boldsymbol{\theta})$ for a given $\boldsymbol{\theta}$ requires $O(m^2N)$ arithmetic operations). Thus, the need for the compromise mentioned when selecting m and N is clearly seen.

7.6. THE RELATIONSHIP BETWEEN THE MUSIC AND ML ESTIMATORS

In this section we assume that conditions A1, A2, AMU, and AML all hold, so that both the MUSIC and ML estimators are usable. We want to investigate possible relationships between MUSIC and MLE.

By invoking the *invariance principle* of the ML estimators, it has been claimed by some authors (see, e.g., [5]) that MUSIC is a realization of the MLE. More precisely, it was claimed that since under the conditions stated, the sample eigenvectors $\{\hat{\mathbf{s}}_i\}$ are ML estimates of the true eigenvectors $\{\mathbf{s}_i\}$ (see, e.g., [28], [31]), the MUSIC estimate $\hat{\boldsymbol{\theta}}$ of $\boldsymbol{\theta}$ obtained from $\{\hat{\mathbf{s}}_i\}$ should be the ML estimate by the invariance principle. However, this line of argument is *not* quite correct. Briefly stated, the reason is as follows: when $\{\boldsymbol{\theta}\}$ span the set of feasible values $\mathbf{D}_\theta$, $\{\mathbf{s}_i\}$ span a set that we denote by $\mathbf{D}_s$. Every point from $\mathbf{D}_s$ can be mapped back to a point in $\mathbf{D}_\theta$ by using the so-called *inverse function.*

Existence of this inverse function is a key condition for the validity of the invariance principle. Now, due to estimation errors, $\{\hat{\mathbf{s}}_i\}$ will in general not belong to $\mathbf{D}_s$ (which is a "thin" set in $C^{m \times 1}$); as a consequence, the inverse function cannot be used to determine the point in $\mathbf{D}_\theta$ that corresponds to $\{\hat{\mathbf{s}}_i\}$ for the simple reason that there is no such point (the mapping from $\{\hat{\mathbf{s}}_i\}$ to $\hat{\boldsymbol{\theta}}$ employed by MUSIC is only an approximation of the correct inverse function). Thus, the invariance principle fails to be applicable. More details on the invariance principle of ML estimators and its failure to apply to the type of problem discussed above, can be found in [29].

In Section 7.3, we have seen that the MUSIC estimator is a large sample (for $N >> 0$) realization of *an* ML estimator obtained from the approximate distribution of the statistic (7.19). This property does not imply any immediate relationship between the MUSIC and *the* MLE of Section 7.5, obtained from the exact distribution of the raw data statistic. However, it suggests that the MUSIC estimator possesses some "optimality" property and, therefore, that a relationship to the MLE is likely to exist. That this is indeed the case is shown in the following theorem.

Theorem 6.1. Under the assumptions stated (A1, A2, AMU, and AML), the MUSIC estimator is a large sample (for $N >> 0$) realization of the MLE of Section 7.5, if and only if the signal covariance matrix $\mathbf{P}$ is diagonal

Proof: See Appendix I.

The result of Theorem 6.1 is pleasantly intuitive. When the n signals are uncorrelated, it should indeed be possible (at least for large N) to decouple the n-dimensional search problem implied by the MLE into the n one-dimensional problems solved by MUSIC. When the signals are correlated, this should not be possible. Note that it is this decoupling that makes the MUSIC estimator much more attractive computationally than the MLE.

We may also remark that the theorem provides a theoretical explanation for the good performance of the MUSIC, observed in many experiments with uncorrelated signals, as well as for a degradation of performance when the signals are highly correlated. More quantitative results on these aspects of the MUSIC performance can be found in the next section (see also Chapter 8).

7.7 AN ANALYTIC AND NUMERICAL STUDY OF PERFORMANCE

Our aim in this section is to study in more detail the MUSIC estimation error variance and to compare it with the CRLB. We will use the asymptotic (for large N) formulas (7.18) and (7.28) for the variance of the MUSIC estimator and the CRLB, respectively. Thus, our results will be valid for a sufficiently large number N of experiments or snapshots.

7.7.1 An Analytic Study

We begin by developing a more convenient formula for the MUSIC error variance. From (7.18) we obtain

$$\begin{aligned}\mathrm{var}_{\mathrm{MU}}(\hat{\omega}_i) &= \frac{\sigma^2}{2N}\left[\sum_{k=1}^{n}\left(\frac{1}{\lambda_k-\sigma^2}+\frac{\sigma^2}{(\lambda_k-\sigma^2)^2}\right)|\mathbf{a}^\dagger(\omega_i)\mathbf{s}_k|^2\right]\Bigg/\left[\sum_{k=1}^{m-n}|\mathbf{d}^\dagger(\omega_i)\mathbf{g}_k|^2\right]\\ &= \frac{\sigma^2}{2N}\left[a^\dagger(\omega_i)(\mathbf{S}\mathring{\Lambda}^{-1}\mathbf{S}^\dagger+\sigma^2\mathbf{S}\mathring{\Lambda}^{-2}\mathbf{S}^\dagger)\mathbf{a}(\omega_i)\right]\Big/[\mathbf{d}^\dagger(\omega_i)\mathbf{G}\mathbf{G}^\dagger\mathbf{d}(\omega_i)], \end{aligned} \qquad (7.40)$$

where

$$\mathring{\Lambda} \triangleq \Lambda - \sigma^2 I \triangleq \begin{bmatrix}\lambda_1 & & \mathbf{0}\\ & \ddots & \\ \mathbf{0} & & \lambda_n\end{bmatrix} - \sigma^2 I$$

Using (7.7) and the eigendecomposition of $\mathbf{R}$, we can write

$$\mathbf{R} = \mathbf{A}\mathbf{P}\mathbf{A}^\dagger + \sigma^2 I = \mathbf{S}\Lambda\mathbf{S}^\dagger + \sigma^2\mathbf{G}\mathbf{G}^\dagger = \mathbf{S}\mathring{\Lambda}\mathbf{S}^\dagger + \sigma^2\mathbf{I}$$

which implies that

$$\mathbf{A}\mathbf{P}\mathbf{A}^\dagger = \mathbf{S}\mathring{\Lambda}\mathbf{S}^\dagger$$

$$\mathbf{A}\mathbf{P}\mathbf{A}^\dagger\mathbf{A}\mathbf{P}\mathbf{A}^\dagger = \mathbf{S}\mathring{\Lambda}^2\mathbf{S}^\dagger$$

and therefore

$$\begin{aligned}(S^\dagger A)P(A^\dagger S) &= \mathring{\Lambda}\\ (S^\dagger A)P(A^\dagger A)P(A^\dagger S) &= \mathring{\Lambda}^2.\end{aligned} \qquad (7.41)$$

Since the columns of A lie in the column space of S, and A has full rank, it follows that the matrix $S^\dagger A$ is nonsingular and

$$SS^{\dagger} = A(A^{\dagger}A)^{-1}A^{\dagger}. \tag{7.42}$$

The nonsingularity of $(S^{\dagger}A)$ and (7.41) give

$$\begin{aligned} (A^{\dagger}S)\mathring{\Lambda}^{-1}(S^{\dagger}A) &= P^{-1} \\ (A^{\dagger}S)\mathring{\Lambda}^{-2}(S^{\dagger}A) &= P^{-1}(A^{\dagger}A)^{-1}P^{-1}. \end{aligned} \tag{7.43}$$

Using (7.42) and (7.43) in (7.40), we get

$$\text{var}_{\text{MU}}(\hat{\omega}_i) = \frac{\sigma^2}{2N}\{[P^{-1}]_{ii} + \sigma^2[P^{-1}(A^{\dagger}A)^{-1}P^{-1}]_{ii}\}/h(\omega_i), \tag{7.44a}$$

where

$$h(\omega) = d^{\dagger}(\omega)[I - A(A^{\dagger}A)^{-1}A^{\dagger}]d(\omega) \tag{7.44b}$$

and $(\cdot)_{ij}$ denotes the i, j element of the matrix in question. The variance (7.44) may take relatively large values if the signals are highly correlated (i.e., P is nearly singular), or they are closely spaced (i.e., $A^{\dagger}A$ is nearly singular), or $d(\omega_i)$ is close to the column space of A for some i (a similar conclusion has been drawn in Section 7.3, using (7.18)).

Evaluation of formula (7.44) for the MUSIC variance can be done directly from the original parameters σ^2, P and $\theta = \{\omega_i\}$ of the problem. The eigendecomposition of R is not necessary, which is in contrast to the evaluation of (7.18). Furthermore, (7.44) can be conveniently used to compare the MUSIC error variance and the CRLB analytically. For this, note from (7.28) that the CRLB is given by

$$\text{var}_{\text{CR}}(\hat{\omega}_i) = \frac{\sigma^2}{2N}[(\{D^{\dagger}[I - A(A^{\dagger}A)^{-1}A^{\dagger}]D\}\odot P^{T})^{-1}]_{ii}. \tag{7.45}$$

First, we consider the case of *uncorrelated signals*. In such a case the matrix P is diagonal and (7.44), (7.45) reduce to

$$\text{var}_{\text{MU}}(\hat{\omega}_i) = \frac{1}{2N\cdot \text{SNR}_i}\left\{1 + \frac{[(A^{\dagger}A)^{-1}]_{ii}}{\text{SNR}_i}\right\}\Big/ h(\omega_i) \tag{7.46a}$$

and

$$\text{var}_{\text{CR}}(\hat{\omega}_i) = \left(\frac{1}{2N\cdot \text{SNR}_i}\right)\Big/ h(\omega_i), \tag{7.46b}$$

where $\text{SNR}_i = P_{ii}/\sigma^2$. Since in the case of uncorrelated signals, the MUSIC is a large sample (for $N >> 0$) realization of the MLE (see Theorem 6.1), it follows that (7.46) also gives the variance of the latter estimator.

From (7.46) we obtain

$$\text{var}_{\text{MU}}(\hat{\omega}_i)/\text{var}_{\text{CR}}(\hat{\omega}_i) = 1 + [(A^{\dagger}A)^{-1}]_{ii}/\text{SNR}_i. \tag{7.47}$$

It is interesting to note that in this case $\text{var}_{\text{MU}}(\hat{\omega}_i)$ decreases *monotonically* with increasing m. This is so since both $[(A^{\dagger}A)^{-1}]_{ii}$ and $\text{var}_{\text{CR}}(\hat{\omega}_i)$ (see Theorem 4.2) monotonically decrease when m increases. For reasonably large values of m and SNR, the ratio (7.47) expressing the efficiency of the MUSIC estimator will be close to one. Furthermore, if the signals are undamped so that (7.39) holds, the ratio will tend to one as m increases. Thus, we rediscover in another way the fact shown in Section 7.5, that *the MLE, which for diagonal* $\boldsymbol{P}$ *is equivalent to MUSIC, achieves the CRLB as* m *becomes large for signal models satisfying* (7.39). If (7.39) is not satisfied and SNR $\neq \infty$, then the ratio (7.47) will remain strictly greater than one, thus reinforcing our claim in Section 7.5 that *for damped signal models the CRLB cannot be attained.*

Next, consider the case of *correlated signals*. In this case, the MUSIC error variance cannot attain the CRLB. Furthermore, if the matrix $\boldsymbol{P}$ is nearly singular, then the differences $\{\text{var}_{\text{MU}}(\hat{\omega}_i) - \text{var}_{\text{CR}}(\hat{\omega}_i)\}$ may take substantial values, even if m and SNR increase without bound. We illustrate these facts by considering the practically important case of $\boldsymbol{a}(\omega)$ given by (7.3). Using the results in Appendix G we can then show that, as m increases, the variances (7.44) and (7.45) tend to the following limits

$$\text{var}_{\text{MU}}(\hat{\omega}_i) = \frac{6\sigma^2}{Nm^3}(\boldsymbol{P}^{-1})_{ii}$$

$$\text{var}_{\text{CR}}(\hat{\omega}_i) = \frac{6\sigma^2}{Nm^3}\Big/(\boldsymbol{P})_{ii}.$$

Thus, the ratio $\text{var}_{\text{MU}}(\hat{\omega}_i)/\text{var}_{\text{CR}}(\hat{\omega}_i) = (\boldsymbol{P})_{ii}(\boldsymbol{P}^{-1})_{ii}$ increases without bound as $\boldsymbol{P}$ approaches a singular matrix.

To conclude, for uncorrelated signals, the MUSIC estimator has an excellent performance for reasonably large values of N, m, and SNR. Furthermore, for undamped uncorrelated signals, the MUSIC error variance attains the CRLB for large N and m (or SNR). For correlated signals, however, the MUSIC cannot achieve the CRLB. If the signals are highly correlated, then the MUSIC estimator may be very inefficient even for large values of N, m, and SNR.

A Numerical Study

Consider the case of two signals ($n = 2$) of equal powers, and let

$$\boldsymbol{P} = \begin{bmatrix} 1 & \rho \\ \rho & 1 \end{bmatrix}, \qquad \rho \in [0, 1).$$

Furthermore, let $\boldsymbol{a}(\omega)$ be given by (7.3). We evaluated the efficiency ratio

$$\textit{eff} = \text{var}_{\text{CR}}(\hat{\omega}_i)/\text{var}_{\text{MU}}(\hat{\omega}_i) \tag{7.48}$$

for $\rho = 0, 0.5, 0.7$, and 0.9; $\sigma^2 = 0.01$ (SNR $= 20$ dB) and 1 (SNR $= 0$ dB); $m = 5, 10, 25$, and 100; and varying (ω_1, ω_2). Note that in this case, when $n = 2$, the efficiency ratio 7.48 can be written as a function of the magnitude of the parameter separation $\Delta\omega = |\omega_1 - \omega_2|$. The results obtained are shown in Figs. 7.1, 7.2, 7.3 and 7.4 as a function of $\Delta\omega/\pi$.

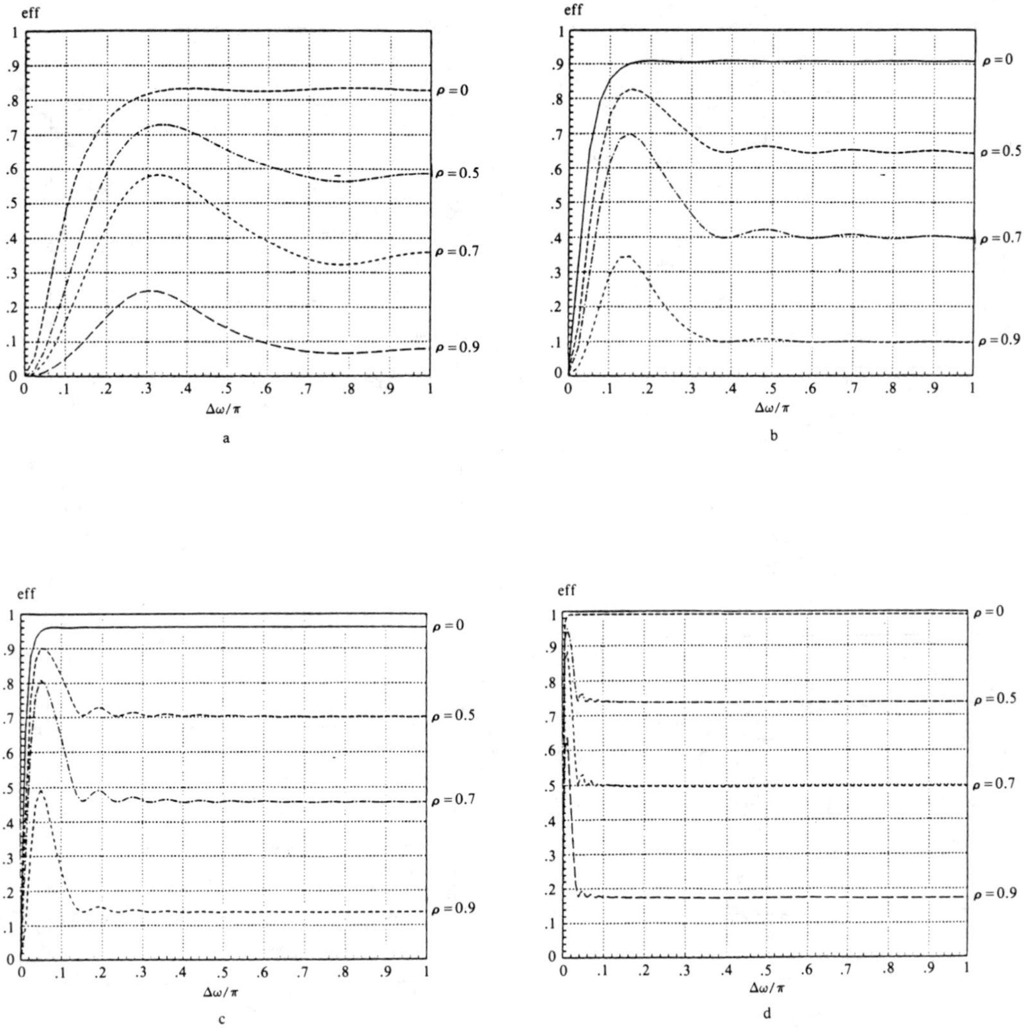

Figure 7.1 The relative efficiency (7.48) of the MUSIC estimator for two equal power signals as a function of signal separation $\Delta\omega = |\omega_1 - \omega_2|$. The results shown are for different values of number of sensors m and correlation factor ρ. SNR = 0 dB. (a) m = 5, (b) m = 10, (c) m = 25, (d) m = 100.

Figures 7.1 and 7.2 show the results for SNR values of 0 dB and 20 dB, respectively. These figures verify our theoretical result that, for uncorrelated and not too closely spaced signals, the MUSIC is statistically efficient for sufficiently large values of m. As shown in the figures, the values of m for which *eff* is close to one decrease when the signal-to-noise ratio or $\Delta\omega$ increases. The figures also demonstrate the degradation of the MUSIC efficiency when the correlation factor ρ increases. Note

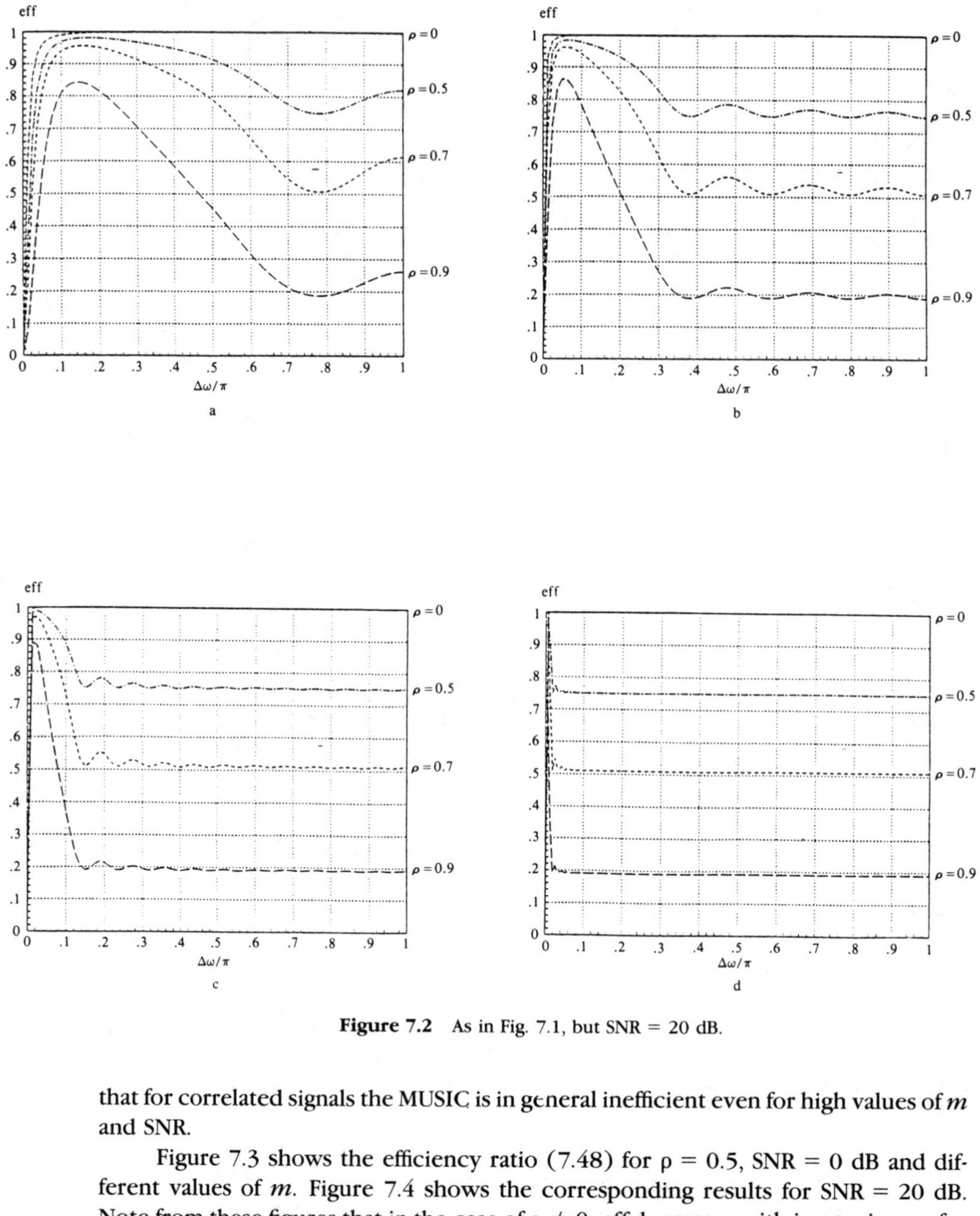

Figure 7.2 As in Fig. 7.1, but SNR = 20 dB.

that for correlated signals the MUSIC is in general inefficient even for high values of m and SNR.

Figure 7.3 shows the efficiency ratio (7.48) for $\rho = 0.5$, SNR = 0 dB and different values of m. Figure 7.4 shows the corresponding results for SNR = 20 dB. Note from these figures that in the case of $\rho \neq 0$, *eff* decreases with increasing m, for some values of $\Delta\omega$.

Next we show how our results can be used to evaluate the resolvability of the MUSIC algorithm quantitatively. The MUSIC algorithm is unlikely to resolve the sig-

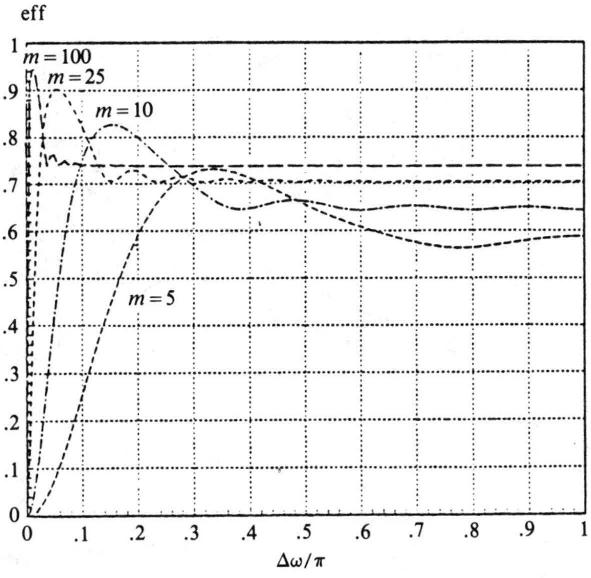

Figure 7.3 The relative efficiency (7.48) of the MUSIC estimator for two equal power signals as a function of signal separation $\Delta\omega = |\omega_1 - \omega_2|$. The results shown are for different values of numbers of sensors m. $\rho = 0.5$ and SNR = 0 dB.

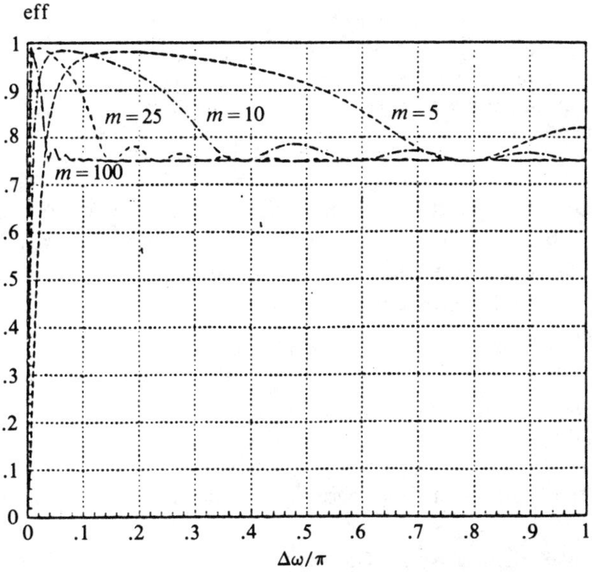

Figure 7.4 As in Fig. 7.3, but for SNR = 20 dB.

nals when 8[st. dev.$_{MU}(\omega_i)$] > $\Delta\omega$ (see [8]). Figure 7.5 shows $8\sqrt{N}$[st. dev.$_{MU}$ (ω_i)] as a function of $\Delta\omega$ for SNR = 0 dB and different values of m and ρ. The straight lines shown in the figure correspond to $\sqrt{N}\Delta\omega$ for $N = 100$, 200, and 500. Figure 7.6 shows similar results for SNR = 20 dB. MUSIC is unlikely to resolve the two signals

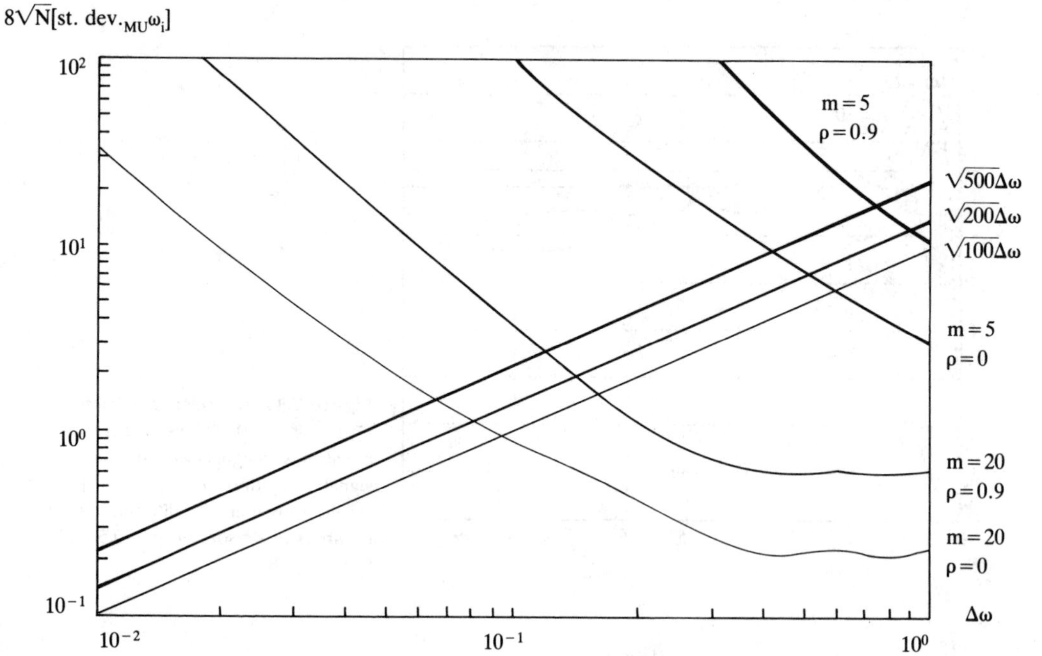

Figure 7.5 Curves of $8\sqrt{N}$[st. dev.$_{MU}(\omega_i)$] as a function of $\Delta\omega$ for two equal power signals. SNR = 0 dB. The straight lines correspond to $\sqrt{N\Delta\omega}$. The MUSIC algorithm is not likely to resolve the signals for values of $\Delta\omega$ smaller than those at the intersection of the standard deviation curves and the straight lines.

for values of $\Delta\omega$ smaller than those at the intersections between the normalized standard deviation curves and the straight lines. It can be seen from the figures that the resolvability of the MUSIC increases with SNR, N or m, as expected. Note also that the resolvability increases much faster with increasing m than with N (as predicted by the developed theory.)

7.8. CONCLUSIONS

There are several new results obtained in this chapter that should be mentioned:

- The MUSIC estimator was shown to be Gaussian distributed for sufficiently large N, and two equivalent explicit formulas, (7.18) and (7.44), for its error variance have been provided. Formula (7.18) can be readily used in practical applications to evaluate the MUSIC accuracy, since consistent estimates of the terms appearing in (7.18) are obtained in the course of estimation of $\boldsymbol{\theta}$. Formula (7.44) is more convenient for theoretical studies of performance, as explained in the previous section. Note that formulas (7.18) or (7.44) can also be used to study the resolvability of the MUSIC algorithm.

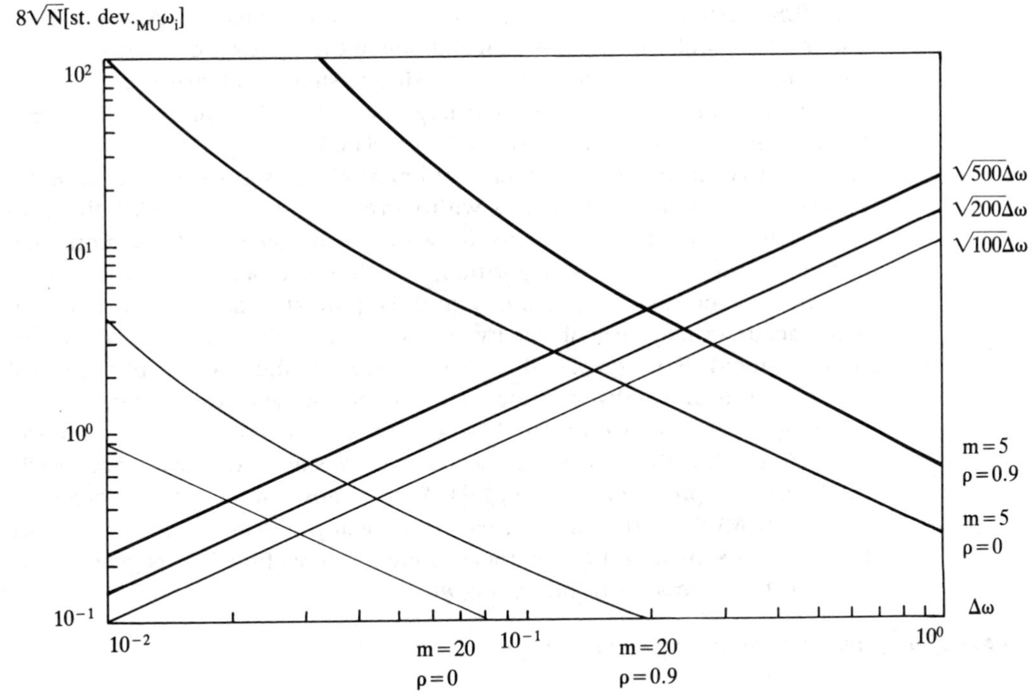

Figure 7.6 As in Fig. 7.5, but for SNR = 20 dB.

- The improved MUSIC estimator introduced in [6] has been shown to perform closely to the (basic) MUSIC estimator for large N, thus providing theoretical justification for the empirically observed results of [6]. In fact, the equivalence (for large N) of MUSIC to a whole class of "improved" MUSIC estimators of the form considered in [6], has been proved. A new MUSIC estimator has been obtained by slightly modifying the estimator of [6]; the new estimator is expected to perform better than other MUSIC-type estimators, for reasonably small values of m and N.
- An explicit formula has been derived for the CRLB on the covariance matrix of any unbiased estimator of $\boldsymbol{\theta}$. It is shown that the CRLB monotonically decreases with increasing m or N. Simple formulas have been presented for the CRLB in the case of large N, and in the case of large m and N and $\mathbf{a}(\omega)$ given by (7.3). The CRLB formulas derived in this chapter should be useful in practical applications and theoretical studies to compare the performance of a given estimator to the ultimate performance corresponding to the CRLB.
- It has been shown that the MLE of the parameter vector $\boldsymbol{\theta}$ does not achieve the CRLB for $N \to \infty$, if $m < \infty$. For undamped signal models, however, the MLE approaches the CRLB performance if m is increased. Based on this type of result, some guidelines for choosing the values of m and N in a given application (when possible) have been provided.

- The MUSIC estimator has been proven to be a large sample (for $N >> 0$) realization of the MLE for any $m > n$, if and only if the signals are uncorrelated. A consequence of this result is that the MUSIC should achieve the CRLB for uncorrelated undamped signals and large m and N. This property has been shown explicitly to hold for general uncorrelated undamped signals. Furthermore, it was shown that in the case of uncorrelated signals the MUSIC error variance monotonically decreases with increasing m. In particular, this provides a theoretical justification to the widespread opinion that the computationally efficient Pisarenko algorithm, which corresponds to MUSIC with $m = n + 1$ (the smallest possible value), is quite statistically inefficient, and better accuracy may be achieved by increasing m (at the expense of additional computations). For correlated signals, however, the MUSIC performance degrades. It is shown that this degradation of performance can be considerable if the signals are highly correlated (as a remedy, in such cases and for uniform linear arrays, the MUSIC based on a subaperture smoothed covariance matrix can be used as proposed, e.g., in [39]). Furthermore, in the case of correlated signals, the MUSIC error variance may occasionally increase when m increases. However, as shown in the numerical examples of Section 7.7, its general tendency is to decrease with increasing m.

APPENDIX A: PROOF OF (7.15) AND (7.16)

Introduce the notation

$$\hat{\Lambda} = \begin{bmatrix} \hat{\lambda}_1 & & 0 \\ & \ddots & \\ 0 & & \hat{\lambda}_n \end{bmatrix}, \qquad \Lambda = \begin{bmatrix} \lambda_1 & & 0 \\ & \ddots & \\ 0 & & \lambda_n \end{bmatrix}$$

$$\hat{\Sigma} = \begin{bmatrix} \hat{\lambda}_{n+1} & & 0 \\ & \ddots & \\ 0 & & \hat{\lambda}_m \end{bmatrix}$$

$$\Delta = G^{\dagger}\hat{R}G - \sigma^2 I$$

$$\Gamma = S^{\dagger}\hat{R}G.$$

For large N, $\hat{S} = S + 0(1/\sqrt{N})$, $\hat{\Lambda} = \Lambda + 0(1/\sqrt{N})$, $\hat{\Sigma} = \sigma^2 I + 0(1/\sqrt{N})$, and $G^{\dagger}\hat{S} = 0(1/\sqrt{N})$. Using these facts and the eigendecomposition of $\hat{R}$, we get

$$(G^{\dagger}\hat{G})(\hat{G}^{\dagger}G) = G^{\dagger}(I - \hat{S}\hat{S}^{\dagger})G = I - (G^{\dagger}\hat{S})(\hat{S}^{\dagger}G) \simeq I, \tag{7.A.1}$$

$$\begin{aligned}\Delta &= G^{\dagger}(\hat{S}\hat{\Lambda}\hat{S}^{\dagger} + \hat{G}\hat{\Sigma}G^{\dagger})G - \sigma^2 I \simeq (G^{\dagger}\hat{G})\hat{\Sigma}(\hat{G}^{\dagger}G) - \sigma^2 I \\ &= (G^{\dagger}\hat{G})(\hat{\Sigma} - \sigma^2 I)(\hat{G}^{\dagger}G) - [I - (G^{\dagger}\hat{G})(\hat{G}^{\dagger}G)]\sigma^2 \simeq (G^{\dagger}\hat{G}) \\ &\qquad (\hat{\Sigma} - \sigma^2 I)(\hat{G}^{\dagger}G)\end{aligned} \tag{7.A.2}$$

$$\begin{aligned}\Gamma &= S^{\dagger}(\hat{S}\hat{\Lambda}\hat{S}^{\dagger} + \hat{G}\hat{\Sigma}\hat{G}^{\dagger})G = (S^{\dagger}\hat{S})\hat{\Lambda}(\hat{S}^{\dagger}G) + (S^{\dagger}\hat{G})\hat{\Sigma}(\hat{G}^{\dagger}G) \\ &\simeq \Lambda(\hat{S}^{\dagger}G) + \sigma^2(S^{\dagger}\hat{G})(\hat{G}^{\dagger}G),\end{aligned} \tag{7.A.3}$$

and

$$(\boldsymbol{S}^{\dagger}\hat{\boldsymbol{G}})(\hat{\boldsymbol{G}}^{\dagger}\boldsymbol{G}) = \boldsymbol{S}^{\dagger}(\boldsymbol{I} - \hat{\boldsymbol{S}}\hat{\boldsymbol{S}}^{\dagger})\boldsymbol{G} = -(\boldsymbol{S}^{\dagger}\hat{\boldsymbol{S}})(\hat{\boldsymbol{S}}^{\dagger}\boldsymbol{G}) \simeq -\hat{\boldsymbol{S}}^{\dagger}\boldsymbol{G}, \qquad (7.A.4)$$

where the terms neglected in the approximations are $0(1/N)$. To proceed we need the following result.

R: The asymptotic distributions of the elements of $\hat{\boldsymbol{S}}^{\dagger}\boldsymbol{G}$ and of $\hat{\boldsymbol{G}}^{\dagger}\boldsymbol{G}$ are independent.

Proof: Define

$$\boldsymbol{u}(t) = \boldsymbol{S}^{\dagger}\boldsymbol{y}(t)$$

$$\boldsymbol{v}(t) = \boldsymbol{G}^{\dagger}\boldsymbol{y}(t) = \boldsymbol{G}^{\dagger}\boldsymbol{e}(t)$$

and observe that

$$\Delta_{ij} = \frac{1}{N}\sum_{t=1}^{N} v_i(t)v_j^*(t) - \sigma^2\delta_{i,j}$$

$$\Gamma_{kp} = \frac{1}{N}\sum_{t=1}^{N} u_k(t)v_p^*(t).$$

Since

$$E\{\boldsymbol{u}(t)\boldsymbol{v}^{\dagger}(s)\} = \boldsymbol{S}^{\dagger}E\{\boldsymbol{y}(t)\boldsymbol{e}^{\dagger}(s)\}\boldsymbol{G} = \sigma^2\boldsymbol{S}^{\dagger}\boldsymbol{G}\delta_{t,s} = 0, \qquad \text{all } t, s$$

$$E\{\boldsymbol{v}(t)\boldsymbol{v}^{\dagger}(s)\} = \boldsymbol{G}^{\dagger}E\{\boldsymbol{e}(t)\boldsymbol{e}^{\dagger}(s)\}\boldsymbol{G} = \sigma^2\boldsymbol{I}\delta_{t,s}$$

$$E\{\boldsymbol{v}(t)\boldsymbol{v}^{T}(s)\} = \boldsymbol{G}^{\dagger}E\{\boldsymbol{e}(t)\boldsymbol{e}^{T}(s)\}\boldsymbol{G}^* = \boldsymbol{0}, \qquad \text{all } t, s$$

$$E\{\boldsymbol{u}(t)\boldsymbol{v}^{T}(s)\} = \boldsymbol{S}^{\dagger}E\{\boldsymbol{y}(t)\boldsymbol{e}^{T}(s)\}\boldsymbol{G}^* = \boldsymbol{0}, \qquad \text{all } t, s,$$

it follows that

$$E\{\Delta_{ij}\} = 0 \text{ and } E\{\Gamma_{kp}\} = 0.$$

Furthermore, using the following formula [38] for the expectation of the product of four complex Gaussian random variables $\{x_i\}_{i=1}^{4}$ of which at least one is of zero mean,

$$E\{x_1x_2x_3x_4\} = (E\{x_1x_2\})(E\{x_3x_4\}) + (E\{x_1x_3\})(E\{x_2x_4\}) + (E\{x_1x_4\})(E\{x_2x_3\})$$

we get

$$E\{\Delta_{ij}\Gamma_{kp}^*\} = \frac{1}{N^2}\sum_{t=1}^{N}\sum_{s=1}^{N} E\{v_i(t)v_j^*(t) - \sigma^2\delta_{i,j}\}\{u_k^*(s)v_p(s)\} = 0 \qquad (7.A.5)$$

$$E\{\Delta_{ij}\Gamma_{kp}\} = \frac{1}{N^2}\sum_{t=1}^{N}\sum_{s=1}^{N} E\{v_i(t)v_j^*(t) - \sigma^2\delta_{i,j}\}\{u_k(s)v_p^*(s)\} = 0. \qquad (7.A.6)$$

Since the elements of $\boldsymbol{\Delta}$ and $\boldsymbol{\Gamma}$ are Gaussian distributed by the multivariate central limit theorem, it follows from (7.A.5) and (7.A.6) that Δ_{ij} and Γ_{kp} are independent (complex) random variables.

Now, observe from (7.A.2) that $\boldsymbol{\Delta}$ determines the asymptotic distribution of $(\hat{\boldsymbol{G}}^{\dagger}\boldsymbol{G})$, and from (7.A.3) and (7.A.4) that $\boldsymbol{\Gamma} \simeq (\boldsymbol{\Lambda} - \sigma^2\boldsymbol{I})(\hat{\boldsymbol{S}}^{\dagger}\boldsymbol{G})$ and, therefore, that $\boldsymbol{\Gamma}$ determines the asymptotic distribution of $(\hat{\boldsymbol{S}}^{\dagger}\boldsymbol{G})$. Since the distributions of $\boldsymbol{\Delta}$ and $\boldsymbol{\Gamma}$ have been shown to be independent, the result follows.

It follows from the result that $(\hat{\boldsymbol{G}}^{\dagger}\boldsymbol{G})$ in (7.A.4) can be considered to be fixed and, therefore, that $\boldsymbol{S}^{\dagger}\hat{\boldsymbol{G}}$ has the same asymptotic distribution as $-\hat{\boldsymbol{S}}^{\dagger}\boldsymbol{G}\boldsymbol{Q}$, where $\boldsymbol{Q}$ is some (fixed) unitary matrix (note from (7.A.1) that $\hat{\boldsymbol{G}}^{\dagger}\boldsymbol{G}$ is asymptotically unitary). However, the columns of $\boldsymbol{G}\boldsymbol{Q}$ form another set of eigenvectors associated with the repeated eigenvalue σ^2. Thus, we conclude that $\boldsymbol{S}^{\dagger}\hat{\boldsymbol{G}}$ and $-\hat{\boldsymbol{S}}^{\dagger}\boldsymbol{G}$ have the same asymptotic distribution.

The important implication of the analysis is that $\boldsymbol{S}\boldsymbol{S}^{\dagger}\hat{\boldsymbol{g}}_i$ and $-\boldsymbol{S}\hat{\boldsymbol{S}}^{\dagger}\boldsymbol{g}_i$ have the same limiting distribution. However, the limiting distribution of the latter is the same as that of $-\boldsymbol{S}(\boldsymbol{\Lambda} - \sigma^2\boldsymbol{I})^{-1}\boldsymbol{\Gamma}_i$, where $\boldsymbol{\Gamma}_i$ is the ith column of $\boldsymbol{\Gamma}$ (see (7.A.3), (7.A.4)). As explained previously, the asymptotic distribution of $\{\boldsymbol{\Gamma}_i\}$ is Gaussian with zero mean. The covariance matrix of this distribution is derived as follows:

$$\begin{aligned}
\lim_{N\to\infty} E\{(\sqrt{N}\boldsymbol{\Gamma}_i \cdot \sqrt{N}\boldsymbol{\Gamma}_j^{\dagger})\} &= \lim_{N\to\infty} \frac{1}{N}\sum_{t=1}^{N}\sum_{s=1}^{N} E\{\boldsymbol{u}(t)v_i^*(t)v_j(s)\boldsymbol{u}^{\dagger}(s)\} \\
&= \sigma^2 \lim_{N\to\infty} \frac{1}{N}\sum_{t=1}^{N}\sum_{s=1}^{N} E\{\boldsymbol{u}(t)\boldsymbol{u}^{\dagger}(s)\}\delta_{t,s}\delta_{i,j} \\
&= \sigma^2 \lim_{N\to\infty} \frac{1}{N}\sum_{t=1}^{N} \boldsymbol{S}^{\dagger}[\boldsymbol{A}\boldsymbol{x}(t)\boldsymbol{x}^{\dagger}(t)\boldsymbol{A}^{\dagger} + \sigma^2\boldsymbol{I}]\boldsymbol{S}\delta_{i,j} \\
&= \sigma^2\boldsymbol{S}^{\dagger}\boldsymbol{R}\boldsymbol{S}\delta_{i,j} = \sigma^2\boldsymbol{\Lambda}\delta_{i,j}
\end{aligned} \tag{7.A.7}$$

and

$$\lim_{N\to\infty} E\{(\sqrt{N}\boldsymbol{\Gamma}_i \cdot \sqrt{N}\boldsymbol{\Gamma}_j^{T})\} = \lim_{N\to\infty} \frac{1}{N}\sum_{t=1}^{N}\sum_{s=1}^{N} E\{\boldsymbol{u}(t)v_i^*(t)v_j^*(s)\boldsymbol{u}^{T}(s)\} = 0. \tag{7.A.8}$$

The results (7.15), (7.16) now follow from (7.A.7), (7.A.8) and the observation that $\boldsymbol{S}\boldsymbol{S}^{\dagger}\hat{\boldsymbol{g}}_i$ and $-\boldsymbol{S}(\boldsymbol{\Lambda} - \sigma^2\boldsymbol{I})^{-1}\boldsymbol{\Gamma}_i$ have the same asymptotic distribution.

APPENDIX B: PROOF OF (7.17)

As $\{\hat{\omega}_i\}$ is a minimum point of $f(\omega)$, we must have

$$f'(\hat{\omega}_i) = 0,$$

where

$$f'(\hat{\omega}_i) \triangleq \left.\frac{df(\omega)}{d\omega}\right|_{\omega = \hat{\omega}_i} = \boldsymbol{d}^\dagger(\hat{\omega}_i)\,\hat{\boldsymbol{G}}\hat{\boldsymbol{G}}^\dagger\boldsymbol{a}(\hat{\omega}_i) + \boldsymbol{a}^\dagger(\hat{\omega}_i)\hat{\boldsymbol{G}}\hat{\boldsymbol{G}}^\dagger\boldsymbol{d}(\hat{\omega}_i)$$

$$= 2\,\mathrm{Re}\,[\boldsymbol{a}^\dagger(\hat{\omega}_i)\hat{\boldsymbol{G}}\hat{\boldsymbol{G}}^\dagger\boldsymbol{d}(\hat{\omega}_i)]\,. \tag{7.B.1}$$

Following the idea of [5], we use the expression (7.13a) of $f(\omega)$, which appears to be more convenient than (7.13b) for the analysis of the distribution of the estimation errors $\{\hat{\omega}_i - \omega_i\}$. Since $\hat{\omega}_i$ is a consistent estimate of ω_i, we can write for sufficiently large N

$$\begin{aligned} 0 &= f'(\hat{\omega}_i) \simeq f'(\omega_i) = f''(\omega_i)(\hat{\omega}_i - \omega_i) \\ &= 2\,\mathrm{Re}[\boldsymbol{a}^\dagger(\omega_i)\hat{\boldsymbol{G}}\hat{\boldsymbol{G}}^\dagger\boldsymbol{d}(\omega_i)] + 2\,\mathrm{Re}[\boldsymbol{d}^\dagger(\omega_i)\hat{\boldsymbol{G}}\hat{\boldsymbol{G}}^\dagger\boldsymbol{d}(\omega_i) + \boldsymbol{a}^\dagger(\omega_i)\hat{\boldsymbol{G}}\hat{\boldsymbol{G}}^\dagger\boldsymbol{d}'(\omega_i)](\hat{\omega}_i - \omega_i) \\ &\simeq 2\,\mathrm{Re}[\boldsymbol{a}^\dagger(\omega_i)\hat{\boldsymbol{G}}\hat{\boldsymbol{G}}^\dagger\boldsymbol{d}(\omega_i)] + 2\,[\boldsymbol{d}^\dagger(\omega_i)\boldsymbol{G}\boldsymbol{G}^\dagger\boldsymbol{d}(\omega_i)](\hat{\omega}_i - \omega_i), \end{aligned} \tag{7.B.2a}$$

where the terms neglected in the approximations are $O(1/N)$. The $\hat{\boldsymbol{G}}^\dagger$ in the first term of (7.B.2a) can be replaced by $\boldsymbol{G}^\dagger$ without affecting the symptotic distribution. Indeed,

$$\begin{aligned} \boldsymbol{A}^\dagger\hat{\boldsymbol{G}}\hat{\boldsymbol{G}}^\dagger &\simeq \boldsymbol{A}^\dagger\boldsymbol{S}\boldsymbol{S}^\dagger\hat{\boldsymbol{G}}\hat{\boldsymbol{G}}^\dagger\boldsymbol{G}\boldsymbol{G}^\dagger = \boldsymbol{A}^\dagger\boldsymbol{S}\boldsymbol{S}^\dagger(\boldsymbol{I} - \hat{\boldsymbol{S}}\hat{\boldsymbol{S}}^\dagger)\hat{\boldsymbol{G}}\hat{\boldsymbol{G}}^\dagger \\ &= -\boldsymbol{A}^\dagger\boldsymbol{S}(\boldsymbol{S}^\dagger\hat{\boldsymbol{S}})(\hat{\boldsymbol{S}}^\dagger\boldsymbol{G})\boldsymbol{G}^\dagger \simeq -\boldsymbol{A}^\dagger\boldsymbol{S}(\hat{\boldsymbol{S}}^\dagger\boldsymbol{G})\boldsymbol{G}^\dagger \end{aligned} \tag{7.B.2b}$$

(where again the terms neglected are $O(1/N)$), and $-\hat{\boldsymbol{S}}^\dagger\boldsymbol{G}$ has been shown in Appendix A to be asymptotically equivalent to $\boldsymbol{S}^\dagger\hat{\boldsymbol{G}}$. Also note that

$$\begin{aligned} \boldsymbol{a}^\dagger(\omega_i)\hat{\boldsymbol{G}}\boldsymbol{G}^\dagger\boldsymbol{d}(\omega_i) &= \boldsymbol{a}^\dagger(\omega_i)\boldsymbol{S}\boldsymbol{S}^\dagger\hat{\boldsymbol{G}}\boldsymbol{G}^\dagger\boldsymbol{d}(\omega_i) \\ &= \boldsymbol{a}^\dagger(\omega_i)[\boldsymbol{S}\boldsymbol{S}^\dagger\hat{\boldsymbol{g}}_1 \;\cdots\; \boldsymbol{S}\boldsymbol{S}^\dagger\hat{\boldsymbol{g}}_{m-n}] \begin{bmatrix} \boldsymbol{g}_1^\dagger\boldsymbol{d}(\omega_i) \\ \vdots \\ \boldsymbol{g}_{m-n}^\dagger\boldsymbol{d}(\omega_i) \end{bmatrix} \\ &= \sum_{k=1}^{m-n} [\boldsymbol{g}_k^\dagger\boldsymbol{d}(\omega_i)][\boldsymbol{a}^\dagger(\omega_i)\boldsymbol{S}\boldsymbol{S}^\dagger\hat{\boldsymbol{g}}_k]\,. \end{aligned} \tag{7.B.3}$$

From (7.B.2) and (7.B.3) we get (neglecting the higher-order terms)

$$\hat{\omega}_i - \omega_i = -\mathrm{Re}\left\{\sum_{k=1}^{m-n} [\boldsymbol{g}_k^\dagger\boldsymbol{d}(\omega_i)][\boldsymbol{a}^\dagger(\omega_i)\boldsymbol{S}\boldsymbol{S}^\dagger\hat{\boldsymbol{g}}_k]\right\} \Big/ h(\omega_i)\,. \tag{7.B.4}$$

The asymptotic zero-mean Gaussian distribution of $\{\hat{\omega}_i - \omega_i\}$ follows from (7.B.4) and Lemma 3.1, part (b). It remains to verify (7.17) for the variances-covariances of the estimation errors $\{\hat{\omega}_i - \omega_i\}$. Toward this end, note that for two scalar-valued complex variables, u and v, we have

$$\mathrm{Re}(u)\cdot\mathrm{Re}(v) = \frac{1}{2}\,[\mathrm{Re}(uv) + \mathrm{Re}(uv^*)]\,. \tag{7.B.5}$$

Let

$$
\begin{aligned}
u &= \sum_{k=1}^{m-n} [g_k^\dagger \boldsymbol{d}(\omega_i)][\boldsymbol{a}^\dagger(\omega_i) \boldsymbol{S}\boldsymbol{S}^\dagger \hat{\boldsymbol{g}}_k] \\
v &= \sum_{p=1}^{m-n} [g_p^\dagger \boldsymbol{d}(\omega_i)][\boldsymbol{a}^\dagger(\omega_j) \boldsymbol{S}\boldsymbol{S}^\dagger \hat{\boldsymbol{g}}_p] .
\end{aligned}
\tag{7.B.6}
$$

Using Lemma 3.1, part (b), we obtain

$$
\mathrm{E}\{uv\} = \sum_{k=1}^{m-n} \sum_{p=1}^{m-n} [g_k^\dagger \boldsymbol{d}(\omega_i)][g_p^\dagger \boldsymbol{d}(\omega_j)] \boldsymbol{a}^\dagger(\omega_i) \mathrm{E}\{(\boldsymbol{S}\boldsymbol{S}^\dagger \hat{\boldsymbol{g}}_k)(\boldsymbol{S}\boldsymbol{S}^\dagger \hat{\boldsymbol{g}}_p)^T\} \boldsymbol{a}^*(\omega_j) = 0
$$

and

$$
\begin{aligned}
\mathrm{E}\{uv^*\} &= \sum_{k=1}^{m-n} \sum_{p=1}^{m-n} [\boldsymbol{g}_k^\dagger \boldsymbol{d}(\omega_i)][\boldsymbol{d}^\dagger(\omega_j) \boldsymbol{g}_p] \boldsymbol{a}^\dagger(\omega_i) \mathrm{E}\left\{ \underbrace{(\boldsymbol{S}\boldsymbol{S}^\dagger \hat{\boldsymbol{g}}_k)(\boldsymbol{S}\boldsymbol{S}^\dagger \hat{\boldsymbol{g}}_p)^\dagger}_{\frac{1}{N} U \delta_{k,p}} \right\} \boldsymbol{a}(\omega_j) \\
&= \frac{1}{N} [\boldsymbol{d}^\dagger(\omega_j) \boldsymbol{G}\boldsymbol{G}^\dagger d(\omega_i)][\boldsymbol{a}^\dagger(\omega_i) \boldsymbol{U} \boldsymbol{a}(\omega_j)] .
\end{aligned}
\tag{7.B.7}
$$

Equation (7.17) now follows from (7.B.4)–(7.B.7), and the proof is finished.

Remark. It is worth noting that the asymptotic distribution of the MUSIC estimation errors could be derived from (7.B.2a), (7.B.2b), and part (a) of Lemma 3.1, thus circumventing the need of using part (b) of the lemma. However, we preferred to base the derivation on part (b) of Lemma 3.1 to motivate inclusion of the results (7.15), (7.16), which appear to be of independent interest. Note, for example, that results (7.15), (7.16) are used in Appendix D to establish the asymptotic distribution of the random variables (7.19).

APPENDIX C: PROOF OF THEOREM 3.2

Let $\hat{\omega}$ denote a generic minimum point of (7.20). A simple Taylor series expansion similar to (7.B.2) gives

$$
0 = \alpha'(\hat{\omega}) \simeq \alpha'(\omega) + \alpha''(\omega)(\hat{\omega} - \omega) , \tag{7.C.1}
$$

where

$$
\alpha'(\omega) = \frac{f'(\omega)r(\omega) - f(\omega)r'(\omega)}{r^2(\omega)} \tag{7.C.2a}
$$

$$
\underline{\alpha''(\omega) = [f''(\omega)r(\omega) - f(\omega)r''(\omega)]r^2(\omega) - [f'(\omega)r(\omega) - f(\omega)r'(\omega)]2r(\omega)r'(\omega)}
$$
$$
r^4(\omega) \, .
$$

(7.C.2b)

Since $f(\omega) = O(1/N)$ and $f'(\omega) = O(1/\sqrt{N})$, it follows from (7.C.1) and (7.C.2) that

$$0 \simeq \alpha'(\omega) + \alpha''(\omega)(\hat{\omega} - \omega) \simeq \frac{f'(\omega)}{r(\omega)} + \frac{f''(\omega)}{r(\omega)}(\hat{\omega} - \omega), \qquad (7.C.3)$$

where the neglected terms go to zero faster than $(\hat{\omega} - \omega)$, when N tends to infinity. From (7.C.3), we get

$$f'(\omega) + f''(\omega)(\hat{\omega} - \omega) \simeq 0,$$

which is exactly (7.B.2) corresponding to MUSIC, and thus the proof is finished.

APPENDIX D: PROOF OF (7.21)

It follows from Lemma 3.1 that the random variables

$$\epsilon_i \triangleq \boldsymbol{a}^\dagger(\omega)\hat{\boldsymbol{g}}_i = \boldsymbol{a}^\dagger(\omega)\boldsymbol{S}\boldsymbol{S}^\dagger\hat{\boldsymbol{g}}_i, \qquad i = 1, \ldots, m-n,$$

are Gaussian distributed with zero means and the following variances-covariances:

$$\begin{aligned} E\{\epsilon_i\epsilon_j^*\} &= \frac{1}{N}\boldsymbol{a}^\dagger(\omega)\boldsymbol{U}\boldsymbol{a}(\omega)\delta_{i,j} \\ E\{\epsilon_i\epsilon_j\} &= 0, \qquad \text{for all } i, j. \end{aligned} \qquad (7.D.1)$$

From (7.D.1) we get

$$E\{\bar{\epsilon}_k\bar{\epsilon}_p\} = \frac{1}{4}E\{\epsilon_k + \epsilon_k^*)(\epsilon_p + \epsilon_p^*)\} = \frac{1}{2N}\boldsymbol{a}^\dagger(\omega)\boldsymbol{U}\boldsymbol{a}(\omega)\delta_{k,p}$$

$$E\{\tilde{\epsilon}_k\tilde{\epsilon}_p\} = -\frac{1}{4}E\{(\epsilon_k - \epsilon_k^*)(\epsilon_p - \epsilon_p^*)\} = \frac{1}{2N}\boldsymbol{a}^\dagger(\omega)\boldsymbol{U}\boldsymbol{a}(\omega)\delta_{k,p}$$

$$E\{\bar{\epsilon}_k\tilde{\epsilon}_p\} = \frac{1}{4_j}E\{(\epsilon_k + \epsilon_k^*)(\epsilon_p - \epsilon_p^*)\} = 0, \qquad \text{for all } k \text{ and } p.$$

Thus, the likelihood function is given by
$L(\epsilon_1, \ldots, \epsilon_{m-n}) \triangleq L(\bar{\epsilon}_1, \ldots, \bar{\epsilon}_{m-n}, \tilde{\epsilon}_1, \ldots, \tilde{\epsilon}_{m-n})$

$$= \frac{1}{(2\pi)^{m-n}[\frac{1}{2N}\boldsymbol{a}^\dagger(\omega)\boldsymbol{U}\boldsymbol{a}(\omega)]^{m-n}} \exp\left\{-\frac{N}{\boldsymbol{a}^\dagger(\omega)\boldsymbol{U}\boldsymbol{a}(\omega)}\sum_{k=1}^{m-n}|\epsilon_k|^2\right\}$$

and, therefore, the log likelihood is

$$\ln L = \text{const} - (m-n)\ln[\boldsymbol{a}^\dagger(\omega)\boldsymbol{U}\boldsymbol{a}(\omega)] - [N\sum_{k=1}^{m-n}|\epsilon_k|^2]\big/[\boldsymbol{a}^\dagger(\omega)\boldsymbol{U}\boldsymbol{a}(\omega)],$$

which proves (7.21).

APPENDIX E: DERIVATION OF THE CRLB

The likelihood function of the data is given by

$$L(y(1), \ldots, y(N)) = \frac{1}{(2\pi)^{mN}\left(\frac{\sigma^2}{2}\right)^{mN}} \exp\left\{-\frac{1}{\sigma^2}\sum_{t=1}^{N}[y(t) - Ax(t)]^\dagger[y(t) - Ax(t)]\right\}.$$

Thus, the log-likelihood function is

$$\ln L = \text{const} - mN\ln\sigma^2 - \frac{1}{\sigma^2}\sum_{t=1}^{N}[y^\dagger(t) - x^\dagger(t)A^\dagger][y(t) - Ax(t)]. \quad (7.E.1)$$

First, we calculate the derivatives of (7.E.1) with respect to σ^2, $\{\bar{x}(t) \triangleq \text{Re } x(t)\}$, $\{\tilde{x}(t) \triangleq \text{Im } x(t)\}$ and $\boldsymbol{\theta}$. We have

$$\frac{\partial \ln L}{\partial \sigma^2} = -\frac{mN}{\sigma^2} + \frac{1}{\sigma^2}\sum_{t=1}^{N} e^\dagger(t)e(t) \quad (7.E.2a)$$

$$\frac{\partial \ln L}{\partial \bar{x}(k)} = \frac{1}{\sigma^2}[A^\dagger e(k) + A^T e^*(k)] = \frac{2}{\sigma^2}\text{Re}[A^\dagger e(k)], \qquad k = 1, \ldots, N \quad (7.E.2b)$$

$$\frac{\partial \ln L}{\partial \tilde{x}(k)} = \frac{1}{\sigma^2}[-jA^\dagger e(k) + jA^T e^*(k)] = \frac{2}{\sigma^2}\text{Im}[A^\dagger e(k)], \qquad k = 1, \ldots, N \quad (7.E.2c)$$

and

$$\frac{\partial \ln L}{\partial \omega_i} = \frac{2}{\sigma^2}\sum_{t=1}^{N}\text{Re}\left[x^\dagger(t)\frac{dA^\dagger}{d\omega_i}e(t)\right] = \frac{2}{\sigma^2}\sum_{t=1}^{N}\text{Re}\,[x_i^\dagger(t)d^\dagger(\omega_i)e(t)],\ i = 1, \ldots, n$$

which can be written more compactly as

$$\frac{\partial \ln L}{\partial \boldsymbol{\theta}} = \frac{2}{\sigma}\sum_{t=1}^{N}\text{Re}[X^\dagger(t)D^\dagger e(t)]. \quad (7.E.2d)$$

To proceed, we need a number of results. These are stated and proved in the following.

$$R1\text{: E}\{e^\dagger(t)e(t)e^\dagger(s)e(s)\} = \begin{cases} m^2\sigma^4, & \text{for } t \neq s \\ m(m+1)\sigma^4, & \text{for } t = s. \end{cases} \quad (7.E.3)$$

Proof: For $t \neq s$,

$$\text{E}\{e^\dagger(t)e(t)e^\dagger(s)e(s)\} = [\text{E}\{\bar{e}^T(t)\bar{e}(t)\} + \text{E}\{\tilde{e}^T(t)\tilde{e}(t)\}]^2 = m^2\sigma^4.$$

For $t = s$,

$$\text{E}\{e^\dagger(t)e(t)e^\dagger(s)e(s)\} = \text{E}\{[\bar{e}^T(t)\bar{e}(t) + \tilde{e}^T(t)\tilde{e}(t)]^2\} =$$
$$= \text{E}\{[\bar{e}^T(t)\bar{e}(t)]^2\} + 2\,\text{E}\{[\bar{e}^T(t)\bar{e}(t)]\}\text{E}\{[\tilde{e}^T(t)\tilde{e}(t)]\}$$

$$+ \mathrm{E}\{[\tilde{e}^T(t)\tilde{e}(t)]^2\} = 2\,\mathrm{E}\{[\bar{e}^T(t)\bar{e}(t)]^2\} + \frac{1}{2}m^2\sigma^4.$$

Since

$$\mathrm{E}\{[\bar{e}^T(t)\bar{e}(t)]^2\} = \mathrm{E}\left\{\sum_{i=1}^{m}\sum_{j=1}^{m}\bar{e}_i^2(t)\bar{e}_j^2(t)\right\}$$

$$= \sum_{i=1}^{m}\sum_{\substack{j=1\\ j\neq i}}^{m}\mathrm{E}\{\bar{e}_i^2(t)\}\mathrm{E}\{\bar{e}_j^2(t)\} + \sum_{i=1}^{m}\mathrm{E}\{\bar{e}_i^4(t)\}$$

$$= (m-1)m\frac{\sigma^4}{4} + 3m\frac{\sigma^4}{4} = m(m+2)\frac{\sigma^4}{4}$$

the proof is finished.

R2: $\mathrm{E}\{\boldsymbol{e}^\dagger(t)\boldsymbol{e}(t)\boldsymbol{e}^T(s)\} = 0$ for all t and s. (7.E.4)

Proof: For $t \neq s$ the result is immediate since $\boldsymbol{e}(t)$ and $\boldsymbol{e}(s)$ are independent. For $t = s$, it follows from the fact that the third-order moments of Gaussian random variables are equal to zero.

R3:

$$\mathrm{Re}\,(\boldsymbol{x})\mathrm{Re}\,(\boldsymbol{y}^T) = \frac{1}{2}[\mathrm{Re}(\boldsymbol{x}\boldsymbol{y}^T) + \mathrm{Re}(\boldsymbol{x}\boldsymbol{y}^\dagger)]$$

$$\mathrm{Im}(\boldsymbol{x})\mathrm{Im}(\boldsymbol{y}^T) = -\frac{1}{2}[\mathrm{Re}(\boldsymbol{x}\boldsymbol{y}^T) - \mathrm{Re}(\boldsymbol{x}\boldsymbol{y}^\dagger)] \tag{7.E.5}$$

$$\mathrm{Re}(\boldsymbol{x})\mathrm{Im}(\boldsymbol{y}^T) = \frac{1}{2}[\mathrm{Im}(\boldsymbol{x}\boldsymbol{y}^T) - \mathrm{Im}(\boldsymbol{x}\boldsymbol{y}^\dagger)].$$

Proof: The result follows from some straightforward calculations.

R4: Let $\boldsymbol{H}$ be a nonsingular complex matrix, and denote its inverse by $\boldsymbol{G} \triangleq \boldsymbol{H}^{-1}$. Then

$$\begin{bmatrix}\bar{\boldsymbol{H}} & -\tilde{\boldsymbol{H}}\\ \tilde{\boldsymbol{H}} & \bar{\boldsymbol{H}}\end{bmatrix}^{-1} = \begin{bmatrix}\bar{\boldsymbol{G}} & -\tilde{\boldsymbol{G}}\\ \tilde{\boldsymbol{G}} & \bar{\boldsymbol{G}}\end{bmatrix}. \tag{7.E.6}$$

Proof: The equality (7.E.6) can equivalently be written as

$$\bar{\boldsymbol{H}}\bar{\boldsymbol{G}} - \tilde{\boldsymbol{H}}\tilde{\boldsymbol{G}} = \boldsymbol{I}$$

$$\bar{\boldsymbol{H}}\tilde{\boldsymbol{G}} + \tilde{\boldsymbol{H}}\bar{\boldsymbol{G}} = \boldsymbol{0},$$

which certainly must hold since

$$\boldsymbol{I} = \boldsymbol{H}\boldsymbol{G} = (\bar{\boldsymbol{H}} + j\tilde{\boldsymbol{H}})(\bar{\boldsymbol{G}} + j\tilde{\boldsymbol{G}}) = (\bar{\boldsymbol{H}}\bar{\boldsymbol{G}} - \tilde{\boldsymbol{H}}\tilde{\boldsymbol{G}}) + j(\bar{\boldsymbol{H}}\tilde{\boldsymbol{G}} + \tilde{\boldsymbol{H}}\bar{\boldsymbol{G}}).$$

Turn now to the evaluation of the CRLB covariance matrix, which is given by

$$\boldsymbol{\Omega} = (\mathrm{E}\{\boldsymbol{\Psi}\boldsymbol{\Psi}^T\})^{-1}, \tag{7.E.7a}$$

where

$$\boldsymbol{\Psi}^T \triangleq \partial \ln L \,/\, \partial[\sigma^2 \bar{\boldsymbol{x}}^T(1)\, \tilde{\boldsymbol{x}}^T(1) \cdots \bar{\boldsymbol{x}}^T(N)\, \tilde{\boldsymbol{x}}^T(N)\, \boldsymbol{\theta}^T]. \tag{7.E.7b}$$

Using R1, we get

$$\begin{aligned} \mathrm{E}\left[\frac{\partial \ln L}{\partial \sigma^2}\right]^2 &= \frac{m^2N^2}{\sigma^4} - 2\frac{mN}{\sigma^6}\sum_{t=1}^{N} \mathrm{E}\{\boldsymbol{e}^\dagger(t)\boldsymbol{e}(t)\} + \frac{1}{\sigma^8}\sum_{t=1}^{N}\sum_{s=1}^{N} \mathrm{E}\{\boldsymbol{e}^\dagger(t)\boldsymbol{e}(t)\boldsymbol{e}^\dagger(s)\boldsymbol{e}(s)\} \\ &= \frac{m^2N^2}{\sigma^4} - 2\frac{m^2N^2}{\sigma^4} + \frac{Nm}{\sigma^4}[(N-1)m + (m+1)] = \frac{mN}{\sigma^4}. \end{aligned} \tag{7.E.8a}$$

Using R2 we note that $\partial \ln L/\partial\sigma^2$ is not correlated with any of the other derivatives in (7.E.2).

Next, we use R3 and the fact that $\mathrm{E}\{\boldsymbol{e}(t)\boldsymbol{e}^T(s)\} = \boldsymbol{0}$ for all t and s (see assumption A2), to obtain

$$\mathrm{E}\left\{\left[\frac{\partial \ln L}{\partial \bar{\boldsymbol{x}}(k)}\right]\left[\frac{\partial \ln L}{\partial \bar{\boldsymbol{x}}(p)}\right]^T\right\} = \frac{4}{\sigma^4}\frac{1}{2}\,\mathrm{Re}[\mathrm{E}\{\boldsymbol{A}^\dagger \boldsymbol{e}(k)\boldsymbol{e}^\dagger(p)\boldsymbol{A}\}] = \frac{2}{\sigma^2}\,\mathrm{Re}[\boldsymbol{A}^\dagger\boldsymbol{A}]\delta_{k,p} \tag{7.E.8b}$$

$$\mathrm{E}\left\{\left[\frac{\partial \ln L}{\partial \bar{\boldsymbol{x}}(k)}\right]\left[\frac{\partial \ln L}{\partial \tilde{\boldsymbol{x}}(p)}\right]^T\right\} = -\frac{4}{\sigma^4}\frac{1}{2}\,\mathrm{Im}[\mathrm{E}\{\boldsymbol{A}^\dagger \boldsymbol{e}(k)\boldsymbol{e}^\dagger(p)\boldsymbol{A}\}] = -\frac{2}{\sigma^2}\,\mathrm{Im}[\boldsymbol{A}^\dagger\boldsymbol{A}]\delta_{k,p} \tag{7.E.8c}$$

$$\mathrm{E}\left\{\left[\frac{\partial \ln L}{\partial \bar{\boldsymbol{x}}(k)}\right]\left[\frac{\partial \ln L}{\partial \boldsymbol{\theta}}\right]^T\right\} = \frac{4}{\sigma^4}\sum_{t=1}^{N}\frac{1}{2}\,\mathrm{Re}[\mathrm{E}\{\boldsymbol{A}^\dagger \boldsymbol{e}(k)\boldsymbol{e}^\dagger(t)\boldsymbol{D}\boldsymbol{X}(t)\}] = \frac{2}{\sigma^2}\mathrm{Re}[\boldsymbol{A}^\dagger\boldsymbol{D}\boldsymbol{X}(k)] \tag{7.E.8d}$$

$$\mathrm{E}\left\{\left[\frac{\partial \ln L}{\partial \tilde{\boldsymbol{x}}(k)}\right]\left[\frac{\partial \ln L}{\partial \tilde{\boldsymbol{x}}(p)}\right]^T\right\} = \frac{4}{\sigma^4}\frac{1}{2}\,\mathrm{Re}[\mathrm{E}\{\boldsymbol{A}^\dagger \boldsymbol{e}(k)\boldsymbol{e}^\dagger(p)\boldsymbol{A}\}] = \frac{2}{\sigma^2}\,\mathrm{Re}[\boldsymbol{A}^\dagger\boldsymbol{A}]\delta_{k,p} \tag{7.E.8e}$$

$$\begin{aligned} \mathrm{E}\left\{\left[\frac{\partial \ln L}{\partial \tilde{\boldsymbol{x}}(k)}\right]\left[\frac{\partial \ln L}{\partial \boldsymbol{\theta}}\right]^T\right\} &= \frac{4}{\sigma^4}\sum_{t=1}^{N}\left(-\frac{1}{2}\right)\mathrm{Im}[\mathrm{E}\{\boldsymbol{X}^\dagger(t)\boldsymbol{D}^\dagger \boldsymbol{e}(t)\boldsymbol{e}^\dagger(k)\boldsymbol{A}\}]^T \\ &= -\frac{2}{\sigma^2}\,\mathrm{Im}[\boldsymbol{X}^\dagger(k)\boldsymbol{D}^\dagger\boldsymbol{A}]^T = \frac{2}{\sigma^2}\mathrm{Im}[\boldsymbol{A}^\dagger\boldsymbol{D}\boldsymbol{X}(k)] \end{aligned} \tag{7.E.8f}$$

$$\begin{aligned} \mathrm{E}\left\{\left[\frac{\partial \ln L}{\partial \boldsymbol{\theta}}\right]\left[\frac{\partial \ln L}{\partial \boldsymbol{\theta}}\right]^T\right\} &= \frac{4}{\sigma^4}\frac{1}{2}\sum_{t=1}^{N}\sum_{s=1}^{N}\mathrm{Re}\,[\mathrm{E}\{\boldsymbol{X}^\dagger(t)\boldsymbol{D}^\dagger \boldsymbol{e}(t)\boldsymbol{e}^\dagger(s)\boldsymbol{D}\boldsymbol{X}(s)\}] \\ &= \frac{2}{\sigma^2}\sum_{t=1}^{N}\mathrm{Re}\,[\boldsymbol{X}^\dagger(t)\boldsymbol{D}^\dagger\boldsymbol{D}\boldsymbol{X}(t)] \triangleq \boldsymbol{\Gamma}. \end{aligned} \tag{7.E.8g}$$

Introduce the following notations

$$\text{var}_{\text{CR}}(\sigma^2) = \sigma^4/mN$$

$$\boldsymbol{H} = \frac{2}{\sigma^2}\boldsymbol{A}^\dagger \boldsymbol{A}$$

$$\boldsymbol{G} = \boldsymbol{H}^{-1}$$

$$\Delta_k = \frac{2}{\sigma^2}\boldsymbol{A}^\dagger \boldsymbol{D}\boldsymbol{X}(k).$$

Observe that since the matrix $\boldsymbol{H}$ is Hermitian, its imaginary part must be skew-symmetric $\tilde{\boldsymbol{H}}^T = -\tilde{\boldsymbol{H}}$. Inserting (7.E.8) into (7.E.7) and using the notation, we get

$$\boldsymbol{\Omega} = \begin{bmatrix} \text{var}_{\text{CR}}^{-1}(\sigma^2) & & & 0 & & \\ & \begin{matrix}\bar{\boldsymbol{H}} & -\tilde{\boldsymbol{H}} \\ \tilde{\boldsymbol{H}} & \bar{\boldsymbol{H}}\end{matrix} & & 0 & & \begin{matrix}\bar{\boldsymbol{\Delta}}_1 \\ \tilde{\boldsymbol{\Delta}}_1\end{matrix} \\ & & \ddots & & & \vdots \\ 0 & & 0 & & \begin{matrix}\bar{\boldsymbol{H}} & -\tilde{\boldsymbol{H}} \\ \tilde{\boldsymbol{H}} & \bar{\boldsymbol{H}}\end{matrix} & \begin{matrix}\bar{\boldsymbol{\Delta}}_N \\ \tilde{\boldsymbol{\Delta}}_N\end{matrix} \\ & \bar{\boldsymbol{\Delta}}_1^T \; \tilde{\boldsymbol{\Delta}}_1^T & \ldots & & \bar{\boldsymbol{\Delta}}_N^T \; \tilde{\boldsymbol{\Delta}}_N^T & \boldsymbol{\Gamma} \end{bmatrix}^{-1}. \tag{7.E.9}$$

The expression (7.24) for $\text{var}_{\text{CR}}(\sigma^2)$ is thus proved. To prove the expression (7.25) for CRB($\boldsymbol{\theta}$) we note that (7.E.9), a standard result on the inverse of a partitioned matrix, and R4 give

$$\text{CRLB}^{-1}(\boldsymbol{\theta}) = \boldsymbol{\Gamma} - [\bar{\boldsymbol{\Delta}}_1^T \, \tilde{\boldsymbol{\Delta}}_1^T \vdots \cdots \vdots \bar{\boldsymbol{\Delta}}_N^T \, \tilde{\boldsymbol{\Delta}}_N^T] \begin{bmatrix} \begin{matrix}\bar{\boldsymbol{G}} & -\tilde{\boldsymbol{G}} \\ \tilde{\boldsymbol{G}} & \bar{\boldsymbol{G}}\end{matrix} & & 0 \\ & \ddots & \\ 0 & & \begin{matrix}\bar{\boldsymbol{G}} & -\tilde{\boldsymbol{G}} \\ \tilde{\boldsymbol{G}} & \bar{\boldsymbol{G}}\end{matrix} \end{bmatrix} \begin{bmatrix} \bar{\boldsymbol{\Delta}}_1 \\ \tilde{\boldsymbol{\Delta}}_1 \\ \vdots \\ \bar{\boldsymbol{\Delta}}_N \\ \tilde{\boldsymbol{\Delta}}_N \end{bmatrix}. \tag{7.E.10}$$

Next, observe that

$$\begin{bmatrix} \bar{\boldsymbol{G}} & -\tilde{\boldsymbol{G}} \\ \tilde{\boldsymbol{G}} & \bar{\boldsymbol{G}} \end{bmatrix} \begin{bmatrix} \bar{\boldsymbol{\Delta}} \\ \tilde{\boldsymbol{\Delta}} \end{bmatrix} = \begin{bmatrix} \bar{\boldsymbol{G}}\bar{\boldsymbol{\Delta}} - \tilde{\boldsymbol{G}}\tilde{\boldsymbol{\Delta}} \\ \tilde{\boldsymbol{G}}\bar{\boldsymbol{\Delta}} + \bar{\boldsymbol{G}}\tilde{\boldsymbol{\Delta}} \end{bmatrix} = \begin{bmatrix} \overline{\boldsymbol{G}\boldsymbol{\Delta}} \\ \widetilde{\boldsymbol{G}\boldsymbol{\Delta}} \end{bmatrix} \tag{7.E.11a}$$

and that

$$[\bar{\Delta}^T \ \tilde{\Delta}^T] \begin{bmatrix} \overline{G\Delta} \\ \widetilde{G\Delta} \end{bmatrix} = \text{Re}[\Delta^\dagger G \Delta]. \tag{7.E.11b}$$

From (7.E.10) and (7.E.11) it follows that

$$\text{CRLB}^{-1}(\boldsymbol{\theta}) = \boldsymbol{\Gamma} - \sum_{t=1}^{N} \text{Re}[\Delta_t^\dagger G \Delta_t] = \frac{2}{\sigma^2} \sum_{t=1}^{N} \text{Re}[X^\dagger(t) D^\dagger D X(t)$$

$$-X^\dagger(t) D^\dagger A (A^\dagger A)^{-1} A^\dagger D X(t)] = \frac{2}{\sigma^2} \sum_{t=1}^{N} \text{Re}\,\{X^\dagger(t) D^\dagger [I - A(A^\dagger A)^{-1} A^\dagger] D X(t)\},$$

which completes the proof.

Remark. A slightly more compressed derivation of the CRLB formula (7.23) can be obtained if one uses the extension of Bangs' formula [40] for the CRLB matrix with information in the mean and covariance, see [41], instead of the standard general formula (7.E.7).

APPENDIX F: PROOF OF (7.25)

We have

$$\text{CRLB}^{-1}(N+1) =$$

$$\text{CRLB}^{-1}(N) + \frac{2}{\sigma^2} \cdot$$

$$\text{Re}\,\{X^\dagger(N+1) D^\dagger [I - A(A^\dagger A)^{-1} A^\dagger] D X(N+1)\}. \tag{7.F.1}$$

The matrix in curly braces is Hermitian positive definite and thus its real part is symmetric positive semidefinite. This observation and (7.F.1) prove (7.25a).

To prove (7.25b), let us introduce for convenience the following notation

$H = (A_m^\dagger A_m)^{-1}$
$G = A_m^\dagger D_m$
$u^\dagger$ = the last row of A_{m+1}
$v^\dagger$ = the last row of D_{m+1}.

Making use of the nested structures of A_{m+1} and D_{m+1},

$$A_{m+1} = \begin{bmatrix} A_m \\ u^\dagger \end{bmatrix}, \qquad D_{m+1} = \begin{bmatrix} D_m \\ v^\dagger \end{bmatrix}$$

and of the *matrix inversion lemma* (see, e.g., [32]), we can write

$$D_{m+1}^{\dagger}[I - A_{m+1}(A_{m+1}^{\dagger}A_{m+1})^{-1}A_{m+1}^{\dagger}]D_{m+1}$$

$$= D_m^{\dagger}D_m + vv^{\dagger} - (G^{\dagger} + vu^{\dagger})(H^{-1} + uu^{\dagger})^{-1}(G + uv^{\dagger})$$

$$= D_m^{\dagger}D_m + vv^{\dagger} - (G^{\dagger} + vu^{\dagger})\left(H - \frac{Huu^{\dagger}H}{1 + u^{\dagger}Hu}\right)(G + uv^{\dagger})$$

$$= D_m^{\dagger}[I - A_m(A_m^{\dagger}A_m)^{-1}A_m^{\dagger}]\, D_m + Q, \tag{7.F.2}$$

where

$$Q \triangleq vv^{\dagger} + \frac{G^{\dagger}Huu^{\dagger}HG}{1 + u^{\dagger}Hu} - G^{\dagger}\left[H - \frac{Huu^{\dagger}H}{1 + u^{\dagger}Hu}\right]uv^{\dagger}$$

$$-vu^{\dagger}\left[H - \frac{Huu^{\dagger}H}{1 + u^{\dagger}Hu}\right]G - vu^{\dagger}\left[H - \frac{Huu^{\dagger}H}{1 + u^{\dagger}Hu}\right]uv^{\dagger}$$

$$= (vv^{\dagger} + vv^{\dagger}u^{\dagger}Hu + G^{\dagger}Huu^{\dagger}HG + G^{\dagger}Huv^{\dagger} - vu^{\dagger}HG - vu^{\dagger}Huv^{\dagger})/(1 + u^{\dagger}Hu)$$

$$= (v - G^{\dagger}Hu)(v - G^{\dagger}Hu)^{\dagger}/(1 + u^{\dagger}Hu).$$

Since the matrix Q is evidently Hermitian positive semidefinite, the inequality (7.25b) follows from (7.F.2) and the expression (7.23) of the CRLB.

APPENDIX G: PROOF OF THEOREM 4.3

Let

$$H \triangleq D^{\dagger}[I - A(A^{\dagger}A)^{-1}A^{\dagger}]D$$

Then, the i, j element of the matrix whose inverse appears in (7.23) can be written as

$$\text{Re}\left\{d^{\dagger}(\omega_i)[I - A(A^{\dagger}A)^{-1}A^{\dagger}]d(\omega_j) \sum_{t=1}^{N} x_i^{\dagger}(t)x_j(t)\right\}$$

$$= N\,\text{Re}\left[H_{ij} \cdot \frac{1}{N}\sum_{t=1}^{N} x_i^{\dagger}(t)x_j(t)\right].$$

Since by definition

$$\lim_{N\to\infty} \frac{1}{N}\sum_{t=1}^{N} x_i^{\dagger}(t)x_j(t) = P_{ji}$$

it readily follows that for sufficiently large N, the CRLB is given by

$$\frac{\sigma^2}{2N}[\mathrm{Re}(\boldsymbol{H} \odot \boldsymbol{P}^T)]^{-1},$$

which proves part (a).

Next consider part (b). To prove (7.29) we make use of the following result [26]

$$\frac{1}{m^{k+1}} \sum_{t=1}^{m} t^k e^{jt(\omega_1 - \omega_2)} \xrightarrow[m \to \infty]{} \begin{cases} 1/(k+1) & \text{for } \omega_1 = \omega_2 \\ 0 & \text{for } \omega_1 \neq \omega_2 . \end{cases} \tag{7.G.1}$$

Using (7.G.1), we can write

$$\frac{1}{m^3}[\boldsymbol{D}^\dagger \boldsymbol{D}]_{kp} = \frac{1}{m^3} \sum_{t=1}^{m-1} t^2 e^{jt(\omega_p - \omega_k)} \xrightarrow[m \to \infty]{} \frac{1}{3}\delta_{k,p}$$

$$\frac{1}{m^2}[\boldsymbol{A}^\dagger \boldsymbol{D}]_{kp} = \frac{j}{m^2} \sum_{t=1}^{m-1} t e^{jt(\omega_p - \omega_k)} \xrightarrow[m \to \infty]{} \frac{j}{2}\delta_{k,p}$$

$$\frac{1}{m}[\boldsymbol{A}^\dagger \boldsymbol{A}]_{kp} = \frac{1}{m} \sum_{t=1}^{m} e^{j(t-1)(\omega_p - \omega_k)} \xrightarrow[m \to \infty]{} \delta_{k,p}$$

which readily give

$$\frac{1}{m^3}\boldsymbol{D}^\dagger[\boldsymbol{I} - \boldsymbol{A}(\boldsymbol{A}^\dagger \boldsymbol{A})^{-1}\boldsymbol{A}^\dagger]\boldsymbol{D}$$

$$= \frac{1}{m^3}\boldsymbol{D}^\dagger \boldsymbol{D} - (\frac{1}{m^2}\boldsymbol{D}^\dagger \boldsymbol{A})(\frac{1}{m}\boldsymbol{A}^\dagger \boldsymbol{A})^{-1}(\frac{1}{m^2}\boldsymbol{A}^\dagger \boldsymbol{D}) \xrightarrow[m \to \infty]{} \frac{1}{12}\boldsymbol{I}. \tag{7.G.2}$$

Inserting (7.G.2) into (7.28), we obtain (7.29), and the proof is finished.

APPENDIX H: PROOF OF (7.37)

A "standard" derivation of $\mathrm{var}_{\mathrm{ML}}(\hat{\omega})$, which begins by developing $\mathrm{d}g(\omega)/\mathrm{d}\omega|_{\omega = \hat{\omega}}$ in a Taylor series around the true frequency, appears to be rather lengthy. In the following, we provide a simpler derivation, which makes use of Theorem 6.1 proven in Appendix I. That result states that

$$\mathrm{var}_{\mathrm{ML}}(\hat{\omega}) = \mathrm{var}_{\mathrm{MU}}(\hat{\omega}) = \frac{\sigma^2}{2N}\frac{\lambda_1}{(\sigma^2 - \lambda_1)^2}|\boldsymbol{a}^\dagger(\omega)\boldsymbol{s}_1|^2/[\boldsymbol{d}^\dagger(\omega)(\boldsymbol{I} - \boldsymbol{s}_1\boldsymbol{s}_1^\dagger)\boldsymbol{d}(\omega)], \tag{7.H.1}$$

where the expression for $\text{var}_{\text{MU}}(\hat{\omega})$ (i.e., the MUSIC variance) follows from (7.18). Since for the case under discussion, $\boldsymbol{R}$ is given by

$$\boldsymbol{R} = P\boldsymbol{a}(\omega)\boldsymbol{a}^\dagger(\omega) + \sigma^2\boldsymbol{I},$$

it readily follows that

$$\begin{aligned} \lambda_1 &= mP + \sigma^2 \\ \boldsymbol{s}_1 &= \boldsymbol{a}(\omega)/\sqrt{m}. \end{aligned} \tag{7.H.2}$$

Inserting (7.H.2) into (7.H.1), we get (see also (7.26))

$$\begin{aligned} \text{var}_{\text{ML}}(\hat{\omega}) &= \frac{\sigma^2}{2N}\frac{mP + \sigma^2}{m^2P^2}m \bigg/ \left[\frac{m(m-1)(2m-1)}{6} - \frac{m(m-1)^2}{4}\right] \\ &= \frac{6\sigma^2}{N}\frac{mP + \sigma^2}{m^2(m^2-1)P^2} \end{aligned}$$

and the proof is finished.

APPENDIX I: PROOF OF THEOREM 6.1

The idea of the proof can be explained as follows. As shown in Section 7.5, the ML estimate $\hat{\boldsymbol{\theta}}$ of $\boldsymbol{\theta}$, which is defined by (see (7.33))

$$g(\hat{\boldsymbol{\theta}}) \triangleq \text{tr}[\hat{\boldsymbol{A}}(\hat{\boldsymbol{A}}^\dagger\hat{\boldsymbol{A}})^{-1}\hat{\boldsymbol{A}}^\dagger\hat{\boldsymbol{R}}] = \max \text{tr}[\boldsymbol{A}(\boldsymbol{A}^\dagger\boldsymbol{A})^{-1}\boldsymbol{A}^\dagger\hat{\boldsymbol{R}}] \tag{7.I.1}$$

converges to the true values when N tends to infinity. Furthermore, the estimation errors $(\hat{\boldsymbol{\theta}} - \boldsymbol{\theta})$ are $0(1/\sqrt{N})$ for large N. According to this fact, we can neglect the terms in (I.1) whose derivative with respect to $\hat{\boldsymbol{\theta}}$ is $0(1/N)$, without affecting the asymptotic (for $N >> 0$) distribution of the MLE that maximizes (7.I.1).

To implement the idea, we need to introduce some additional notation. Let

$$\boldsymbol{\Lambda} = \begin{bmatrix} \hat{\lambda}_1 & & \boldsymbol{0} \\ & \ddots & \\ \boldsymbol{0} & & \hat{\lambda}_n \end{bmatrix}$$

$$\boldsymbol{\Sigma} = \begin{bmatrix} \hat{\lambda}_{n+1} & & \boldsymbol{0} \\ & \ddots & \\ \boldsymbol{0} & & \hat{\lambda}_m \end{bmatrix}.$$

Then, the eigendecomposition of $\hat{\boldsymbol{R}}$ can be written as follows

$$\hat{\boldsymbol{R}} = \hat{\boldsymbol{S}}\boldsymbol{\Lambda}\hat{\boldsymbol{S}}^\dagger + \hat{\boldsymbol{G}}\boldsymbol{\Sigma}\hat{\boldsymbol{G}}^\dagger. \tag{7.I.2}$$

Inserting (7.I.2) into (7.I.1), we obtain

$$g(\hat{\boldsymbol{\theta}}) = \text{tr}[(\hat{\boldsymbol{S}}^\dagger\hat{\boldsymbol{A}})(\hat{\boldsymbol{A}}^\dagger\hat{\boldsymbol{A}})^{-1}(\hat{\boldsymbol{A}}^\dagger\hat{\boldsymbol{S}})\boldsymbol{\Lambda}] + \text{tr}[(\hat{\boldsymbol{A}}^\dagger\hat{\boldsymbol{A}})^{-1}(\hat{\boldsymbol{A}}^\dagger\hat{\boldsymbol{G}})\boldsymbol{\Sigma}(\hat{\boldsymbol{G}}^\dagger\hat{\boldsymbol{A}})]. \tag{7.I.3}$$

Since, for large N

$$\hat{A}^{\dagger}\hat{G} = O(1/\sqrt{N}) \tag{7.I.4}$$

and

$$\Sigma = \sigma^2 I + O(1/\sqrt{N}),$$

it follows that we can replace Σ in (7.I.3) by $\sigma^2 I$

$$\begin{aligned} g(\hat{\theta}) &\cong \mathrm{tr}[(\hat{S}^{\dagger}\hat{A})(\hat{A}^{\dagger}\hat{A})^{-1}(\hat{A}^{\dagger}\hat{S})\Lambda] + \mathrm{tr}[(\hat{A}^{\dagger}\hat{A})^{-1}\hat{A}^{\dagger}\hat{G}\hat{G}^{\dagger}\hat{A}]\sigma^2 \\ &= \mathrm{tr}(\Lambda) + \mathrm{tr}[(\hat{S}^{\dagger}\hat{A})(\hat{A}^{\dagger}\hat{A})^{-1}(\hat{A}^{\dagger}\hat{S}) - I]\Lambda - \mathrm{tr}[(\hat{A}^{\dagger}\hat{A})^{-1}\hat{A}^{\dagger}\hat{S}\hat{S}^{\dagger}\hat{A} - I]\sigma^2 \\ &= \mathrm{tr}(\Lambda) + \mathrm{tr}[(\hat{S}^{\dagger}\hat{A})(\hat{A}^{\dagger}\hat{A})^{-1}(\hat{A}^{\dagger}\hat{S}) - I][\Lambda - \sigma^2 I]. \end{aligned} \tag{7.I.5}$$

Define

$$\Delta = \hat{G}\hat{G}^{\dagger}\hat{A} = \hat{A} - \hat{S}\hat{S}^{\dagger}\hat{A}$$

and observe from (7.I.4) that $\Delta = O(1/\sqrt{N})$. Since $\hat{S}^{\dagger}\Delta = \mathbf{0}$, we can write

$$\hat{A}^{\dagger}\hat{A} = [\Delta^{\dagger} + \hat{A}^{\dagger}\hat{S}\hat{S}^{\dagger}][\Delta + \hat{S}\hat{S}^{\dagger}\hat{A}] = \Delta^{\dagger}\Delta + (\hat{A}^{\dagger}\hat{S})(\hat{S}^{\dagger}\hat{A}). \tag{7.I.6}$$

Next, note that the matrix $S^{\dagger}A$ is nonsingular. This follows from the fact that $A = SQ$ for some matrix Q, and Q must be nonsingular since A has full rank by assumption. Thus, the matrix $\hat{S}^{\dagger}\hat{A}$, which, for large N, is close to $S^{\dagger}A$, must also be nonsingular. Using this observation and (7.I.6), we get

$$\begin{aligned} (\hat{S}^{\dagger}\hat{A})(\hat{A}^{\dagger}\hat{A})^{-1}(\hat{A}^{\dagger}\hat{S}) - I &= [I + (\hat{A}^{\dagger}\hat{S})^{-1}\Delta^{\dagger}\Delta(\hat{S}^{\dagger}\hat{A})^{-1}]^{-1} - I \\ &\simeq -(\hat{A}^{\dagger}\hat{S})^{-1}\Delta^{\dagger}\Delta(\hat{S}^{\dagger}\hat{A})^{-1}, \end{aligned} \tag{7.I.7}$$

where we also used the fact that for some matrix Γ of subunitary norm $(I + \Gamma)^{-1} = I - \Gamma + \Gamma^2 - \Gamma^3 + \ldots$. Inserting (7.I.7) into (7.I.5), we obtain

$$g(\hat{\theta}) \simeq \mathrm{const} - \mathrm{tr}[\Delta^{\dagger}\Delta(\hat{S}^{\dagger}\hat{A})^{-1}[\Lambda - \sigma^2 I](\hat{A}^{\dagger}\hat{S})^{-1}. \tag{7.I.8}$$

To complete the proof we only need to show that the matrix that multiplies $\Delta^{\dagger}\Delta$ in (7.I.8) can be replaced by P. Since

$$R = APA^{\dagger} + \sigma^2 I = \hat{A}P\hat{A}^{\dagger} + \sigma^2 I + O(1/\sqrt{N}) = \hat{S}\Lambda\hat{S}^{\dagger} + \hat{G}\Sigma\hat{G}^{\dagger} + O(1/\sqrt{N})$$

it follows that

$$(\hat{S}^{\dagger}\hat{A})P(\hat{A}^{\dagger}\hat{S}) + \sigma^2 I = \Lambda + O(1/\sqrt{N})$$

or yet

$$P = (\hat{S}^{\dagger}\hat{A})^{-1}(\Lambda - \sigma^2 I)(\hat{A}^{\dagger}\hat{S})^{-1} + O(1/\sqrt{N}). \tag{7.I.9}$$

From (7.I.8) and (7.I.9) we obtain the following large sample (for $N >> 0$) approximation to $\boldsymbol{g}(\hat{\boldsymbol{\theta}})$:

$$\boldsymbol{g}(\hat{\boldsymbol{\theta}}) \simeq \text{const} - \text{tr}\,[\hat{\boldsymbol{A}}^{\dagger}\hat{\boldsymbol{G}}\hat{\boldsymbol{G}}^{\dagger}\hat{\boldsymbol{A}}][\boldsymbol{P}]. \qquad (7.I.10)$$

Maximization of (7.I.10) with respect to $\hat{\boldsymbol{\theta}}$ is equivalent to minimization of the MUSIC function (7.13) if and essentially only if the matrix $\boldsymbol{P}$ is diagonal. Thus, the proof is finished.

ACKNOWLEDGMENT

The authors are grateful to M. Kaveh for very helpful comments on a first version of this paper. The work of A. Nehorai was supported by the Air Force Office of Scientific Research, under Grant No. AFOSR-88-0080.

REFERENCES

1. R. O. Schmidt, "Multiple Emitter Location and Signal Parameter Estimation," *Proc. RADC Spectral Estimation Workshop,* Rome, New York, 1979, pp. 243–258.
2. G. Bienvenu and L. Kopp, "Adaptivity to Background Noise Spatial Coherence for High Resolution Passive Methods," *Proc. IEEE Int. Conf. on Acoustics, Speech, and Signal Processing,* Denver, Colorado, April 1980, pp. 307–310.
3. G. Bienvenu and L. Kopp, "Optimality of High Resolution Array Processing," *IEEE Trans. Acoustics, Speech, and Signal Processing,* **31** (October 1983), 1235–1248.
4. M. Wax and T. Kailath, "Determining the Number of Signals by Information Theoretic Criteria," *Proc. IEEE Int. Conf. Acoustics, Speech, and Signal Processing,* San Diego, California, March 1984, 6.3.1–6.3.4.
5. K. Sharman, T. S. Durrani, M. Wax, and T. Kailath, "Asymptotic Performance of Eigenstructure Spectral Analysis Methods," *Proc. IEEE Int. Conf. Acoustic, Speech, and Signal Processing,* San Diego, California, March 1984, 45.5.1–45.5.4.
6. K. Sharman and T. S. Durrani, "A Comparative Study of Modern Eigenstructure Methods for Bearing Estimation—A New High Performance Approach," *Proc. 25th IEEE Conf. Dec. Contr.,* Athens, Greece, December. 1986, pp. 1737–1742.
7. D. J. Jeffries and D. R. Farrier, "Asymptotic Results for Eigenvector Methods," *IEE Proc.,* pt. F, **132** (1985), pp. 589–594.
8. M. Kaveh and A. J. Barabell, "The Statistical Performance of the MUSIC and the Minimum-Norm Algorithms in Resolving Plane Waves in Noise," *IEEE Trans. Acoustics, Speech, and Signal Processing,* **34** (April 1986), pp. 331–341, corrections in *IEEE Trans. Acoustics, Speech, and Signal Processing,* **34** (June 1986), pg. 633.
9. A. M. Bruckstein, T.-J. Shan, and T. Kailath, "The Resolution of Overlapping Echoes," *IEEE Trans. Acoustics, Speech, and Signal Processing,* **33** (December 1985), pp. 1357–1367.

10. R. Kumaresan and A. K. Shaw, "High Resolution Bearing Estimation Without Eigendecomposition," *Proc. IEEE Int. Conf. Acoustics, Speech, and Signal Processing,* Tampa, Florida, March 1985, pp. 576–579.

11. R. Kumaresan, L. L. Scharf, and A. K. Shaw, "An Algorithm for Pole-Zero Modeling and Spectral Analysis," *IEEE Trans. Acoustics, Speech, and Signal Processing,* **34,** June 1986, pp. 637–640.

12. Y. Bresler and A. Macovski, "Exact Maximum Likelihood Parameter Estimation of Superimposed Exponential Signals in Noise," *IEEE Trans. Acoustics, Speech, and Signal Processing,* **34** (October 1986), pp. 1081–1089.

13. D. W. Tufts and R. Kumaresan, "Estimation of Frequencies of Multiple Sinusoids: Making Linear Prediction Perform Like Maximum Likelihood," *Proc. IEEE,* **70,** (September 1982), pp. 975–989.

14. S. Parthasarathy and D. W. Tufts, "Maximum-Likelihood Estimation of Parameters of Exponentially Damped Sinusoids," *Proc. IEEE,* **73** (October 1985), pp. 1528–1530.

15. S. J. Orfanidis, "A Reduced MUSIC Algorithm," *Third ASSP Workshop on Spectrum Estimation and Modeling,* Boston, November 1986, pp. 165–167.

16. R. Kumaresan and D. W. Tufts, "Estimating the Angles of Arrival of Multiple Plane Waves," *IEEE Trans. Aerosp. Electron. Syst.*, **19** (January 1983), pp. 134–139.

17. G. Xu and Y.-H. Pao, "Single Vector Approaches to Eigenstructure Analysis for Harmonic Retrieval," *Proc. IEEE Int. Conf. on Acoustics, Speech, and Signal Processing,* Tokyo, April 1986, pp. 173–176.

18. K. C. Sharman, "Adaptive Algorithms for Estimating the Complete Covariance Eigenstructure," *Proc. IEEE Int. Conf. Acoustics, Speech, and Signal Processing,* Tokyo, April 1986, pp. 1401–1404.

19. R. Schreiber, "Implementation of Adaptive Array Algorithms," *IEEE Trans. Acoustics, Speech, and Signal Processing,* **34** (October 1986), pp. 1038–1045.

20. D. W. Tufts and C. D. Melissinos, "Simple, Effective Computation of Principal Eigenvectors and Their Eigenvalues and Application to High-Resolution Estimation of Frequencies," *IEEE Trans. Acoustics, Speech, and Signal Processing*, **34** (October 1986), pp. 1046–1053.

21. H. Wang and M. Kaveh, "On the Performance of Signal-Subspace Processing," Part I: Narrow-Band Systems," *IEEE Trans. Acoustics, Speech, and Singal Processing,* **34** (October 1986), pp. 1201–1209.

22. F. Tuteur and Y. Rockah, "The Covariance Difference Method in Signal Detection," *Third ASSP Workshop on Spectrum Estimation and Modeling,* Boston, November 1986, pp. 120–122.

23. R. Roy, A. Paulraj, and T. Kailath, "ESPRIT—A Subspace Rotation Approach to Estimation of Parameters of Cisoids in Noise," *IEEE Trans. Acoustics, Speech, and Signal Processing*, **34** (October 1986), pp. 1340–1342.

24. D. C. Rife and R. R. Boorstyn, "Single-Tone Parameter Estimation from Discrete-Time Observations," *IEEE Trans. Inform. Theory,* **20** (September 1974), pp. 591–598.

25. D. C. Rife and R. R. Boorstyn, "Multiple Tone Parameter Estimation from Discrete-Time Observations," *Bell Syst. Tech. J.,* **55** (November 1976), pp. 1389–1410.

26. P. Stoica, R. Moses, B. Friedlander, and T. Söderström, "Maximum Likelihood Estimation of the Parameters of Multiple Sinusoids from Noisy Measurements," *IEEE Trans. Acoustics, Speech, and Signal Processing*, **37** (March 1989), pp. 378–392.

27. M. G. Kendall and A. Stuart, *The Advanced Theory of Statistics* (New York: Hafner Publ. Co., 1961).

28. T. W. Anderson, *An Introduction to Multivariate Statistical Analysis* (New York: John Wiley, 1984).

29. P. Stoica and T. Söderström, "*On Reparametrization of Loss Functions Used in Estimation and the Invariance Principle*," Technical Report UPTEC 87113R, Uppsala University, Sweden, 1987; also *Signal Processing*, **17** (August 1989).

30. A. A. Beex, "Fast Recursive/Iterative Toeplitz Eigenspace Decomposition," in *Signal Processing III: Theories and Applications*, eds. I. T. Young et al. (Amsterdam: North-Holland, 1986), pp. 1001–1004.

31. T. W. Anderson, "Asymptotic Theory for Principal Component Analysis," *Ann. Math. Stat.*, **34**, no. 1 (1963), pp. 122–148.

32. T. Söderström and P. Stoica, *System Identification* (London: Prentice-Hall, 1989).

33. K. Sharman, "Maximum Likelihood Parameter Estimation by Simulated Annealing," *Proc. IEEE Int. Conf. Acoustics, Speech, and Signal Processing*, New York, April 1988, pp. 2741–2744.

34. M. Feder and E. Weinstein, "Parameter Estimation of Superimposed Signals Using the EM Algorithm," *IEEE Trans. Acoustics, Speech, and Signal Processing*, **36** (April 1988), pp. 477–489.

35. I. Ziskind and M. Wax, "Maximum Likelihood Estimation via the Alternating Projection Maximization Algorithm," *Proc. IEEE Int. Conf. Acoustics, Speech, and Signal Processing*, Dallas, Texas, March 1987, pp. 2280–2283.

36. P. Stoica and K. Sharman, "Maximum Likelihood Methods for Direction-of-Arrival Estimation," to appear in *IEEE Trans. Acoustics, Speech, and Signal Processing*, 1989.

37. P. Stoica and A. Nehorai, "Statistical Efficiency Study of Direction Estimation Methods," Part II: "Analysis of Weighted MUSIC and MLM," Chapter 8 of this book.

38. P. Janssen and P. Stoica, "*On the Expectation of the Product of Four Matrix-Valued Gaussian Random Variables,*" EUT Report 87-E-178, Eindhoven University of Technology, Eindhoven, The Netherlands, 1987; also *IEEE Trans. Automatic Control,* **33** (September 1988), pp. 867–870.

39. T.-J. Shan, M. Wax and T. Kailath, "On Spatial Smoothing for Direction-of-Arrival Estimation of Coherent Signals," *IEEE Trans. Acoustics, Speech, and Signal Processing*, **33** (August 1985), pp. 806–811.

40. W. J. Bangs, "Array Processing with Generalized Beamformers," Ph. D. dissertation, Yale University, New Haven, Conneticut, 1971.

41. H. Hung and M. Kaveh, "On the Statistical Sufficiency of the Coherently Averaged Covariance Matrix for the Estimation of the Parameters of Wideband Sources," *Proc. IEEE Int. Conf. Acoustics, Speech, and Signal Processing*, Dallas, Texas, March 1987, pp. 33–36.

42. D. Starer and A. Nehorai, "Maximum Likelihood Estimation of Exponential Signals in Noise Using a Newton Algorithm," *The Fourth ASSP Workshop on Spectrum Estimation and Modeling*, Minneapolis, Minnesota, August 1988, pp. 240–245.

8 Statistical Efficiency Study of Direction Estimation Methods Part II: Analysis of Weighted MUSIC and MLM

Petre Stoica and Arye Nehorai

8.1 INTRODUCTION

In Chapter 7 we presented a number of results on the performance of the *MUltiple SIgnal Characterization* (*MUSIC*) [8], [9] and the *Maximum-Likelihood* (*ML*) [10], [11], [12] estimates of the direction parameters $\boldsymbol{\theta}$. (Note: The problem formulation, data model and notation are the same as those of Chapter 7.) We also derived the *Cramér-Rao lower bound* (*CRLB*) on the covariance matrix of any unbiased estimate of $\boldsymbol{\theta}$, and provided a comparison of the MUSIC performance to the ultimate performance corresponding to the CRLB. In this companion chapter, we continue this study and establish a number of additional results on MUSIC, ML method, and CRLB, which complete the picture on the estimation methods in question and, in particular, on their performances. In the following, the results of Chapter 7 which are relevant to this chapter are presented in a compact *matrix* form, for ease of reference.

(a) MUSIC Covariance Matrix

The MUSIC estimation error $\{\hat{\omega}_i - \omega_i\}$ are asymptotically (for large N) jointly Gaussian distributed with zero means and the following covariance matrix

$$C_{MU} = \frac{\sigma^2}{2N}(\boldsymbol{H} \odot \boldsymbol{I})^{-1}\text{Re}\{\boldsymbol{H} \odot (\boldsymbol{A}^{\dagger}\boldsymbol{U}\boldsymbol{A})^T\}(\boldsymbol{H} \odot \boldsymbol{I})^{-1}, \tag{8.1}$$

where Re(x) denotes the real part of x,

$$\boldsymbol{H} = \boldsymbol{D}^{\dagger}\boldsymbol{G}\boldsymbol{G}^{\dagger}\boldsymbol{D} = \boldsymbol{D}^{\dagger}[I - A(\boldsymbol{A}^{\dagger}\boldsymbol{A})^{-1}\boldsymbol{A}^{\dagger}]\boldsymbol{D} \tag{8.2}$$

and $\boldsymbol{U}$ is implicitly defined by

$$\boldsymbol{A}^{\dagger}\boldsymbol{U}\boldsymbol{A} = \boldsymbol{P}^{-1} + \sigma^2\boldsymbol{P}^{-1}(\boldsymbol{A}^{\dagger}\boldsymbol{A})^{-1}\boldsymbol{P}^{-1}. \tag{8.3}$$

The diagonal elements of $\boldsymbol{C}_{MU}$ give the variances of estimation errors. An equivalent but less compact expression for these variance elements has been presented in [1] for the special case of uncorrelated sources (i.e., diagonal $\boldsymbol{P}$ matrices).

(b) CRB Covariance Matrix

The Cramér-Rao lower bound on the covariance matrix of any (asymptotically) unbiased estimator of $\boldsymbol{\theta}$ is, for large N, given by

$$\boldsymbol{C}_{CR} = \frac{\sigma^2}{2N}\{\text{Re}[\boldsymbol{H} \odot \boldsymbol{P}^T]\}^{-1}. \tag{8.4}$$

A formula for the CRB matrix that is valid for any value of N was also derived in [2]. However, the commonly used estimators for $\boldsymbol{\theta}$, such as MUSIC and MLE, are complicated nonlinear functions of the data vector, and their statistical behavior for "small" N appears difficult to establish. In the study of the statistical efficiency of these estimators we will thus use the (asymptotic) formula (8.4) for the CRB.

(c) Relationship Between MUSIC and MLE

The MUSIC estimator is a large-sample (for $N >> 0$) realization of the MLE if and only if the covariance matrix $\boldsymbol{P}$ is *diagonal.*

(d) Statistical Efficiency of MUSIC

For diagonal $\boldsymbol{P}$, it holds that

$$[\boldsymbol{C}_{MU}]_{ii} \geq [\boldsymbol{C}_{CR}]_{ii} \tag{8.5}$$

where *equality* holds in the limit as m increases, if and only if

$$\boldsymbol{a}^*(\omega)\boldsymbol{a}(\omega) \to \infty, \qquad \text{as } m \to \infty. \tag{8.6}$$

For nondiagonal $\boldsymbol{P}$, $[\boldsymbol{C}_{MU}]_{ii}$ is strictly larger than $[\boldsymbol{C}_{CR}]_{ii}$.

There are a number of directions in which the results of Chapter 7 should be extended to obtain a fairly complete description of the topic under discussion:

1. A conceptually simple generalization of the MUSIC estimator is obtained by considering the following function

$$f_W(\omega) = \boldsymbol{a}^\dagger(\omega)\hat{\boldsymbol{G}}\boldsymbol{W}\hat{\boldsymbol{G}}^\dagger\boldsymbol{a}(\omega), \tag{8.7}$$

where $\boldsymbol{W}$ is a positive definite (weighting) matrix. Minimization of (8.7) with respect to ω gives the so-called *weighted MUSIC estimators* (see [3], [4]). The statistical behavior of the weighted MUSIC estimators needs to be established, and their performance compared to that of the nonweighted MUSIC.

2. The study in Chapter 7 of the statistical (in)efficiency of MUSIC was concerned only with the diagonal elements $[\boldsymbol{C}_{MU}]_{ii}$ and $[\boldsymbol{C}_{CR}]_{ii}$. It should be of interest to compare the whole covariance matrices $\boldsymbol{C}_{MU}$ and $\boldsymbol{C}_{CR}$. *Inter alia*, this comparison will provide more exact results on the statistical efficiency of the MUSIC estimator.

3. The covariance matrix of the ML estimator was not derived in Chapter 7 (except for the special case of diagonal $\boldsymbol{P}$ matrices, when $\boldsymbol{C}_{ML} = \boldsymbol{C}_{MU}$. Derivation of the covariance matrix $\boldsymbol{C}_{ML}$ of the MLE appears to be of significant interest. Among other things, the availability of an expression for $\boldsymbol{C}_{ML}$ will make it possible to study the statistical efficiency of the MLE, and to compare the performances of the MUSIC and MLE in the general case of nondiagonal $\boldsymbol{P}$ matrices.

The aim of this chapter is to provide a number of results along the lines just outlined. More exactly, in Section 8.2 we establish the covariance matrix of the *weighted MUSIC estimator* and show that the nonweighted MUSIC possesses minimum variance. Thus, we provide theoretical support to the empirically observed fact that use of a weighting matrix $\boldsymbol{W} \neq \boldsymbol{I}$ in (8.7) cannot improve the accuracy of the MUSIC estimator (see, e.g., [3]).

In Section 8.3 we prove that $\boldsymbol{C}_{MU} \geq \boldsymbol{C}_{CR}$. Furthermore, we show that for diagonal $\boldsymbol{P}$ and vectors $\boldsymbol{a}(\omega)$ satisfying (8.6), the equality $\boldsymbol{C}_{MU} = \boldsymbol{C}_{CR}$ holds in the limit, as m increases. For nondiagonal $\boldsymbol{P}$ we show that the inequality is strict and that the difference $\boldsymbol{C}_{MU} - \boldsymbol{C}_{CR}$ may be substantial if the elements of $\boldsymbol{x}(t)$ are nearly colinear. These results again provide theoretical support to a fact observed in simulations: the performance of MUSIC for diagonal $\boldsymbol{P}$ is excellent, but it degrades if $\boldsymbol{P}$ approaches a (nearly) singular matrix.

In Section 8.4 we derive the covariance matrix of the ML estimator. Also in that section we prove that $\boldsymbol{C}_{ML} \geq \boldsymbol{C}_{CR}$, and that for vectors $\boldsymbol{a}(\omega)$ that satisfy (8.6), the equality $\boldsymbol{C}_{ML} = \boldsymbol{C}_{CR}$ holds in the limit as m increases. For vectors $\boldsymbol{a}(\omega)$ that do not satisfy (8.6), the inequality $\boldsymbol{C}_{ML} \geq \boldsymbol{C}_{CR}$ is shown to be strict. These results were anticipated in [2] using simple examples and argumentation based on the general properties of the ML estimators. Here, we provide direct algebraic proofs of them.

In Section 8.5 we compare the covariance matrices $\boldsymbol{C}_{MU}$ and $\boldsymbol{C}_{ML}$ of the MUSIC and ML estimators. For diagonal $\boldsymbol{P}$ we prove that $\boldsymbol{C}_{MU} = \boldsymbol{C}_{ML}$, thus rediscovering in another way the result of [2] that asserts that MUSIC and MLE coincide asymptotically (for large N) in this case. For nondiagonal $\boldsymbol{P}$, we show that a generally valid order relation between $\boldsymbol{C}_{MU}$ and $\boldsymbol{C}_{ML}$ does not exist. Quite often, the MLE is expected to offer better performance, ***but*** in certain rare cases MUSIC can be superior to the MLE.

In Section 8.6 we present the results of a numerical study of performance, whose aim is to provide a more quantitative comparison of $\boldsymbol{C}_{MU}$, $\boldsymbol{C}_{ML}$, and $\boldsymbol{C}_{CR}$ in the case of a two-direction finding application.

Finally, in the appendices we include some useful results on the Hadamard matrix product and proofs of the theorems in the paper.

8.2 WEIGHTED AND NONWEIGHTED MUSIC

First, we derive the distribution of the weighted MUSIC estimator.

Theorem 2.1. The estimation errors $\{\hat{\omega}_k - \omega_k\}$ associated with the weighted MUSIC estimator which minimizes (8.7), are asymptotically (for large N) jointly Gaussian distributed with zero means and the following covariance matrix

$$\boldsymbol{C}_{WMU} = \frac{\sigma^2}{2N}(\bar{\boldsymbol{H}} \odot \boldsymbol{I})^{-1}\mathrm{Re}\{\tilde{\boldsymbol{H}} \odot (\boldsymbol{A}^{\dagger}\boldsymbol{U}\boldsymbol{A})^T\}(\bar{\boldsymbol{H}} \odot \boldsymbol{I})^{-1} \tag{8.8}$$

where

$$\bar{\boldsymbol{H}} = \boldsymbol{D}^{\dagger}\boldsymbol{G}\boldsymbol{W}\boldsymbol{G}^{\dagger}\boldsymbol{D} \tag{8.9a}$$

$$\tilde{\boldsymbol{H}} = \boldsymbol{D}^{\dagger}\boldsymbol{G}\boldsymbol{W}^2\boldsymbol{G}^{\dagger}\boldsymbol{D} \tag{8.9b}$$

and all the other quantities have been defined before.

Proof: See Appendix B.

Next, we show that the minimum variance in the class of weighted MUSIC estimators is achieved by the nonweighted MUSIC.

Theorem 2.2 The diagonal elements of $\boldsymbol{C}_{WMU}$ are greater than the corresponding diagonal elements of

$$\boldsymbol{C}_{MU} = \boldsymbol{C}_{WMU}|_{W = I}, \text{ or}$$

$$\boldsymbol{C}_{WMU} \odot \boldsymbol{I} \geq \boldsymbol{C}_{MU} \odot \boldsymbol{I}. \tag{8.10}$$

Proof: See Appendix C.

It is an open question whether the inequality (8.10) extends to the whole covariance matrices C_{WMU} and C_{MU}. However, the answer to this question cannot change the conclusion that use of weighting matrix $W \neq I$ in (8.7) will, for large N, worsen the accuracy of the estimates, instead of improving it.

8.3 MUSIC AND CRLB

First, we prove that the covariance matrix C_{MU} is bounded from below by the CRLB covariance matrix C_{CR}. We present a simple proof of this expected result, which has the virtue of revealing some ways of doing more accurate comparisons between C_{MU} and C_{CR}.

Theorem 3.1 The covariance matrix of the MUSIC estimator can be decomposed additively as follows

$$C_{MU} = \tilde{C}_{MU} + \bar{C}_{MU}, \tag{8.11}$$

where (see (8.1), (8.3))

$$\tilde{C}_{MU} = \frac{\sigma^2}{2N}(H \odot I)^{-1}\text{Re}[H \odot P^{-T}](H \odot I)^{-1} \tag{8.12}$$

$$\bar{C}_{MU} = \frac{\sigma^4}{2N}(H \odot I)^{-1}\text{Re}\{H \odot [P^{-1}(A^\dagger A)^{-1}P^{-1}]^T\}(H \odot I)^{-1}. \tag{8.13}$$

The matrices $\tilde{C}_{MU}$ and $\bar{C}_{MU}$ satisfy

$$\tilde{C}_{MU} \geq C_{CR} \tag{8.14}$$

$$\bar{C}_{MU} \geq 0. \tag{8.15}$$

Proof: The results follow immediately from Lemmas A.1 and A.2 in Appendix A.

Next, assume that P is *diagonal*. Then, we get

$$\tilde{C}_{MU} = \frac{\sigma^2}{2N}(H \odot I)^{-1}P^{-1} = C_{CR}. \tag{8.16}$$

If we additionally assume that $a(\omega)$ *satisfies* (8.6) then it is not difficult to see that C_{MU} tends to $\tilde{C}_{MU}$, as m increases. Indeed, under assumption (8.6), the matrix $(A^\dagger A)^{-1}$ tends to zero as $m \to \infty$, which implies that $\bar{C}_{MU}$ tends to zero faster than $\tilde{C}_{MU}$ (compare the expressions of the two matrices); thus the contribution of $\bar{C}_{MU}$ in the decomposition (8.11) vanishes as m increases. As an example, consider the case of the steering vector

$$a(\omega) = [1 \; e^{j\omega} \; \cdots \; e^{j(m-1)\omega}]^T, \tag{8.17}$$

which appears in direction estimation applications using uniform linear arrays. For (8.17), we have $\boldsymbol{a}^\dagger(\omega)\boldsymbol{a}(\omega) = m$ and, therefore, the condition (8.6) is satisfied. Some straightforward calculations show that the limits, for $m \to \infty$, of the matrices $\boldsymbol{H}$ and $(\boldsymbol{A}^\dagger\boldsymbol{A})$ corresponding to (8.17), are given by (see [2])

$$\frac{1}{m}(\boldsymbol{A}^\dagger\boldsymbol{A}) \to \boldsymbol{I} \tag{8.18a}$$

$$\frac{1}{m^3}\boldsymbol{H} \to \frac{1}{12}\boldsymbol{I}. \tag{8.18b}$$

Thus, as m increases,

$$\tilde{\boldsymbol{C}}_{MU} \to \frac{6\sigma^2}{Nm^3}\boldsymbol{I} \odot \boldsymbol{P}^{-1} \tag{8.19}$$

and

$$\bar{\boldsymbol{C}}_{MU} \to \frac{6\sigma^4}{Nm^4}\boldsymbol{I} \odot P^{-2}. \tag{8.20}$$

In conclusion, under the preceding assumptions ($\boldsymbol{P}$ is diagonal and $\boldsymbol{a}(\omega)$ satisfies (8.6)) $\boldsymbol{C}_{MU}$ approaches $\boldsymbol{C}_{CR}$ as m increases, which means that the MUSIC estimator is asymptotically (for large *N and m*) efficient in this case. This result was also obtained in [2] by different reasoning.

If either the matrix $\boldsymbol{P}$ is *nondiagonal or* the vector $\boldsymbol{a}(\omega)$ *does not satisfy (8.6)*, then $\boldsymbol{C}_{MU}$ cannot attain $\boldsymbol{C}_{CR}$, and the MUSIC estimator is statistically inefficient. This is so since the equality in (8.14) cannot hold if $\boldsymbol{P}$ is not diagonal, and $\bar{\boldsymbol{C}}_{MU}$ will be strictly bounded from zero if $\boldsymbol{a}(\omega)$ does not satisfy (8.6).

For nondiagonal $\boldsymbol{P}$, it is worth noting that the difference $\tilde{\boldsymbol{C}}_{MU} - \boldsymbol{C}_{CR}$ (and, hence, $\boldsymbol{C}_{MU} - \boldsymbol{C}_{CR}$) may be quite large if $\boldsymbol{P}$ is nearly singular. We illustrate this fact by considering once more the case of the steering vector $\boldsymbol{a}(\omega)$ given by (8.17). The corresponding matrices $\boldsymbol{C}_{MU}$ and $\boldsymbol{C}_{CR}$ behave, for large m, as follows (see (8.18)–(8.20)):

$$\boldsymbol{C}_{MU} \to \frac{6\sigma^2}{Nm^3}(\boldsymbol{I} \odot \boldsymbol{P}^{-1}) \tag{8.21}$$

$$\boldsymbol{C}_{CR} \to \frac{6\sigma^2}{Nm^3}(\boldsymbol{I} \odot \boldsymbol{P})^{-1}. \tag{8.22}$$

Thus, the difference matrix $\boldsymbol{C}_{MU} - \boldsymbol{C}_{CR}$ may indeed take very large values if $\boldsymbol{P}$ is nearly singular. As a simple example, for

$$\boldsymbol{P} = \begin{bmatrix} 1 & \rho \\ \rho & 1 \end{bmatrix}, \qquad |\rho| < 1,$$

we obtain from (8.21) and (8.22)

$$C_{MU} - C_{CR} \rightarrow \frac{6\sigma^2}{Nm^3} \frac{\rho^2}{1 - \rho^2} I,$$

which increases without bound as $|\rho|$ approaches one.

In the foregoing analysis it was assumed that the *signal-to-noise ratio (SNR)* is finite (the SNR for the ith signal is defined as P_{ii}/σ^2). If the SNRs of all signals are very large, then $\bar{C}_{MU}$ is much less than $\tilde{C}_{MU}$ and can be neglected in (8.11). If in addition P is diagonal then it follows from (8.16) that as SNR $\rightarrow \infty$, $C_{MU} \rightarrow \tilde{C}_{MU} = C_{CR}$. This observation confirms a result established in [1] by a different method. The case of a very large SNR appears, however, of little practical interest as the user may have no control on this quantity.

8.4 MLE AND CRLB

First, we derive the asymptotic (for large N) distribution of the ML estimator. Derivation of the MLE distribution from its cost function, by the usual technique of expanding the corresponding gradient equation $F'(\hat{\theta}) = 0$ in a Taylor series around the true parameter values, appears somewhat difficult. To obtain a simpler derivation we make use of a result proven in [2] which states that a large-sample (for $N >> 0$) realization of the MLE is given by the minimizer of the following function

$$h(\theta) = \text{tr}[A^{\dagger}\hat{G}\hat{G}^{\dagger}AP]. \tag{8.23}$$

Observe that $h(\theta)$ depends on θ in a simpler way than $F(\theta)$ does.

Theorem 4.1. The estimation errors $\{\hat{\omega}_i - \omega_i\}$ of the ML estimator are asymptotically (for large N) jointly Gaussian distributed, with zero means and a covariance matrix given by

$$C_{ML} = \frac{\sigma^2}{2N} [\text{Re}(H \odot P^T)]^{-1}\{\text{Re}[H \odot (PA^{\dagger}UAP)^T]\}[\text{Re}(H \odot P^T)]^{-1} \tag{8.24}$$

(All the quantities appearing in (8.24) have been defined previously.)

Proof. See Appendix D.

Next, we show a simple relationship between the covariance matrix of the MLE and the CRB.

Theorem 4.2. The covariance matrix C_{ML} can be decomposed additively as follows:

$$C_{ML} = C_{CR} + \bar{C}_{ML}, \tag{8.25}$$

where

$$\bar{C}_{ML} = \frac{\sigma^4}{2N}[\mathrm{Re}(H \odot P^T)]^{-1}\{\mathrm{Re}[H \odot (A^\dagger A)^{-T}]\}[\mathrm{Re}(H \odot P^T)]^{-1}. \tag{8.26}$$

Proof. Inserting the expression (8.3) of $A^\dagger UA$ into (8.24), we readily get the additive decomposition (8.25), (8.26).

The simple result of Theorem 4.2 has several immediate implications.

1. Since the matrix $\bar{C}_{ML}$ is positive semidefinite (see, Lemma A.1), it follows that $C_{ML} \geq C_{CR}$ (as expected).
2. For vectors $a(\omega)$ that do not satisfy (8.6) and for finite SNR, $\bar{C}_{ML}$ is strictly bounded from zero, and, therefore, the MLE cannot achieve the CRB.
3. For vectors $a(\omega)$ that satisfy (8.6), $\bar{C}_{ML}$ goes to zero faster than C_{CR} when m increases (compare the expressions of these two matrices). Thus, in this case the MLE is asymptotically (for large N *and* m) statistically efficient. Note, however, that for "small or medium" values of m (and finite SNR), $\bar{C}_{ML}$ is different from zero and, thus, the MLE is statistically inefficient. Also note that C_{CR} is inversely proportional to SNR, while $\bar{C}_{ML}$ is inversely proportional to (SNR)^2. Thus, for a given finite m, the C_{ML} approaches the C_{CR} as the signal-to-noise ratio increases. These properties of the MLE are similar to the analogous properties of the MUSIC but here they hold for arbitrary P matrices.

The results reinforce some conclusions of the analysis of the MLE, obtained in Chapter 7 by less direct argumentation. The statistical inefficiency of the MLE in the case of finite m (and finite SNR), appears to be a rather unusual result that leaves open the question whether estimators better than MLE could exist. In this light, comparison between the MLE and the (much) simpler computationally MUSIC estimator becomes of significant interest. This aspect is addressed in the next section.

8.5 MUSIC AND MLE

For *diagonal* P matrices we simply have

$$C_{ML} = \frac{\sigma^2}{2N}(H \odot I)^{-1}\mathrm{Re}[H \odot (A^\dagger UA)^T](H \odot I)^{-1} = C_{MU}. \tag{8.27}$$

The equality between C_{ML} and C_{MU} was expected in view of the fact that, for diagonal P, the MLE and MUSIC are known to coincide as N increases [2] (we stress that the equality (8.27) holds for all $m > n$).

Next, consider the more general case of *nondiagonal* P matrices. If either the SNR is "large enough" or the vector $a(\omega)$ satisfies (8.6) and m is "sufficiently large," then we have $C_{ML} = C_{CR} < C_{MU}$. However, in all other cases $C_{ML} > C_{CR}$, which means that the inequality $C_{ML} \geq C_{MU}$ is not a priori excluded. In fact, this inequality can really hold true, as we show in the following.

From (8.11) and (8.25), we get

$$\frac{2N}{\sigma^2}(C_{MU} - C_{ML}) = \frac{2N}{\sigma^2}(\tilde{C}_{MU} - C_{CR}) + \frac{2N}{\sigma^2}(\overline{C}_{MU} - \overline{C}_{ML}). \tag{8.28}$$

The first term on the right side of (8.28) is positive semidefinite and independent of σ^2, while the second term is proportional to σ^2. We thus conclude that a sufficient *and* necessary condition for the inequality $C_{MU} \geq C_{ML}$ to hold is

$$\overline{C}_{MU} \geq \overline{C}_{ML}. \tag{8.29}$$

The sufficiency of (8.29) is evident. The necessity follows from the observation that the second term in (8.28) is proportional to σ^2, while the first does not depend on σ^2. Thus, if (8.29) does not hold then one can choose σ^2 sufficiently large such that $C_{MU} \geq C_{ML}$ does not hold either. In the following we provide a simple counterexample to (8.29).

Let $P = (A^\dagger A)^{-1/2}$ [the (Hermitian) positive definite square root of the inverse matrix $(A^\dagger A)^{-1}$].* Then we obtain

$$\overline{C}_{MU} = \frac{\sigma^4}{2N}(H \odot I)^{-1}$$

and

$$\overline{C}_{ML} = \frac{\sigma^4}{2N}[\text{Re}(H \odot P^T)]^{-1}\,[\text{Re}(H \odot P^T P^T)]\,[\text{Re}(H \odot P^T)]^{-1},$$

which satisfy the following inequality (see Lemma A.2 in Appendix A)

$$\overline{C}_{ML} \geq \overline{C}_{MU}.$$

Thus, for the choice of the matrix P (which is special but is not peculiar), the MUSIC estimator will be more accurate than the MLE for "sufficiently large" values of σ^2! From a theoretical standpoint, this result is of significant interest (we are not aware of a similar result in the literature, for an estimation problem of the complexity of that treated here). From a practical standpoint, the result makes a detailed study of the performances of MUSIC and MLE, of considerable interest. In the next section we present a numerical study of performance for the case of a two-direction finding application.

---

*Let H be a (Hermitian) positive definite matrix. Then we can write $H = Q^\dagger \Lambda Q$, where Q is unitary and Λ is a positive definite real-valued diagonal matrix. Define $P = Q^\dagger \Lambda^{1/2} Q$ and observe that $P^2 = H$. Thus, P is a (Hermitian) positive definite square root of H.

8.6 MUSIC, MLE, AND CRLB: A NUMERICAL COMPARISON

Consider the situation of two narrow-band plane waves impinging on a uniform linear array of m sensors. Then $n = 2$ and $\boldsymbol{a}(\omega)$ is given by (8.17) (see [3], [5], [4]). Let

$$\boldsymbol{P} = \begin{bmatrix} 1 & \rho \\ \rho & 1 \end{bmatrix}, \qquad |\rho| < 1.$$

The variance elements $[\boldsymbol{C}_{MU}]_{11} = [\boldsymbol{C}_{MU}]_{22}$, $[\boldsymbol{C}_{ML}]_{11} = [\boldsymbol{C}_{ML}]_{22}$ and $[\boldsymbol{C}_{CR}]_{11} = [\boldsymbol{C}_{CR}]_{22}$ have been evaluated for several values of ρ, m, and σ, and for varying (ω_1, ω_2). Note that in the present case these variance elements depend on $\Delta\omega = |\omega_1 - \omega_2|$ only. Thus they have been evaluated for $\Delta\omega \in (0, \pi)$. However, only the variance values for $\Delta\omega \in (0, 0.2)$ will be shown since the case of practical interest is of closely spaced sources.

Figure 8.1 shows the results obtained as outlined. These results confirm that for highly correlated sources, the performance of MUSIC degrades significantly compared to MLE and CRLB; whereas for weakly correlated sources, MUSIC and MLE provide similar performance that is close to the CRLB for reasonably high SNR. Figure 8.1 also shows the extent to which a degradation of estimation accuracy caused by a decrease in SNR or $\Delta\omega$ may be compensated for by an increase in m. This aspect is further studied in Fig. 8.2 where the ratios

$$\text{st. dev. (CRLB)/st. dev. (MUSIC)} \quad \text{and} \quad \text{st. dev. (CRLB)/st. dev. (ML)} \tag{8.30}$$

are plotted versus $\Delta\omega$, for $\rho = 0.9$, $\sigma = 0.1$ (SNR $= 10$ dB) and several values of m. It is seen from this figure that ML achieves CRLB as m increases (the smaller $\Delta\omega$ (or SNR), the larger m required for MLE to achieve CRLB). However, MUSIC variance for correlated sources ($\rho \neq 0$) cannot achieve CRLB by increasing m and, in fact, MUSIC statistical efficiency does not necessarily increase when m increases.

8.7 CONCLUSIONS

The present chapter and its companion chapter 7 provide a farily complete picture on the performance and statistical efficiency of the MUSIC and ML estimators. The main contributions of this chapter were outlined in Section 8.1 and will not be repeated herein. In this concluding section we only reemphasize the importance of the explicit expressions for the covariance matrices of the ML and MUSIC estimators and of the CRB, derived in Chapters 7 and 8. Using these expressions we were able to perform general analytic comparisons between the methods under discussion, as well as rapid numerical evaluations of performance in specific situations. A main result that emerged from the present study concerns the fact that for small or medium values of m and SNR (which is the case of practical interest), the MLE is *not* the most accurate estimator. This fact opens the possibility that other estimators which are computationally simpler *and* statistically more accurate than the MLE could exist.

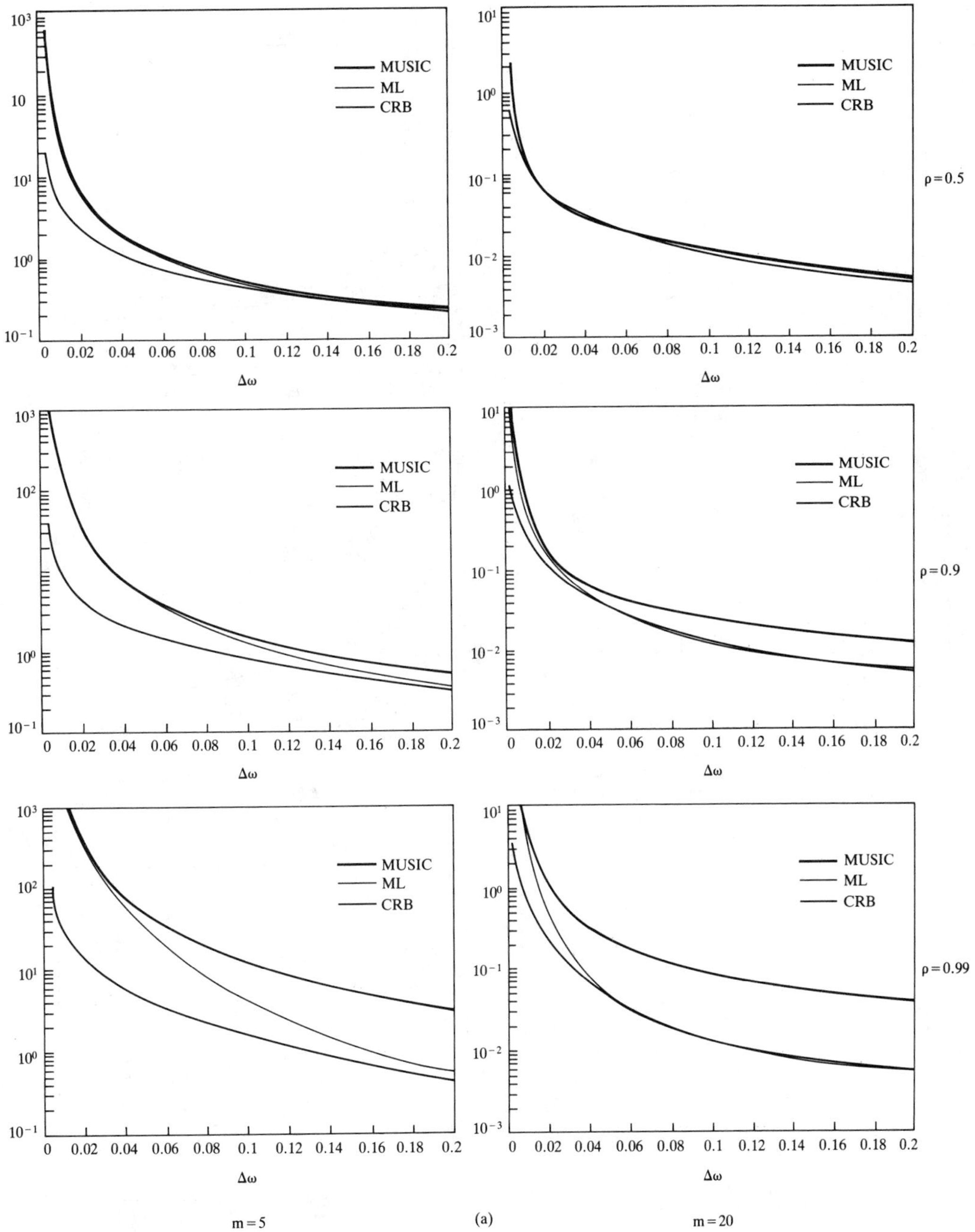

Figure 8.1 Comparison of the estimation error standard deviations of MUSIC, MLE, and CRLB for the shown values of ρ, m, and SNR, and for varying $\Delta\omega$. (a) $\sigma^2 = 0.01$ (SNR = 20 dB), (b) $\sigma^2 = 1$ (SNR = 0 dB).

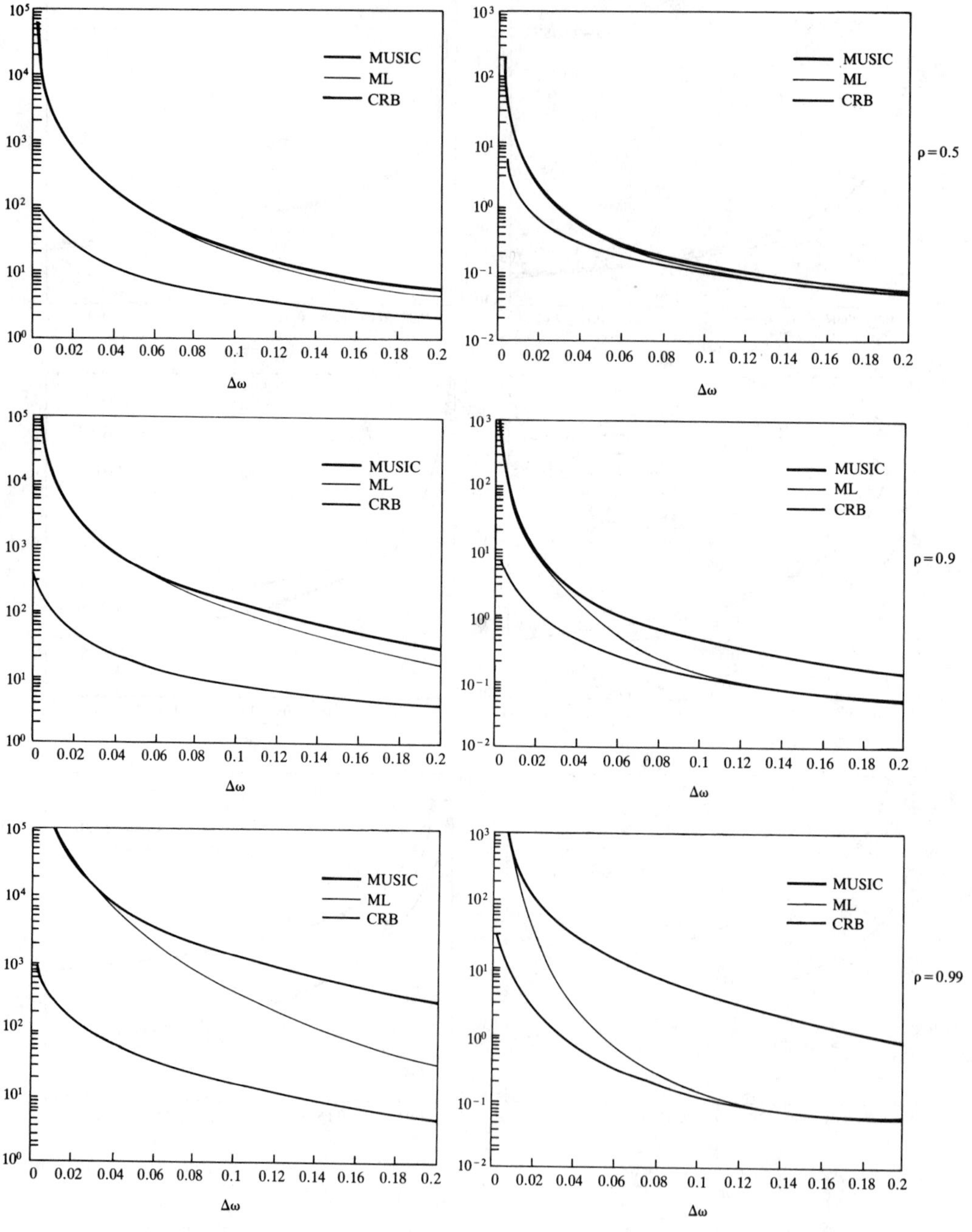

m=5 (b) m=20

Figure 8.1 continued

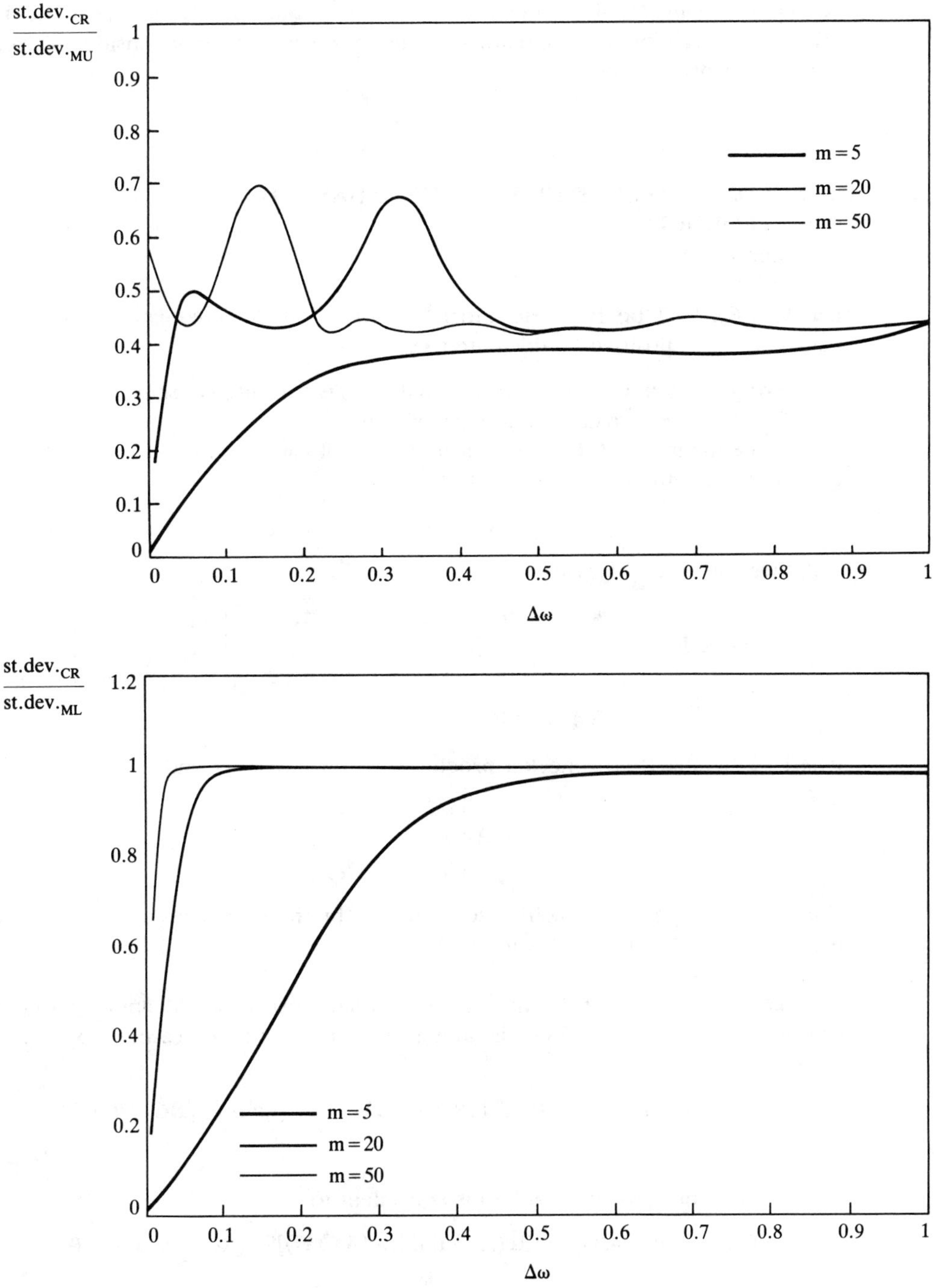

Figure 8.2 The ratios (8.30) for $\rho = 0.9$, $\sigma^2 = 0.1$ (SNR = 10 dB); $m = 5$, 20, and 50; and varying $\Delta\omega$.

(Computationally, MUSIC is such a simple estimator but is rarely more accurate than MLE. The search for such estimators remains a research topic of considerable practical and theoretical interest.

APPENDIX A: SOME USEFUL RESULTS ON THE HADAMARD PRODUCT

Lemma A.1

Let A, $B \in C^{n \times n}$ be two (Hermitian) positive semidefinite matrices. Then the matrix $A \odot B$ is positive semidefinite too.

Proof: This result (for the real-valued case) is attributed to Schur (see [6]). For completeness, we provide a simple proof of it.

Since the matrix B is positive semidefinite, it can be written as $B = W^{\dagger}W$. Let ω_k denote the kth column of W. Then

$$(A \odot B)_{ij} = A_{ij}\omega_i^{\dagger}\omega_j$$

and, therefore, we can write

$$\begin{aligned}(A \odot B) &= \begin{bmatrix} w_1^* & & 0 \\ & \ddots & \\ 0 & & w_n^* \end{bmatrix} \begin{bmatrix} A_{11}I & \cdots & A_{1n}I \\ \vdots & & \vdots \\ A_{n1}I & \cdots & A_{nn}I \end{bmatrix} \begin{bmatrix} w_1 & & 0 \\ & \ddots & \\ 0 & & w_n \end{bmatrix} \\ &= \overline{W}^{\dagger}(A \otimes I)\overline{W},\end{aligned}$$

where $\otimes$ denotes the Kronecker product (see, e.g., [6], [7]), and

$$\overline{W} \triangleq \begin{bmatrix} w_1 & & 0 \\ & \ddots & \\ 0 & & w_n \end{bmatrix}.$$

Since the matrix $A \otimes I$ is positive semidefinite (by the properties of the Kronecker product, e.g., [6]), the proof is finished.

Lemma A.2 Let A, B, and C be (Hermitian) positive semidefinite matrices (B can be Hermitian only). Then, assuming that the inverses appearing next exist, it holds that

$$[\mathrm{Re}(A \odot B)]^{-1}\,[\mathrm{Re}(A \odot C)]\,[\mathrm{Re}(A \odot B)]^{-1} \geq \{\mathrm{Re}[A \odot (BC^{-1}B)]\}^{-1}. \tag{8.A.1}$$

Proof: The inequality (8.A.1) is equivalent to

$$\mathrm{Re}[A \odot (BC^{-1}B)] - [\mathrm{Re}(A \odot B)]\,[\mathrm{Re}(A \odot C)]^{-1}\,[\mathrm{Re}(A \odot B)] \geq 0,$$

which, in turn, is equivalent to

$$\mathrm{Re}\begin{bmatrix} A \odot BC^{-1}B & A \odot B \\ A \odot B & A \odot C \end{bmatrix} = \mathrm{Re}\left\{\begin{bmatrix} A & A \\ A & A \end{bmatrix} \odot \begin{bmatrix} BC^{-1}B & B \\ B & C \end{bmatrix}\right\} \geq 0.$$

Since the matrices

$$\begin{bmatrix} A & A \\ A & A \end{bmatrix} = \begin{bmatrix} I \\ I \end{bmatrix} A[I \quad I]$$

and

$$\begin{bmatrix} BC^{-1}B & B \\ B & C \end{bmatrix} = \begin{bmatrix} BC^{-1} \\ I \end{bmatrix} C[C^{-1}B \quad I]$$

are both positive semidefinite, the assertion of the lemma follows from Lemma A.1 and the fact that $\mathrm{Re}(H) \geq 0$ if $H \geq 0$.

Corollary. Let P be a (Hermitian) positive definite matrix. Then

$$(I \odot P^{-1}) \geq (I \odot P)^{-1}.$$

Proof: Set $A = B = I$ and $C = P^{-1}$ in Lemma A.2.

APPENDIX B: PROOF OF THEOREM 2.1

To prove the theorem, we will use the following result proved in Chapter 7 (also, see [3]).

Lemma B.1

The orthogonal projections of $\{\hat{g}_i\}$ onto the column space of S are asymptotically (for large N) jointly Gaussian distributed with zero means and the following variances-covariances

$$E\left\{(SS^\dagger\hat{g}_i)(SS^\dagger\hat{g}_j)^\dagger\right\} = \frac{\sigma^2}{N}U\delta_{ij}$$

$$E\left\{(SS^\dagger\hat{g}_i)(SS^\dagger\hat{g}_j)^T\right\} = 0, \qquad \text{for all } i, j,$$

where the matrix U is defined (implicitly) by (8.3).

Now we turn to the proof of Theorem 2.1, which will closely follow the proof for the nonweighted case in chapter 7. As $\hat{\omega}_i$ is a minimum point of $f_W(\omega)$, we must have

$$f'_W(\hat{\omega}_i) = 2\mathrm{Re}[a^\dagger(\hat{\omega}_i)\hat{G}W\hat{G}^\dagger d(\hat{\omega}_i)] = 0.$$

Since $\hat{\omega}_i$ is a consistent estimate of ω_i ([2], [3]) we can write (for large N)

$$
\begin{aligned}
0 = f'_{\boldsymbol{W}}(\hat{\omega}_i) &\simeq f'_{\boldsymbol{W}}(\omega_i) + f''_{\boldsymbol{W}}(\omega_i)(\hat{\omega}_i - \omega_i) \\
&= 2\mathrm{Re}[\boldsymbol{a}^{\dagger}(\omega_i)\hat{\boldsymbol{G}}\boldsymbol{W}\hat{\boldsymbol{G}}^{\dagger}\boldsymbol{d}(\omega_i)] + 2\mathrm{Re}[\boldsymbol{d}^{\dagger}(\omega_i)\hat{\boldsymbol{G}}\boldsymbol{W}\hat{\boldsymbol{G}}^{\dagger}\boldsymbol{d}(\omega_i) \\
&\quad + \boldsymbol{a}^{\dagger}(\omega_i)\hat{\boldsymbol{G}}\boldsymbol{W}\hat{\boldsymbol{G}}^{\dagger}\boldsymbol{d}'(\omega_i)](\hat{\omega}_i - \omega_i) \simeq 2\mathrm{Re}[\boldsymbol{a}^{\dagger}(\omega_i)\hat{\boldsymbol{G}}\boldsymbol{W}\boldsymbol{G}^{\dagger}\boldsymbol{d}(\omega_i)] \\
&\quad + 2[\boldsymbol{d}^{\dagger}(\omega_i)\boldsymbol{G}\boldsymbol{W}\boldsymbol{G}^{\dagger}\boldsymbol{d}(\omega_i)](\hat{\omega}_i - \omega_i).
\end{aligned}
\tag{8.B.1}
$$

Remark: The replacement of $\hat{\boldsymbol{G}}^{\dagger}$ by $\boldsymbol{G}^{\dagger}$ in the foregoing calculation is a subtle issue discussed in some detail in chapter 7. The complication is caused by the fact that $\hat{\boldsymbol{G}}$ is unique (with probability one), whereas $\boldsymbol{G}$ is not (any multiplication of $\boldsymbol{G}$ on the right by a unitary matrix produces another set of noise eigenvectors, i.e., another $\boldsymbol{G}$ matrix). Thus, the problem is which $\boldsymbol{G}$ should be used instead of $\hat{\boldsymbol{G}}$. Consider the orthogonal projection of $\hat{\boldsymbol{G}}$ on the subspace spanned by the columns of $\boldsymbol{G} : \boldsymbol{G}(\boldsymbol{G}^{\dagger}\hat{\boldsymbol{G}})$. The distribution of $\boldsymbol{G}^{\dagger}\hat{\boldsymbol{G}}$ can be shown to be independent of the distribution of the other terms in (8.B.1) (see [2]). Thus, the matrix $\boldsymbol{G}^{\dagger}\hat{\boldsymbol{G}}$ can be treated as given, and since it is asymptotically unitary (as can be readily verified, see [2]), it follows that asymptotically $\boldsymbol{G}(\boldsymbol{G}^{\dagger}\hat{\boldsymbol{G}})$ forms a matrix of noise eigenvectors. This matrix can thus be redenoted by $\boldsymbol{G}$ and used in (8.B.1) instead of $\hat{\boldsymbol{G}}$. Note that the difference $\hat{\boldsymbol{G}} - \boldsymbol{G}\boldsymbol{G}^{\dagger}\hat{\boldsymbol{G}} = (\boldsymbol{I} - \boldsymbol{G}\boldsymbol{G}^{\dagger})\hat{\boldsymbol{G}} = \boldsymbol{S}\boldsymbol{S}^{\dagger}\hat{\boldsymbol{G}}$ tends to zero as $N \to \infty$.

Next note that

$$
\begin{aligned}
\boldsymbol{a}^{\dagger}(\omega_i)\hat{\boldsymbol{G}}\boldsymbol{W}\boldsymbol{G}^{\dagger}\boldsymbol{d}(\omega_i) &= \boldsymbol{a}^{\dagger}(\omega_i)\boldsymbol{S}\boldsymbol{S}^{\dagger}\hat{\boldsymbol{G}}\boldsymbol{W}\boldsymbol{G}^{\dagger}\boldsymbol{d}(\omega_i) \\
&= \sum_{k=1}^{m-n} [\tilde{\boldsymbol{g}}_k^{\dagger}\boldsymbol{d}(\omega_i)][\boldsymbol{a}^{\dagger}(\omega_i)\boldsymbol{S}\boldsymbol{S}^{\dagger}\hat{\boldsymbol{g}}_k]
\end{aligned}
\tag{8.B.2}
$$

where $[\tilde{\boldsymbol{g}}_1 \cdots \tilde{\boldsymbol{g}}_{m-n}] \triangleq \boldsymbol{G}\boldsymbol{W}$. Thus, from (8.B.1) and (8.B.2) we obtain (neglecting the higher-order terms)

$$
(\hat{\omega}_i - \omega_i) = -\mathrm{Re}\left\{\sum_{k=1}^{m-n} [\tilde{\boldsymbol{g}}_k^{\dagger}\boldsymbol{d}(\omega_i)][\boldsymbol{a}^{\dagger}(\omega_i)\boldsymbol{S}\boldsymbol{S}^{\dagger}\hat{\boldsymbol{g}}_k]\right\} \Big/ (\overline{\boldsymbol{H}} \odot \boldsymbol{I})_{ii}. \tag{8.B.3}
$$

The asymptotic zero-mean joint Gaussian distribution of $\{\hat{\omega}_i - \omega_i\}$ follows from (8.B.3) and Lemma B.1. To evaluate the covariance matrix of the distribution, we make use of the following simple result: for two scalar variables u and v, it holds that

$$
\mathrm{Re}(u) \cdot \mathrm{Re}(v) = \frac{1}{2}[\mathrm{Re}(uv) + \mathrm{Re}(uv^{\dagger})].
$$

Let

$$
u = \sum_{k=1}^{m-n} [\tilde{\boldsymbol{g}}_k^{\dagger}\boldsymbol{d}(\omega_i)][\boldsymbol{a}^{\dagger}(\omega_i)\boldsymbol{S}\boldsymbol{S}^{\dagger}\hat{\boldsymbol{g}}_k]
$$

and let v be defined similarly to u but with ω_i replaced by ω_j. Using Lemma B.1, we obtain

$$E\{uv\} = \sum_{k=1}^{m-n}\sum_{p=1}^{m-n} [\tilde{\boldsymbol{g}}_k^\dagger \boldsymbol{d}(\omega_i)][\tilde{\boldsymbol{g}}_p^\dagger \boldsymbol{d}(\omega_j)] \cdot E\{[\boldsymbol{a}^\dagger(\omega_i)\boldsymbol{S}\boldsymbol{S}^\dagger \hat{\boldsymbol{g}}_k] \cdot [\boldsymbol{a}^\dagger(\omega_j)\boldsymbol{S}\boldsymbol{S}^\dagger \hat{\boldsymbol{g}}_p]^T\} = 0$$

and

$$\begin{aligned} E\{uv^\dagger\} &= \sum_{k=1}^{m-n}\sum_{p=1}^{m-n} [\boldsymbol{d}^\dagger(\omega_j)\tilde{\boldsymbol{g}}_p \tilde{\boldsymbol{g}}_k^\dagger \boldsymbol{d}(\omega_i)][\boldsymbol{a}^\dagger(\omega_i) E\{(\boldsymbol{S}\boldsymbol{S}^\dagger \hat{\boldsymbol{g}}_k)(\boldsymbol{S}\boldsymbol{S}^\dagger \hat{\boldsymbol{g}}_p)^\dagger\}\boldsymbol{a}(\omega_j)] \\ &= \frac{\sigma^2}{N}\sum_{k=1}^{m-n} \boldsymbol{d}^\dagger(\omega_j)\tilde{\boldsymbol{g}}_k \tilde{\boldsymbol{g}}_k^\dagger \boldsymbol{d}(\omega_i)\boldsymbol{a}^\dagger(\omega_i)\boldsymbol{U}\boldsymbol{a}(\omega_j) \\ &= \frac{\sigma^2}{N}[\boldsymbol{d}^\dagger(\omega_j)\boldsymbol{G}\boldsymbol{W}^2\boldsymbol{G}^\dagger \boldsymbol{d}(\omega_i)][\boldsymbol{a}^\dagger(\omega_i)\boldsymbol{U}\boldsymbol{a}(\omega_j)]. \end{aligned}$$

Thus, we have

$$[\boldsymbol{C}_{WMU}]_{ij} = \frac{\sigma^2}{2N}\mathrm{Re}[\tilde{H}_{ij}(\boldsymbol{A}^\dagger\boldsymbol{U}\boldsymbol{A})_{ji}] \Big/ [(\bar{\boldsymbol{H}} \odot \boldsymbol{I})_{ii}(\bar{\boldsymbol{H}} \odot \boldsymbol{I})_{jj}],$$

and the proof is complete.

APPENDIX C: PROOF OF THEOREM 2.2

Since

$$\boldsymbol{C}_{WMU} \odot \boldsymbol{I} - \boldsymbol{C}_{MU} \odot \boldsymbol{I} = \frac{\sigma^2}{2N}\{[(\bar{\boldsymbol{H}} \odot \boldsymbol{I})^{-1}(\tilde{\boldsymbol{H}} \odot \boldsymbol{I})(\bar{\boldsymbol{H}} \odot \boldsymbol{I})^{-1} - (\boldsymbol{H} \odot \boldsymbol{I})^{-1}] \odot (\boldsymbol{A}^\dagger\boldsymbol{U}\boldsymbol{A})^T\},$$

the inequality (8.10) will follow from Lemma A.1 if we can prove that

$$(\boldsymbol{I} \odot \bar{\boldsymbol{H}})^{-1}(\boldsymbol{I} \odot \tilde{\boldsymbol{H}})(\boldsymbol{I} \odot \bar{\boldsymbol{H}})^{-1} \geq (\boldsymbol{I} \odot \boldsymbol{H})^{-1},$$

or, equivalently,

$$(\boldsymbol{I} \odot \bar{\boldsymbol{H}})(\boldsymbol{I} \odot \tilde{\boldsymbol{H}})^{-1}(\boldsymbol{I} \odot \bar{\boldsymbol{H}}) \leq (\boldsymbol{I} \odot \boldsymbol{H}),$$

or yet

$$\begin{aligned} \begin{bmatrix} \boldsymbol{I} \odot \boldsymbol{H} & \boldsymbol{I} \odot \bar{\boldsymbol{H}} \\ \boldsymbol{I} \odot \bar{\boldsymbol{H}} & \boldsymbol{I} \odot \tilde{\boldsymbol{H}} \end{bmatrix} &= \begin{bmatrix} \boldsymbol{I} & \boldsymbol{I} \\ \boldsymbol{I} & \boldsymbol{I} \end{bmatrix} \odot \begin{bmatrix} \boldsymbol{H} & \bar{\boldsymbol{H}} \\ \bar{\boldsymbol{H}} & \tilde{\boldsymbol{H}} \end{bmatrix} \\ &= \left\{\begin{bmatrix} \boldsymbol{I} \\ \boldsymbol{I} \end{bmatrix} [\boldsymbol{I} \;\; \boldsymbol{I}]\right\} \odot \left\{\begin{bmatrix} \boldsymbol{D}^\dagger\boldsymbol{G} \\ \boldsymbol{D}^\dagger\boldsymbol{G}\boldsymbol{W} \end{bmatrix} [\boldsymbol{G}^\dagger\boldsymbol{D} \quad \boldsymbol{W}\boldsymbol{G}^\dagger\boldsymbol{D}]\right\} \geq 0 \end{aligned} \tag{8.C.1}$$

However, using Lemma A.1 once more we conclude readily that (8.C.1) holds, and the proof is finished.

APPENDIX D: PROOF OF THEOREM 4.1

It follows from (8.23) and the consistency of the MLE [2] that, for large N, we can write (neglecting higher-order terms)

$$0 = h'(\hat{\boldsymbol{\theta}}) = \mathrm{h}'(\boldsymbol{\theta}) + h''(\boldsymbol{\theta})(\hat{\boldsymbol{\theta}} - \boldsymbol{\theta}). \tag{8.D.1}$$

Let p_{ij} denote the i, j element and $\boldsymbol{P}_i$ the ith column of the matrix $\boldsymbol{P}$. Some straightforward calculations then give,

$$\begin{aligned}\frac{\partial h(\boldsymbol{\theta})}{\partial \omega_i} &= \mathrm{tr}\left[\frac{\partial \boldsymbol{A}^\dagger}{\partial \omega_i}\hat{\boldsymbol{G}}\hat{\boldsymbol{G}}^\dagger \boldsymbol{A}\boldsymbol{P}\right] + \mathrm{tr}\left[\boldsymbol{P}\boldsymbol{A}^\dagger\hat{\boldsymbol{G}}\hat{\boldsymbol{G}}^\dagger\frac{\partial \boldsymbol{A}}{\partial \omega_i}\right] \\ &= 2\mathrm{Re}\{\boldsymbol{d}^\dagger(\omega_i)\hat{\boldsymbol{G}}\hat{\boldsymbol{G}}^\dagger\boldsymbol{A}\boldsymbol{P}_i\}\end{aligned} \tag{8.D.2}$$

and

$$\begin{aligned}\frac{\partial^2 h(\boldsymbol{\theta})}{\partial \omega_i \partial \omega_j} = {} & 2\mathrm{Re}\{\boldsymbol{d}^\dagger(\omega_i)\hat{\boldsymbol{G}}\hat{\boldsymbol{G}}^\dagger\boldsymbol{d}(\omega_j)p_{ji}\} \\ & +2\mathrm{Re}\left\{\frac{\partial \boldsymbol{d}^\dagger(\omega_i)}{\partial \omega_j}\hat{\boldsymbol{G}}\hat{\boldsymbol{G}}^\dagger\boldsymbol{A}\boldsymbol{P}_i\right\}.\end{aligned} \tag{8.D.3}$$

The sample projector matrix $\hat{\boldsymbol{G}}\hat{\boldsymbol{G}}^\dagger$ that appears in the expression of the Hessian matrix $\boldsymbol{h}''(\boldsymbol{\theta})$ can be replaced by the true projector matrix $\boldsymbol{G}\boldsymbol{G}^\dagger$, without affecting the dominant term in (8.D.1). This observation implies that the second term in (8.D.3) can be neglected, since $\boldsymbol{G}^\dagger\boldsymbol{A} = \boldsymbol{0}$ (as can readily be verified).

Similarly, we can replace the first $\hat{\boldsymbol{G}}$ in (8.D.2) by $\boldsymbol{G}$ (see the remark after (8.B.1). This replacement will only introduce a higher-order term in (8.D.1), since

$$\hat{\boldsymbol{G}}^\dagger\boldsymbol{A} = \boldsymbol{G}^\dagger\boldsymbol{A} + (\hat{\boldsymbol{G}} - \boldsymbol{G})^\dagger\boldsymbol{A} = (\hat{\boldsymbol{G}} - \boldsymbol{G})^\dagger\boldsymbol{A}.$$

Using the observation, we get from (8.D.1)–(8.D.3)

$$\hat{\boldsymbol{\theta}} - \boldsymbol{\theta} = -[\mathrm{Re}(\boldsymbol{H} \odot \boldsymbol{P}^T)]^{-1}\mathrm{Re}(\boldsymbol{\mu}), \tag{8.D.4}$$

where the ith component of the vector $\boldsymbol{\mu}$ is given by

$$\mu_i = [(\boldsymbol{A}\boldsymbol{P})_i^\dagger\hat{\boldsymbol{G}}\boldsymbol{G}^\dagger\boldsymbol{d}(\omega_i)] \tag{8.D.5}$$

(here $(\boldsymbol{AP})_i$ denotes the ith column of the matrix $\boldsymbol{AP}$).

Next, observe that μ_i is similar to the quantity (8.B.2) (with $\boldsymbol{W} = \boldsymbol{I}$) in Appendix B. This similarity and the calculations in Appendix B can be exploited to conclude readily that $\mathrm{Re}(\boldsymbol{\mu})$ is asymptotically (for large N) Gaussian distributed with zero mean and the following covariance matrix

$$\mathrm{E}\{\mathrm{Re}(\boldsymbol{\mu}) \cdot \mathrm{Re}(\boldsymbol{\mu}^\dagger)\} = \frac{\sigma^2}{2N}\mathrm{Re}\{\boldsymbol{H} \odot (\boldsymbol{P}\boldsymbol{A}^\dagger\boldsymbol{U}\boldsymbol{A}\boldsymbol{P})^T\}. \tag{8.D.6}$$

The assertion of the theorem now follows from (8.D.4) and (8.D.6).

ACKNOWLEDGMENT

The work of A. Nehorai was supported by the Air Force Office of Scientific Research, under Grant. No. AFOSR-88-0080.

REFERENCES

1. B. Porat and B. Friedlander, "Analysis of the Asymptotic Relative Efficiency of the MUSIC Algorithm," *IEEE Trans. Acoustics, Speech, and Signal Processing,* **36** (April 1988), pp. 532–544.
2. P. Stoica and A. Nehorai, "Statistical Efficiency Study of Direction Estimation Methods," Part I: "Analysis of MUSIC and Preliminary Study of MLM," Chapter 7 of this book
3. K. Sharman and T. S. Durrani, "A Comparative Study of Modern Eigenstructure Methods for Bearing Estimation — A New High Performance Approach," *Proc. 25th IEEE Conf. Dec. Contr.,* Athens, Greece, December 1986, pp. 1737–1742.
4. D. Johnson, "The Application of Spectral Estimation Methods to Bearing Estimation Problems," *Proc. IEEE,* **70** (September 1982), pp. 1018–1028.
5. H. Wang and M. Kaveh, "On the Performance of Signal- Subspace Processing–Part I: Narrow-Band Systems," *IEEE Trans. Acoustics, Speech, and Signal Processing,* **34**, (October 1986), pp. 1201–1209.
6. R. Bellman, *Introduction to Matrix Analysis,* 2nd ed. (New York: McGraw-Hill, 1972).
7. T. Söderström and P. Stoica, *System Identification* (London: Prentice-Hall, 1989).
8. R. O. Schmidt, "Multiple Emitter Location and Signal Parameter Estimation," *Proc. RADC Spectral Estimation Workshop,* Rome, NY, pp. 243–258, 1979.
9. G. Bienvenu and L. Kopp, "Optimality of High Resolution Array Processing Using the Eigensystem Approach," *IEEE Trans. Acoust., Speech, Signal Processing,* Vol. ASSP-31, pp. 1235–1248, Oct. 1983.
10. Y. Bresler and A. Macovski, "Exact Maximum Likelihood Parameter Estimation of Superimposed Exponential Signals in Noise," *IEEE Trans. Acoust., Speech, Signal Processing,* Vol. ASSP-34, pp. 1081–1089, Oct. 1986.
11. R. Kumaresan, L. L. Scharf and A. K. Shaw, "An Algorithm for Pole-Zero Modeling and Spectral Analysis," *IEEE Trans. Acoust., Speech, Signal Processing,* Vol. ASSP-34, pp. 637–640, June 1986.
12. D. Starer and A. Nehorai, "Maximum Likelihood Estimation of Exponential Signals in Noise Using a Newton Algorithm," *The 4th ASSP Workshop on Spectrum Estimation and Modeling,* pp. 240–245, Minneapolis, MN, Aug. 1988.

9

Estimation of Direction of Arrival of Signals: Asymptotic Results

Z. D. Bai, B. Q. Miao, and C. Radhakrishna Rao

9.1. INTRODUCTION

The problem of estimation of the directions of arrivals of signals from a number of sources has been considered by several of authors, the latest in the series being papers by Wax, Shan, and Kailath [1] and Paulraj and Kailath [2]. The reader is referred to these papers and Chapters 1–8 of this book for a list of earlier papers on the subject.

The problem considered by these authors is based on the model

$$\boldsymbol{x}(t) = \boldsymbol{A}\boldsymbol{s}(t) + \boldsymbol{n}(t), \qquad t = 1, \ldots, N, \tag{9.1}$$

where at time t,

$\boldsymbol{x}(t)$ is a p-complex vector of observations received by p sensors uniformly spaced
$\boldsymbol{s}(t)$ is a q-complex vector of unobservable signals emitted from q sources
$\boldsymbol{n}(t)$ is the additive noise p-vector

It is assumed that $[\boldsymbol{n}(t), \boldsymbol{s}(t)]$, $t = 1, \ldots, N$, are identically and independently distributed random variables, such that $\boldsymbol{s}(t)$ and $\boldsymbol{n}(t)$ are independent and

$$\begin{aligned} E\{\boldsymbol{n}(t)\} &= 0, \qquad E\{\boldsymbol{n}(t)\boldsymbol{n}(t)^{\dagger}\} = \sigma^2 \boldsymbol{I} \\ E\{\boldsymbol{s}(t)\} &= 0, \qquad E\{\boldsymbol{s}(t)\boldsymbol{s}(t)^{\dagger}\} = \Gamma \end{aligned} \tag{9.2}$$

where † indicates complex conjugate. Then we have

$$\Sigma = E\{\boldsymbol{x}(t)\boldsymbol{x}(t)^{\dagger}\} = \boldsymbol{A}\Gamma\boldsymbol{A}^{\dagger} + \sigma^2 I. \tag{9.3}$$

The $p \times q$ matrix $\boldsymbol{A}$ has the special structure

$$\begin{aligned} \boldsymbol{A} &= (\boldsymbol{a}_1 : \ldots : \boldsymbol{a}_q) \\ \boldsymbol{a}_k &= a(\tau_k) = (1, e^{-j\omega_0\tau_k}, \ldots, e^{-j\omega_0(p-1)\tau_k})^T \end{aligned} \tag{9.4}$$

where

$\tau_k = c^{-1}\Delta \sin \theta_k$, c = speed of propagation

Δ = spacing between the sensors

θ_k = *direction of arrival* (*DOA*) of the signal from the kth source.

Usually ω_0 is assumed to be known; we can take $\omega_0 = 1$ without loss of generality.

There are two important problems associated with the model (9.1), (9.2). One is the detection of the number of sources q and another is the estimation of $\tau_1, \ldots, \tau_k$. The first problem is approached through model selection criteria such as *AIC* (*an information criterion* due to Akaike), *BIC* (*Bayes information criterion* due to Schwarz), and cross-validation. Recently Zhao, Krishnaiah, and Bai [3], [4] proposed a *GIC* (*general information criterion*), which provides a strongly consistent estimate of q. The GIC is a more flexible criterion that can be adopted in small samples by a suitable choice of the multiplier involved in the criterion (see Bai, Krishnaiah, and Zhao [5] and Rao and Wu [6]). See the appendix for the general definition and application of the GIC.

The second problem is solved through what is called the eigen structure method, the principal contributors being Bienvenu [7], Schmidt [8], and Wax, Shan, and Kailath [1] in the case when the error vector is white noise and Paulraj and Kailath [2] when the error vector is colored noise. The solution obtained by Kailath and his coworkers depends on the minimization of a Hermitian form of exponential functions. The solution is obtained basically by a search method, and it is extremely complicated to obtain the limiting distribution of the estimates in such a case. In the light of this, Bai, Krishnaiah, and Zhao [9] proposed a new approach and established strong consistency of the estimators. The purpose of the present chapter is to derive the limiting distribution of the estimates of DOA based on the method of Bai, Krishnaiah, and Zhao.

9.2. ESTIMATION OF DOA

For given $q = \text{rank}\, \boldsymbol{A\Gamma A}^{\dagger} = \text{rank}\, \boldsymbol{A}$ under the condition that Γ is positive definite, let us denote the eigenvalues of $\boldsymbol{\Sigma} = \boldsymbol{A\Gamma A}^{\dagger} + \sigma^2 \boldsymbol{I}$ by

$$\lambda_1 \geq \lambda_2 \geq \cdots \geq \lambda_q > \lambda_{q+1} = \cdots = \lambda_p = \sigma^2$$

and eigenvectors by

$$\boldsymbol{e}_1, \ldots, \boldsymbol{e}_q, \boldsymbol{e}_{q+1}, \ldots, \boldsymbol{e}_p.$$

Define the matrices of eigenvectors

$$\boldsymbol{E}_s = (\boldsymbol{e}_1 : \cdots : \boldsymbol{e}_q), \qquad \boldsymbol{E}_n = (\boldsymbol{e}_{q+1} : \cdots : \boldsymbol{e}_p),$$

and the spaces spanned by the vectors in $\boldsymbol{E}_s$ and $\boldsymbol{E}_n$ as the signal eigenspace and the noise eigenspace, respectively. It is well known that

$$\boldsymbol{a}(\tau)^{\dagger}\boldsymbol{E}_n = \boldsymbol{0} \Longleftrightarrow \boldsymbol{D}(\tau) \overset{\Delta}{=} \boldsymbol{a}(\tau)^{\dagger}\boldsymbol{E}_n\boldsymbol{E}_n^{\dagger}\boldsymbol{a}(\tau) = 0 \tag{9.5}$$

when $\tau = \tau_i$, $i = 1, \ldots, q$ defined in (9.4). In practice, we have only an estimate of $\boldsymbol{\Sigma}$, as shown by

$$\hat{\boldsymbol{\Sigma}} = N^{-1} \sum_{t=1}^{N} \boldsymbol{x}(t)\boldsymbol{x}(t)^{\dagger}$$

with eigenvalues, say,

$$\hat{\lambda}_1 \geq \hat{\lambda}_2 \geq \ldots \geq \hat{\lambda}_{q+1} \geq \hat{\lambda}_p$$

and the corresponding eigenvectors

$$\hat{\boldsymbol{e}}_1, \ldots, \hat{\boldsymbol{e}}_q, \hat{\boldsymbol{e}}_{q+1}, \ldots, \hat{\boldsymbol{e}}_p.$$

Then $\hat{\boldsymbol{E}}_n = (\hat{\boldsymbol{e}}_{q+1} : \ldots : \hat{\boldsymbol{e}}_p)$ is an estimate of $\boldsymbol{E}_n = (\boldsymbol{e}_{q+1} : \ldots : \boldsymbol{e}_p)$ in the sense of spaces generated by their columns. The function

$$\hat{\boldsymbol{D}}(\tau) = \boldsymbol{a}(\tau)^{\dagger}\, \hat{E}_n\hat{E}_n^{\dagger}\, \boldsymbol{a}(\tau) \tag{9.6}$$

may not vanish for any τ, but estimates of $\tau_1, \ldots, \tau_q$ may be formed by plotting the function $1/\hat{D}(\tau)$ and locating the peaks, which is the basis of the MUSIC algorithm. As the solutions are obtained essentially by a search method, their analytical characterization is complicated so that the statistical properties of the estimates are difficult to study.

Bai, Krishnaiah, and Zhao [9] suggested a different algorithm for estimating $\tau_1, \ldots, \tau_q$, which involves a specified sequence of computations. The estimates have a simple representation from which their asymptotic properties can be studied. The motivation for the suggested algorithm may be explained as follows.

We note that there exists a matrix $\boldsymbol{G}$ of order $p \times (p - q)$

$$\boldsymbol{G} = (\gamma_{q+1} : \ldots : \gamma_p) \tag{9.7}$$

where

$$\gamma_{q+1} = (g_1, \ldots, g_{q+1}, 0, \ldots, 0)^T$$

$$\gamma_{q+2} = (0, g_1, \ldots, g_{q+1}, 0, \ldots, 0)^T$$

$$\cdots$$

$$\gamma_p = (0, 0, \ldots, g_1, \ldots, g_{q+1})'$$

with $\Sigma_i |g_i|^2 = 1$ such that $\boldsymbol{a}(\tau)^\dagger \boldsymbol{G} = \boldsymbol{0}$, for $\tau = \tau_1, \ldots, \tau_q$ and $e^{j\tau_1}, \ldots, e^{j\tau_q}$ are the solutions of the polynomial equation

$$\mathrm{B}(z) = g_{q+1}z^q + \ldots + g_1 = 0. \tag{9.8}$$

The strategy is then to obtain γ_{q+1} and use the estimated components of $g_1, \ldots, g_{q+1}$ in (9.8) to obtain the roots that provide the estimates of $\tau_1, \ldots, \tau_q$. Let us first observe that we can choose $\boldsymbol{E}_n$ as the Gram-Schmidt orthogonalized version of $\boldsymbol{G}$ defined in (9.7) with the first column γ_{q+1} fixed. Let us now take $\hat{\boldsymbol{E}}_n$ as defined in (9.6) and by *Householder transformation,* that is, multiplying by an unitary matrix $\boldsymbol{O}_{p-q}$ of order $p - q$, convert it into the form

$$\hat{\boldsymbol{E}}_n \boldsymbol{O}_{p-q} = \boldsymbol{U} = (\hat{u}_{ij}), \qquad i = 1, \ldots, p, \;\; j = q + 1, \ldots, p, \tag{9.9}$$

where $\hat{u}_{ii} \geq 0$ for $i = q + 1, \ldots, p$ and $\hat{u}_{ij} = 0$ for $i > j$. (Note that the first column of $\boldsymbol{U}$ has zero elements beyond the $(q + 1)$-th row.) We estimate

$$(g_1, \ldots, g_{q+1}) \text{ by } (\hat{u}_{1,q+1}, \ldots, \hat{u}_{q+1,q+1})$$

the first $q + 1$ components in the first column of $\boldsymbol{U}$.

Then, we solve the equation

$$\hat{B}(z) = \hat{u}_{q+1,q+1}z^q + \cdots + \hat{u}_{2,q+1}z + \hat{u}_{1,q+1} = 0,$$

obtain the roots in the form

$$\hat{\rho}_k e^{j\hat{\tau}_k}, \qquad k = 1, \ldots, q \tag{9.10}$$

and choose $\hat{\tau}_1, \ldots, \hat{\tau}_q$ as estimates of $\tau_1, \ldots, \tau_q$. The method of estimation is thus straightforward.

In Section 9.3, we give the proof of strong consistency of $\hat{\tau}_i$ as an estimate of τ_i, $i = 1, \ldots, q$ as given by Bai, Krishnaiah, and Zhao [9]. In Section 9.4, we derive the asymptotic distribution of $\hat{\tau}_i$, $i = 1, \ldots, q$.

Remark 2.1. Sometimes $\hat{u}_{q+1,q+1}$ may be zero. In such a case there may be less than q roots for $\hat{B}(z)$ and all τ_k cannot be estimated. However, it can be proved that, with probability one, $\hat{u}_{q+1,q+1} > 0$ for all large N.

Remark 2.2. In the estimated noise eigenspace, that is, the space spanned by the columns of $\hat{\boldsymbol{E}}_n$, there is a unique vector of the type γ_{q+i} defined in (9.7) for each $i = 1, \ldots, p - q$, for all large N, providing estimates of the coefficients $g_1, \ldots, g_{q+1}$. There are various ways of estimating these unique vectors, which are discussed in Bai and Rao [10]. Here we only point out in that practice the first column of U can be obtained by Gaussian elimination, which is much simpler than using Householder transformation.

Remark 2.3. In our discussion, we assumed that q is known or determined by the GIC (general information criterion given by Zhao, Krishnaiah, and Bai [3], [4]).

9.3. STRONG CONSISTENCY OF DOA ESTIMATES

In this section, we establish the strong consistency of the estimates $\hat{\tau}_1, \ldots, \hat{\tau}_q$ as defined in (9.10). We need the following lemma.

Lemma 3.1. Let $\boldsymbol{A} = (a_{ik})$ and $\boldsymbol{B} = (b_{ik})$ be two Hermitian $p \times p$ matrices with spectral decompositions

$$\boldsymbol{A} = \sum_{i=1}^{p} \delta_i \boldsymbol{u}_i \boldsymbol{u}_i^{\dagger}, \ \delta_1 \geq \delta_2 \geq \cdots \geq \delta_p,$$

and

$$B = \sum_{i=1}^{p} \lambda_i \boldsymbol{v}_i \boldsymbol{v}_i^{\dagger}, \ \lambda_1 \geq \lambda_2 \geq \cdots \geq \lambda_p,$$

where δ's and λ's are eigenvalues of $\boldsymbol{A}$ and $\boldsymbol{B}$, respectively, $\boldsymbol{u}$'s and $\boldsymbol{v}$'s are orthogonal unit (or orthonormal) eigenvectors associated with δ's and λ's, respectively. Further, we assume that

$$\lambda_{n_{b-1}+1} = \cdots = \lambda_{n_b} = \tilde{\lambda}_b, \ n_0 = 0 < n_1 < \ldots < n_s = p, \ b = 1, \ldots, s,$$
$$\tilde{\lambda}_1 > \tilde{\lambda}_2 > \cdots > \tilde{\lambda}_s,$$

and that

$$|a_{ik} - b_{ik}| < \alpha, \qquad i, k = 1, \ldots, p.$$

Then there is a constant M independent of α, such that

(a) $|\delta_i - \lambda_i| < M\alpha, \qquad i = 1, \ldots, p.$

(b) $\Sigma_{i=n_{b-1}+1}^{n_b} \boldsymbol{u}_i \boldsymbol{u}_i^{\dagger} = \Sigma_{i=n_{b-1}+1}^{n_b} \boldsymbol{v}_i \boldsymbol{v}_i^{\dagger} + \boldsymbol{C}^{(b)},$

$$\boldsymbol{C}^{(b)} = (C_{lk}^{(b)}), \ |C_{lk}^{(b)}| \leq M\alpha, \qquad l, k = 1, \ldots, p, \ b = 1, \ldots, s.$$

Remark 3.1. Note that $\Sigma_{i=n_{b-1}+1}^{n_b} \boldsymbol{u}_i \boldsymbol{u}_i^{\dagger}$ and $\Sigma_{i=n_{b-1}+1}^{n_b} \boldsymbol{v}_i \boldsymbol{v}_i^{\dagger}$ are projection operators onto the subspaces spanned by the eigenvectors $\{\boldsymbol{u}_i, i = n_{b-1} + 1, \ldots, n_b\}$ and $\{\boldsymbol{v}_i, i = n_{b-1} + 1, \ldots, n_b\}$, respectively. So the assersion (b) describes the closeness of the eigenspaces of two matrices when these two matrices are close to each other.

Proof. By von Neumann's [11] inequality, one can easily obtain

$$\sum_{i=1}^{p} (\delta_i - \lambda_i)^2 \leq \mathrm{tr}[(\boldsymbol{A} - \boldsymbol{B})^2],$$

which implies (a) with $M = p$.

For simplicity, we denote by $\boldsymbol{D} = O(\alpha)$ for an $m \times n$ matrix $\boldsymbol{D} = (d_{ik})$, the fact that $|d_{ik}| \leq M\alpha, i = 1, \ldots, m, k = 1, \ldots, n$. To prove (b), without loss of generality, we can assume

$$\boldsymbol{A} = \sum_{h=1}^{s} \tilde{\lambda}_h \sum_{i \in L_h} \boldsymbol{u}_i \boldsymbol{u}_i^{\dagger},$$

where $L_h = (n_{h-1} + 1, \ldots, n_h)$. When $s = 1$, (b) is trivial. Now we assume (b) to be true for $s = t - 1$ and proceed to prove (b) for $s = t$. When $s = t$,

$$\sum_{h=1}^{t-1} (\tilde{\lambda}_h - \tilde{\lambda}_t) \sum_{i \in L_h} \boldsymbol{u}_i \boldsymbol{u}_i^{\dagger} = \sum_{h=1}^{t-1} (\tilde{\lambda}_h - \tilde{\lambda}_t) \sum_{i \in L_h} \boldsymbol{v}_i \boldsymbol{v}_i^{\dagger} + O(\alpha). \tag{9.11}$$

Multiply from right by $\boldsymbol{v}_k$, $k \in L_t$, on the two sides of (9.11), we get

$$\sum_{h=1}^{t-1} (\tilde{\lambda}_h - \tilde{\lambda}_t) \sum_{i \in L_h} \boldsymbol{u}_i (\boldsymbol{u}_i^{\dagger} \boldsymbol{v}_k) = O(\alpha),$$

which implies that

$$\boldsymbol{u}_i^{\dagger} \boldsymbol{v}_k^{\dagger} = O(\alpha), \qquad i \notin L_t, k \in L_t.$$

Thus, we have

$$\boldsymbol{U}_1^{\dagger} \boldsymbol{V}_2 = O(\alpha), \qquad \boldsymbol{V}_1^{\dagger} \boldsymbol{U}_2 = O(\alpha), \tag{9.12}$$

where

$$\boldsymbol{U}_1 = (\boldsymbol{u}_1, \ldots, \boldsymbol{u}_{n_{t-1}}), \boldsymbol{U}_2 = (\boldsymbol{u}_{n_{t-1}+1}, \ldots, \boldsymbol{u}_{n_t}), n_t = p,$$

$$\boldsymbol{V}_1 = (\boldsymbol{v}_1, \ldots, \boldsymbol{v}_{n_{t-1}}), \boldsymbol{V}_2 = (\boldsymbol{v}_{n_{t-1}+1}, \ldots, \boldsymbol{v}_{n_t}).$$

Put $\boldsymbol{U}_2 = \boldsymbol{V}_1\boldsymbol{G}_1 + \boldsymbol{V}_2\boldsymbol{G}_2$, where $\boldsymbol{G}_1 : n_{t-1} \times (p - n_{t-1})$, $\boldsymbol{G}_2 : (p - n_{t-1}) \times (p - n_{t-1})$. By (9.12),

$$\boldsymbol{V}_2^{\dagger} \boldsymbol{U}_2 \boldsymbol{U}_2^{\dagger} \boldsymbol{V}_2 = \boldsymbol{V}_2^{\dagger}(\boldsymbol{I}_p - \boldsymbol{U}_1 \boldsymbol{U}_1^{\dagger})\boldsymbol{V}_2 = \boldsymbol{V}_2^{\dagger} \boldsymbol{V}_2 + O(\alpha) = \boldsymbol{I}_{p - n_{t-1}} + O(\alpha), \tag{9.13}$$

By (9.11) and (9.12), we get

$$O(\alpha) = \boldsymbol{V}_1^{\dagger} \boldsymbol{U}_2 = \boldsymbol{G}_1 + \boldsymbol{V}_1^{\dagger} \boldsymbol{V}_2 \boldsymbol{G}_2 = \boldsymbol{G}_1,$$

which implies that

$$U_2 = V_2 G_2 + O(\alpha). \tag{9.14}$$

By (9.13) and (9.14)

$$\begin{aligned} G_2 G_2^\dagger = V_2^\dagger V_2 G_2 G_2^\dagger V_2^\dagger V_2 = V_2^\dagger U_2 U_2^\dagger V_2 + O(\alpha) \\ = I_{p - n_{t-1}} + O(\alpha). \end{aligned} \tag{9.15}$$

From (9.14 and (9.15), it follows that

$$\sum_{i \in L_t} u_i u_i^\dagger = U_2 U_2^\dagger = V_2 V_2^\dagger + O(\alpha) = \sum_{i \in L_t} v_i v_i^\dagger + O(\alpha) \tag{9.16}$$

and that

$$\begin{aligned} &\sum_{h=1}^{t-2} \tilde{\lambda}_h \sum_{i \in L_h} u_i u_i^\dagger + \tilde{\lambda}_{t-1} \sum_{i \in L_{t-1} + L_t} u_i u_i^\dagger \\ &\quad = \sum_{h=1}^{t-2} \tilde{\lambda}_h \sum_{i \in L_h} v_i v_i^\dagger + \tilde{\lambda}_{t-1} \sum_{i \in L_{t-1} + L_t} v_i v_i^\dagger + O(\alpha). \end{aligned}$$

By the induction assumption,

$$\sum_{i \in L_h} u_i u_i^\dagger = \sum_{i \in L_h} v_i v_i^\dagger + O(\alpha), \qquad h = 1, \ldots, t-2, \tag{9.17}$$

and

$$\sum_{i \in L_{t-1}} u_i u_i^\dagger = \sum_{i \in L_{t-1}} v_i v_i^\dagger + O(\alpha). \tag{9.18}$$

Thus, (b) is true for $s = t$ by (9.16)–(9.18). Lemma 3.1 is proved.

We have the following theorem on the strong consistency of the estimators:

Theorem 3.1. Suppose the second moments of $s(t)$ and $n(t)$ are finite. Then the estimates $\hat{\tau}_k$'s are strongly consistent.

Proof. Let $\mathbf{g} = (g_1, \ldots, g_{q+1}, 0, \ldots, 0)^{\mathrm{T}}$ be the $p \times 1$ vector whose elements $g_1, \ldots, g_{q+1}$ are the coefficients of the polynomial $g_{q+1} \prod_{k=1}^{q} (z - e^{j\tau_k}) \triangleq B(z)$ with restrictions $\Sigma_{k=1}^{q+1} |g_k|^2 = 1$ and $g_{q+1} > 0$.

Let $\gamma_{q+1} = \mathbf{g}, \gamma_{q+2} = (0, g_1, \ldots, g_{q+1}, 0, \ldots, 0)^T, \ldots, \gamma_p = (0, \ldots, g_1, \ldots, g_{q+1})^T$ be the p-vectors as defined in (9.7). From $a_k^\dagger \gamma_{q+l} = exp(\mathrm{j}(l - 1)_k)\, \mathrm{B}(exp[j\tau_k]) = 0, k = 1, \ldots, q, l = 1, \ldots, p - q$, it follows that $\gamma_{q+1}, \ldots, \gamma_p$ are all eigenvectors of $A\Psi A^\dagger$ associated with zero eigenvalue. Since they are linearly independent, they span the eigensubspace of $A\Psi A^\dagger$ associated with zero eigenvalue. Let V_2 denote this subspace and consider the vectors $\hat{u}_{q+i}$ as derived in (9.9) from $\hat{E}_n$, the matrix generating the estimated noise space. Let $P_2(\hat{u}_{q+1}) = \Sigma_{k=q+1}^{p} \beta_k^{(N)} \gamma_k$ denote the projection of $\hat{u}_{q+1}$ on V_2. By the *strong law of large numbers*, we have

$$\hat{\Sigma} \to \Sigma = A\Psi A^{\dagger} + \sigma^2 I_p, \quad \text{almost surely as } N \to \infty.$$

By Lemma 3.1, it is easy to see that

$$\hat{u}_{q+1} = P_2(\hat{u}_{q+1}) + o(1), \quad \text{almost surely as } N \to \infty. \tag{9.19}$$

Since $\hat{u}_{q+1,l} = 0$ for $l = q+2, \ldots, p$, we see that the last $p - q - 1$ components of $P_2(\hat{u}_{q+1}) = \Sigma_{k=q+1}^{p} \beta_k^{(N)} \gamma_k$ tend to zero almost surely. From this and the expressions for γ_k, $k = q+1, \ldots, p$, we get, noting that g_{q+1} is a positive constant,

$$\lim_{N \to \infty} \beta_k^{(N)} = 0, \quad \text{almost surely,} \qquad \text{for } k = q+2, \ldots, p,$$

and

$$\lim_{N \to \infty} \beta_{q+1}^{(N)} = 1, \quad \text{almost surely,}$$

which imply that

$$\hat{u}_{q+1} \to \gamma_{q+1} = g, \quad \text{almost surely as } N \to \infty. \tag{9.20}$$

By the definition of g, we know that $e^{j\tau_k}$, $k = 1, \ldots, q$ are the roots of the polynomial equation

$$\sum_{k=1}^{q+1} g_k z^{k-1} = 0. \tag{9.21}$$

Hence, after suitable rearrangement,

$$\hat{\rho}_k e^{j\hat{\tau}_k} \to e^{j\tau_k}, \quad \text{almost surely,} \quad k = 1, \ldots, q.$$

and, consequently,

$$\hat{\rho}_k \to 1 \quad \text{almost surely}$$

$$\hat{\tau}_k \to \tau_k, \qquad k = 1, \ldots, q \quad \text{almost surely as } N \to \infty, \tag{9.22}$$

which proves the theorem.

Remark 3.2. If q is known, to ensure the strong consistency of $\hat{\tau}_k$'s, we need only assume the existence of the second moments of $s(t)$'s and $n(t)$'s. But in the GIC procedure, to guarantee the strong consistency of the estimate of the signal number, the existence of the fourth moments of $s(t)$ and $n(t)$ was assumed (see Zhao, Krishnaiah, and Bai[3]). The conclusion of Theorem 3.1 remains true when q is first estimated by the GIC and then the relevant DOAs are estimated provided the existence of the fourth moments is assumed.

9.4. ASYMPTOTIC DISTRIBUTION OF THE DOA ESTIMATES

9.4.1. Some Lemmas

We prove some lemmas that are needed in the sequel.

Lemma 4.1. Let

$$f(\lambda) = a_k\lambda^k + \cdots + a_o, \qquad a_k \neq 0$$
$$g_n(\lambda) = a_{nm}\lambda^m + \cdots + a_{no}, \qquad a_{nm} \neq 0, m \geq k$$

be polynomials such that $g_n(\lambda) \to f(\lambda)$ as $n \to \infty$. Denote the roots of $f(\lambda)$ by $\lambda_1, \ldots, \lambda_k$ and those of $g_n(\lambda)$ by $\lambda_{n1}, \ldots, \lambda_{nm}$. Then under certain rearrangement

$$\lambda_{in} \to \begin{cases} \lambda_i, & i \leq k \\ \infty, & i > k. \end{cases}$$

Proof. The reader is referred to Bai [12].

We need the following notations to prove the other lemmas. We group the eigenvalues of $\boldsymbol{\Sigma}$ by their multiplicities as

$$\lambda_1 = \cdots = \lambda_{p_1} = \tilde{\lambda}_1 > \lambda_{p_1+1} = \cdots = \lambda_{p_2} = \tilde{\lambda}_2 > \cdots$$
$$> \lambda_{p_{t-1}+1} = \cdots = \lambda_{p_t} = \tilde{\lambda}_t > \lambda_{q+1} = \cdots = \lambda_p = \sigma^2 \tag{9.23}$$

where $p_t = q$ and $p_{t+1} = p$. In the sequel, we consider the case $t = 2$ for simplicity, observing that the general case can be studied along the same lines. Corresponding to the grouping (9.23) with $t = 2$, we have the grouping of the associated eigenvectors $\boldsymbol{e}_1, \ldots, \boldsymbol{e}_p$ of $\boldsymbol{\Sigma}$ as

$$\tilde{\boldsymbol{V}}_1 = (\boldsymbol{e}_1 : \cdots : \boldsymbol{e}_{p1}), \tilde{\boldsymbol{V}}_2 = (\boldsymbol{e}_{p_1+1} : \cdots : \boldsymbol{e}_q), \tilde{\boldsymbol{V}}_3 = (\boldsymbol{e}_{q+1} : \cdots : \boldsymbol{e}_p) \tag{9.24}$$

$$\boldsymbol{V}_1 = (\tilde{\boldsymbol{V}}_1 : \tilde{\boldsymbol{V}}_2), \quad \boldsymbol{V}_2 = \tilde{\boldsymbol{V}}_3, \boldsymbol{V} = (\boldsymbol{V}_1 : \boldsymbol{V}_2). \tag{9.25}$$

Denote the corresponding groupings of the eigenvectors $\hat{e}_i$'s of $\hat{\boldsymbol{\Sigma}}$ by

$$\tilde{\boldsymbol{U}}_{N1} = (\hat{\boldsymbol{e}}_1 : \cdots : \hat{\boldsymbol{e}}_{p1}), \tilde{\boldsymbol{U}}_{N2} = (\hat{\boldsymbol{e}}_{p_1+1} : \cdots : \hat{\boldsymbol{e}}_q), \tilde{\boldsymbol{U}}_{N3} = (\hat{\boldsymbol{e}}_{q+1} : \cdots : \hat{\boldsymbol{e}}_p) \tag{9.26}$$

$$\boldsymbol{U}_{N1} = (\tilde{\boldsymbol{U}}_{N1} : \tilde{\boldsymbol{U}}_{N2}), \quad \boldsymbol{U}_{N2} = \tilde{\boldsymbol{U}}_{N3}, \boldsymbol{U}_N = (\boldsymbol{U}_{N1} : \boldsymbol{U}_{N2}) \tag{9.27}$$

$$\hat{\boldsymbol{R}}_N = \sqrt{N}(\hat{\boldsymbol{\Sigma}}_N - \boldsymbol{\Sigma}) \tag{9.28}$$

$$\hat{\boldsymbol{H}}_N = \boldsymbol{V}^\dagger \hat{\boldsymbol{R}} \boldsymbol{V} = (\hat{H}_{ij}), \qquad i,j = 1,2,3. \tag{9.29}$$

Further, denote

$$\boldsymbol{D} = \text{diag}[\tilde{\lambda}_1 \boldsymbol{I}_{p_1}, \tilde{\lambda}_2 \boldsymbol{I}_{q-p_1}, \sigma^2 \boldsymbol{I}_{p-q}] \tag{9.30}$$

$$\boldsymbol{\Lambda}_N = \text{diag}[\boldsymbol{\Lambda}_{N1}, \boldsymbol{\Lambda}_{N2}, \boldsymbol{\Lambda}_{N3}] = \text{diag}[\hat{\lambda}_1, \ldots, \hat{\lambda}_p] \tag{9.31}$$

where $\hat{\lambda}_i$'s are the eigen values of $\hat{\boldsymbol{\Sigma}}$ with $\boldsymbol{\Lambda}_{Ni}$ corresponding to the groupings of $\hat{\lambda}_i$'s into three sets associated with the sets of eigenvectors (9.26). We can write

$$\hat{\boldsymbol{\Sigma}}_N = \boldsymbol{U}_N \boldsymbol{\Lambda}_N \boldsymbol{U}_N^{\dagger}, \; \boldsymbol{\Sigma} = \boldsymbol{VDV}^{\dagger}. \tag{9.32}$$

For the sake of convenience, we omit the subscript N associated with the random variables.

By the central limit theorem, it is seen that there are two random Hermitian matrices $\boldsymbol{R}$ and $\boldsymbol{H}$ with $\boldsymbol{H} = \boldsymbol{V}^{\dagger} \boldsymbol{RV}$ and that

$$\hat{\boldsymbol{R}} \xrightarrow{\mathcal{D}} \boldsymbol{R} = (r_{ij}), \; \hat{\boldsymbol{H}} \xrightarrow{\mathcal{D}} \boldsymbol{H} = (h_{ij}) \tag{9.33}$$

where the elements h_{ij}, $i \le j$ of $\boldsymbol{H}$ have $[p(p+1)/2]$-variate normal distribution with mean zero and variance-covariance matrix same as that of $V^* x(1) x^*(1) V$. The details are given in Appendix A.

By Skorohod's [13] *representation theorem,* we can change the underlying probability space and assume without loss of generality that (9.33) is true pointwise.

Lemma 4.2. Let

$$\eta_{p_{i-1}+1} > \cdots > \eta_{p_i}, \qquad i = 1,2,3,(p_0 = 0)$$

be the eigenvalues of $\boldsymbol{H}_{ii}$ and define $\hat{\eta}_k = \sqrt{n}(\hat{\lambda}_k - \lambda_k)$; then

$$\hat{\eta}_k \to \eta_k, \qquad k = 1, \ldots, p.$$

Proof. Let

$$\hat{\zeta}_k = \sqrt{N}(\hat{\lambda}_k - \sigma^2) = \sqrt{N}(\hat{\lambda}_k - \tilde{\lambda}_3), \qquad k = 1, \ldots, p.$$

It is easy to see that $\hat{\zeta}_1, \ldots, \hat{\zeta}_p$ are the roots of the pth-degree polynomial

$$\det\begin{bmatrix} N^{-1/2}\hat{\boldsymbol{H}}_{11} + (\tilde{\lambda}_1 - \tilde{\lambda}_3 - N^{-1/2}\zeta)\boldsymbol{I}_{p_1} & N^{-1/2}\hat{\boldsymbol{H}}_{12} & N^{-1/2}\hat{\boldsymbol{H}}_{13} \\ N^{-1/2}\hat{\boldsymbol{H}}_{21} & N^{-1/2}\hat{\boldsymbol{H}}_{22} + (\tilde{\lambda}_2 - \tilde{\lambda}_3 - N^{-1/2}\zeta)I_{p_2-p_1} & N^{-1/2}\hat{\boldsymbol{H}}_{23} \\ N^{-1/2}\hat{\boldsymbol{H}}_{31} & N^{-1/2}\hat{\boldsymbol{H}}_{32} & N^{-1/2}\hat{\boldsymbol{H}}_{33} - N^{-1/2}\zeta\boldsymbol{I}_{p-q} \end{bmatrix},$$

or equivalently of the polynomial

$$g_n(\zeta) = \det\begin{bmatrix} N^{-1/2}\hat{\boldsymbol{H}}_{11} + (\tilde{\lambda}_1 - \tilde{\lambda}_3 - \tilde{\boldsymbol{N}}^{-1/2}\zeta)\boldsymbol{I}_{p_1} & N^{-1/2}\hat{\boldsymbol{H}}_{12} & N^{-1/4}\hat{\boldsymbol{H}}_{13} \\ N^{-1/2}\hat{\boldsymbol{H}}_{21} & N^{-1/2}\hat{\boldsymbol{H}}_{22} + (\tilde{\lambda}_2 - \tilde{\lambda}_3 - N^{-1/2}\zeta)I_{p_2-p_1} & N^{-1/4}\hat{\boldsymbol{H}}_{23} \\ N^{-1/4}\hat{\boldsymbol{H}}_{31} & N^{-1/4}\hat{\boldsymbol{H}}_{32} & \hat{\boldsymbol{H}}_{33} - \zeta I_{p-q} \end{bmatrix},$$

which tends to

$$f(\zeta) = \det \begin{bmatrix} (\tilde{\lambda}_1 - \tilde{\lambda}_3)\boldsymbol{I}_{p_1} & 0 & 0 \\ 0 & (\tilde{\lambda}_2 - \tilde{\lambda}_3)\boldsymbol{I}_{p_2 - p_1} & 0 \\ 0 & 0 & \mathrm{H}_{33} - \zeta \boldsymbol{I}_{P - q} \end{bmatrix}.$$

It is obvious that $\hat{\zeta}_k \to \infty$ for $k = 1, \ldots, q$ and $\hat{\zeta}_k = \hat{\eta}_k$ for $k = q + 1, \ldots, p$. Thus by Lemma 4.1, we have $\hat{\zeta}_k = \hat{\eta}_k \to \eta_k$ for $k = q + 1, \ldots, p$. By the same approach we can prove that $\hat{\eta}_k \to \eta_k$ for all $k \leq p$, which proves Lemma 4.2.

Lemma 4.3. We have

$$(a)\ \tilde{\boldsymbol{V}}_i^\dagger \tilde{\boldsymbol{U}}_i \to \boldsymbol{\Phi}_i,\ i = 1,2,3, \tag{9.34}$$

$$(b)\ \sqrt{N}\tilde{\boldsymbol{V}}_i^\dagger \tilde{\boldsymbol{U}}_j \to (\tilde{\lambda}_j - \tilde{\lambda}_i)^{-1}\boldsymbol{H}_{ij}\boldsymbol{\Phi}_j, \qquad 1 \leq i \neq j \leq 3, \tag{9.35}$$

where $\boldsymbol{\Phi}_i$ is $(p_i - p_{i-1}) \times (p_i - p_{i-1})$ unitary matrix, which, in fact, is the matrix whose columns are the eigenvectors of $\boldsymbol{H}_{ii}$.

Proof. By the definition of $\hat{\lambda}_k$ and $\hat{\mathbf{e}}_k$, we have

$$(\hat{\boldsymbol{\Sigma}} - \hat{\lambda}_k \boldsymbol{I}_p)\hat{\boldsymbol{e}}_k = 0,$$

or equivalently

$$(N^{-1/2}\hat{\boldsymbol{H}} + \mathrm{diag}[\tilde{\lambda}_1 \boldsymbol{I}_{p_1}, \tilde{\lambda}_2 \boldsymbol{I}_{q - p_1}, \tilde{\lambda}_3 \boldsymbol{I}_{p - q}] - \hat{\lambda}_k \boldsymbol{I}_p)\boldsymbol{V}^\dagger \hat{\boldsymbol{e}}_k = \boldsymbol{0}. \tag{9.36}$$

Consider the case $k \leq p_1$ then (9.36) is equivalent to

$$(\hat{\boldsymbol{H}}_{11} - \hat{\eta}_k \boldsymbol{I}_{p_1})\tilde{\boldsymbol{V}}_1^\dagger \hat{\boldsymbol{e}}_k + \hat{\boldsymbol{H}}_{12}\tilde{\boldsymbol{V}}_2^\dagger \hat{\boldsymbol{e}}_k + \hat{\boldsymbol{H}}_{13}\tilde{\boldsymbol{V}}_3^\dagger \hat{\boldsymbol{e}}_k = \boldsymbol{0} \tag{9.37}$$

$$\hat{\boldsymbol{H}}_{21}\tilde{\boldsymbol{V}}_1^\dagger \hat{\boldsymbol{e}}_k + (\hat{\boldsymbol{H}}_{22} - N^{1/2}(\hat{\lambda}_k - \tilde{\lambda}_2)\boldsymbol{I}_{q - p})\tilde{\boldsymbol{V}}_2^\dagger \boldsymbol{e}_k + \hat{\boldsymbol{H}}_{23}\tilde{\boldsymbol{V}}_3^\dagger \hat{\boldsymbol{e}}_k = \boldsymbol{0} \tag{9.38}$$

$$\hat{\boldsymbol{H}}_{31}\tilde{\boldsymbol{V}}_1^\dagger \hat{\boldsymbol{e}}_k + \hat{\boldsymbol{H}}_{32}\tilde{\boldsymbol{V}}_2^\dagger \hat{\boldsymbol{e}}_k + (\hat{\boldsymbol{H}}_{33} - N^{1/2}(\hat{\lambda}_k - \tilde{\lambda}_3)\boldsymbol{I}_{p - q})\tilde{\boldsymbol{V}}_3^\dagger \hat{\boldsymbol{e}}_k = \boldsymbol{0}. \tag{9.39}$$

By (9.38) and (9.39), we have

$$\tilde{\boldsymbol{V}}_i^\dagger \hat{\boldsymbol{e}}_k = \boldsymbol{O}(N^{-1/2}),\ i = 2,3. \tag{9.40}$$

Substituting (9.40) into (9.37) and using the facts that $\hat{\boldsymbol{H}}_{11} \to \boldsymbol{H}_{11}$, $\hat{\eta}_k \to \eta_k$ and with probablity 1, $\eta_1, \ldots, \eta_{p_1}$ are distinct we find that $\tilde{\boldsymbol{V}}_1^\dagger \hat{\boldsymbol{e}}_k$ tends to the eigenvector $\boldsymbol{\phi}_k$ of $\boldsymbol{H}_{11}$ associated with η_k. Thus (9.34) is proved for $i = 1$. The proof for $i = 2$ and 3 is similar and is omitted.

Substituting $\boldsymbol{V}_1^\dagger \hat{\boldsymbol{e}}_k \to \boldsymbol{\phi}_k$ into (9.38) or (9.39) we get

$$\sqrt{N}\boldsymbol{V}_2^\dagger \hat{\boldsymbol{e}}_k \to \frac{1}{\tilde{\lambda}_1 - \tilde{\lambda}_2}\boldsymbol{H}_{21}\boldsymbol{\phi}_k$$

$$\sqrt{N}\boldsymbol{V}_3^\dagger \hat{\boldsymbol{e}}_k \to \frac{1}{\tilde{\lambda}_1 - \tilde{\lambda}_3}\boldsymbol{H}_{31}\boldsymbol{\phi}_k,$$

which proves (9.35) for $j = 1$ and $i = 2, 3$. The proof for the other results in (9.35) is the same and is omitted.

Lemma 4.4. We have

$$\sqrt{N}(\hat{U}_2 - V_2) \to -V_1(\Lambda_1 - \sigma^2 I_q)^{-1}B + V_2 W_0 \text{ in distribution} \tag{9.41}$$

where $\Lambda_1 = \text{diag}\,[\tilde{\lambda}_1 I_{p_1}, \tilde{\lambda}_2 I_{q-p_1}]$, $B' = (H'_{13} : H'_{23})$, W_0 is a $(p-q)\times(p-q)$ anti-Hermitian matrix whose elements below the diagonal are those of $V_{22}^{-1}V_{12}(\Lambda_1 - \sigma^2 I_q)^{-1}B$ and the diagonal elements are the imaginary parts of those of $V_{22}^{-1}V_{12}(\Lambda_1 - \sigma^2 I_q)^{-1}B$, and $V_i^T = (V_{i1}^T : V_{i2}^T)$, $i = 1, 2$.

Proof. By using the fact that

$$\tilde{V}_1\tilde{V}_1^\dagger + \tilde{V}_2\tilde{V}_2^\dagger + \tilde{V}_3\tilde{V}_3^\dagger = I_p,$$

from Lemma 4.3, we get

$$\begin{aligned}\sqrt{N}(U_2 - V_2V_2^\dagger U_2) &= \sqrt{N}(\tilde{V}_1\tilde{V}_1^\dagger\tilde{U}_3 + \tilde{V}_2\tilde{V}_2^\dagger\tilde{U}_3)\\ &\to \tilde{V}_1(\tilde{\lambda}_3 - \tilde{\lambda}_1)^{-1}H_{13}\Phi_3 + \tilde{V}_2(\tilde{\lambda}_3 - \tilde{\lambda}_2)^{-1}H_{23}\Phi_3\\ &= -V_1(\Lambda_1 - \sigma^2 I_q)^{-1}B\,\Phi_3.\end{aligned} \tag{9.42}$$

Recalling $\hat{U}_2 = U_2O_2$, $\hat{U}_2 \to V_2$ and $V_2^\dagger U_2 \to \Phi_3$, we have $O_2 \to \Phi_3^\dagger$. This and (9.42) implies that

$$\sqrt{N}(\hat{U}_2 - V_2V_2^\dagger\hat{U}_2) \to -V_1(\Lambda_1 - \sigma^2 I_q)^{-1}B. \tag{9.43}$$

Write $X = V_2^\dagger\hat{U}_2 - I_{p-q}$ and $W = (1/2)(X - X^\dagger)$. We proceed to show that

$$\sqrt{N}(\hat{U}_2 - V_2 - V_2W) \to -V_1(\Lambda_1 - \sigma^2 I_q)^{-1}B. \tag{9.44}$$

To prove (9.44), it is sufficient to show that

$$X + X^\dagger = O(1/N). \tag{9.45}$$

By Lemma 4.3, we have

$$\begin{aligned}I + X + X^\dagger + XX^\dagger &= V_2^\dagger\hat{U}_2\hat{U}_2^\dagger V_2 = V_2^\dagger U_2U_2^\dagger V_2\\ &= I - V_2^\dagger U_1U_1^\dagger V_2 = I + O(1/N).\end{aligned} \tag{9.46}$$

Thus we only need to prove that

$$X = O(1/\sqrt{N}). \tag{9.47}$$

By Householder transformation, there exists a unitary matrix $C : (p-q)\times(p-q)$ such that $G = \hat{U}_2^\dagger V_2 C$ is an upper triangular matrix with positive diagonal elements. Then (9.46) implies that

$$G^\dagger G = I + O(1/N)$$

and hence by the fact that G is an upper triangular matrix,

$$G = I + O(1/N).$$

Thus we have

$$V_2^{\dagger} \hat{U}_2 = C + O(1/N). \tag{9.48}$$

Substituting (9.48) into (9.43), we have

$$\hat{U}_2 - V_2 C = O(1/\sqrt{N}). \tag{9.49}$$

Since $\hat{U}_2$, V_2 are upper triangular matrices with positive diagonal elements and V_2 is independent of N, (9.49) implies that

$$\begin{aligned} C_{ij} &= O(1/\sqrt{N}), \qquad \text{for } q < i < j \le p, \\ Im\,(C_{ii}) &= O(1/\sqrt{N}), \qquad \text{for } q < i \le p, \end{aligned} \tag{9.50}$$

where Im stands for the imaginary part and C_{ij} are elements of C. Since C is unitary, (9.50) implies that

$$C = I + O(1/\sqrt{N}),$$

which together with (9.48) implies (9.47). Thus we have proved (9.44).

Noting the structures of $\hat{U}_2$ and V_2, it follows from (9.44) that $\sqrt{N}\,W$ has a limit, and we can obtain this limit from (9.44). In fact, if we denote

$$V_1 = \begin{pmatrix} V_{11} \\ V_{12} \end{pmatrix}, \qquad V_2 = \begin{pmatrix} V_{21} \\ V_{22} \end{pmatrix}$$

where V_{11} and V_{21} have q rows and V_{12} and V_{22} have $(p - q)$ rows, and

$$L = -V_{12}(\Lambda_1 - \sigma^2 \mathbf{I}_q)^{-1} B,$$

then the limits of the elements below the diagonal of $\sqrt{N}\,W$ are those of $V_{22}^{-1}L$ and the limits of diagonal elements of $\sqrt{N}\,W$ are the imaginary parts of the diagonal elements of $V_{22}^{-1}L$. The limits of other elements of $\sqrt{N}\,W$ can be obtained by the fact that W is anti-Hermitian. Hence Lemma 4.4 is proved.

9.4.2. The Main Theorem

Let $\alpha = (\alpha_1, \ldots, \alpha_p)^T$ denote the first column of the matrix $-V_1(\Lambda_1 - \sigma^2 I_q)^{-1}B$, which is defined in Lemma 4.4. Denote

$$D(\tau) = j \frac{d}{d\tau} B(e^{j\tau})$$

where $B(z) = g_{q+1} z^q + \cdots + g_1$ and

$$G = \text{diag}[D(\tau_1), \ldots, D(\tau_q)]. \tag{9.51}$$

Write

$$T_{Nk} = \sqrt{N}(\hat{\rho}_k - 1), \qquad \boldsymbol{T}_N = (T_{N1}, \ldots, T_{Nq})^T$$
$$\Delta_{Nk} = \sqrt{N}(\hat{\tau}_k - \tau_k), \qquad \boldsymbol{\Delta}_N = (\Delta_{N1}, \ldots, \Delta_{Nq})^T,$$

where $\hat{\rho}_k$, $\hat{\tau}_k$ are defined in Section 9.2. Then we have the following theorem.

Theorem 4.1. Under the assumptions (9.2) and the existence of the fourth moments of $\boldsymbol{s}(t)$ and $\boldsymbol{n}(t)$, we have

$$\boldsymbol{T}_N + j\boldsymbol{\Delta}_N \to \boldsymbol{G}^{-1}\boldsymbol{A}^{\dagger}\boldsymbol{\alpha} \text{ in distribution,} \tag{9.52}$$

where $j = \sqrt{-1}$.

Remark 4.1. In Section 9.4.1, the lemmas are stated for $t = 2$. For the general case of t, it only needs to write

$$\boldsymbol{B} = \begin{bmatrix} H_{1,t+1} \\ \vdots \\ H_{t,t+1} \end{bmatrix}, \qquad \boldsymbol{\Lambda}_1 = \begin{bmatrix} \tilde{\lambda}\boldsymbol{I}_{p_1} & \tilde{\lambda}_2\boldsymbol{I}_{p_2 - p_1} & & \\ & & \ddots & \\ & & & \tilde{\lambda}_t\boldsymbol{I}_{q - p_{t-1}} \end{bmatrix}$$

$$\boldsymbol{V}_1 = (\boldsymbol{v}_1, \ldots, \boldsymbol{v}_q) = (\tilde{\boldsymbol{V}}_1, \ldots, \tilde{\boldsymbol{V}}_t).\ \boldsymbol{V}_2 = \tilde{\boldsymbol{V}}_{t+1} = (\boldsymbol{v}_{q+1}, \ldots, \boldsymbol{v}_p).$$

$\boldsymbol{W}_0$ is similarly defined.

Proof. Write $\boldsymbol{\alpha} = (\alpha_1, \ldots, \alpha_p)^T$ and $\boldsymbol{\beta} = (\beta_1, \ldots, \beta_p)^T$, where $\boldsymbol{\beta}$ is the first column vector of $\boldsymbol{V}_2\boldsymbol{W}_0$.

By the definition of $B(z)$ as in (9.8), we have

$$B(e^{j\tau_k}) = 0, \qquad \textit{for } k = 1, \ldots, q.$$

Then for each k, we have,

$$O = \sqrt{N}\,[\hat{B}_n(\hat{\rho}_k e^{j\hat{\tau}}k) - B(e^{j\tau_k})]$$

$$= \sqrt{N} \sum_{l=0}^{q} (\hat{u}_{l+1,\,q+1}\hat{\rho}_k^l e^{jl\hat{\tau}_k} - b_{l+1}e^{jl\tau_k})$$

$$= \sum_{l=0}^{q} \{[\sqrt{N}\,(\hat{u}_{l+1,\,q+1} - b_{l+1})]\hat{\rho}_k^l e^{jl\hat{\tau}_k} + b_{l+1}[\sqrt{N}(\hat{\rho}_k^l e^{jl\hat{\tau}_k} - e^{jl\tau_k})]\}$$

$$= \sum_{l=0}^{q} (\alpha_{l+1} + \beta_{l+1})e^{jl\tau_k} + \sum_{l=0}^{q} b_{l+1}\, l e^{jl\tau_k}\,(T_{Nk} + j\Delta_{Nk})[1 + o(1)] + o(1)$$

$$= a_k^{\dagger}(\boldsymbol{\alpha} + \boldsymbol{\beta}) - D(\tau_k)(\,T_{Nk} + j\Delta_{Nk}\,)[1 + o(1)] + o(1).$$

Here we have used the facts that $\alpha_l + \beta_l = 0$ for $l = q + 2, \ldots, p$ and $\hat{\rho}_k \to 1, \hat{\tau}_k \to \tau_k$. By the orthogonality of $\boldsymbol{a}_k$ to $\boldsymbol{V}_2$ and hence to $\boldsymbol{\beta}$ (i.e., $\boldsymbol{a}_k^\dagger \boldsymbol{\beta} = 0$). So we obtain

$$T_N + j\Delta_N = G^{-1}A^\dagger\alpha + o(1).$$

This completes the proof of Theorem 4.1.

9.5. GENERAL AND CONCLUDING REMARKS

Remark 5.1. The limiting distribution of $\hat{\tau}_1, \ldots, \hat{\tau}_k$ is independent of the choice of the base of the eigenspace of $\boldsymbol{\Sigma}$, that is, the choice of $\tilde{V}_1, \ldots, \tilde{V}_t$. In fact, this can be seen from (9.41). Suppose there are two possible choices V_l and $\hat{V}_1$, there must exist a unitary matrix

$$O = \text{diag}[\hat{O}_1, \ldots, \hat{O}_t],\ \hat{O}_k : (p_k - p_{k-1}) \times (p_k - p_{k-1}),$$

such that $V_1 = \hat{V}_1 O$.

Note that

$$B = \begin{bmatrix} H_{1,t+1} \\ \vdots \\ H_{t,t+1} \end{bmatrix} = V_1{}^\dagger R V_2$$

$$\begin{aligned} V_1(\Lambda_1 - \sigma^2 I_q)^{-1}B &= V_1(\Lambda_1 - \sigma^2 I_q)^{-1}V_1{}^\dagger R V_2 \\ &= \hat{V}_1 O(\Lambda_1 - \sigma^2 I_q)^{-1}O^\dagger \hat{V}_1{}^\dagger R V_2 \\ &= \hat{V}_1(\Lambda_1 - \sigma^2 I_q)^{-1}\hat{V}_1{}^\dagger R V_2. \end{aligned}$$

Since W_0 is a function of $V_1(\Lambda_1 - \sigma^2 I_q)^{-1}B$, the distribution of W_0 is also independent of the choice of V_1. So we proved that not only the limiting distribution of $\hat{\tau}_k$'s, but also that of $\sqrt{N}(\hat{U}_2 - V_2)$, is independent of the choice of V_1.

Remark 5.2. When $s(t)$ and $n(t)$ are independently distributed as multivariate complex normal, the elements above or on the diagonal of H are independent and of mean zero, the diagonal elements of H_{ii} have distribution $RN(0, \tilde{\lambda}_i^2)$ and the off-diagonal elements of H_{ij} have distribution $CN(0, \tilde{\lambda}_i\tilde{\lambda}_j)$, where RN and CN denote *real* and *complex normal*, respectively.

Remark 5.3. Since $G^{-1}A^\dagger\,\alpha$ is a vector of linear combinations of the elements of B whose elements have multivariate normal distribution, the limiting distribution of Δ_N is multivariate normal with mean zero. But the asymptotic variances and covariances are very complicated and involve many nuisance parameters, even when the underlying distribution of $s(t)$ and $n(t)$ are independent complex multivariate normal as mentioned in Remark 5.2. For example, the expression for the limiting distribution contains V_1, though basically independent of its choice, but one can not use U_1 to approximate it since U_1 is random. So, for the purpose of testing hypotheses of the DOA, the bootstrapping approach may have to be adopted.

Suppose $m = m_N$ such that $m_N \to \infty$ and m_N / N is bounded. Choose randomly m samples $Y_{N1}, \ldots, Y_{Nm}$ from $X_1, \ldots, X_N$ (may be repeated). Then use the resampling samples $Y_{N1}, \ldots, Y_{Nm}$ and the same algorithm to estimate the DOAs. The estimates are denoted by $\hat{\tau}_{1m}, \ldots, \hat{\tau}_{km}$. By similar arguments as in proving the main theorem, one can show that with probability one (with respect to $X_1, \ldots, X_N, \ldots$) the limiting distribution of $(\hat{\tau}_{m1}, \ldots, \hat{\tau}_{km})$, is the same as that of $(\hat{\tau}_1, \ldots, \hat{\tau}_k)$. Repeat this procedure μ ($\mu = \mu_N$ is comparable with N) times. We obtain μ points $(\hat{\tau}_{1ms}, \ldots, \hat{\tau}_{kms})$. $s = 1, \ldots, \mu$, of the resampling estimates of the DOA. Then the empirical distribution of the μ points can provide an approximation to the limiting distribution of $(\hat{\tau}_1, \ldots, \hat{\tau}_k)$. We may use it to test hypotheses or make other statistical inferences about the DOA like interval estimation and so on.

ACKNOWLEDGMENTS

*This work is supported by Contract No. 0014-85-K-0292 of the Office of Naval Research. The U.S. government is authorized to reproduce and distribute reprints for governmental purposes notwithstanding any copyright notation hereon.

APPENDIX A. THE COVARIANCES OF ELEMENTS OF *H* DEFINED IN (9.33)

Write $y = V^\dagger x(1) = (y_1, \ldots, y_p)^T$. Then we have

$$E\{y_k\} = 0, \quad E\{y_k y_g^*\} = 0, \qquad \text{for } k \neq g,$$

$$E\{|y_k|^2\} = \lambda_k = \tilde{\lambda}_i, \text{ if } k \in \{n_{i-1} + 1, \ldots, n_i\}, \qquad i = 1, \ldots, t + 1,$$

where y_g denotes the conjugate of y_g^*. Since the covariances of elements of H can be obtained by writing h_{kg} as $y_k y_g^*$, we get

$$\begin{aligned} cov(h_{kk}, h_{gg}) &= E\{|y_k|^2 |y_g|^2\} - \lambda_k \lambda_g \\ &= E\{|y_k|^2 |y_h|^2\} - \tilde{\lambda}_i \tilde{\lambda}_{i'}, \end{aligned}$$

if $k \in \{n_{i-1} + 1, \ldots, n_i\}$, $g \in \{n_{i'-1} + 1, \ldots, n_{i'}\}$, $k, g = 1, \ldots, t + 1$ and

$$cov(h_{kg}, h_{k'g'}) = E\{y_k y_g^* y_{k'}^* y_{g'}\}$$

if $k < g$, $k' \leq g'$ or if $k = g$, $k' < g'$.

APPENDIX B. EXACT LIMITING DISTRIBUTION OF THE ESTIMATE OF DOA

In Theorem 5.1, we have already proved that the joint limiting distribution of $T_N + j\Delta_N$ is given by that of $G^{-1}A^\dagger\alpha$, which is a linear combination of normal variables. So the limiting distribution is also normal, and hence we need only to compute its second moment.

For simplicity of computation, we make the following additional assumptions that are usually met in practice.

(A1). $\text{cov}[\text{Re}\boldsymbol{s}(t)] = \text{cov}[\text{Im}\boldsymbol{s}(t)] = (1/2)\,\text{Re}(\boldsymbol{\Psi})$

$$E\text{Re}\boldsymbol{s}(t)(\text{Im}\boldsymbol{s}(t))^T = -(\text{E Im}\boldsymbol{s}(t)(\text{Re}\boldsymbol{s}(t))^T)^T = 2^{-1}\,\text{Im}(\boldsymbol{\Psi}).$$

(A2). $\boldsymbol{n}(t) = \boldsymbol{n}_1(t) + j\boldsymbol{n}_2(t)$

$$= [n_{11}(t), \ldots, n_{1p}(t)]^T + j[n_{21}(t), \ldots, n_{2p}(t)]^T,$$

$$E\{n_{gk}^4(t)\} = (3/4)\sigma^4,\ E\{n_{gk}^2(t)n_{g'k'}^2(t)\} = (1/4)\sigma^4 \text{ if } (g, k) \neq (g', k').$$

$$E\{n_{gk}(t)n_{g'k'}(t)n_{g''k''}(t)n_{g'''k'''}(t)\} = 0, \text{ for all cases other than above.}$$

Under the preceding assumptions, it is not difficult to verify that

$$(1/\sqrt{N}) \sum_{t=1}^{N} [\boldsymbol{n}(t)\boldsymbol{n}^\dagger(t) - \sigma^2\boldsymbol{I}_p] \xrightarrow{\mathcal{D}} \boldsymbol{M}_p,$$

$$(1/\sqrt{N}) \sum_{t=1}^{N} \boldsymbol{s}(t)[\boldsymbol{n}^\dagger(t)\boldsymbol{v}_{q+1}] \xrightarrow{\mathcal{D}} \boldsymbol{m}_q,$$

where $\boldsymbol{M}_p$ is a complex Gaussian matrix, with variance of each entry equal to σ^4, $\boldsymbol{m}_q$ is a complex multivariate normal vector with mean zero and covariance matrix $\sigma^2\boldsymbol{\Psi}$, and $\boldsymbol{M}_p$, $\boldsymbol{m}_q$ are independent of each other.

Note that $\boldsymbol{B} = \boldsymbol{V}_1^\dagger \boldsymbol{R}\boldsymbol{V}_2$, $\boldsymbol{\alpha} = -\boldsymbol{V}_1(\boldsymbol{\Lambda}_1 - \boldsymbol{\sigma}^2\boldsymbol{I}_q)^{-1}\boldsymbol{V}_1^\dagger \boldsymbol{R}\boldsymbol{v}_{q+1}$. Then we have

$$\begin{aligned}
\boldsymbol{G}^{-1}\boldsymbol{A}^\dagger\boldsymbol{\alpha} &= -\lim_{N\to\infty} \boldsymbol{G}^{-1}\boldsymbol{A}^\dagger\boldsymbol{V}_1(\boldsymbol{\Lambda}_1 - \sigma^2\boldsymbol{I}_q)^{-1}\boldsymbol{V}_1^\dagger\,(1/\sqrt{N}) \sum_{t=1}^{N} \{\boldsymbol{A}(\boldsymbol{s}(t)\boldsymbol{s}^\dagger(t) - \boldsymbol{\Psi})\boldsymbol{A}^\dagger \\
&\quad + [\boldsymbol{n}(t)\boldsymbol{n}^\dagger(t) - \sigma^2\boldsymbol{I}_p] + \boldsymbol{A}\boldsymbol{s}(t)\boldsymbol{n}^\dagger(t) + \boldsymbol{n}(t)\boldsymbol{s}^\dagger(t)\boldsymbol{A}^\dagger\}\boldsymbol{v}_{q+1} \\
&= -\lim_{n\to\infty} \boldsymbol{G}^{-1}\boldsymbol{A}^\dagger\boldsymbol{V}_1(\boldsymbol{\Lambda}_1 - \sigma^2\boldsymbol{I}_q)^{-1}\boldsymbol{V}_1^\dagger[1/\sqrt{N}] \sum_{t=1}^{N} [\boldsymbol{n}(t)\boldsymbol{n}^\dagger(t) - \sigma^2\boldsymbol{I}_p]\boldsymbol{v}_{q+1} \\
&\quad + (1/\sqrt{N}) \sum_{t=1}^{N} [\boldsymbol{A}\boldsymbol{s}(t)\boldsymbol{n}^\dagger(t)\boldsymbol{v}_{q+1}] \\
&= -\boldsymbol{G}^{-1}\boldsymbol{A}^\dagger\boldsymbol{V}_1(\boldsymbol{\Lambda}_1 - \sigma^2\boldsymbol{I}_q)^{-1}\boldsymbol{V}_1^\dagger(\boldsymbol{M}_p\boldsymbol{v}_{q+1} + \boldsymbol{A}\boldsymbol{m}_q) \\
&= -\boldsymbol{G}^{-1}\boldsymbol{A}^\dagger\boldsymbol{V}_1(\boldsymbol{\Lambda}_1 - \sigma^2\boldsymbol{I}_q)^{-1}(\boldsymbol{\gamma} + \boldsymbol{V}_1^\dagger\boldsymbol{A}\boldsymbol{m}_q)
\end{aligned}$$

where $\boldsymbol{\gamma} = \boldsymbol{V}_1^\dagger \boldsymbol{M}_p\boldsymbol{v}_{q+1}$ is a q-vector of iid complex normal components, of mean zero and variance σ^4, by the reason that $\boldsymbol{M}_p$ and $\boldsymbol{V}^\dagger\boldsymbol{M}_p\boldsymbol{V}$ are identically distributed.

Therefore $\boldsymbol{G}^{-1}\boldsymbol{A}^\dagger\boldsymbol{\alpha}$ is a multivariate complex normal vector whose covariance matrix is given by

$$\begin{aligned}
\Sigma_0 &= \boldsymbol{G}^{-1}[\sigma^4\,\boldsymbol{A}^\dagger\boldsymbol{V}_1(\boldsymbol{\Lambda}_1 - \sigma^2\boldsymbol{I}_q)^{-2}\boldsymbol{V}_1^\dagger\boldsymbol{A} \\
&\quad + \sigma^2\,\boldsymbol{A}^\dagger\boldsymbol{V}_1(\boldsymbol{\Lambda}_1 - \sigma^2\boldsymbol{I}_q)^{-1}\boldsymbol{V}_1^\dagger\boldsymbol{A}\boldsymbol{\Psi}\boldsymbol{A}^\dagger\boldsymbol{V}_1(\boldsymbol{\Lambda}_1 - \sigma^2\boldsymbol{I}_q)^{-1}\boldsymbol{V}_1^\dagger\boldsymbol{A}](\boldsymbol{G}^{-1})^\dagger.
\end{aligned}$$

By elementary computation, we find that

$$A^\dagger V_1(\Lambda_1 - \sigma^2 I_q)^{-1} V_1^\dagger A = \Psi^{-1}$$

and

$$A^\dagger V_1(\Lambda_1 - \sigma^2 I_q)^{-2} V_1^\dagger A = \Psi^{-1}(A^\dagger A)^{-1}\Psi^{-1}.$$

Consequently, we get

$$\Sigma_0 = G^{-1}[\sigma^4 \Psi^{-1}(A^\dagger A)^{-1}\Psi^{-1} + \sigma^2\Psi^{-1}](G^{-1})^\dagger.$$

Finally we obtain that the covariance matrix of the limiting joint distribution of $\Delta_N = \sqrt{N}(\hat{\tau}_1 - \tau_1, \ldots, \hat{\tau}_q - \tau_q)^T$ as

$$(1/2)\mathrm{Re}[G^{-1}(\sigma^4\Psi^{-1}(A^\dagger A)^{-1}\Psi^{-1} + \sigma^2\Psi^{-1})(G^{-1})^\dagger].$$

Remark A.1. From the above expression for the asymptotic covariance matrix of Δ_N, we can see the following:

1. If the number of sensors p increases, then $(A^\dagger A)^{-1}$ decreases and hence Σ_0 decreases. This is intuitively correct since more sensors provide more information and hence the estimation can be improved.
2. Note the $G = \mathrm{diag}[D(\tau_1), \ldots, D(\tau_q)]$ and

$$D(\tau_k) = j\frac{d}{d\tau} f(e^{j\tau})|_{\tau = \tau_k} = -b_{q+1}\prod_{\substack{i \neq k \\ i=1}}^{q}(e^{j\tau_k} - e^{j\tau_i})e^{j\tau_k}, \qquad k = 1, \ldots, q.$$

If q is large and τ_k's are approximately uniformly distributed, then we can show that $D(\tau_k)$'s have the same order of q.

This fact opens up a new possibility of obtaining improved estimates of the unknown DOA parameters $\tau_1, \ldots, \tau_q$, when $p \gg q$ (p being the number of sensors and q being the number of signals). We introduce an additional number, say, r, of (artificial) signals with known DOAs, $\tau_{q+1}, \ldots, \tau_{q+r}$, and observe the combined response of $q + r$ signals at the p sensors. Using these observations we estimate $(q + r)$ parameters $\tau_1, \ldots, \tau_q, \tau_{q+1}, \ldots, \tau_{q+r}$ assuming that $\tau_{q+1}, \ldots, \tau_{q+r}$ are also unknown. If $\tilde{\tau}_i$ is the estimate of τ_i, $i = 1, \ldots, q + r$, then we choose the subset $\tilde{\tau}_1, \ldots, \tilde{\tau}_q$ as estimates of $\tau_1, \ldots, \tau_q$. It is interesting to note, which may appear as a paradox, that $\tilde{\tau}_1, \ldots, \tilde{\tau}_q$ are more efficient as estimates of $\tau_1, \ldots, \tau_q$ than $\hat{\tau}_1, \ldots, \hat{\tau}_q$ obtained from the observations without introducing the artificial signals.

APPENDIX C. INFORMATION CRITERIA FOR MODEL SELECTION

Let $X = (X_1, \ldots, X_N)'$ be a sample of N independent observations with an unknown joint density function g. We wish to find an approximation to g through a suitable parametric model. Let us suppose that there are k alternative models

$$M_i = \{f_i(X|\theta_i),\ \theta_i \in \Theta_i \subset R^{d_i}\}, \qquad i = 1, \ldots, k, \tag{9.C.1}$$

where f_i denotes the density function of X depending on an unknown d_i-vector parameter θ_i. Let the best parametric model approximation to g correspond to $i = i_0$ in (9.C.1). Our problem is to estimate i_0 on the basis of the observed X.

Define

$$L_N^{(i)}(\hat{\theta}_i) = \max_{\theta_i} f_i\,(X|\theta_i) \tag{9.C.2}$$

the maximum likelihood under the model i. While the method of maximum likelihood is valuable in selecting a value of θ_i given the model i, it is not, by itself, an appropriate criterion for selection between models since the magnitudes in (9.C.2) are not comparable for different values of i because of different dimensions of θ_i. However, it is found that a general information criterion (GIC) for model selection can be developed by using (9.C.2) with a penalty function correcting for the dimensions of the parameter. Let us define for the model i

$$\text{GIC}(i) = 2\,L_N^{(i)}\,(\hat{\theta}_i) - d_i C_N \tag{9.C.3}$$

where d_i is the dimension of θ_i and C_N satisfies the conditions

$$\frac{C_N}{N} \to 0 \quad \text{and} \quad \frac{C_N}{\log\log N} \to \infty. \tag{9.C.4}$$

The decision rule is to choose the model $i = i_*$ where

$$\text{GIC}(i_*) = \max_i \text{GIC}(i). \tag{9.C.5}$$

It is shown by Zhao, Krishnaiah, and Bai [3, 4] that under the condition (9.C.4), the procedure (9.C.5) is strongly consistent, that is,

$$i_* \to i_0, \quad \text{almost surely as } n \to \infty. \tag{9.C.6}$$

Criteria of the kind (9.C.5) for model selection have been used with special choices: $C_N = 2$ by Akaike [14], $C_N = \log N$ by Schwarz [15] and Rissanen [16], and $C_N = K \log\log N$ with $K >$ by Hannan and Quinn [17]. All these choices except Akaike's give a consistent estimate of i_0 but not necessarily a strongly consistent estimate as ensured by the condition (9.B.4).

In signal detection and estimation, we have the problem of deciding on the number of signals present. This, in the context of the problem we are discussing, is equivalent to deciding on the rank of the matrix Γ in the decomposition of Σ in (9.3)

$$\Sigma = A\Gamma A^{\dagger} + \sigma^2 I. \tag{9.C.6}$$

Based on the statistic

$$S = N^{-1} \sum_{t=1}^{N} \boldsymbol{x}(t)\boldsymbol{x}(t)'$$

where $\boldsymbol{x}(t)$ are observed p-vectors as in (9.1), the log likelihood is, apart from a constant,

$$\log L_N^{(i)}(\Sigma) = -\frac{N}{2}\log|\Sigma| - \frac{N}{2}\operatorname{tr}[\Sigma^{-1}S] \tag{9.C.7}$$

where i represents the rank of Γ. The maximum of (9.C.7) with respect to Σ under the constraint (9.B.5) for given i is

$$\log L_N^{(i)}(\hat{\Sigma}) = -\frac{N}{2}\left[\sum_{j=1}^{i}\log\delta_j + (p-i)\log\frac{\delta_{i+1}+\cdots+\delta_p}{p-\text{i}}\right] \tag{9.C.8}$$

where $\delta_1 > \cdots > \delta_p$ are the eigenvalues of S, assumed to be distinct with probability 1. The criterion for model selection (for estimating the number of signals) is

$$\text{GIC}(i) = 2\log L_N^{(i)}(\hat{\Sigma}) - i(2p-i) + 1\,C_N, \tag{9.C.9}$$

where C_N is chosen to satisfy the conditions (9.C.4).

How does one choose C_N subject to the conditions (9.C.4) in a given problem? Usually the larger C_N is, the faster is the convergence rate of wrong detection probability to zero. Some rules have been recently developed in Bai, Krishnaiah, and Zhao [5] and Rao and Wu [6] for an optimum choice of C_N in finite samples observed data based on itself. The criterion $\mathrm{C_N}$ is, indeed, flexible unlike other criteria, and we suggest that an appropriate should be made in each individual case.

REFERENCES

1. M. Wax, T. J. Shan, and T. Kailath, "Spatio-temporal Spectral Analysis by Eigenstructure Methods," *IEEE Trans. Acoustics, Speech, and Signal Processing,* **32** (1984), 817–827.
2. A. Paulraj, and T. Kailath, "Eigenstructure Methods for Direction of Arrival Estimation in the Presence of Unknown Noise Fields, *IEE Trans. Acoustics, Speech, and Signal Processing,* **34,** no. 1 (1986), 13–20.
3. L. C. Zhao, P. R. Krishnaiah, and Z. D. Bai, "On Detection of the Number of Signals in Presence of White Noise," *J. Multivariate Analy.* **20** (1986), 1–25.
4. L. C. Zhao, P. R. Krishnaiah, and Z. D. Bai, "On Detection of the Number of Signals When the Noise Covariance Matrix is Arbitrary," *J. Multivariate Anal.* **20** (1986), 26–49.
5. Z. D. Bai, P. R. Krishnaiah, and L. C. Zhao, "On rates of convergence of efficient detection criteria in signal processing with white noise," *IEEE Transactions, Information Theory,* **35** (1989), 380–388.
6. C. R. Rao, and Y. Wu, "A Strongly Consistent Procedure for Model Selection," *Biometrika* 76 (1989), 369–74.
7. G. Bienvenu, "Influence of the Spatial Coherence of the Background Noise on High Resolution Passive Methods," *Proc. IEEE ICASSP* (1979), 306–309.
8. R. O. Schmidt, "A Signal Subspace Approach to Multiple Source Location and Spectral Estimation," Ph.D. dissertation, Stanford University, Stanford, California, 1981.

9. Z. D. Bai, P. R. Krishnaiah, and L. C. Zhao, "On the Direction of Arrival Estimation," *Tech. Report No. 87–12,* Center for Multivariate Analysis, University of Pittsburgh, 1987.

10. Z. D. Bai, and C. Radhakrishna Rao, (1989). "spectral analytical methods for the estimation of number of signals and directions of arrival," To appear in *Proc. Indo-US Workshop: Spectral Analysis in One or Two Dimensions.*

11. J. Von Neumann, "Some Matrix Inequalities and Metrization of Metric Space," *Tomsk Univ. Rev.* **1** (1937), 386–300.

12. Z. D. Bai, "A Note on the Asymptotic Joint Distribution of the Eigenvalues of a Noncentral Multivariate *F* Matrix," *J. Math. Research and Exposition,* **15** (1986), 113–118.

13. A. V. Skorohod, "Limit Theorems for Stochastic Process," *Theory Probab. Appl.* **1**, no. 3 (1956), 261–290.

14. H. Akaike, "Information Theory and an Extension of the Maximum Likelihood Principle," *Proc. 2nd Internat. Symp. on Information Theory,* eds. B. N. Petrov et al., Akademia Kiado, Budapest, 1973, pp. 267–281.

15. G. Schwarz, "Estimating the Dimension of a Model," *Ann. Statist.,* **6** (1978), 461–464.

16. J. Rissanen, "Modelling by Shortest Data Description," *Automatica,* **14** (1978), 465–471.

17. E. J. Hannan and B. G. Quinn, "The Determination of the Order of an Autoregression," *J. Roy. Statist. Soc.,* **B 41** (1979), 190–195.

10

Self-Calibration for High-Resolution Array Processing

Benjamin Friedlander and Anthony J. Weiss

10.1. INTRODUCTION

Direction-finding techniques based on the eigen decomposition of the covariance matrix of the received signals have been discussed extensively in the literature since the early 1980s. Computer simulations and a relatively limited number of experimental systems have demonstrated that in certain cases involving multiple closely spaced sources, these techniques have superior performance compared to conventional direction-finding techniques.

In spite of the potential advantages of eigenstructure-based methods, their application to real systems has been very limited. One of the reasons for this situation are the difficulties associated with calibrating the data collection system. Eigenstructure-based direction-finding techniques such as MUSIC [1], [2] require precise knowledge of the signals received by the sensor array from a standard source located at any given direction. The collection of the received signal vectors for all possible directions is often called the *array manifold*. The performance of the eigenstructure-based system depends strongly on the accuracy of this array manifold.

The process of measuring the array manifold can be time consuming and expensive. Calibrating an antenna array designed for two-dimensional (azimuth and elevation) direction finding with the accuracy required by these *superresolution* techniques poses numerous practical difficulties. The amount of memory required for storing the array manifold once it has been measured may also increase the size and cost of the system, even if interpolation techniques are used to reduce the number of points that need to be stored [3].

In addition to the problem of *initial* array calibration, there is the problem of *maintaining* array calibration. Many factors contribute to changing the response of the sensor array over time: gradual changes in the behavior of the sensor itself and of the electronic circuitry between the sensor and the output of the digitizer (due to thermal effects, aging of components, moisture, etc.), changes due to the environment around the sensor array (e.g., the effect of metal objects near an antenna array on its beam pattern), and changes in the location of the sensors (e.g., an antenna array located on the vibrating wing of an aircraft or a hydrophone array towed behind a ship). In many practical situations it is impossible to maintain array calibration to the accuracy required for the proper operation of these eigenstructure-based techniques. This results in significant degredation in system performance, sometimes to the point where these *superresolution* techniques perform no better (or worse) than conventional direction-finding methods.

A useful approach to alleviating the problems introduced by imprecise array calibration is to use the received signals to adjust or fine-tune the array calibration. *Self-calibrating* or *self-cohering antenna arrays* have been developed and tested by Steinberg [4] and others. Rockah and Schultheiss [5], [6] have studied in detail the self-calibration issue in the context of passive sensor arrays with imprecisely known sensor locations. Self-calibration techniques for eigenstructure-based array processing techniques seem to have received little attention. Lo and Marple [7] discussed a calibration technique that requires calibrating sources whose directions are known (at least two sources are required), and therefore their technique is not a true self-calibrating technique. Paulraj and Kailath [8] presented a method for *directions of arrival (DOA)* estimation by eigenstructure methods for an array with unknown sensor gains and phases. Their method does not require calibrating sources with known directions, but is limited to uniformly spaced *linear* arrays.

In an attempt to address this important problem we have recently developed a general class of self-calibration techniques for sensor arrays having arbitrary geometries. These techniques are based on simultaneous estimation of the DOA of plane waves impinging on the array, and of the unknown array parameters. We addressed specifically the cases of unknown sensor gains and phases, unknown mutual coupling coefficients, and unknown sensor locations. We were able to show that given a sufficient number of sensors and sources it is theoretically possible to obtain unique and accurate estimates of both the directions of arrival and of the array parameters. We were able to test and validate our algorithms to a limited extent on real data. However, it is too early to draw any conclusions about the role of self-calibration techniques in making high-resolution techniques practically useful.

In this chapter we present a summary of our work to date on self-calibration. Section 10.2 introduces the problem and the necessary notation and discusses the errors arising in the modeling of the array outputs. In Section 10.3 we briefly review eigenstructure-based direction finding algorithms, focusing on the MUSIC algorithm. In Section 10.4 we present some results on the sensitivity of these algorithms to modeling errors. Section 10.5 presents the basic idea underlying our self-calibration technique and an asymptotic analysis of its performance. In Section 10.6 we outline several specific estimation algorithms for different types of modeling errors, and illustrate their performance by numerical examples. Section 10.7 has some concluding remarks.

10.2. PROBLEM FORMULATION

Direction finding is an important problem arising in a wide range of applications, including surveillance, navigation, and communications. The challenging part of the direction-finding problem is to estimate the parameters of interest in the presence of *multiple* closely spaced sources. The case of a *single* source can be treated by "conventional" techniques such as time-of-arrival estimation (or phase estimation in the narrow-band case) [9]. The case of multiple well-separated sources (i.e., separated by more than a beamwidth) can be handled by beamforming [10]. The difficult case of multiple closely spaced sources requires more sophisticated signal processing techniques such as the MUSIC algorithm [1], [2] or other so-called "high-resolution" techniques. In this section we formulate the multisource direction-finding problem and discuss the type of modeling errors that arise in the "standard" formulation used in the literature. In this chapter we address only the narrow-band case, although the same ideas can be extended to broad-band signals.

10.2.1 The Multisource Estimation Problem

Consider N radiating sources observed by an arbitrary array of M sensors, and assume that $M > N$. The signal at the output of the mth sensor can be described by

$$x_m(t) = \sum_{n=1}^{N} \alpha_m s_n(t - \tau_{mn} - \psi_m) + v_m(t),$$

$$-T/2 \leq t \leq T/2, \qquad m = 1, 2, \ldots, M, \tag{10.1}$$

where $\{s_n(t)\}_{n=1}^{N}$ are the *radiated signals*, $\{v_m(t)\}_{m=1}^{M}$ are sample waveforms from additive *noise* processes and T is the observation interval. The parameters

$\{\tau_{mn}\}$ are *delays* associated with the signal propagation time from the nth source to the mth sensor. These parameters are of interest since they contain information about the source locations relative to the array. Finally, the parameters α_m and ψ_m are the *gain* and the *delay* associated with the mth sensor.

A convenient separation of the parameters to be estimated is obtained by representing the signals using *Fourier coefficients* defined by

$$X_m(\omega_l) = \frac{1}{\sqrt{T}} \int_{-T/2}^{T/2} x_m(t) e^{-j\omega_l t} dt, \tag{10.2}$$

where $\omega_l = \frac{2\pi}{T}(l_1 + l)$, $l = 1, 2, \ldots, L$, and l_1 is a constant. In principle, the number of coefficients required to capture all the signal information is infinite. However, if we consider signals with energy concentrated in a finite spectral band, we can use a finite (perhaps small) number of coefficients. Here we consider narrow-band signals with spectrum concentrated around some angular frequency ω_0, with a bandwidth that is small compared to $2\pi/T$. In other words, we are using *single* Fourier coefficient ($L = 1$). Taking the relevant Fourier coefficient of (10.1) and suppressing the dependence on ω_0 we obtain,

$$X_m = \sum_{n=1}^{N} \alpha_m e^{-j\phi_m} \cdot e^{-j\omega_0 \tau_{mn}} S_n + V_m, \qquad m = 1, 2, \ldots, M, \tag{10.3}$$

where S_n and V_m are the Fourier coefficients of $s_n(t)$ and $v_m(t)$, respectively, and where $\phi_m = \omega_0 \psi_m$ is the *phase shift* introduced by the mth sensor. Equation (10.3) may be expressed using vector notation as follows,

$$\boldsymbol{X}(j) = \Gamma \cdot A \cdot \mathbf{S}(j) + \mathbf{V}(j), \qquad j = 1, 2, \ldots, J, \tag{10.4}$$

where j is the index of different (independent) samples and

$$\begin{aligned}
\mathbf{X}(j) &= [X_1(j), \ldots, X_2(j), X_M(j)]^T, \\
\mathbf{S}(j) &= [S_1(j), S_2(j), \ldots, S_N(j)]^T, \\
\mathbf{V}(j) &= [V_1(j), V_2(j), \ldots, V_M(j)]^T, \\
\Gamma &= \text{diag}\,\{\alpha_1 e^{-j\phi_1}, \alpha_2 e^{-j\phi_2}, \ldots, \alpha_M e^{-j\phi_M}\}. \\
A_{mn} &= e^{-j\omega_0 \tau_{mn}}, \qquad m = 1, 2, \ldots, M; \quad n = 1, 2, \ldots, N.
\end{aligned}$$

To simplify the exposition further, we assume that the sensors and sources are coplanar and that the sources are far enough from the observing array so that the signal wavefronts are effectively planar over the array. It is easy to verify that the delays τ_{mn} are given by

$$\begin{aligned}
\tau_{mn} &= -d_{mn}/c, \\
d_{mn} &= \tilde{x}_m \sin \gamma_n + \tilde{y}_m \cos \gamma_n,
\end{aligned} \tag{10.5}$$

where c is the *propagation velocity*, d_{mn} is the *distance* from sensor m to sensor number one (reference sensor) in the direction of the nth source, $(\tilde{x}_m, \tilde{y}_m)$ are the coordinates of the mth sensor, γn is the DOA of the nth source relative to the $\tilde{y}$ axis, and the origin of the Cartesian coordinate system coincides with sensor number one—see Figure 10.1.

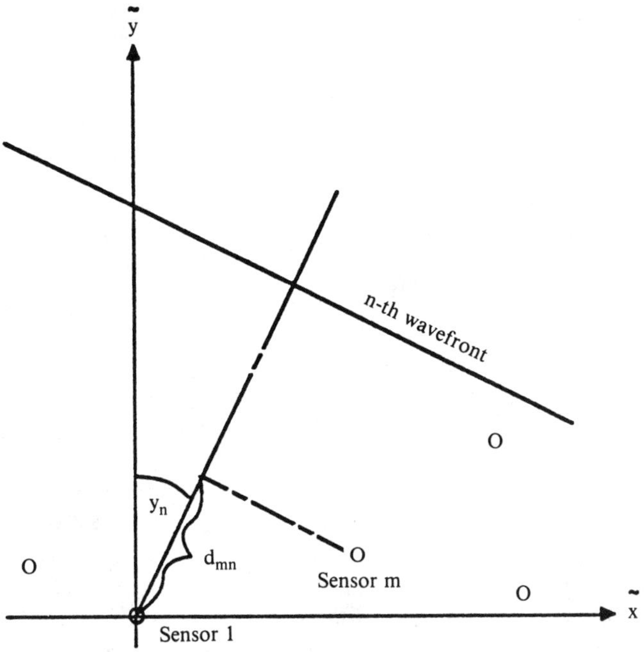

Figure 10.1. Problem geometry.

From (10.4) and (10.5), it follows that the elements of the matrix A are given by

$$A_{mn} = e^{j(\omega_0/c)(\tilde{x}_m \sin \gamma_n + \tilde{y}_m \cos \gamma_n)}. \tag{10.6}$$

It also follows that $\{A_{1n}\}_{n=1}^{N} = 1$ and that only the nth column of A depends on γ_n.

We are now ready to address the mutual coupling between the array elements. In the foregoing theory it was assumed that each element in the array acts independently of all the others. This assumption is often invalid in practice. Reflected radiation from one element *couples* to its neighbors, as do currents that propagate along the surface of the array. The output voltage of each array element is the sum of the primary voltage due to the incident radiation, plus all the contributions from various coupling sources from each of its neighbors. Hence, the actual voltage at the output of the array is given by the following modification of (10.4),

$$\mathbf{X}(j) = C \cdot \Gamma \cdot \boldsymbol{A} \cdot \mathbf{S}(j) + \mathbf{V}(j); \qquad j = 1, 2, \ldots, \boldsymbol{J}, \tag{10.7}$$

where $\boldsymbol{C}$ is an $M \times M$ complex matrix.

In the following we sometimes rewrite (10.7) as

$$\mathbf{X}(j) = \mathbf{A} \cdot \mathbf{S}(j) + \mathbf{V}(j), \qquad j = 1, 2, \ldots, \boldsymbol{J}, \tag{10.8}$$

where

$$\mathbf{A} = [\mathbf{a}(\gamma_1), \mathbf{a}(\gamma_2), \ldots, \mathbf{a}(\gamma_N)] \triangleq C \cdot \Gamma \cdot \boldsymbol{A}.$$

Each column of $\mathbf{A}$ represents the signals that would appear at the output of the array for a single-unit source located at a particular direction.

In general, the matrix $\boldsymbol{C}$ has no special structure. However, for a linear uniform array that is well balanced, a *banded matrix* provides an excellent model. The rationale behind this model is the fact that the mutual coupling coefficients are inversely proportional to the distance between the elements. Therefore, the mutual coupling between two elements that are far enough from each other, can often be approximated as zero. Moreover, we expect that an "ideal" linear uniform arrays will exhibit a *banded Toeplitz mutual coupling matrix* (that is, the coupling between any two equally spaced sensors is the same).

Using the same kind of reasoning leads to the conclusion that the mutual coupling matrix for a uniform *circular* array consists of three bands; a center band, a band at the upper right corner and a band at the lower left corner. Moreover, it is expected that an "ideal" circular uniform array will exhibit a circulant *mutual coupling matrix* (*MCM*). Exploiting the special structure of the MCM we can reduce the number of parameters needed to characterize the coupling matrix. A parsimonious parametrization is essential for the self-calibration algorithms discussed later.

10.2.2 Modeling Errors

The formulation just given is idealized in several respects. It assumes that the sensor gains and phases (α_i, ϕ_i), the mutual coupling coefficients (the entries of C), and the sensor locations ($\tilde{x}_m$, $\tilde{y}_m$) are precisely known. In practice, the actual values of these parameters will differ from their assumed values, and will in fact change over time due to environmental and other effects.

In other words, the actual vector of received signals, may differ from the assumed model for these signals due to various effects, including the ones mentioned in the introduction. In the next section we investigate the effect of these differences on the performance of direction-finding algorithms based on (10.8). But first, we briefly review the MUSIC algorithm [1], [2], which is perhaps the best known eigenstructure-based direction-finding algorithm.

10.3. REVIEW OF THE MUSIC ALGORITHM

The literature dealing with MUSIC and related algorithms is quite extensive. The reader interested in learning more about eigenstructure-based array processing techniques can find papers on this topic in almost every issue of the *IEEE Transactions on Acoustics, Speech, and Signal Processing* (*ASSP*) that appeared in the past few years, and in recent proceedings of the International Conference on ASSP. In this section we review the MUSIC algorithm only briefly, assuming that the reader is familiar with this algorithm.

The *MUltiple SIgnal Classification* (*MUSIC*) algorithm and related algorithms are based on the properties of the covariance matrix of the vector of received signals. The covariance matrices of the signal, noise, and observation vectors are given by

$$\begin{aligned} \boldsymbol{R}_s &= E\{\mathbf{S}\mathbf{S}^\dagger\}, \ \sigma^2\Sigma_0 = E\{\mathbf{V}\mathbf{V}^\dagger\}, \\ \boldsymbol{R}_x &= E\{\mathbf{X}\mathbf{X}^\dagger\} = \mathbf{A}\boldsymbol{R}_s\mathbf{A}^\dagger + \sigma^2\Sigma_0, \end{aligned} \tag{10.9}$$

where $(\cdot)^\dagger$ represents the Hermitian (conjugate) transpose operation.

We make the standard assumptions underlying the MUSIC algorithm and other eigenstructure-based methods for direction finding:

1. The signals and the noise processes are stationary over the observation interval.
2. The columns of $\mathbf{A}$ are linearly independent.
3. The signals are not perfectly correlated.
4. The noise is uncorrelated with the signals and its covariance matrix is full rank and is known except for a multiplicative constant σ^2.

The following theorem forms the basis for the eigenstructure approach.

Theorem 3.1. Let λ_i and $\mathbf{u}_i$, $i = 1, 2, \ldots, M$ be the *eigenvalues* and corresponding *eigenvectors* of the matrix pencil $(\boldsymbol{R}_x, \Sigma_0)$, (that is, the solutions of $\boldsymbol{R}_x\mathbf{u} = \lambda\Sigma 0\mathbf{u}$), where the λ_i's are listed in descending order. Then,

(a) $\lambda_{N+1} = \lambda_{N+2} = \ldots = \lambda_M = \sigma^2$.

(b) Each of the columns of $\mathbf{A}$ is orthogonal to the matrix

$$U = [\mathbf{u}_{N+1}, \mathbf{u}_{N+2}, \ldots, \mathbf{u}_M].$$

Proof. See [1], [2] for details. Here we present an outline of the proof.

Define the $M \times M$ matrix $\boldsymbol{Z} = [\mathbf{u}_1, \mathbf{u}_2, \ldots, \mathbf{u}_M]$ and the $M \times M$ diagonal matrix $\Lambda = \text{diag}\{\lambda_1, \lambda_2, \ldots, \lambda_M\}$. By definition.

$$\boldsymbol{R}_x\mathbf{u}_i = \Sigma_0\mathbf{u}_i\lambda_i, \qquad \text{for } i = 1, 2, \ldots, M, \tag{10.10}$$

or equivalently

$$R_x Z = \Sigma_0 \mathbf{Z} \Lambda. \tag{10.11}$$

Substituting the expression given in (10.9) for R_x and subtracting $\sigma^2 \Sigma_0 Z$ from the result we obtain

$$\mathbf{A} R_s \mathbf{A}^\dagger \mathbf{Z} = \Sigma_0 Z(\Lambda - \sigma^2 I). \tag{10.12}$$

The left side of (10.12) has rank $N < M$ according to our assumptions, but Σ_0 and Z are full-rank matrices, and hence $M - N$ elements of Λ are equal to σ^2. Note that the λ_i's that are equal to σ^2 are the smallest, since $\mathbf{A}R_s\mathbf{A}^\dagger$ is the signal covariance matrix and is therefore nonnegative definite. Thus, part 1 of the theorem is proved. Now, since by definition

$$R_x U = (\mathbf{A}R_s\mathbf{A}^\dagger + \sigma^2\Sigma_0)\, U = \sigma^2 \Sigma_0 U, \tag{10.13}$$

then

$$\mathbf{A}R_s\mathbf{A}^\dagger U = \mathbf{0}. \tag{10.14}$$

But $\mathbf{A}R_s$ is an $M \times N$ matrix of rank N (guaranteed by our assumptions,) and therefore we can find an $N \times M$ matrix, D, with rank N so that

$$D\mathbf{A}R_s = I_N \qquad (I_N \text{ is the } N \times N \text{ identity matrix}). \tag{10.15}$$

Multiplying (10.14) on the left by D we get

$$D\mathbf{A}R_s\mathbf{A}^\dagger U = \mathbf{A}^\dagger U = \mathbf{0}, \tag{10.16}$$

which proves the second part of the theorem.

This theorem suggests that we should first estimate R_x and then use the estimates of λ_i to determine the number of signals. Once N is known, reasonable estimates of $\{\gamma_n\}_{n=1}^{N}$ may be obtained by searching for the N smallest local minima of $\tilde{J} = \|\hat{U}^\dagger A(\gamma)\|^2$, or equivalently, searching for the N largest peaks of $1/\|\hat{U}^\dagger A(\gamma)\|^2$, where $\hat{U}$ is an estimate of U.

If $\hat{U}$ were a perfect estimate of U (that is, $\hat{U} = U$) then the minimum value of $\tilde{J}$ ($\tilde{J} = 0$) will be achieved for the true $\{\hat{\gamma}_n\}_{n=1}^{N}$. When $\hat{U}$ is an imperfect estimate of U, the minimization of $\tilde{J}$ will provide estimates $\{\hat{\gamma}_n\}_{n=1}^{N}$ of the true DOAs. The accuracy of these estimates has been investigated extensively by simulations [11] and analysis [12]. In [12] it was shown that the variance of the DOA estimates provided by MUSIC approach asymptotically (as the number of snapshots increases) the Cramér-Rao lower bound, for the case of a single source. The same holds for the case of multiple sources, as the signal-to-noise ratio increases to infinity. Simulation results have demonstrated the good performance of the algorithm for practical values of the signal-to-noise ratio and the number of snapshots.

10.4. THE SENSITIVITY OF MUSIC TO MODELING ERRORS

In this section we study the effect of differences between the true and assumed array manifolds caused by modeling errors, on the MUSIC algorithm. First, we need to introduce some notation. Let $A(\gamma;\boldsymbol{\theta})$ denote the array manifold, that is, the vector of received signals for a unit-energy far-field source at direction γ, for a sensor array with parameters $\boldsymbol{\theta}$. The parameter vector $\boldsymbol{\theta}$ contains the array parameters which are the subject of the sensitivity analysis. For example, $\boldsymbol{\theta}$ may contain the sensor gains ($\boldsymbol{\theta} = [\alpha_1, \ldots, \alpha_M]^T$), phases ($\boldsymbol{\theta} = [\phi_1, \ldots, \phi_M]^T$), the sensor locations ($\boldsymbol{\theta} = [(\tilde{x}_1, \tilde{y}_1), \ldots, (\tilde{x}_M, \tilde{y}_M)]^T$), or other parameters such as the mutual coupling coefficients of the array elements.

Next we consider the case where the array manifold used by the MUSIC algorithm $A(\gamma; \boldsymbol{\theta}_0)$, corresponding to a nominal value $\boldsymbol{\theta}_0$ of the array parameters, differs from the true array manifold $A(\gamma; \boldsymbol{\theta})$, where $\boldsymbol{\theta}$ is slightly different from $\boldsymbol{\theta}_0$. We refer to the difference between the true and assumed array parameters as a *modeling error.*

We study the effect of such modeling errors on the MUSIC algorithm in the "ideal" case where we are given the exact covariance matrix of the received signal vector. This corresponds to having an infinite (or very large) number of independent "snapshots" of the received signal vector. In other words, we are given the covariance matrix

$$R_x(\boldsymbol{\theta}) = \mathbf{A}(\boldsymbol{\theta})R_s\mathbf{A}(\boldsymbol{\theta})^\dagger + \sigma^2\Sigma_0, \tag{10.17}$$

where

$$\mathbf{A}(\boldsymbol{\theta}) = [A(\gamma_1; \boldsymbol{\theta}), A(\gamma_2; \boldsymbol{\theta}), \ldots, A(\gamma_N; \boldsymbol{\theta})]. \tag{10.18}$$

The MUSIC algorithm involves searching for the N largest peaks of

$$S(\gamma; \boldsymbol{\theta}) = 1/\|U_n^\dagger (\boldsymbol{\theta})\, A(\gamma; \boldsymbol{\theta}_0)\|^2, \tag{10.19}$$

where $U_n\,(\theta)$ is the "noise subspace" obtained by eigen decomposition of $R_x\,(\boldsymbol{\theta})$. Alternatively, we could search for the N smallest minima of $F(\gamma; \boldsymbol{\theta})$. We note the following alternative expression for F:

$$F(\gamma; \boldsymbol{\theta}) = A^\dagger\,(\gamma; \boldsymbol{\theta}_0)P_n(\boldsymbol{\theta})A(\gamma; \boldsymbol{\theta}_0), \tag{10.20}$$

where

$$P_n(\theta) = I - \mathbf{A}(\boldsymbol{\theta})(\mathbf{A}^\dagger\,(\boldsymbol{\theta})\mathbf{A}(\boldsymbol{\theta}))^{-1}\,\mathbf{A}^\dagger\,(\boldsymbol{\theta}). \tag{10.21}$$

In the absence of modeling errors, $\theta_0 = \theta$, in which case the peaks of $S(\gamma,\theta)$ will provide exact estimates of the directions of arrival. However, if $\theta_0 \neq \theta$, the peaks of $S(\gamma,\theta)$ will no longer coincide with the true DOAs. In fact, even small modeling errors can lead to significant degradation in the performance of the MUSIC algorithm, as is illustrated by the following example.

Figure 10.2 shows the MUSIC "spectrum" for a circular array of four uniformaly spaced sensors, half a wavelength apart. The algorithm is given the exact data covariance matrix for two sources at directions 13.5° and 16.5°. Modeling errors were introduced in the sensor phases that is, $\boldsymbol{\theta} = [\phi_1, \phi_2, \phi_3, \phi_4]$. The phases were assumed to be zero and the gains to be unity. However, the true phases were $\boldsymbol{\theta} = \beta[-0.5829°, -0.0329°, -0.7036°, -0.4051°]$. The parameter β determines the magnitude of the modeling error in each case. The MUSIC "spectrum" is plotted for DOAs between 7.5° and 22.5°. Examination of this figure shows the gradual degradation of the "spectrum" as the modeling errors increase and the transition from a "spectrum" containing two peaks to a flat single peak "spectrum."

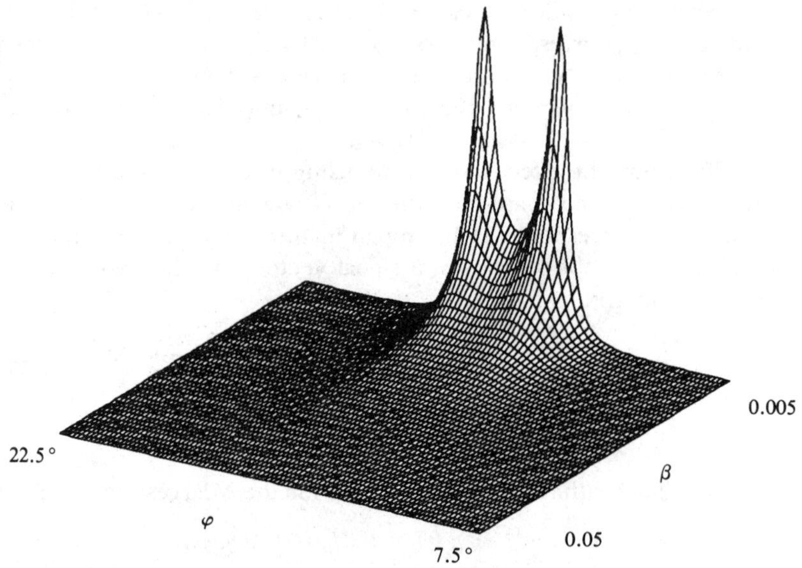

Figure 10.2 The MUSIC "spectrum" in the presence of phase errors, $0.005 < \beta < 0.05$.

For small modeling errors the MUSIC spectrum has two distinct peaks. The DOA estimates are provided by the locations of the peaks in this "spectrum." Due to modeling errors the peak locations are different from the true source DOAs. The locations of the peaks are characterized by the fact that the first derivative of $S(\gamma;\theta)$ equals zero at the peaks.

Alternatively, we could look at the partial derivatives of $F(\gamma;\theta) = 1/S(\gamma;\theta)$. Let us denote

$$F_1(\gamma;\boldsymbol{\theta}) = \frac{\partial F(\gamma,\boldsymbol{\theta})}{\partial \gamma}. \tag{10.22}$$

In the following we will deal with F rather than with S, since its derivatives are somewhat simpler to compute.

To see how much the peak locations shift due to modeling errors we perform a first-order Taylor expansion of $F_1(\gamma;\theta)$ around the true source DOAs γ_1, γ_2.

$$F_1(\gamma;\boldsymbol{\theta}) = F_1(\gamma_i;\boldsymbol{\theta}_0) + \frac{\partial F_1(\gamma;\boldsymbol{\theta}_0)}{\partial \gamma}\Big|_{\gamma = \gamma_i}(\gamma - \gamma_i) + \frac{\partial F_1(\gamma_i;\boldsymbol{\theta})}{\partial \boldsymbol{\theta}}\Big|_{\boldsymbol{\theta} = \boldsymbol{\theta}_0}(\boldsymbol{\theta} - \boldsymbol{\theta}_0), \tag{10.23}$$

where $i = 1,2$. Note that $F_1(\gamma_i;\boldsymbol{\theta}_0) = 0$.

By setting $F_1(\gamma;\boldsymbol{\theta}) = 0$ we get the following relationship between the change in the DOA estimate and the modeling error:

$$\tilde{\gamma}_i \triangleq \gamma - \gamma_i = -\frac{\frac{\partial F_1(\gamma_i;\boldsymbol{\theta})}{\partial \theta}\Big|_{\boldsymbol{\theta} = \boldsymbol{\theta}_0}\tilde{\boldsymbol{\theta}}}{\frac{\partial F_1(\gamma;\boldsymbol{\theta}_0)}{\partial \gamma}\Big|_{\gamma = \gamma_i}}, \tag{10.24}$$

where

$$\tilde{\boldsymbol{\theta}} \triangleq \boldsymbol{\theta} - \boldsymbol{\theta}_0. \tag{10.25}$$

Let us assume that

$$\tilde{\boldsymbol{\theta}} = \sigma_\theta \mathbf{u}, \tag{10.26}$$

where $\mathbf{u}$ is a random vector whose elements are zero-mean unit-variance random variables and where σ_θ is a positive scalar. It is straightforward to show that

$$E\{\tilde{\gamma}_i\} = 0 \tag{10.27}$$

and

$$\text{std}\{\tilde{\gamma}_i\} = \frac{\left\|\frac{\partial F_1(\gamma_i;\boldsymbol{\theta})}{\partial \theta}\Big|_{\boldsymbol{\theta} = \boldsymbol{\theta}_0}\right\|\sigma_\theta}{\left\|\frac{\partial F_1(\gamma;\boldsymbol{\theta}_0)}{\partial \gamma}\Big|_{\gamma = \gamma_i}\right\|}, \tag{10.28}$$

where std$\{\cdot\}$ is the standard deviation of $\{\cdot\}$, and where $\|\cdot\|$ is the vector norm. Let us define

$$\sigma_i = \frac{\text{std}\{\tilde{\gamma}_i\}}{\sigma_\theta} = \frac{\left\|\frac{\partial F_1(\gamma_i;\boldsymbol{\theta})}{\partial \theta}\Big|_{\boldsymbol{\theta} = \boldsymbol{\theta}_0}\right\|}{\left\|\frac{\partial F_1(\gamma;\boldsymbol{\theta}_0)}{\partial \gamma}\Big|_{\gamma = \gamma_i}\right\|}, \tag{10.29a}$$

and

$$\sigma_{\text{DOA}} = \sqrt{\frac{1}{N}\sum_{i=1}^{N} \sigma_i^2}\,. \tag{10.29b}$$

This quantity is a measure of the sensitivity of the MUSIC algorithm, since it is the ratio of the DOA errors (averaged over all sources) and the modeling error that caused it.

To evaluate (10.29) we need to compute the derivatives of F_1 for the particular array and the particular parameter vector θ under consideration. In [13] we present the formulas needed to compute these derivatives for arbitrary array geometries and for three different choices of the array parameters (θ): sensor gains, sensor phases, and element locations.

To gain some insight into the sensitivity properties of the MUSIC algorithm we evaluated σ_{DOA} for different array geometries and different DOAs [13]. Here we give only one example: We consider a circular arrays with 4, 8, and 16 elements uniformly spaced around the circle, half a wavelength apart. There are two equal-power sources whose midpoint was fixed at 15° and whose separation was allowed to vary. The nominal gains are unity, the nominal phases are zero, and there is no mutual coupling between the array elements (i.e., C = identity matrix). Figure 10.3 depicts the sensitivity parameter versus source separation (in degrees). The solid lines depict the results for a 4-element array, the dashed lines for an 8-element array, and the dotted lines for a 16-element array.

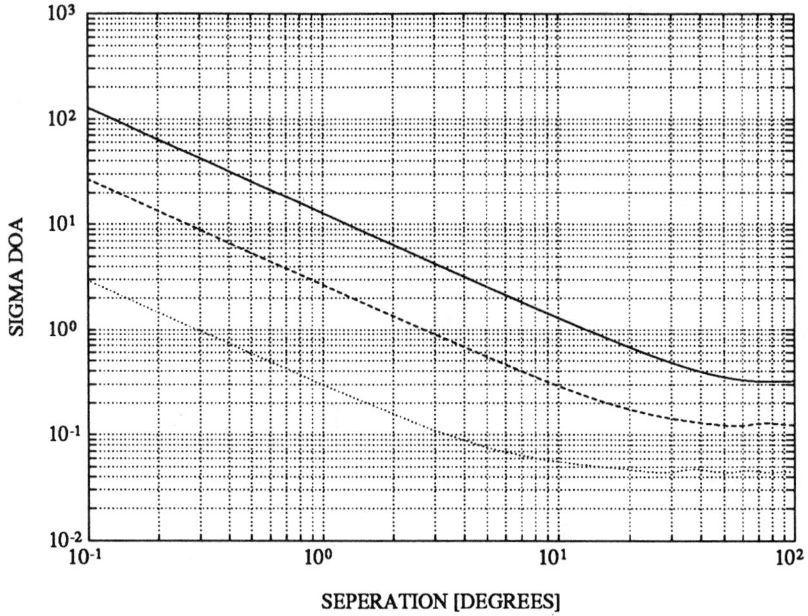

Figure 10.3 σ_{DOA} for phase errors versus source separation, circular array, 0.5λ spacing, for 4, 8, and 16 element arrays.

10.4.1 The Failure Threshold of MUSIC

We analyzed the effect of modeling errors that are sufficiently small so that the MUSIC "spectrum" has two distinct peaks. Next we consider the case where these errors are sufficiently large so that the algorithm fails to resolve the two sources. To understand our approach to analyzing this problem we examine again Fig. 10.2. Note that between the two "spectral" peaks there is a valley. As the modeling errors increase, the valley becomes more and more flat, until finally the two surrounding peaks merge into one, and the valley disappears.

Let us denote the location of the valley bottom by γ^*. This point is characterized by the fact that $\frac{\partial S(\gamma,\boldsymbol{\theta})}{\partial \gamma}|_{\gamma = \gamma^*} = 0$, and $\frac{\partial^2 S(\gamma,\boldsymbol{\theta})}{\partial \gamma^2}|_{\gamma = \gamma^*} > 0$. At the transition point where the valley disappears as the algorithm fails to resolve the two sources we have $\frac{\partial^2 S(\gamma,\boldsymbol{\theta})}{\partial \gamma^2}|_{\gamma = \gamma^*} = 0$.

Alternatively, we could look at the partial derivatives of $F(\gamma;\boldsymbol{\theta}) = 1/S(\gamma;\boldsymbol{\theta})$. Again, we denote

$$F_2(\gamma;\boldsymbol{\theta}) = \frac{\partial^2 F(\gamma,\boldsymbol{\theta})}{\partial \gamma^2}. \tag{10.30}$$

Then clearly $F_1(\gamma^*;\theta) = 0$ and $F_2(\gamma^*;\boldsymbol{\theta}) < 0$ until the transition point at which $F_2(\gamma^*;\boldsymbol{\theta}) = 0$. In the following we deal with F rather than with S, since its derivatives are somewhat simpler to compute. These observations suggest that by studying how $F_2(\gamma^*;\boldsymbol{\theta})$ changes as a function of $\boldsymbol{\theta}$ it should be possible to quantify the magnitude of the phase errors that cause the MUSIC algorithm to fail.

In general, the location of the valley bottom, γ^*, will change with the modeling error $\boldsymbol{\theta}$. This fact complicates the analysis considerably. To simplify matters we assume for the moment that γ^* is constant and that $\gamma^* = \gamma_0$, where $\gamma_0 \stackrel{\Delta}{=} (\gamma_1 + \gamma_2)/2$. In other words, we assume that the valley bottom is fixed at the midpoint between the two source directions. This enables us to study $F_2(\gamma_0;\boldsymbol{\theta})$ rather than $F_2(\gamma^*;\boldsymbol{\theta})$.

We approximate $F_2(\gamma_0;\boldsymbol{\theta})$ by a first-order Taylor expansion around θ_0

$$F_2(\gamma_0;\boldsymbol{\theta}) \approx F_2(\gamma_0;\boldsymbol{\theta}_0) + \frac{\partial F_2(\gamma_0;\boldsymbol{\theta})}{\partial \boldsymbol{\theta}}|_{\boldsymbol{\theta} = \boldsymbol{\theta}_0}(\boldsymbol{\theta} - \boldsymbol{\theta}_0). \tag{10.31}$$

Let us denote

$$\boldsymbol{\theta} - \boldsymbol{\theta}_0 = ||\boldsymbol{\theta} - \boldsymbol{\theta}_0||\boldsymbol{u}, \tag{10.32}$$

where $\boldsymbol{u}$ is a unit length vector and where $||\cdot||$ is the vector norm. The biggest change in F_2 for a given value of $||\boldsymbol{\theta} - \boldsymbol{\theta}_0||$ will occur when $\boldsymbol{u}$ is collinear with $\frac{\partial F_2(\gamma_0;\boldsymbol{\theta})}{\partial \boldsymbol{\theta}}|\boldsymbol{\theta} = \boldsymbol{\theta}_0$ in which case,

$$F_2(\gamma_0;\boldsymbol{\theta}) \approx F_2(\gamma_0;\boldsymbol{\theta}_0) \pm \left|\left| \frac{\partial F_2(\gamma_0;\boldsymbol{\theta})}{\partial \boldsymbol{\theta}}|_{\boldsymbol{\theta} = \boldsymbol{\theta}_0} \right|\right| \, ||\boldsymbol{\theta} - \boldsymbol{\theta}_0||. \tag{10.33}$$

We are interested in the point where $F_2(\gamma_0;\boldsymbol{\theta}) = 0$, which will occur when

$$||\boldsymbol{\theta} - \boldsymbol{\theta}_0|| = |F_2(\gamma_0;\boldsymbol{\theta}_0)|/|| \frac{\partial F_2(\gamma_0;\boldsymbol{\theta})}{\partial \boldsymbol{\theta}} |_{\boldsymbol{\theta} = \boldsymbol{\theta}_0}||. \tag{10.34}$$

In other words, the *smallest norm* modeling error that will cause the MUSIC algorithm to fail is given by

$$\boldsymbol{\theta} - \boldsymbol{\theta}_0 = \pm F_2(\gamma_0;\boldsymbol{\theta}_0) \frac{\partial F_2(\gamma_0;\boldsymbol{\theta})}{\partial \boldsymbol{\theta}} |_{\boldsymbol{\theta} = \boldsymbol{\theta}_0} /|| \frac{\partial F_2(\gamma_0;\boldsymbol{\theta})}{\partial \boldsymbol{\theta}} |_{\boldsymbol{\theta} = \boldsymbol{\theta}_0}||^2. \tag{10.35}$$

We refer to

$$\sigma_{\text{fail}} = ||\boldsymbol{\theta} - \boldsymbol{\theta}_0||/\sqrt{M}, \tag{10.36}$$

as the *failure threshold* of MUSIC. We divide the total phase error by $\sqrt{M}$ to get a per-sensor error. Formulas for computing the partial derivatives appearing in (10.34) can be found in [13].

To gain some insight into the failure properties of the MUSIC algorithm, we evaluated σ_{fail} for the case considered earlier. In Fig. 10.4 we plot the failure threshold versus source separation for the case of phase errors for element spacing of half a wavelength. Note the quadratic dependence of the failure threshold on source separation. A more comprehensive set of numerical examples can be found in [13].

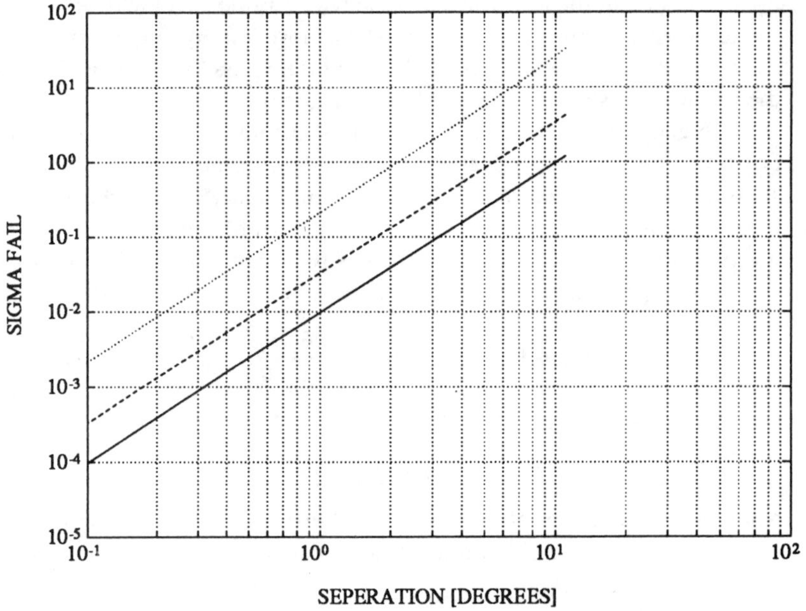

Figure 10.4: σ_{fail} for phase errors versus source separation, circular array, 0.5λ spacing, for 4, 8, and 16 element arrays.

It should be emphasized again that these results are for the infinite data case. In the finite data case, MUSIC is expected to fail in the presence of smaller modeling errors. Of course, if the signal-to-noise ratio or the number of "snapshots" used by the algorithm are too small, MUSIC will fail even in the absence of modeling errors. This failure mode was analyzed in [14].

10.5. PERFORMANCE ANALYSIS OF SELF-CALIBRATION

The high sensitivity of eigenstructure algorithms to modeling errors is a serious problem that needs to be solved if these techniques are to be practically useful. Here we propose a solution technique that is conceptually simple, although its implementation may not be straightforward. Let the elements of $\boldsymbol{\theta}$ be the array parameters that are most difficult to calibrate and that account for most of the sensitivity problem. Use the received signals to estimate *simultaneously* the directions of arrival $\{\gamma_n\}$ and the array parameters $\boldsymbol{\theta}$.

The idea of jointly estimating the DOAs and the array parameters raises a number of questions such as (1) *uniqueness*—under what conditions is it possible to uniquely determine all of the estimated parameters? (2) *accuracy*—how accurate are the estimates of the DOA and of the array parameters? To what extent is the accuracy of the DOA estimates degraded due to the need to estimate the array parameters? (3) *estimation algorithms*—are there estimation algorithms that are robust and that do not require an excessive amount of computation? In this section we address briefly the first two questions, while the third is addressed in Section 10.6.

The uniqueness issue can be studied in several ways, but a completely satisfactory answer is not available at this time. By counting the number of equations and the number of unknown parameters we are able to obtain necessary (but not sufficient) conditions for a unique solution. For example, in the case where $\boldsymbol{\theta} = [\alpha_1, \ldots, \alpha_M, \psi_1, \ldots, \psi_M]^T$, it is necessary to have at least two sources present. In the case of a *linear* array, it is also necessary to know the direction to one of the sources; see [15]–[17] for a more detailed discussion of the uniqueness issue.

The accuracy issue can be addressed by means of the *Cramér-Rao lower bound* (*CRLB*). Let $\{\mathbf{X}(j), j = 1, \ldots, T\}$ be independent, Gaussian, zero-mean random variables with a covariance matrix $\boldsymbol{R}_x = E\{\mathbf{X}\mathbf{X}^\dagger\}$ that depends on an unknown parameter vector β. The CRLB provides a lower bound for any unbiased estimate $\hat{\boldsymbol{\beta}}$ of $\boldsymbol{\beta}$. More specifically,

$$\operatorname{cov}\{\hat{\beta}\} \geq \boldsymbol{J}^{-1}, \qquad \boldsymbol{J} = [\mathrm{J}_{mn}], \tag{10.37}$$

$$J_{mn} = T \cdot \operatorname{tr}\left[\boldsymbol{R}_x^{-1} \frac{\partial \boldsymbol{R}_x}{\partial \beta_m} \boldsymbol{R}_x^{-1} \frac{\partial \boldsymbol{R}_x}{\partial \beta_n}\right]. \tag{10.38}$$

In the cases considered in this paper, $\boldsymbol{\beta} = [\gamma, \theta]$, in which case the ***Fisher information matrix J*** can be divided up as follows:

$$\boldsymbol{J} = \begin{bmatrix} J_{\gamma\gamma} & J_{\gamma\theta} \\ J_{\theta\gamma} & J_{\theta\theta} \end{bmatrix}. \tag{10.39}$$

The inverse of the Fisher information matrix can be similarly divided:

$$\boldsymbol{J}^{-1} = \begin{bmatrix} \boldsymbol{C}_{\gamma\gamma} & \boldsymbol{C}_{\gamma\theta} \\ \boldsymbol{C}_{\theta\gamma} & \boldsymbol{C}_{\theta\theta} \end{bmatrix}. \tag{10.40}$$

Note that $\boldsymbol{C}_{\gamma\gamma}$ is the CRLB for the DOA estimates, when the array parameters θ are being jointly estimated, while $\boldsymbol{J}_{\gamma\gamma}^{-1}$ is the CRLB for the DOA estimates when the array parameters are perfectly known. It is straightforward to show that $\boldsymbol{C}_{\gamma\gamma} \geq \boldsymbol{J}_{\gamma\gamma}^{-1}$, as expected. A comparison of the diagonal elements of $\boldsymbol{C}_{\gamma\gamma}$ and $\boldsymbol{J}_{\gamma\gamma}^{-1}$ indicates how much accuracy is lost in estimating the DOAs, due to the need to estimate the array parameters. The diagonal elements of $\boldsymbol{C}_{\theta\theta}$ indicate the accuracy with which the unknown array parameters can be estimated.

The detailed derivations of the CRLB for various choices of the array parameter vector $\boldsymbol{\theta}$ can be found in Appendix B. To get some insight into what can be learned from examination of the Cramér-Rao lower bound, we present some numerical examples.

The following examples consist of plots of the CRLB given in Appendix B, evaluated for particular system parameters. In each case we have a circular array of sensors, half a wavelength apart. Each figure contains four curves:

Solid line: CRLB for DOA γ_1, all the array parameters are known.
Dashed line: CRLB for DOA γ_1, sensor gains are unknown.
Dotted line: CRLB for DOA γ_1, sensor phases are unknown.
Dot-dash line: CRLB for DOA γ_1, sensor gains and phases are unknown.

Each curve depicts the standard deviation of the DOA estimate (actually 10 times the logarithm of the standard deviation) for the first source, as function of the signal-to-noise ratio.

Figures 10.5 to 10.8 were computed for a six-sensor array, for different source configurations. In these and the following examples, we see quite clearly that lack of knowledge about the sensor gains has very little effect on performance. Most of the loss in performance is due to the need to estimate the sensor phases. In these examples we note that increasing the number of sources increases the estimation error, but reduces the difference between the known and unknown phase cases.

Figures 10.9 to 10.11 were computed for the case of two sources at 0° and 20°, for different numbers of sensors. As expected, increasing the number of sensors improves estimation accuracy and reduces the loss of performance due to the need to estimate unknown parameters. Note in particular the curve for the unknown DOA-gain-phase case in Fig. 10.9. This figure seems to indicate that attempting to

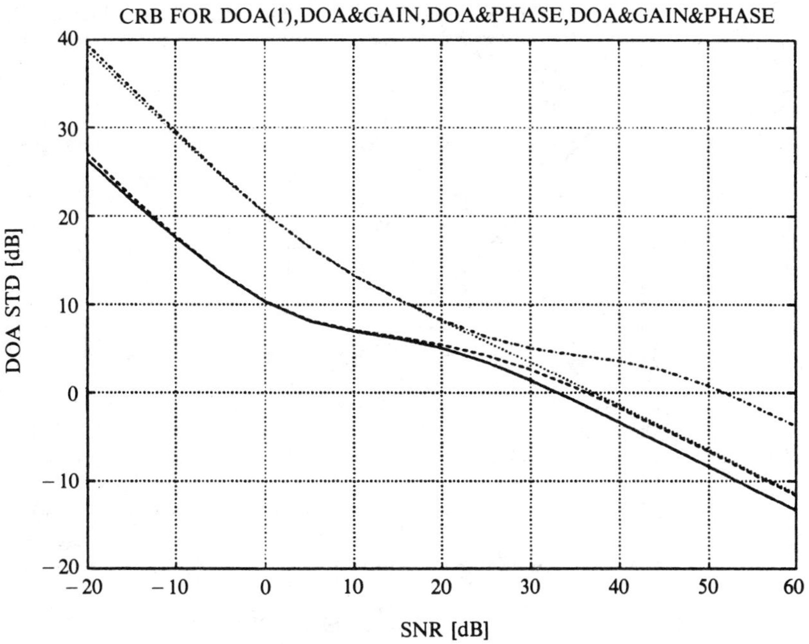

Figure 10.5 The CRLB for the estimated DOA, two sources at 10° and 20°.

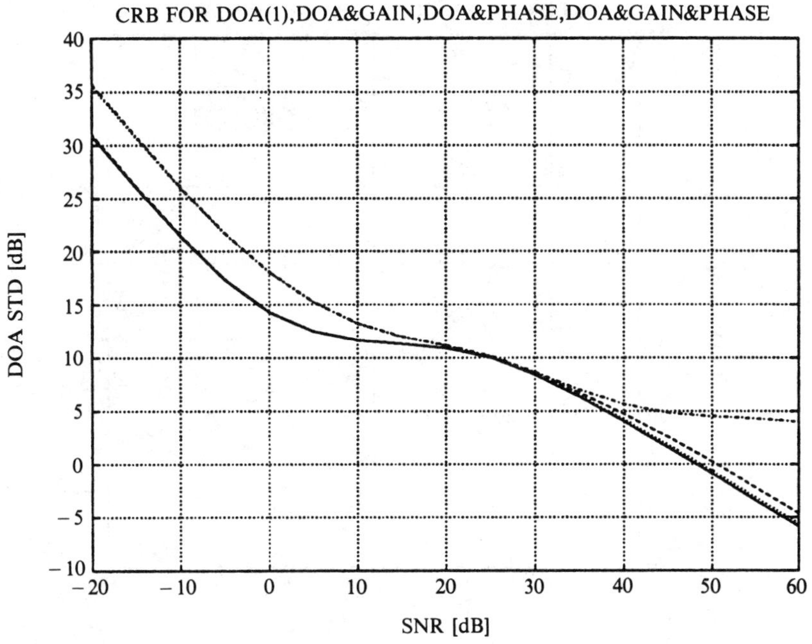

Figure 10.6 The CRLB for the estimated DOA, three sources at 10°, 0°, and 20°.

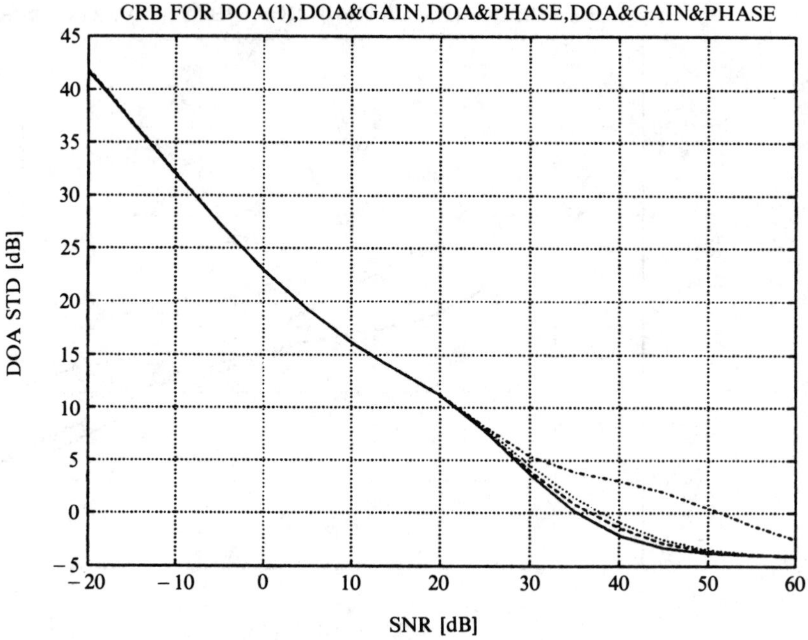

Figure 10.7 The CRLB for the estimated DOA, three sources at 10°, 11°, and 20°.

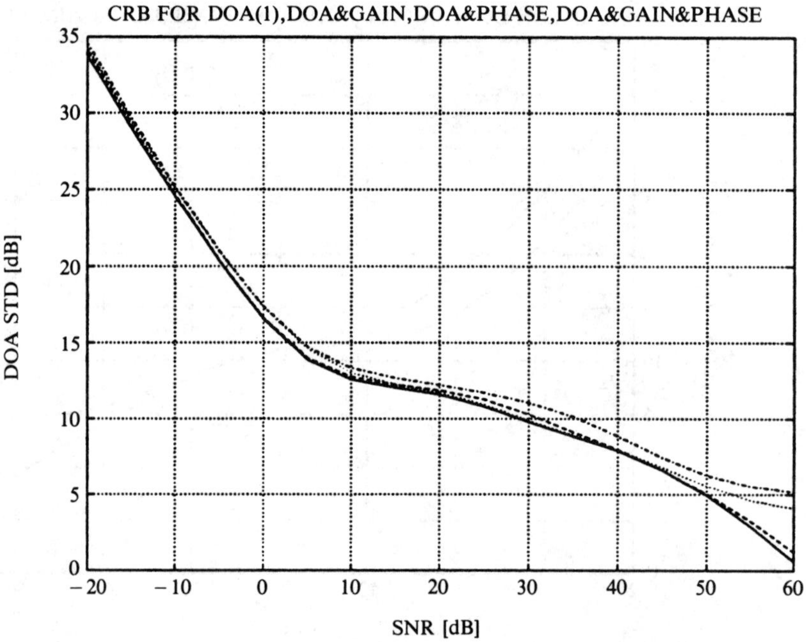

Figure 10.8 The CRLB for the estimated DOA, four sources at 10°, 0°, 20°, and 30°.

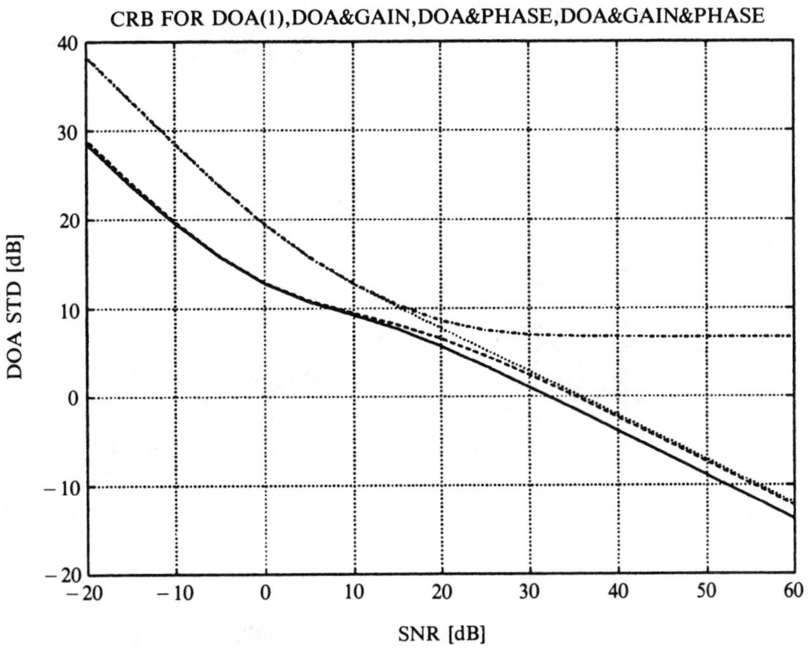

Figure 10.9 The CRLB for the estimated DOA, four sensors.

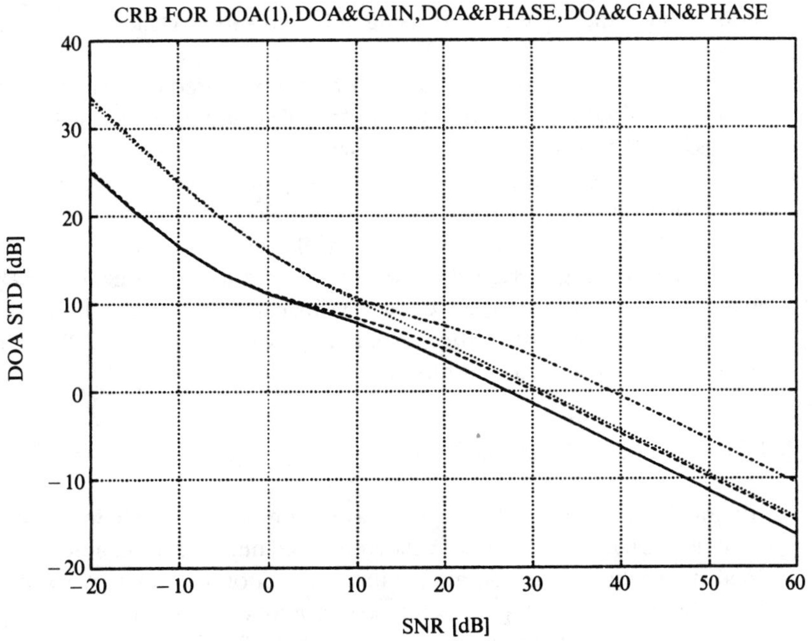

Figure 10.10 The CRLB for the estimated DOA, six sensors.

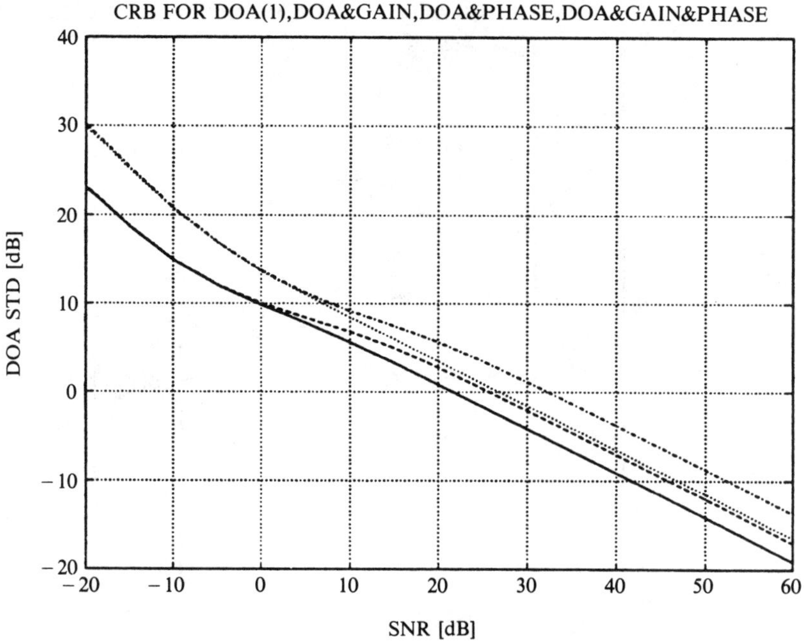

Figure 10.11 The CRLB for the estimated DOA, eight sensors.

estimate all of these parameters using only four sensors leads to a very ill-conditioned problem.

Figures 10.10 and 10.12 to 10.14 were computed for a six-sensor array, for two sources centered at 10°, with different spatial separations. These figures indicate that the loss in performance due to estimating the phase increases significantly as the source separation decreases. Thus, it is very desirable to have the sources used for self-calibration well separated.

By studying such plots for a variety of scenarios we observed many interesting phenomena. As a general rule it seems that the joint estimation of DOA, gains and phases, has only a minor effect on performance when the number of sources is sufficiently large and when they are spatially widely separated. This lends credibility to the idea of array self-calibration.

10.6. SELF-CALIBRATION ALGORITHMS

The joint estimation of DOAs and array parameters is simple in principle but may be difficult in practice due to the relatively large number of parameters involved. "Optimal" techniques, such as the maximum-likelihood estimator, involve global search in a high-dimensional parameter space, and are computationally prohibitive. In [15]–[17] we developed and tested a number of self-calibration algorithms based on the idea of iterating between estimating the DOAs and estimating the array parameters. We start with a nominal value of the array parameters ($\boldsymbol{\theta}_0$). Given the array

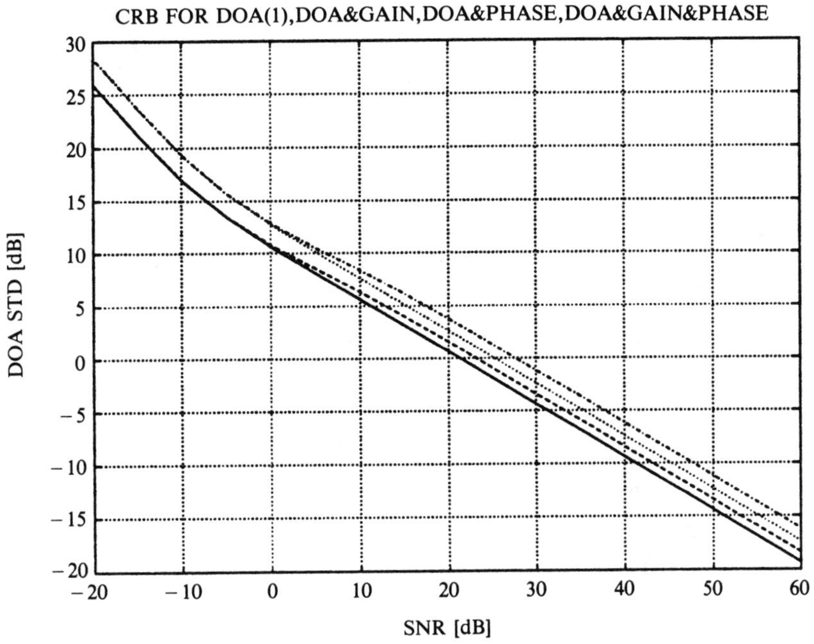

Figure 10.12 The CRLB for the estimated DOA, source separation 40° ($\gamma_1 = -10°$, $\gamma_2 = 30°$).

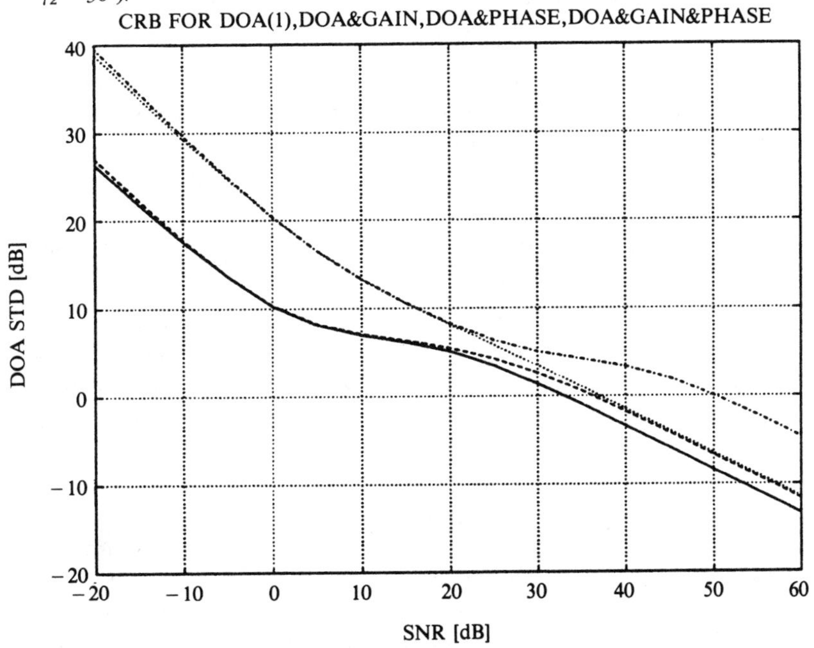

Figure 10.13 The CRLB for the estimated DOA, source separation 10° ($\gamma_1 = 5°$, $\gamma_2 = 15°$).

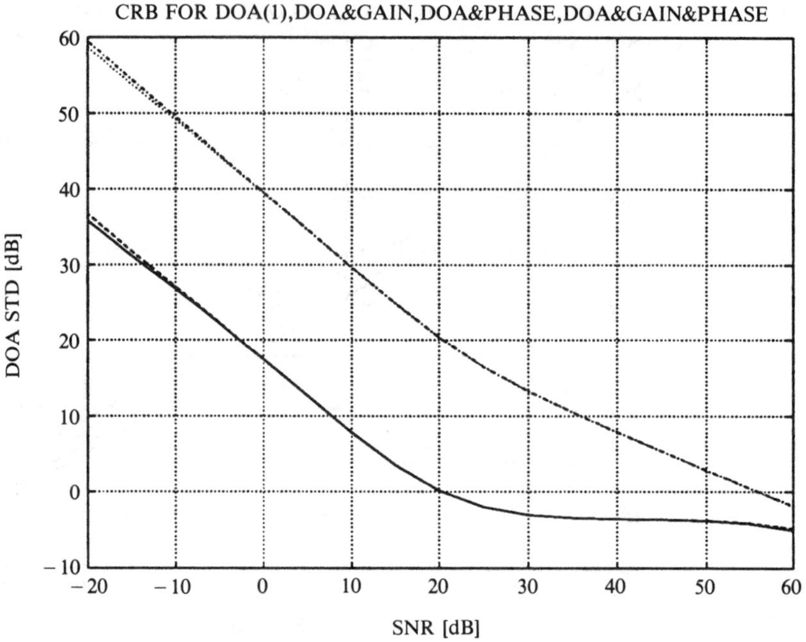

Figure 10.14. The CRLB for the estimated DOA, source separation 1° ($\gamma_1 = 9.5°$, $\gamma_2 = 10.5°$).

parameters, the DOAs can be estimated by "conventional" means, such as the MUSIC algorithm. Having obtained an estimate of the DOAs, we estimate the array parameters by minimizing an appropriate cost function. For example, in the case of [15] and [16] we minimize the following cost function over the unknown array parameters $\boldsymbol{\theta}$,

$$\tilde{J} = \sum_{n=1}^{N} \|\hat{U}^{\dagger}A(\gamma_n;\boldsymbol{\theta})\|^2 = \sum_{n=1}^{N} A(\gamma_n;\boldsymbol{\theta})^{\dagger}\hat{U}\hat{U}^{\dagger}A(\gamma_n;\boldsymbol{\theta}). \qquad (10.41)$$

In the case where the unknown parameters are the sensor gains and phases, the cost function $\tilde{J}$ is a quadratic function of the optimization parameters $\{g_m \triangleq \alpha_m e^{-j\phi_m}\}$. It is therefore straightforward to derive a closed-form expression for the minimizing parameter values [15] [18]. In other words, the self-calibration algorithm iterates between the MUSIC algorithm and a simple update of the array parameters. This is described in more detail in Section 10.6.1.

Similar ideas were explored in [16] for the case where the array parameters include the mutual coupling matrix as well as the gains and phases. In Section 10.6.3 we describe a three-step algorithm that iterates between estimating the DOAs, the gain/phase parameters, and the mutual coupling coefficients.

A somewhat different approach was presented in [17] for the joint estimation of DOAs and the sensor locations, using an iterative maximum-likelihood estimator. This is described briefly in Section 10.6.5.

We performed a large number of simulations to study the performance of these algorithms. We were able to obtain accurate DOA estimates (corresponding to the accuracy predicted by the CRLB discussed earlier) in the presence of significant modeling errors, which caused the "conventional" MUSIC algorithm to fail completely. Some typical numerical examples are presented in Sections 10.6.2 and 10.6.4.

10.6.1 An Algorithm for Estimating Both DOAs and Gain-Phase Parameters

In this section we consider the case where the unknown array parameters are the sensor gains and phases, that is, the elements of $\boldsymbol{\Gamma}$. The mutual coupling matrix is assumed to be perfectly known and assumed to be $\boldsymbol{C} = \boldsymbol{I}$. However, the algorithm can be easily modified to handle an arbitrary (but known) mutual coupling matrix, as will become obvious in Section 10.6.3.

Based on a discussion of the uniqueness of the solution to the joint estimation problem, which was presented in [15], we make the following nonrestrictive assumptions:

5. The array configuration is nonlinear.
6. The relation between the number of sources N and the number of sensors M is $2 \leq N < M$.

These assumptions are in addition to the standard assumptions made in Section 10.3.

Theorem 3.1 suggests that we should first estimate $\boldsymbol{R}_x$ and then use the estimates of λ_i to determine the number of signals. Once N is known, reasonable estimates of $\{\gamma_n\}_{n=1}^N$ and $\boldsymbol{\Gamma}$ may be obtained by minimizing:

$$\tilde{J} = \sum_{n=1}^{N} \|\hat{\boldsymbol{U}}^{\dagger}\boldsymbol{\Gamma}\mathbf{a}(\gamma_n)\|^2 = \sum_{n=1}^{N} \mathbf{a}(\gamma_n)^{\dagger}\boldsymbol{\Gamma}^{\dagger}\hat{\mathbf{U}}\hat{\mathbf{U}}^{\dagger}\boldsymbol{\Gamma}\mathbf{a}(\gamma_n) \qquad (10.42a)$$

where $\hat{\boldsymbol{U}}$ is an estimate of $\boldsymbol{U}$ and $\mathbf{a}(\gamma_n)$ is the nth column of $\boldsymbol{A}$ (see (10.28)). If $\hat{\boldsymbol{U}}$ were a perfect estimate of $\boldsymbol{U}$ (i.e., $\hat{\boldsymbol{U}} = \boldsymbol{U}$), then the minimum value of $\tilde{J}$ ($\tilde{J} = 0$) will be achieved for the true $\boldsymbol{\Gamma}$ and $\{\gamma_n\}_{n=1}^N$. When $\hat{\boldsymbol{U}}$ is an imperfect estimate of $\boldsymbol{U}$, the minimization of $\tilde{J}$ will provide estimates $\hat{\boldsymbol{\Gamma}}$and $\{\hat{\gamma}_n\}_{n=1}^N$ of the true DOAs and gain/phase parameters. The accuracy of these estimates has been investigated by simulations (Sections 10.6.2), but a detailed error analysis has not been carried out as yet.

Before we turn to describe the proposed procedure, we briefly discuss the conditions for well posedness following [1, p. 84]. Referring to the basic equations (10.9) we observe that $\boldsymbol{R}_x$ can be perfectly described by $2MN - N^2 + 1$ parameters. These parameters are the $N + 1$ different (real) eigenvalues and $2M\,N - N^2 - N$ parameters that define the N complex eigenvectors describing the signal subspace, that satisfy $N(N + 1)/2$ complex orthogonality constraints. On the other hand, we have $r \cdot N$ unknown location parameters ($r = 1$ for azimuth only system, $r = 2$ for azimuth and elevation system, etc.), N^2 unknown parameters that define the Hermitian matrix $\boldsymbol{R}_s$, a single unknown parameter σ^2, and $2(M-1)$ parameters associated with $\boldsymbol{\Gamma}$. Thus, the problem is not strictly well posed unless

$$2MN - N^2 + 1 \geq r \cdot N + N^2 + 1 + 2(M-1)$$
$$M \geq \frac{2N^2 + r \cdot N - 2}{2(N-1)}. \tag{10.42b}$$

The proposed minimization algorithm is based on a two-step procedure. First, we assume that the gain/phase parameters $\{\Gamma_{mm}\}_{m=2}^{M}$ are known, and we estimate $\{\gamma_n\}_{n=1}^{N}$ in the usual way (see the description of the MUSIC algorithm in [2]). Given estimates of $\{\gamma_n\}_{n=1}^{N}$ we then minimize $\tilde{J}$ over the gain/phase parameters $\{\Gamma_{mm}\}_{m=2}^{M}$. Before presenting the algorithm for minimizing $\tilde{J}$ note that

$$\boldsymbol{\Gamma}\mathbf{a}(\gamma_n) = \tilde{\boldsymbol{a}}(\gamma_n)\boldsymbol{\delta}, \tag{10.42c}$$

where $\tilde{\boldsymbol{a}}(\gamma_n)$ is a diagonal matrix given by

$$\tilde{\boldsymbol{a}}(\gamma_n) = \text{diag}\,\{\mathbf{a}(\gamma_n)\} \tag{10.42d}$$

and $\boldsymbol{\delta}$ is a vector given by

$$\boldsymbol{\delta} = [\Gamma_{11}, \Gamma_{22}, \ldots, \Gamma_{MM}]^T. \tag{10.42e}$$

The proposed algorithm can then be summarized as follows:

1. Initialization: Set $k = 0$; select $\boldsymbol{\Gamma}^{(k)} = \boldsymbol{\Gamma}_0$; $\boldsymbol{\Gamma}_0$ may be based on the nominal gain and phase values or on any recent calibration information.
2. Search for the N highest peaks of the spatial spectrum defined by

$$P(\gamma|\boldsymbol{\Gamma}^{(k)}) = \|\hat{U}^{\dagger}\boldsymbol{\Gamma}^{(k)}\mathbf{a}(\gamma)\|^{-2}. \tag{10.42f}$$

These peaks are associated with the N column vectors $\{\mathbf{a}(\gamma_n^{(k)})\}_{n=1}^{N}$.

3. Substituting (10.42c) in (10.42a) we have

$$\tilde{J} = \boldsymbol{\delta}^{\dagger}\left\{\sum_{n=1}^{N} \tilde{\boldsymbol{a}}(\gamma_n^{(k)})^{\dagger}\hat{U}\hat{U}^{\dagger}\tilde{\boldsymbol{a}}(\gamma_n^{(k)})\right\}\boldsymbol{\delta}. \tag{10.42g}$$

Hence, we want to minimize (10.42g) with respect to $\boldsymbol{\delta}$ under the constraint $\boldsymbol{\delta}^{\dagger}\mathbf{w} = 1$ where $\mathbf{w} = [1, 0, 0, \ldots, 0]^T$. The result of this minimization problem is well known and is given by

$$\boldsymbol{\delta}^{(k+1)} = \boldsymbol{Q}_k^{-1}\mathbf{w}/(\mathbf{w}^T\boldsymbol{Q}_k^{-1}\mathbf{w}), \tag{10.42h}$$

where $\boldsymbol{Q}_k$ is the matrix

$$\boldsymbol{Q}_k = \sum_{n=1}^{N} \tilde{\boldsymbol{a}}(\gamma_n^{(k)})^{\dagger}\hat{U}\hat{U}^{\dagger}\tilde{\boldsymbol{a}}(\gamma_n^{(k)}). \tag{10.42i}$$

4. Compute $\boldsymbol{\Gamma}^{(k+1)}$ from $\boldsymbol{\delta}^{(k+1)}$.

5. Compute

$$\tilde{J}_k = (\delta^{k+1})^\dagger Q_k \delta^{k+1} \tag{10.42j}$$

If $\tilde{J}_{k-1} - \tilde{J}_k > \varepsilon$ (a preset threshold), then set $k = k + 1$, and go to (2). If $\tilde{J}_{k-1} - \tilde{J}_k \leq \varepsilon$, done.

The algorithm performs the iterations until $\tilde{J}$ converges. Note that at each updating step (i.e., steps 2 and 3), we decrease the cost function $\tilde{J}$ defined in (10.42a). Since $\tilde{J} \geq 0$ the algorithm will converge at least to a local minimum of $\tilde{J}$. Depending on the initial estimate of $\mathbf{\Gamma}$ and on the structure of $\tilde{J}$, the local minimum may or may not coincide with the global minimum. However, Monte Carlo simulations indicate that a convergence to a local minimum that is not also the global minimum is seldom encountered.

10.6.2 Numerical Examples for the DOA-Gain-Phase Algorithm

In this section we present some examples that illustrate the behavior of the proposed algorithm.

Consider a uniform circular array of six sensors separated by half a wavelength of the actual narrow-band source signals, as shown in Fig. 10.15. Three equal-power narrow-band sources are located in the far field of the array at directions: $\gamma_1 = -30°$, $\gamma_2 = -5°$, $\gamma_3 = +35°$. Additive uncorrelated sensor noise is injected with SNR of 30

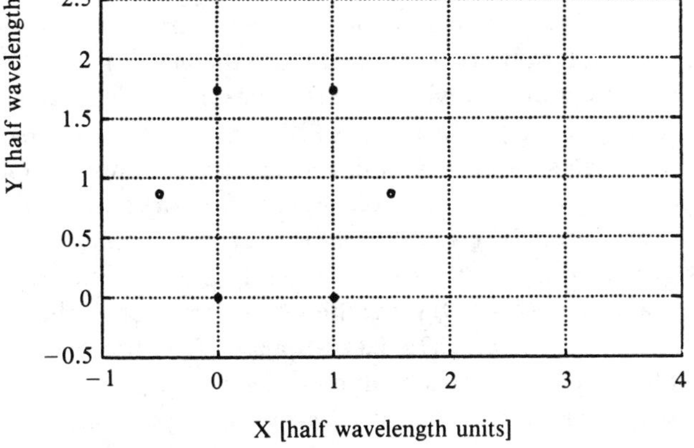

Figure 10.15 Uniform circular array configuration.

dB referenced to the signal sources. One hundred snapshots of array data are accumulated prior to the application of the algorithm. The nominal gain of the sensors is one and the nominal phase is zero. The actual gain is perturbed from the nominal value by up to 80 percent and the actual phase is perturbed from the nominal value by up to 80 degrees.

Figure 10.16 shows the performance of the traditional MUSIC algorithm with and without knowledge of the sensor characteristics (gain and phase). We observe that when the MUSIC algorithm has no knowledge of the gain and phase parameters, no reliable estimates of the DOAs can be extracted from the plot.

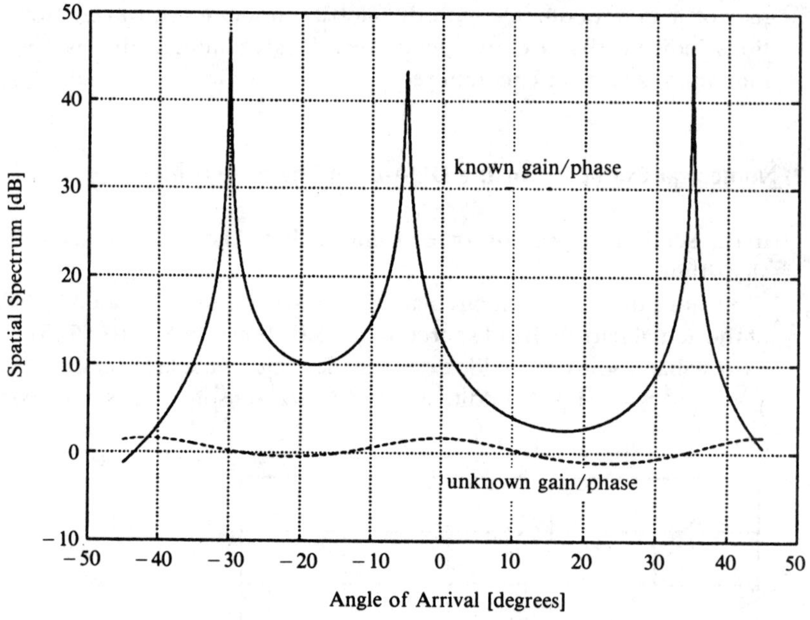

Figure 10.16 Spatial spectrum of the MUSIC algorithm.

Figure 10.17 shows the spatial spectrum generated in iterations 1, 3, 7, and 15 of the new technique. Note that iteration one is equivalent to the application of the traditional MUSIC algorithm without any knowledge of the sensor gains and phases. Iteration 15 is already very close to the performance of MUSIC when the gain and phase are known. Table 10.1 contains the DOA estimates in all the iterations from 1 to 15. It is clear that the algorithm has been able to correct DOA errors of more than 10 degrees. In Table 10.1 we also listed the estimates of the gain and the phase of one of the sensors. The actual gain of this sensor was 1.795 and its actual phase was -41.384° referenced to sensor 1. Figure 10.18 plots the phase errors of all the sensors as a function of the number of iterations. The largest phase error after iteration 15 was 1.06°. Figure 10.19 plots the gain errors of all the sensors as a percentage of their actual gain. The largest gain error after 15 iterations was 1.8 percent.

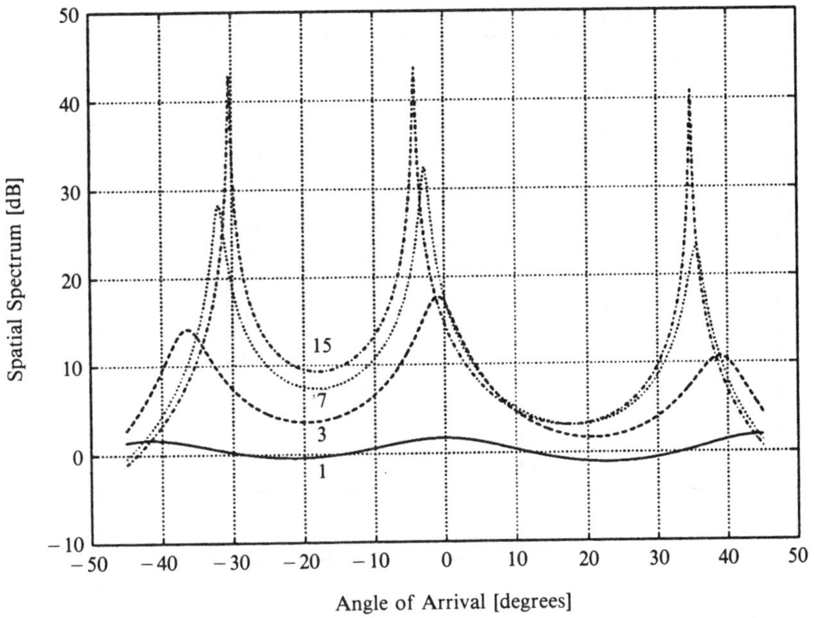

Figure 10.17 Spatial spectrum of the proposed algorithm for iterations 1, 3, 7, and 15.

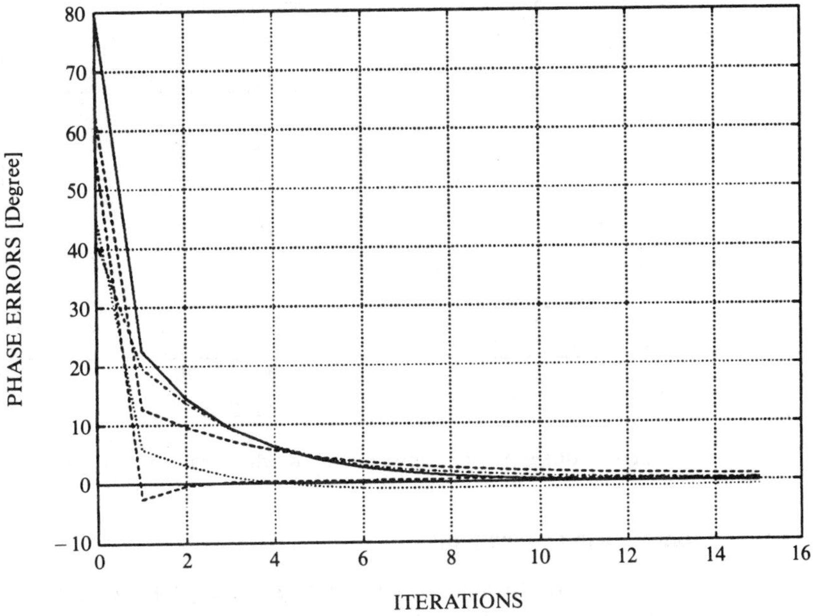

Figure 10.18 Sensor phase errors as a function of the number of iterations.

TABLE 10.1. The Estimated DOAs and Phase-Gain Parameters

Iteration	$\hat{\gamma}_1$	$\hat{\gamma}_2$	$\hat{\gamma}_3$	$\lvert\Gamma_{22}\rvert$	Phase Γ_{22}
1	−41.6	−0.3	44.5	1.30	−21.74
2	−38.7	−0.5	41.4	1.54	−27.72
3	−36.4	−1.0	38.9	1.66	−32.15
4	−34.7	−1.5	37.5	1.72	−35.12
5	−33.4	−2.0	36.5	1.76	−37.16
6	−32.5	−2.4	35.9	1.77	−38.45
7	−31.9	−2.8	35.5	1.78	−39.23
8	−31.5	−3.1	35.2	1.79	−39.71
9	−31.2	−3.4	35.0	1.79	−40.04
10	−30.9	−3.8	34.8	1.79	−40.40
11	−30.7	−3.9	34.8	1.79	−40.61
12	−30.6	−3.9	34.8	1.79	−40.73
13	−30.5	−4.0	34.8	1.79	−40.84
14	−30.4	−4.1	34.8	1.80	−40.85
15	−30.4	−4.2	34.8	1.80	−40.93

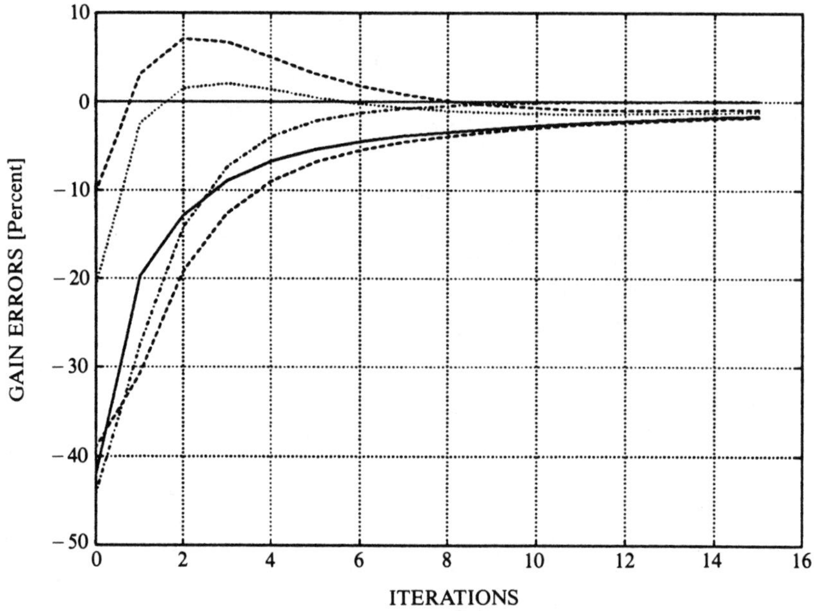

Figure 10.19 Sensor gain errors as a function of the number of iterations.

Monte Carlo Experiments

To demonstrate the statistical efficiency of the proposed procedure, under different conditions, we performed three sets of experiments.

Experiment 1

Consider the circular array of six sensors described in the previous section. The sources are two narrow-band emitters located in the far field of the array. The gains $\{\alpha_m\}$ and phases $\{\phi_m\}$ of the sensors were chosen according to

$$\alpha_m = 1 + \sqrt{12} \cdot \sigma_\alpha \cdot \beta_m \tag{10.43a}$$
$$\phi_m = \sqrt{12} \cdot \sigma_\phi \cdot \eta_m, \tag{10.43b}$$

where the β_m and η_m are independent and identically distributed random numbers distributed uniformly over the interval $[-0.5, 0.5]$, and σ_α and σ_ϕ are the standard deviation of α_m and ϕ_m, respectively. The signals, $\mathbf{S}(j)$, and the noise, $\mathbf{V}(j)$, are random complex Gaussian vectors with covariance matrices $\sigma_s^2 \cdot \boldsymbol{I}$, and $\sigma_n^2 \cdot \boldsymbol{I}$, respectively.

In the first experiment, the following DOAs were chosen: $\gamma_1 = -60°$, $\gamma_2 = +60°$. Using the following definition for SNR,

$$\text{SNR} \triangleq 20 \log(\sigma_s/\sigma_n) \text{ dB}$$

we performed 200 experiments for each SNR = 0, 3, 6,..., 30 dB. In each experiment 100 snapshots of data were collected and processed according to the proposed algorithm. The gains and phases of the sensors were chosen according to (10.43a), (10.43b) with $\sigma_\alpha = 0.05$, $\sigma_\phi = 5°$, and were kept constant throughout the entire simulations. For each SNR value we used 200 DOA results to calculate the estimated bias and estimated root mean square error (RMSE) of the algorithm. Figures 10.20, 10.21, 10.22, and 10.23 display these results together with the Cramér-Rao lower bound. The derivation of the CRLB is detailed in Appendix B. It is clear from Figs. 10.20 and 10.21 that the proposed method is statistically efficient even for fairly low SNR, for the test case used in this experiment.

Experiment 2

To explore the sensitivity of the algorithm to closely spaced sources we repeated experiment no 1 with the following parameters: $\gamma_1 = -5°$, $\gamma_2 = 5°$, $\gamma_3 = 90°$, $\sigma_\alpha = 0.01$, $\sigma_\phi = 1°$.

For each SNR = 20, 30, 40 dB we performed 100 experiments, and in each experiment we collected 500 data snapshots. The RMSE, bias, and CRLB are shown in Figs. 10.24–10.29. It is clear from the figures that the RMSE for the source at 90° achieves the CRLB for the entire range of SNR values that were considered in this experiment. However, the RMSE for the closely spaced sources achieve the CRLB only at relatively high SNR. Whether this is due to the fact that the CRLB is not a tight bound for the finite data case, or due to a shortcoming of the algorithm, is still an open research problem.

Figures 10.30 through 10.33 describe the RMSE and bias in estimating the phase of sensor 2 relative to sensor 1. Note that the variance of the estimated sensor

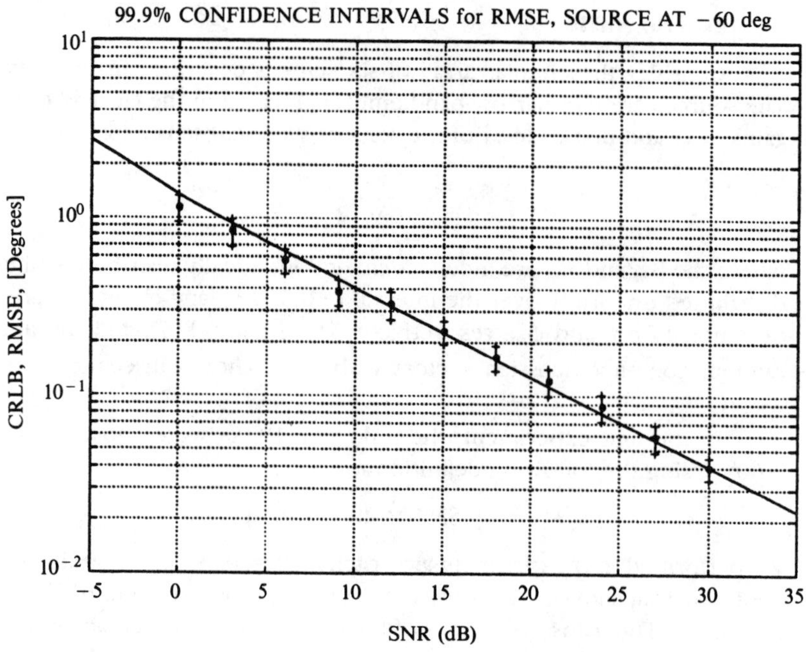

Figure 10.20 RMSE for a source at −60%; experiment 1.

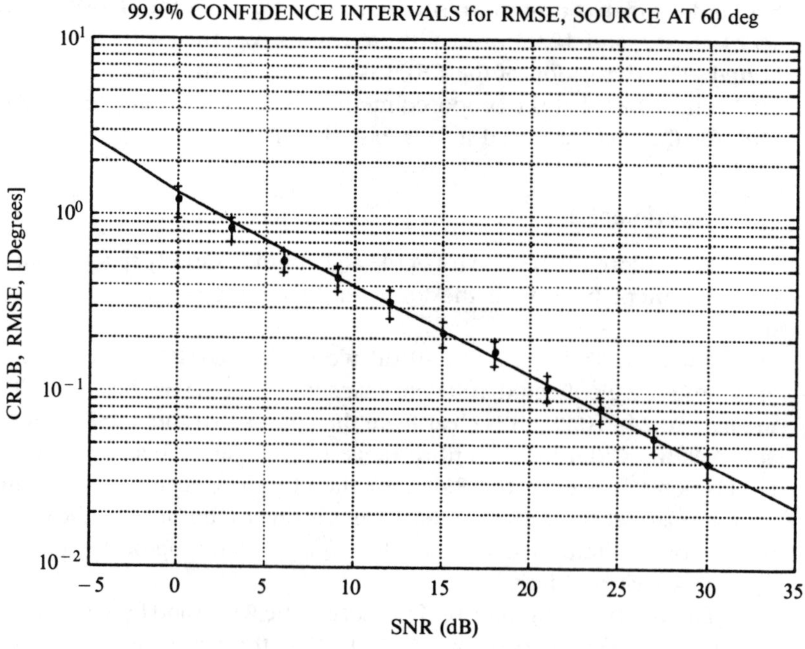

Figure 10.21 RMSE for a source at 60°; experiment 1.

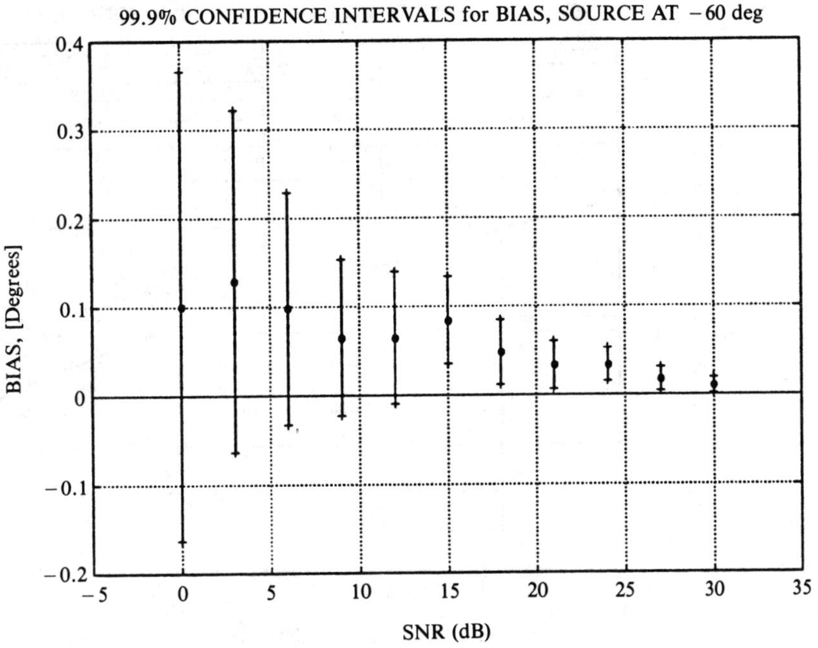

Figure 10.22 Bias for a source at $-60°$; experiment 1.

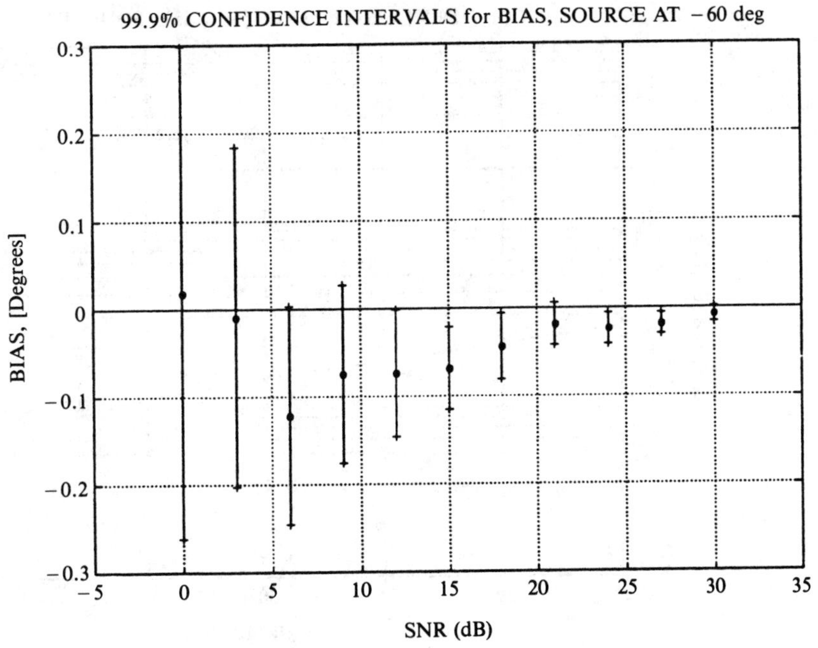

Figure 10.23 Bias for a source at $60°$; experiment 1.

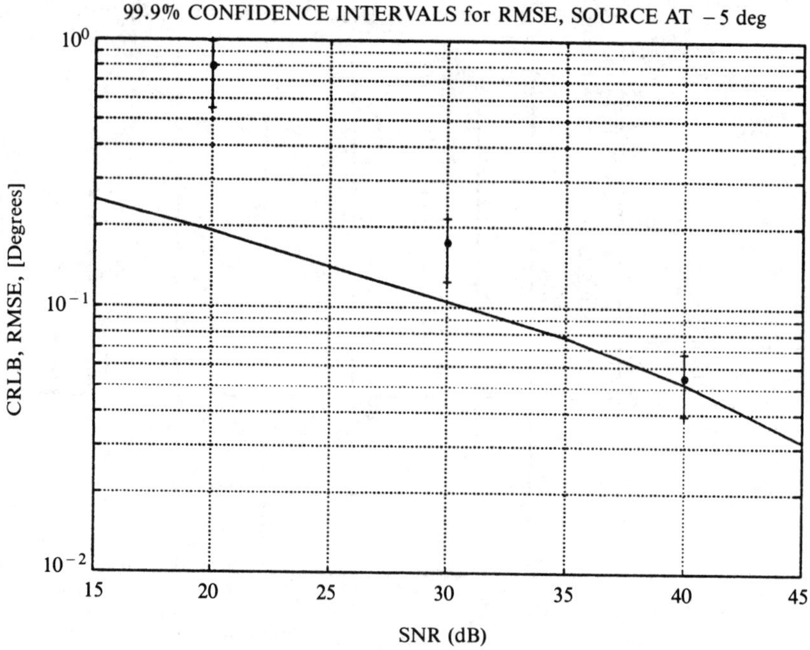

Figure 10.24 RMSE for source at −5°; experiment 2.

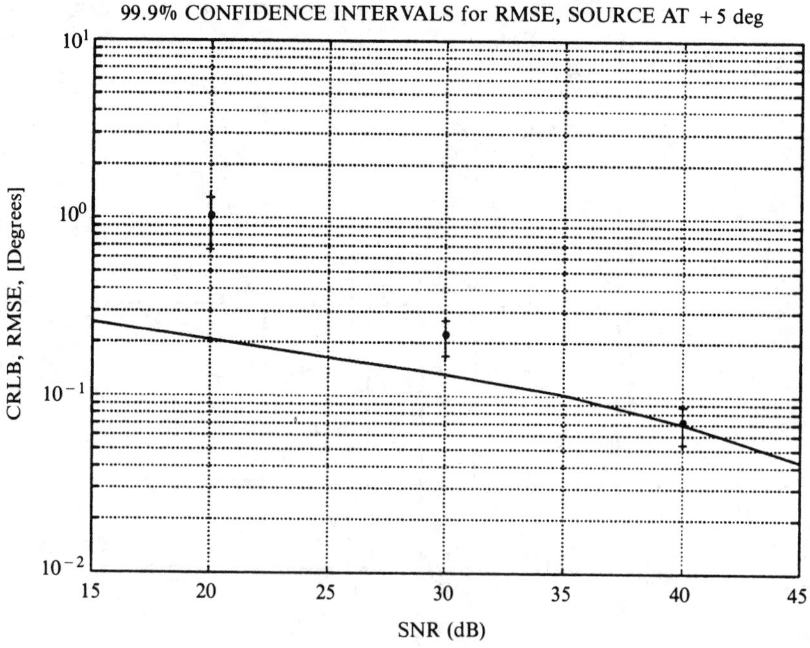

Figure 10.25 RMSE for source at 5°; experiment 2.

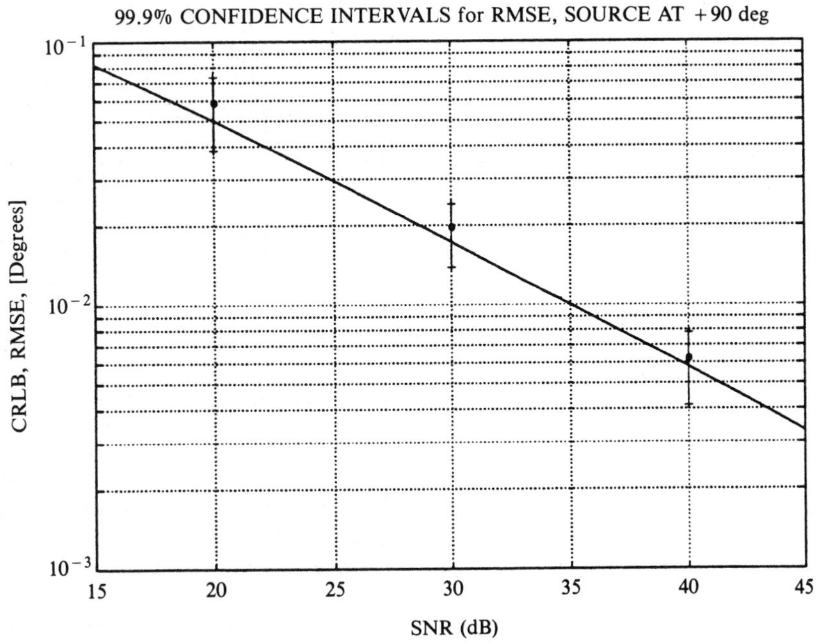

Figure 10.26 RMSE for source at 90°; experiment 2.

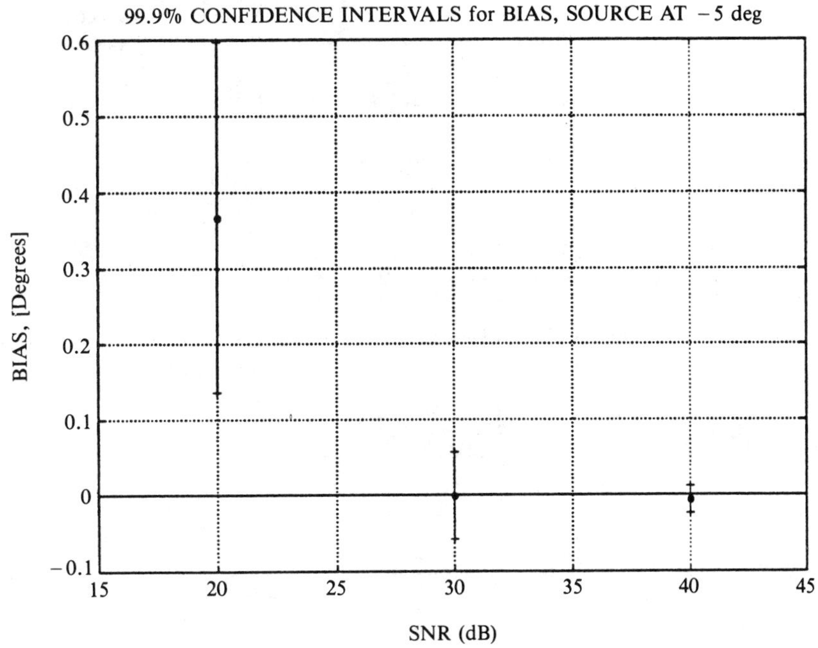

Figure 10.27 Bias for source at −5°; experiment 2.

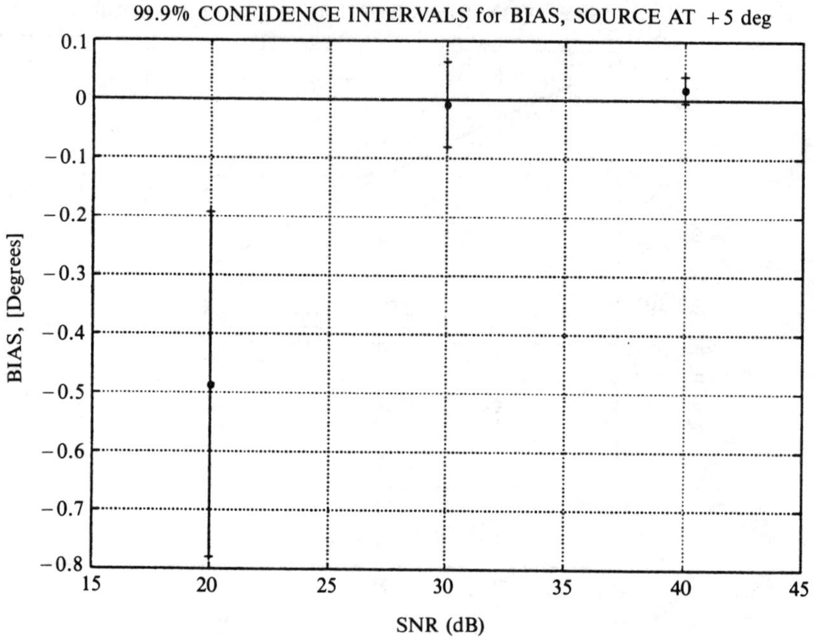

Figure 10.28 Bias for source at 5°; experiment 2.

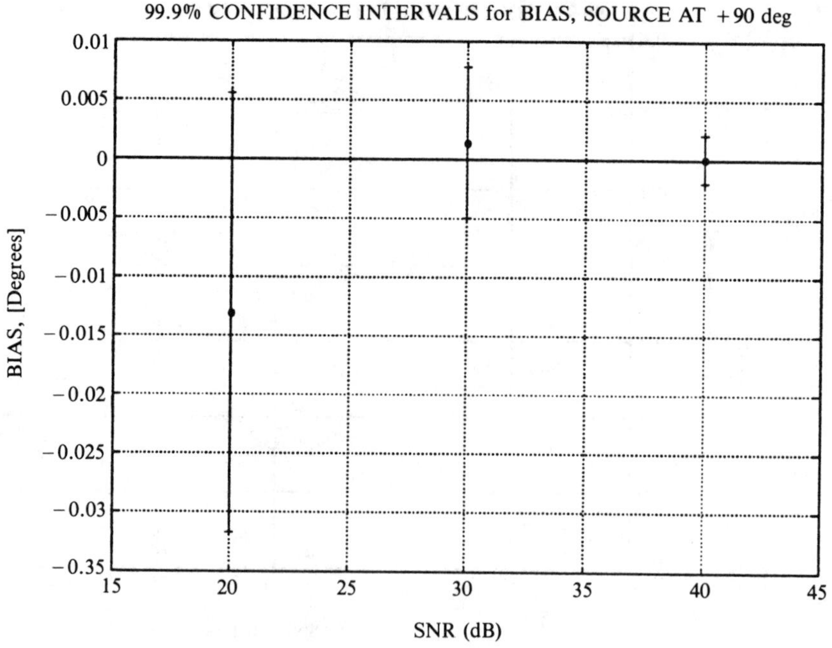

Figure 10.29 Bias for source at 90°; experiment 2.

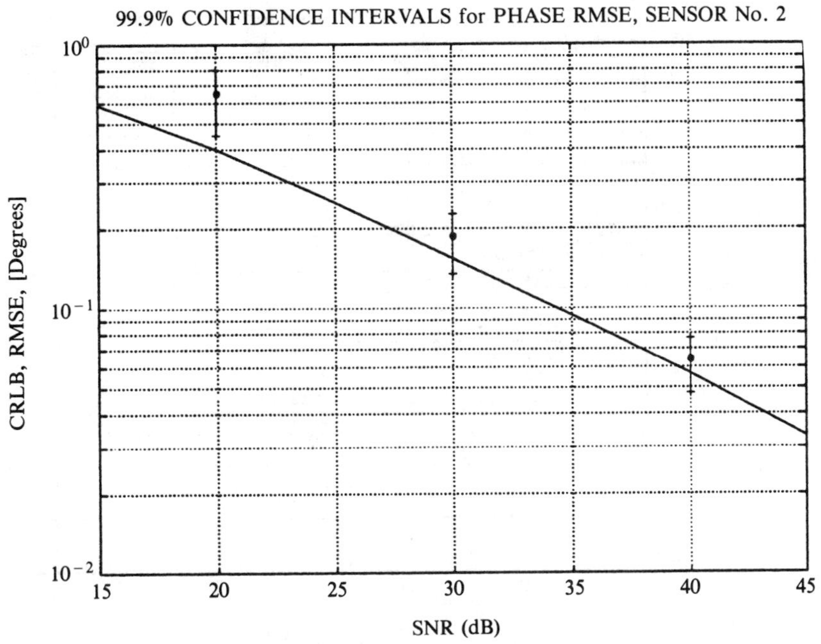

Figure 10.30 RMSE of the estimated phase of sensor 2.

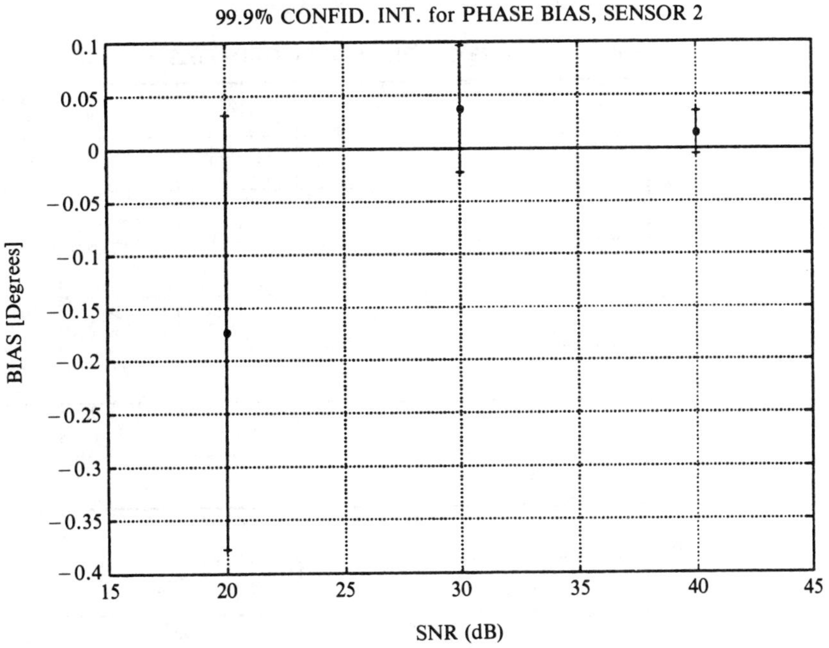

Figure 10.31 Bias of the estimated phase of sensor 2.

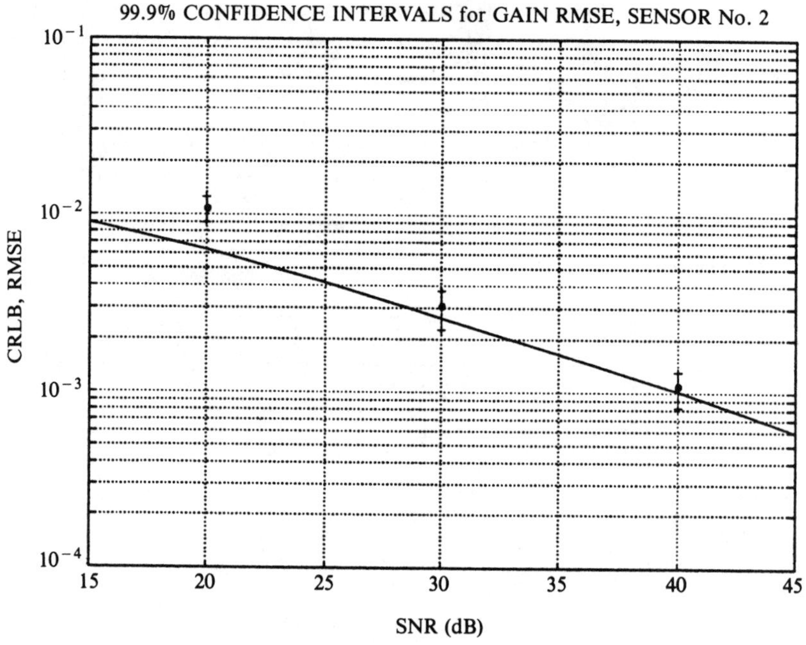

Figure 10.32 RMSE of the estimated gain of sensor 2.

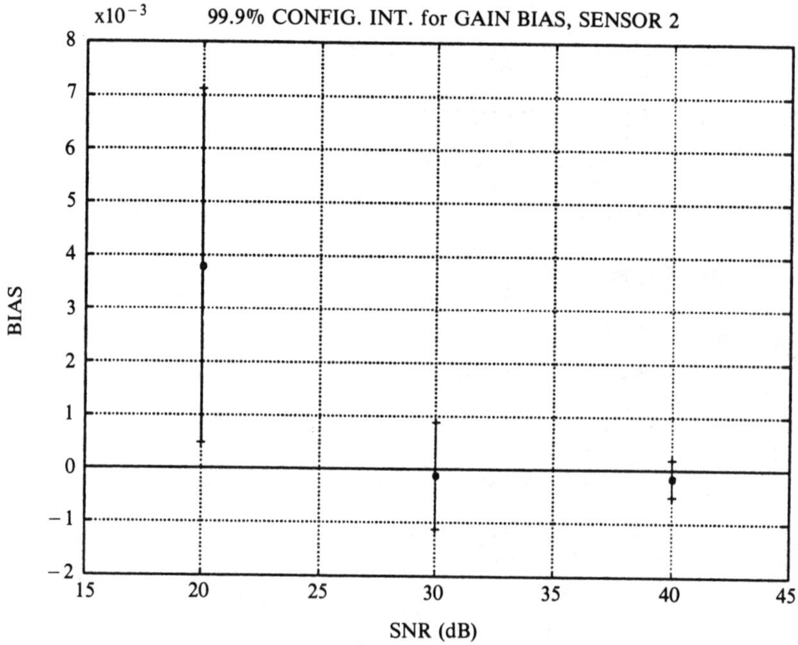

Figure 10.33 Bias of the estimated gain of sensor 2.

parameters is very close to the CRLB, and that the bias is negligible. The figures corresponding to sensors 3–6 are very similar and have, therefore, been omitted.

Experiment 3

To check the sensitivity of the algorithm to the particular selection of the gain/phase of the sensors, we performed the following test. Three sources at DOAs $\gamma_1 = -5°$, $\gamma_2 = +5°$, $\gamma_3 = +90°$, were placed in the far field of a six-element circular array. We performed 100 experiments. In each experiment different (random) gain and phase were selected according to (10.43a) and (10.43b) with $\sigma_\alpha = 0.015$, $\sigma_\phi = 1.5°$. We assumed that the number of snapshots is large enough so that the estimated covariance $\hat{\boldsymbol{R}}_x$ is given precisely by $\boldsymbol{R}_x$. Based on this ideal covariance matrix, the algorithm estimated the DOAs and the gain/phase parameters.

In each of the 100 experiments the standard MUSIC failed to resolve the closely spaced sources. Hence, the initial DOAs were selected, arbitrarily, as $\gamma_0 \pm 15°$, where γ_0 is the MUSIC single result for the two closely spaced sources. In each case the algorithm converged to the correct result with a small residual error, which is partly due to the limited number of iterations. It appears that if $\hat{\boldsymbol{R}}_x$ is accurate enough, the algorithm converges to the correct result and is not "trapped" by a local minimum.

10.6.3 Estimating the DOAs, Gains, Phases, and MCM

In this section we consider the case where the *mutual coupling matrix* (*MCM*) needs to be estimated as well as the sensor gains and phases. In addition to the standard assumptions underlying the MUSIC algorithm we assume that the following relation is satisfied,

$$M \geq \frac{2N^2 + r \cdot N + P - 2}{2(N - 1)} \tag{10.44a}$$

where P is the number of parameters estimated for the MCM. This relationship can be obtained by an obvious modification of (10.42b).

As before, we first estimate $\boldsymbol{R}_x$ and use the estimates of the eigenvectors to estimate the number of signals. Once N is known, reasonable estimates of $\{\gamma_n\}$, $\boldsymbol{\Gamma}$ and $\boldsymbol{C}$ may be obtained by minimizing the cost function

$$J_c = \sum_{n=1}^{N} \|\hat{\boldsymbol{U}}^{\dagger}\boldsymbol{C}\boldsymbol{\Gamma}\mathbf{a}(\gamma_n)\|^2, \tag{10.44b}$$

which is the squared Euclidean norm of the matrix $\hat{\boldsymbol{U}}^{\dagger}\boldsymbol{C}\boldsymbol{\Gamma}\mathrm{a}(\gamma_n)$. Here $\hat{\boldsymbol{U}}$ stands for the estimate of the matrix $\boldsymbol{U}$. If $\hat{\boldsymbol{U}}$ were a perfect estimate of $\boldsymbol{U}$ (i.e., $\hat{\boldsymbol{U}} = \boldsymbol{U}$), then the minimum value of J_c ($J_c = 0$) will be achieved for the true, $\boldsymbol{C}$, $\boldsymbol{\Gamma}$, and $\{\gamma_n\}$. When $\hat{\boldsymbol{U}}$ is an imperfect estimate of $\boldsymbol{U}$, the minimization of J_c will provide estimate of $\boldsymbol{C}$, $\boldsymbol{\Gamma}$ and $\{\gamma_n\}$, the true MCM, gain-phase parameters, and DOAs. The accuracy of these estimates has been investigated by simulations. A detailed error analysis has not been carried out as yet.

The proposed minimization algorithm is based on a three-step procedure. First, we assume that the gain/phase and mutual coupling coefficients are (approximately) known, and we estimate $\{\gamma_n\}_{n=1}^{N}$ using the principles of the standard MUSIC algorithm. Given estimates of $\{\gamma_n\}$, we then minimize J_c over the gain/phase parameters. Given $\{\gamma_n\}$ and Γ, we minimize J_c over the MCM components. These minimization steps can be repeated until J_c converges.

Before presenting the algorithm for minimizing J_c we introduce three useful lemmas.

Lemma 6.3.1 For any $M \times 1$ complex vector $\mathbf{X}$ and any $M \times M$ complex diagonal matrix $\boldsymbol{D}$ we have

$$\boldsymbol{D} \cdot \mathbf{X} = \boldsymbol{Q}_1(\mathbf{X}) \cdot \mathbf{d}$$

where the components of the $M \times 1$ vector $\mathbf{d}$ and the $M \times M$ matrix $\boldsymbol{Q}_1(\mathbf{X})$ are given by

$$d_i = D_{ii}, \qquad i = 1, 2, \ldots, M$$

$$[Q_2(\mathbf{X})]_{ij} = X_i \cdot \delta_{ij}, \qquad i, j = 1, 2, \ldots, M.$$

Lemma 6.3.2: For any $M \times 1$ complex vector $\mathbf{X}$ and any $M \times M$ complex symmetric circulant matrix $\boldsymbol{A}$ we have

$$\boldsymbol{A} \cdot \mathbf{X} = \boldsymbol{Q}_2(\mathbf{X}) \cdot \mathbf{a},$$

where the components of the $L \times 1$ vector $\mathbf{a}$ are given by

$$a_i = A_{1i}, \qquad i = 1, 2, \ldots, L,$$

where $L = M/2 + 1$ when M is even and $L = M/2 + 1/2$ when M is odd.

The $M \times L$ matrix $\boldsymbol{Q}_2(\mathbf{X})$ is the sum of the four $M \times L$ following matrices:

$$[W_1]_{pq} = \begin{cases} X_{p+q-1} & \text{for } p + q \le M + 1 \\ 0 & \text{otherwise} \end{cases}$$

$$[W_2]_{pq} = \begin{cases} X_{p-q+1} & p \ge q \ge 2 \\ 0 & \text{otherwise} \end{cases}$$

$$[W_3]_{pq} = \begin{cases} X_{M+1+p-q} & p < q \le \ell \\ 0 & \text{otherwise} \end{cases}$$

$$[W_4]_{pq} = \begin{cases} X_{p+q-M-1} & 2 \le q \le \ell, p + q \ge M + 2 \\ 0 & \text{otherwise} \end{cases}$$

$\ell = M/2$ for even M and $\ell = (M + 1)/2$ for odd M.

Lemma 6.3.3: For any $M \times 1$ complex vector $\mathbf{X}$ and any $M \times M$ banded complex symmetric Toeplitz matrix $\boldsymbol{A}$ we have

$$\boldsymbol{A} \cdot \mathbf{X} = \boldsymbol{Q}_3(\mathbf{X}) \cdot \mathbf{a}$$

where the $L \times 1$ vector $\mathbf{a}$ is given by

$$a_i = A_{1i}, \qquad i = 1, 2, \ldots, L,$$

and L is the highest superdiagonal that is different from zero. The $M \times L$ matrix $Q_3(\mathbf{X})$ is given by the sum of the two $M \times L$ following matrices

$$[W_1]_{pq} = \begin{cases} X_{p+q-1} & \text{for } p + q \leq M + 1 \\ 0 & \text{otherwise} \end{cases}$$

$$[W_2]_{pq} = \begin{cases} X_{p-q+1} & p \geq q \geq 2 \\ 0 & \text{otherwise} \end{cases}$$

Proof: The proof of the lemmas is based on the special properties of diagonal matrices, circulant matrices and Toeplitz matrices. (See Appendix A.)

The proposed algorithm for minimizing the cost function may be described as follows.

Initialization

1. Set the iteration counter to zero: $k = 0$.
2. Select initial values for the gain-phase matrix $\mathbf{\Gamma}$ and initial value for the MCM (i.e., C). Usually the initial values are based on any previous knowledge (e.g., last measured values or predictions based on idealized model).
3. Use all available data vectors to compute the data covariance matrix estimate

$$\hat{R}_x = \frac{1}{J} \sum_{j=1}^{J} \mathbf{X}(j)\mathbf{X}(j)^{\dagger}. \tag{10.44c}$$

4. Perform eigenanalysis and construct $\hat{U}$ according to Theorem 3.1.

Step 1: Estimating DOAs

1. Search for the N highest peaks of the spatial spectrum defined by

$$P^{(k)}(\gamma) = \|\hat{U}^{\dagger} C^{(k)} \mathbf{\Gamma}^{(k)} \mathbf{a}(\gamma)\|^{-2}. \tag{10.44d}$$

These peaks are associated with the N DOAs $\{\gamma_n\}_{n=1}^{N}$. Also note that substituting $\{\hat{\gamma}_n^{(k)}\}$, so estimated, in J_c (given by (10.44b)) guarantees that $J_c^{(k)}$ is minimized for given $C^{(k)}$ and $\mathbf{\Gamma}^{(k)}$.

Step 2: Estimating Gain-Phase This step is an obvious modification of the second step of the algorithm described in Section 10.6.1.

1. Fixing the DOAs and the MCM, minimize J_c with respect to the gain and phase of each of the sensors. Using Lemma 6.3.1 in expression (10.44b), we thus obtain

$$J_c = \sum_{n=1}^{N} \mathbf{a}(\gamma_n)^\dagger \Gamma^\dagger C^\dagger \hat{U}\hat{U}^\dagger C \Gamma \mathbf{a}(\gamma_n)$$

$$= \boldsymbol{\delta}^H \{ \sum_{n=1}^{N} Q_1^\dagger(n) C^\dagger \hat{U}\hat{U}^\dagger C Q_1(n) \} \boldsymbol{\delta}, \tag{10.44e}$$

where

$$\boldsymbol{\delta} = [\Gamma_{11}, \Gamma_{22}, \ldots, \Gamma_{MM}]^T$$

$$Q_1(n) = \text{diag}\{\mathbf{a}(\gamma_n)\}.$$

Hence, we want to minimize (10.44e) with respect to $\boldsymbol{\delta}$ under the constraint $\boldsymbol{\delta}^\dagger \mathbf{w} = 1$, where $\mathbf{w} = [1, 0, 0, \ldots, 0]^T$. The result of this quadratic minimization problem under linear constraints is well known; it is given by

$$\hat{\boldsymbol{\delta}} = Z_k^{-1}\mathbf{w}/(\mathbf{w}^T Z_k^{-1}\mathbf{w}), \tag{10.44f}$$

where Z_k is the matrix

$$Z_k \triangleq \sum_{n=1}^{N} Q_1^\dagger(n) C^\dagger \hat{U}\hat{U}^\dagger C Q_1(n). \tag{10.44g}$$

2. Compute the gain-phase matrix Γ from the vector $\hat{\boldsymbol{\delta}}$ given by (10.44f)

$$\Gamma^{k+1} = \text{diag}\,(\hat{\boldsymbol{\delta}}). \tag{10.44h}$$

Thus, we minimized the cost function J_c by holding the MCM and the DOAs fixed and searching over the space of the sensor gain-phase parameters.

Step 3: Estimating the Mutual Coupling Matrix

1. In this step we hold the DOAs and sensor gain-phase fixed and find the MCM that minimizes the cost function J_c. The minimization step capitalizes on Lemma 6.3.2 if C is circulant (circular array) and on Lemma 6.3.3 if C is Toeplitz (linear array). In the following we assume that C is a complex symmetric circulant matrix.

   Using Lemma 6.3.2 in expression (10.44a) for the cost function J_c we obtain

$$J_c = \sum_{n=1}^{N} \tilde{\mathbf{a}}(\gamma_n)^\dagger C^\dagger \hat{U}\hat{U}^\dagger C \tilde{\mathbf{a}}(\gamma_n)$$

$$= \mathbf{c}^\dagger \{ \sum_{n=1}^{N} Q_2(n)^\dagger \hat{U}\hat{U}^\dagger Q_2(n) \} \mathbf{c} \tag{10.44i}$$

where we used the following notation:

$$\tilde{\mathbf{a}}(\gamma_n) = \mathbf{\Gamma a}(\gamma_n)$$

$$\mathbf{Q}_2(n) = Q_2(\tilde{\mathbf{a}}(\gamma_n))$$

$$c_i = C_{1i}, \qquad i = 1, 2, \ldots, L$$

Note that Q_2 $(\mathbf{X})$ and L are defined in Lemma 6.3.2. Relation (10.44i) represents (again) a quadratic minimization problem under linear constraints. The linear constraints represent the assumed model of C (e.g., $C_{11} = 1$). Hence if the constraint equation is $W_1^T\,\mathbf{c} = \mathbf{u}$, then

$$\hat{\mathbf{c}} = G^{-1}W_1(W_1^T G^{-1}W_1)^{-1}\mathbf{u} \tag{10.44j}$$

where G is the matrix

$$G \triangleq \sum_{n=1}^{N} Q_2(n)^{\dagger}\hat{U}\hat{U}^{\dagger}Q_2(n) \tag{10.44k}$$

and W_1 represents the linear constraints.

2. Reconstruct the MCM matrix from the vector $\hat{\mathbf{c}}$ given by (10.44j).

Convergence Check

Compute J_c^{k+1} using the estimated DOAs, sensor gain-phase and the MCM. If

$$J_c^{\mathrm{k}} - J_c^{\mathrm{k}+1} > \varepsilon \quad \text{(a preset threshold)}$$

then update the iteration counter $k = k + 1$ and go back to step 1.
If

$$J_c^{\mathrm{k}} - J_c^{k+1} \leq \varepsilon$$

stop.

The algorithm performs the iterations until J_c converges. Note that at each step the cost function reduces so that

$$J_c^{(0)} > J_c^{(1)} > \cdots > J_c^{(k)} \geq 0.$$

Hence, $J_c^{(k)}$ is a convergent series and convergence is guaranteed.

10.6.4 Numerical Examples for the DOA-Gain-Phase/MCM Algorithm

To illustrate the behavior of the algorithm, consider a circular uniform array of six omnidirectional sensors separated by half a wavelength of the actual narrowband source signals. We used simulated signal vectors $\mathbf{S}(j)$ and noise vectors $\mathbf{V}(j)$ drawn from a complex Gaussian distribution with zero mean and covariance matrices $\sigma_s^2 \cdot I$ and $\sigma_n^2 \cdot I$, respectively. We assumed that each sensor is significantly coupled with its nearest neighbors while the coupling with other sensors can be ignored. This assumption reduces the MCM to 6×6 matrix with 5 nonzero diagonals. The gain of the sensors was selected using the following relation:

$$\alpha_i = [(\beta_i - 0.5)\sigma_\alpha \cdot \sqrt{12} + 1] \quad i = 1,2, \cdots, M$$

where β_i is a uniformly distributed random number between 0 and 1, and σ_α^2 is the variance of the gain. The phase of each of the sensors was selected according to:

$$\phi_i = [(\eta_i - 0.5)\sigma_\phi \cdot \sqrt{12}] \quad i = 1,2, \cdots, M$$

where η_i is a uniformly distributed random number between zero and one and σ_ϕ^2 is the variance of the sensors phase.

Figures 10.34–10.38 describe an experiment with the following parameters: $SNR = 10 \log_{10}(\sigma_s^2/\sigma_n^2) = 30\text{dB}$, $J = 500$ snapshots, $\sigma_\alpha = 0.2$, $\sigma_\phi = 20°$, coupling coefficient $= 0.2 + j \cdot 0$, $DOA_1 = -30°$, $DOA_2 = -5°$, $DOA_3 = 35°$, Gains = 1.000, 0.5800, 0.8215, 1.077, 0.7970, 0.9620, Phases = 0°, −3.8°, 14.6°, 34.4°, −3.4°, 41.4°.

Figure 10.34 shows the spatial spectrum of the proposed procedure at the first, second, fourth, and thirtieth iteration. It is clear that significant DOA errors are corrected.

Figure 10.35 shows the reduction of the cost function value during the iterations, until convergence is obtained. Figure 10.36 shows $(\|\Gamma_i - \Gamma_t\|/\|\Gamma_t\|) \cdot 100$,

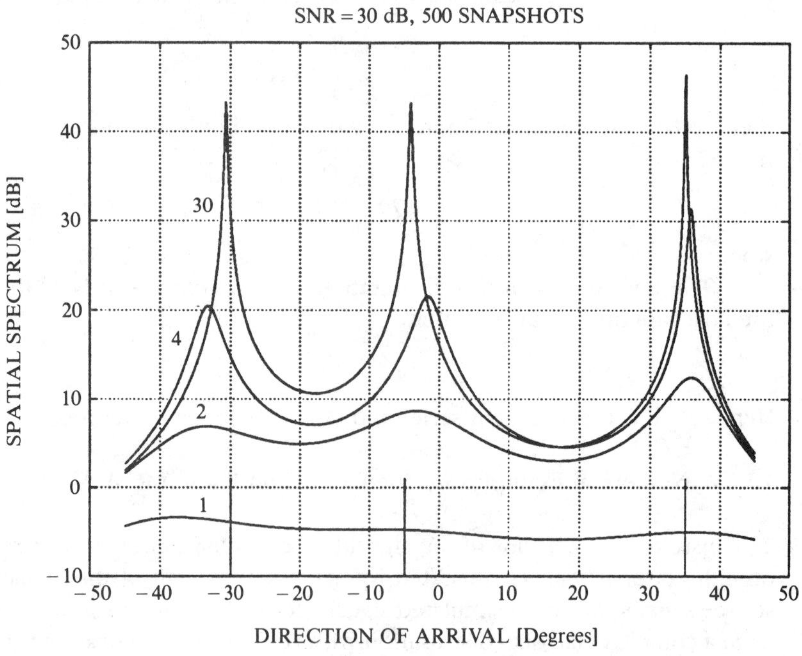

Figure 10.34 Spatial spectrum of the proposed algorithm after iterations 1, 2, 4, 30.

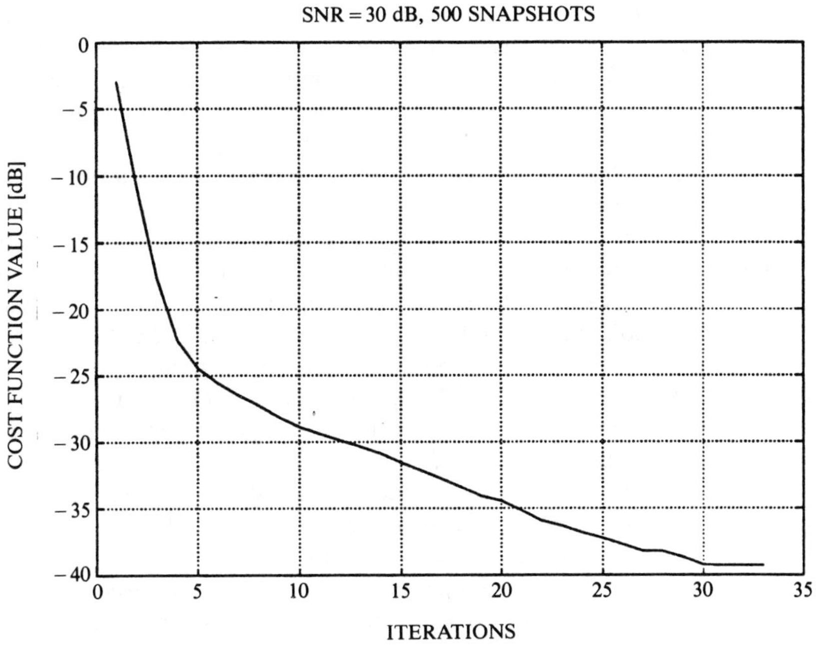

Figure 10.35 Value of the cost function versus iteration number.

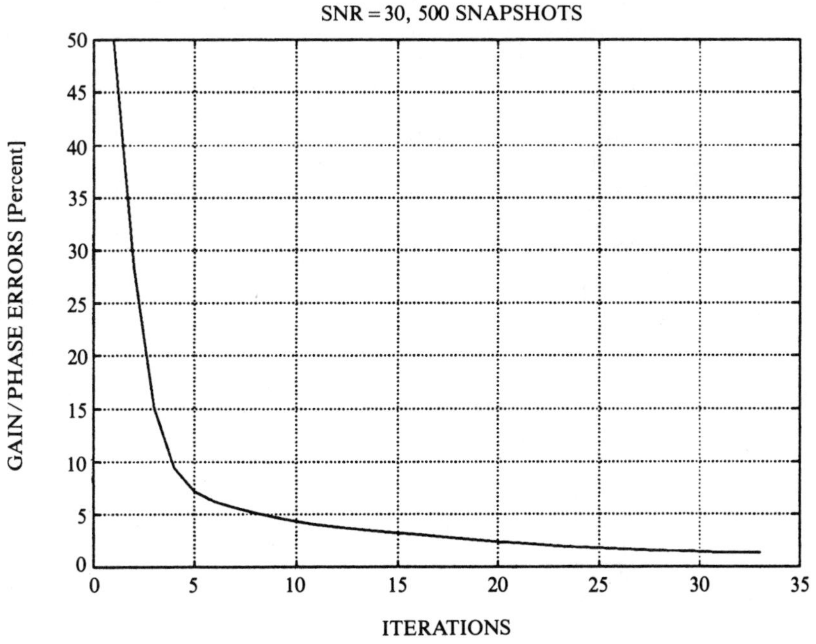

Figure 10.36 Relative gain-phase errors versus iteration number.

which is a measure of the relative gain-phase errors of all the sensors. Γ_i is the gain/phase matrix at iteration i while Γ_t is the true gain-phase matrix.

Figure 10.37 shows the relative coupling coefficient error as a function of iterations. Before the first iteration the error is 100 percent and it reduces to 1.9 percent.

Figure 10.38 shows the DOA errors for the three sources as a function of the iteration number.

To demonstrate the statistical efficiency of the proposed procedure we performed the following Monte Carlo experiments. The six sensor circular array described was used, with three far field narrow-band emitters. The gains and phases were selected as before, with $\sigma_\alpha = 0.02$ and $\sigma_\phi = 2°$. The DOAs were $\gamma_1 = 0°$, $\gamma_2 = 120°$, $\gamma_3 = 240°$. The coupling coefficient between any two adjacent sensors was $cc = 0.2(x + iy)$ where x and y are two i.i.d. random variables with uniform distribution over the interval $[-0.5, 0.5]$. The coupling coefficient for any nonadjacent sensors was assumed to be zero.

We performed 30 experiments for each signal-to-noise ratio, for SNR = 10 dB, 20 dB, 30 dB. In each experiment 500 snapshots of data were collected and processed by the algorithm. The values of the sensor gains and phases were kept constant throughout these simulations. For each SNR we used the results of the 30 experiments to compute the estimated root mean square error and the bias. Figures 10.39–10.52 depict these results and compare them to the corresponding CRLB

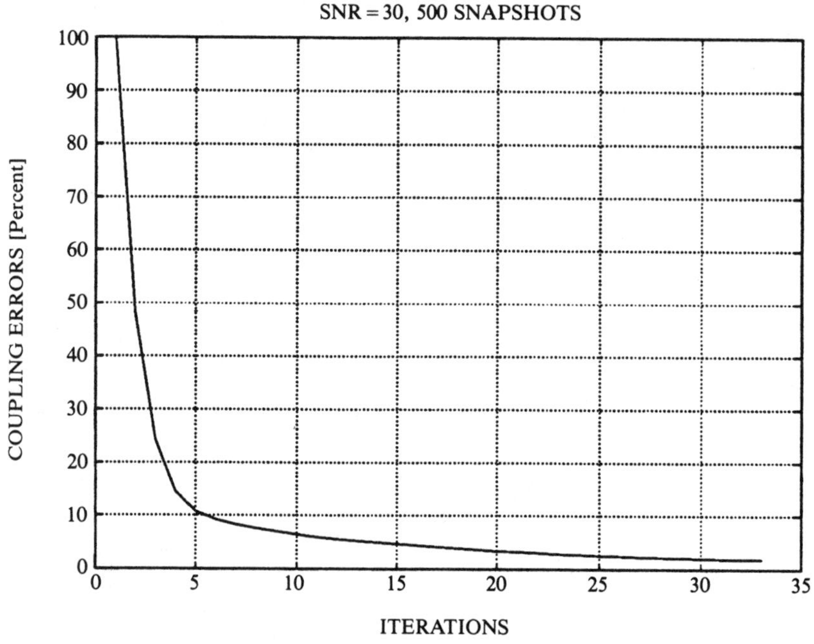

Figure 10.37 Coupling coefficient error versus iteration number.

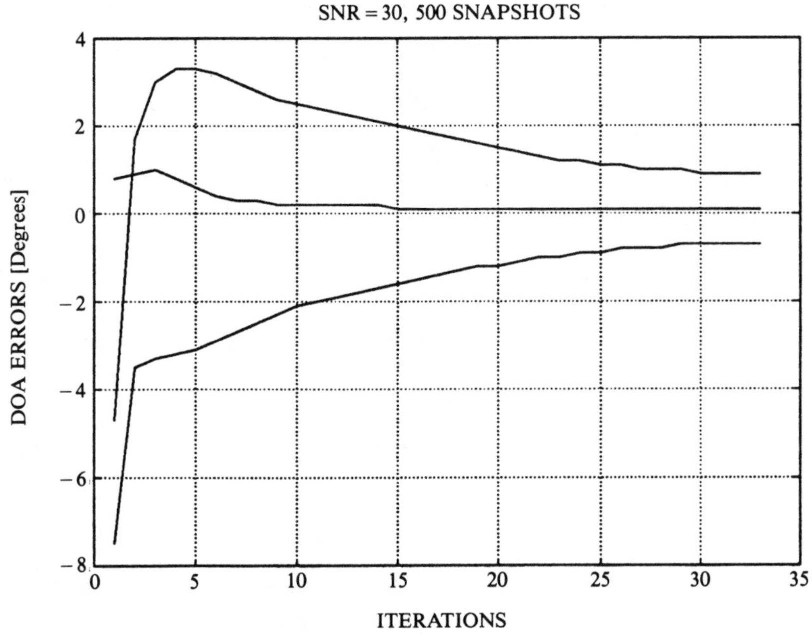

Figure 10.38 DOA errors versus iteration number.

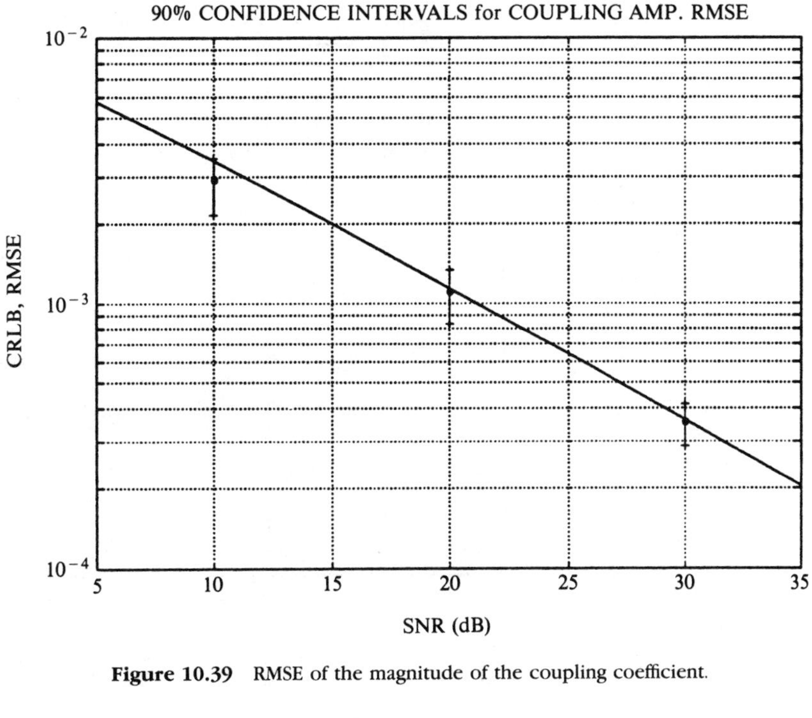

Figure 10.39 RMSE of the magnitude of the coupling coefficient.

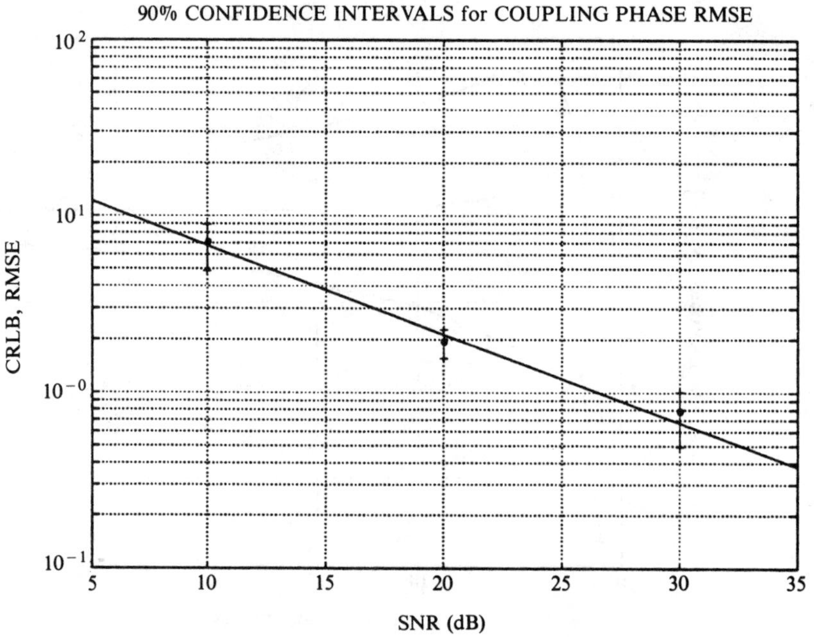

Figure 10.40 RMSE of the phase of the coupling coefficient.

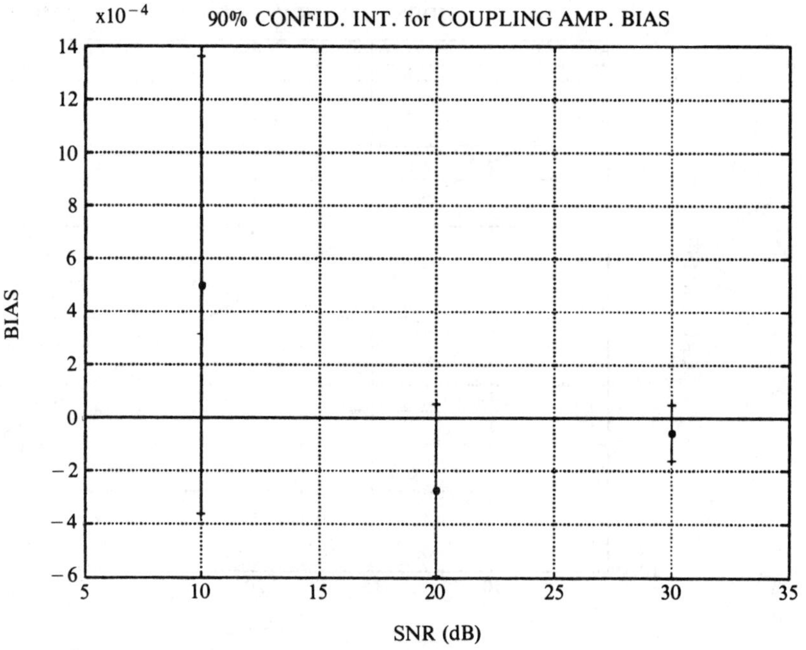

Figure 10.41 BIAS of the amplitude of the coupling coefficient.

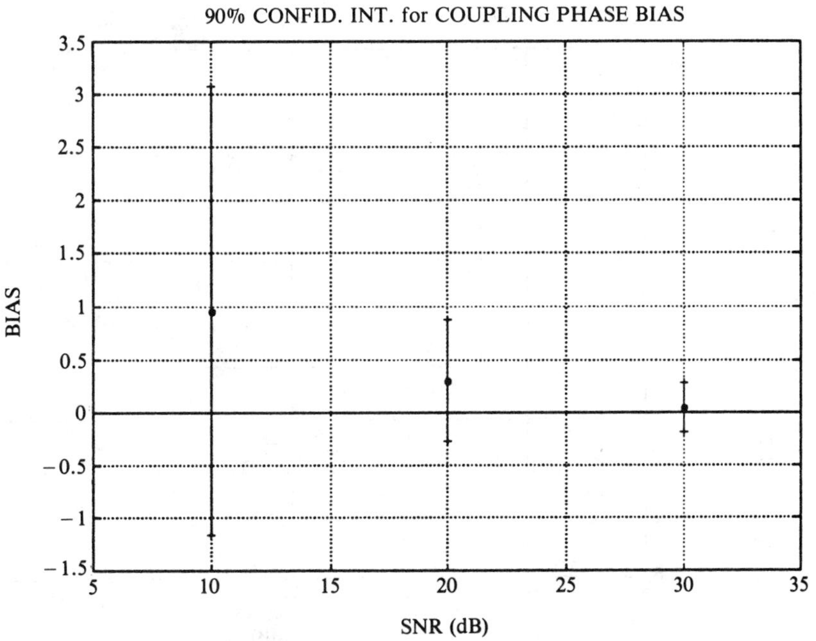

Figure 10.42 BIAS of the phase of the coupling coefficient.

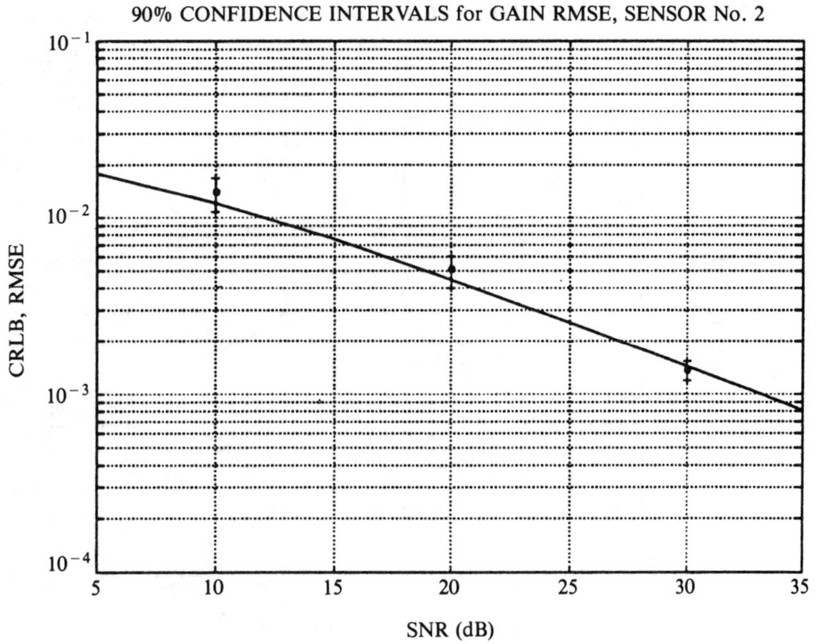

Figure 10.43 RMSE of the gain of sensor 2.

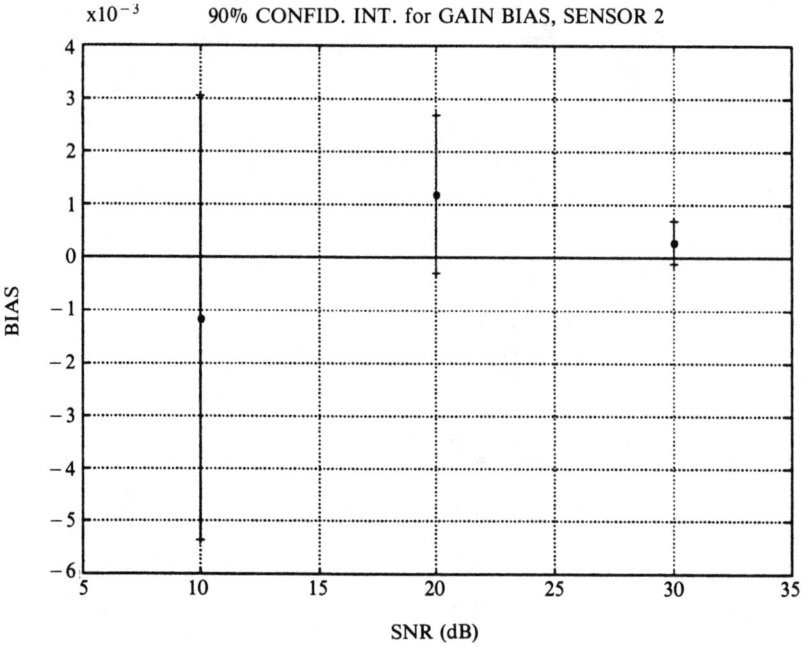

Figure 10.44 BIAS of the gain of sensor 2.

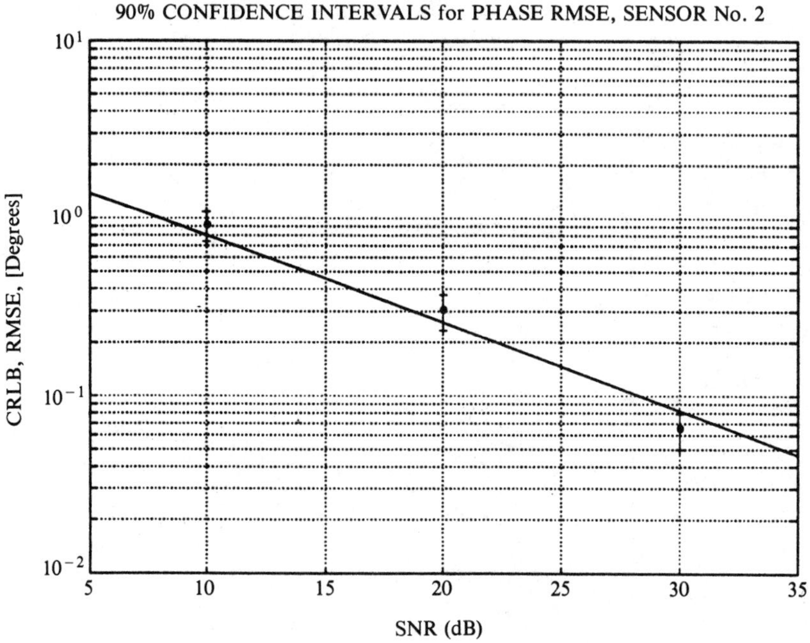

Figure 10.45 RMSE of the phase of sensor 2.

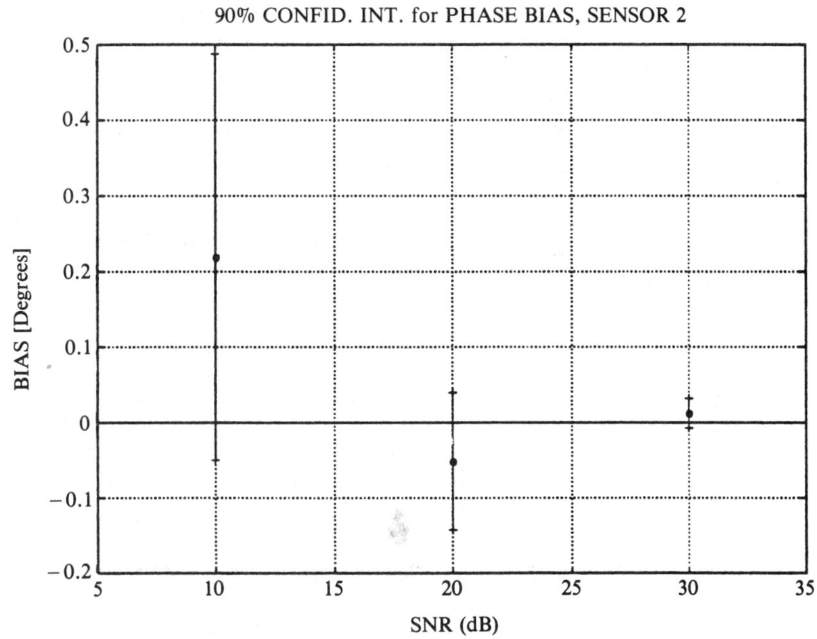

Figure 10.46 BIAS of the phase of sensor 2.

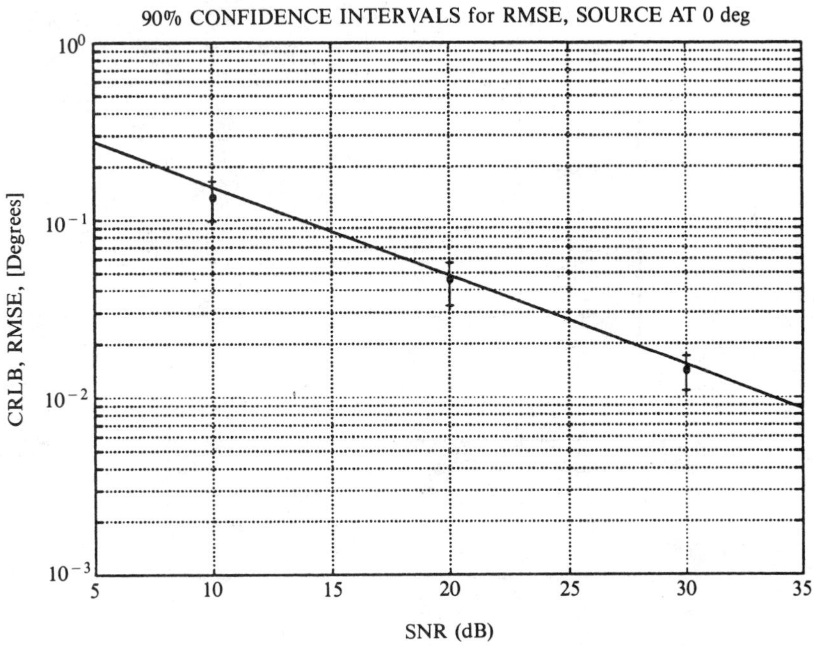

Figure 10.47 RMS for DOA of source at 0°.

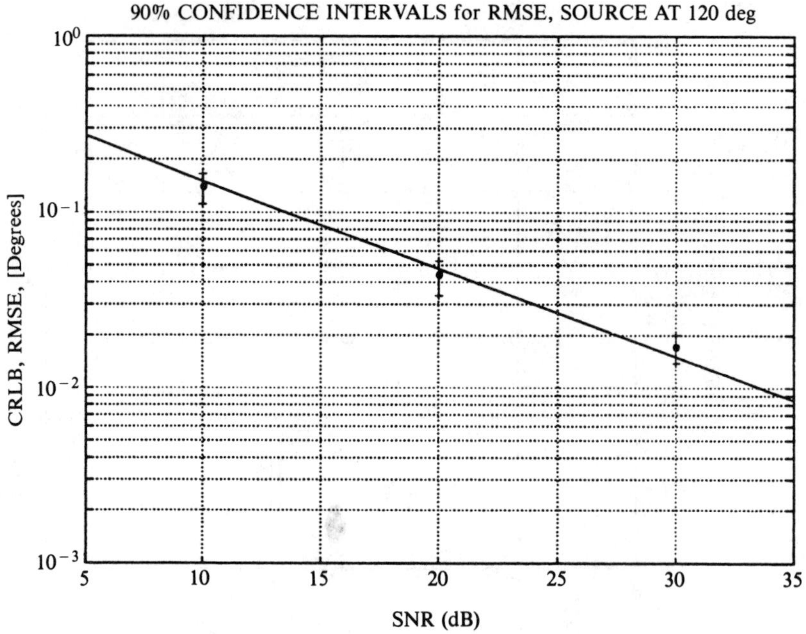

Figure 10.48 RMS for DOA of source at 120°.

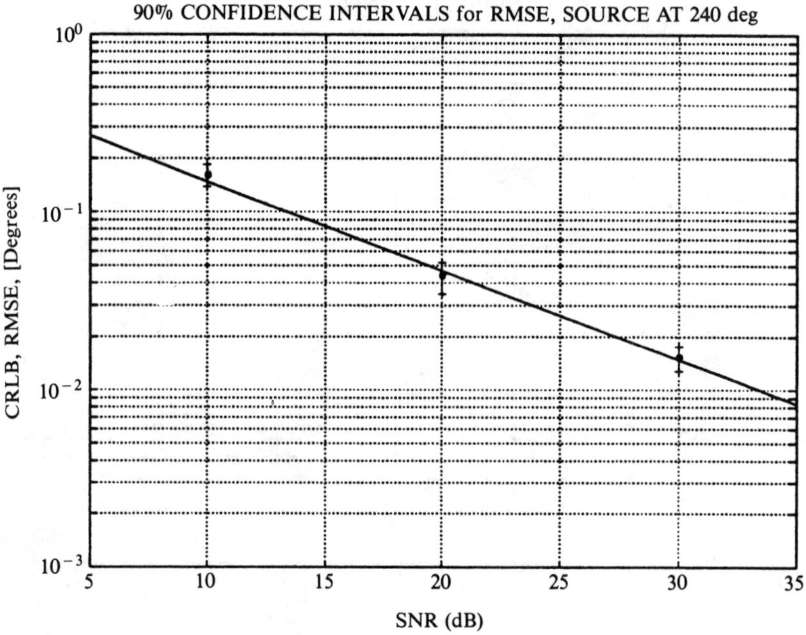

Figure 10.49 RMS for DOA of source at 240°.

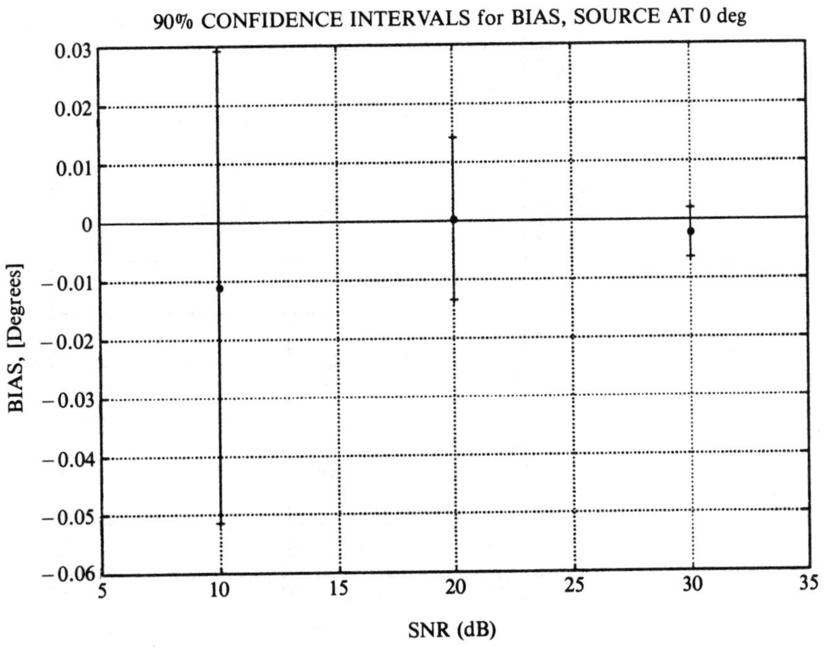

Figure 10.50 RMS for DOA of source at 0°.

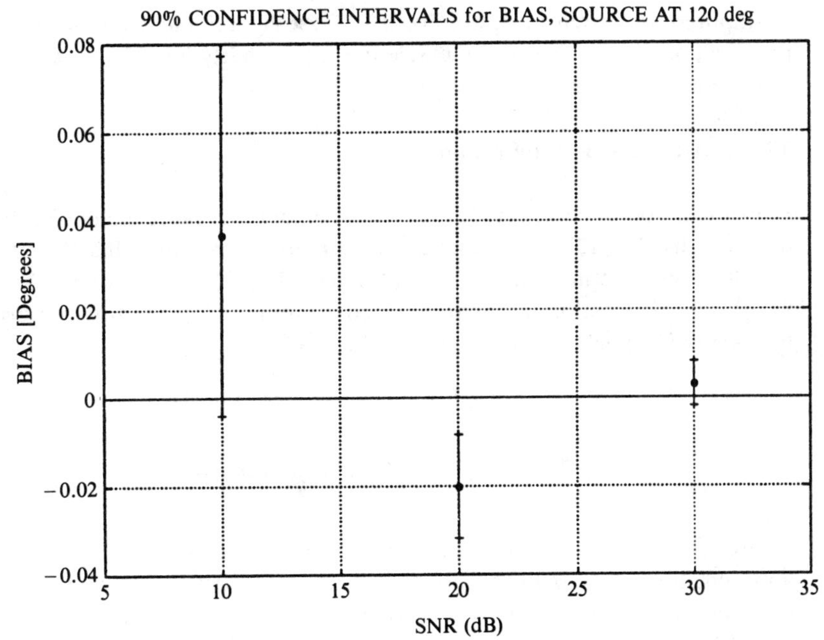

Figure 10.51 RMS for DOA of source at 120°.

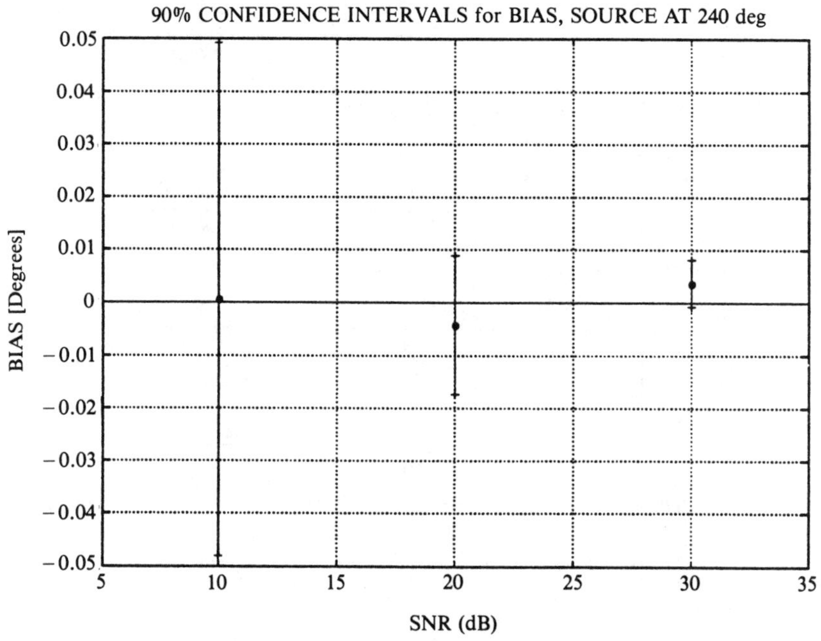

Figure 10.52 RMS for DOA of source at 240°.

(computed as shown in Appendix B.) These figures clearly indicate that the proposed algorithm is statistically efficient even for fairly low SNRs, at least for the test case considered here.

10.6.5 Array Shape Calibration

Sensor location uncertainty can severely degrade the performance of direction-finding systems. In [17] we presented an iterative maximum-likelihood method for simultaneously estimating directions of arrival $\{\gamma_n\}_{n=1}^{N}$ and sensor locations $\{(x_m, y_m)\}_{m=1}^{M}$. The estimation is performed by minimizing the squared error between the received signals and their assumed model:

$$\{\hat{\gamma}_n\}, \{(\hat{x}_m, \hat{y}_m)\} = \arg\min \boldsymbol{Q} \tag{10.45a}$$

$$\boldsymbol{Q} \triangleq \sum_{j=1}^{J} \|\mathbf{X}(j) - \boldsymbol{A}\mathbf{S}(j)\|^2 \tag{10.45b}$$

where $\|\cdot\|$ denotes the Euclidean norm. Equation (10.45a) also represents the maximum-likelihood estimates under the assumption that the noise vectors $\{\mathbf{V}(j)\}$ are i.i.d. zero-mean Gaussian with covariance $\sigma^2 I$.

The minimization required in (10.45a) is not trivial since the vectors $\mathbf{S}(j)$ and the matrix A are not known to the observer. However, whenever A is known, Q is minimized by choosing

$$\hat{\mathbf{S}}(j) = (\mathbf{A}^{\dagger}\mathbf{A})^{-1}\mathbf{A}^{\dagger}\mathbf{X}(j) \tag{10.45c}$$

as the estimates of $\mathbf{S}(j)$ for $j = 1, 2, \ldots, J$. Inserting (10.45c) into (10.45b) eliminates $\mathbf{S}(j)$ from the optimization problem.

The minimization of Q is performed in two separate steps. In the first step we minimize over $(\hat{\gamma}_n)$ using at first the nominal sensor locations and later using estimates provided by the second step. During the second step we minimize over $(\hat{x}_m, \hat{y}_m)$, while fixing $(\hat{\gamma}_n)$ at the values provided by the first step. The algorithm iterates between the two steps until convergence is achieved. A more detailed description of the algorithm and numerical examples illustrating its behavior are presented in [17].

Perhaps the most distinctive feature of the algorithm is its ability to locate the sensors accurately, without deploying calibration sources at known locations. The algorithm does not require disjoint sources or sources that can be turned on and off as proposed elsewhere in the literature. The proposed technique can be used to improve the initial DOA estimates provided by an algorithm that does not take into account sensor location uncertainties.

10.7 CONCLUSIONS

The sensitivity of high-resolution direction-finding techniques to errors in the calibration of the array manifold has made the practical implementation of these techniques a difficult task. In this chapter we explored in some detail one particular approach, which we call array self-calibration, aimed at making these techniques more robust to calibration errors.

The approach is based on modeling the calibration errors by means of a parametric model in which the sensor gains, phases, mutual coupling coefficients, or locations are considered to be imprecisely known parameters. The assumption is that if the "true" values of these parameters were known, then our model will provide the exact values of the array manifold (i.e., the array will be perfectly calibrated). Using signals received by the array we then attempt to estimate the "true" values of these parameters and, thus, self-calibrate the array.

In this chapter we considered the most challenging form of the self-calibration problem: multiple sources are present at the same time and in the same frequency band, and the directions of these sources are unknown. We show that given a sufficient number of sources, it is possible to estimate both the source directions and the unknown array parameters (subject to some mild assumptions). We provide some specific examples of algorithms for performing the joint direction/array-parameter estimation for different choices of the array parameters that are to be estimated. It is possible, of course, to develop other self-calibration algorithms.

Next we presented an accuracy analysis of the self-calibration technique, using the Cramér-Rao lower bound as the principal tool. We studied the asymptotic variance of the DOA estimation errors under two different situations: (1) the case where all the array parameters are known and only the DOAs need to be estimated and (2) the case where both the DOAs and some array parameters need to be jointly estimated. This theoretical analysis, which was also verified by computer simulations, indicates that it is indeed feasible to jointly estimate DOAs and unknown array parameters, and to achieve accuracies that are only slightly less than the accuracy that would be achieved by a perfectly calibrated array.

The results developed in this chapter seem to indicate that self-calibration is a potentially useful technique for making high-resolution array processing techniques implementable in practice. It should be noted that in many situations of practical interest it is possible to develop and apply self-calibration techniques that are somewhat simpler than the class of techniques developed here. Consider, for example, the following situations:

1. Calibration sources are at known locations and can be separated at the receiver either in time or in frequency (that is, only one source is turned on at any given time, or the sources are at different frequency bands).
2. Calibration sources are at unknown locations, but can be separated at the receiver either in time or in frequency.

Both these situations are easier than the one considered in this chapter since the estimation algorithm has to deal with a single source. Various self-calibration algorithms have been and can be developed for these cases.

We believe that self-calibration is likely to become an essential ingredient of practical implementations of high-resolution array processing systems. However, the work reported here represents only the first few steps in the development of self-calibration techniques. A considerable amount of work remains to be done before these techniques are fully developed and validated. In particular, these techniques need to be validated on real data collected by experimental arrays.

REFERENCES

1. R. O. Schmidt, "A Signal Subspace Approach to Multiple Emitter Location and Spectral Estimation," Ph.D. dissertation, Stanford University, Stanford, California, 1981.
2. R. O. Schmidt, "Multiple Emitter Location and Signal Parameter Estimation," *Proc. RADC Spectrum Estimation Workshop*," Griffiths AFB, New York, 1979, pp. 243–258. Reprinted in *IEEE Trans. Antennas and Propag.*, **34**, 3 (March 1986), 276–280.
3. R. O. Schmidt, "Multilinear Array Manifold Interpolation," *Technical Memo ESL-TM1661*, ESL, Inc., Sunnyvale, California, September 1983.

4. B. D. Steinberg, *Principles of Aperture and Array System Design Including Random and Adaptive Array*, (New York: John Wiley, 1976).
5. Y. Rockah and P. M. Schultheiss, "Array Shape Calibration Using Sources in Unknown Locations Part I: Far Field Sources," *IEEE Trans. Acoustics, Speech, and Signal Processing*, **35,** no. 3 (March 1987), 286–299.
6. Y. Rockah and P. M. Schultheiss, "Array Shape Calibration Using Sources in Unknown Locations Part II: Near-Field Sources and Estimator Implementation," *IEEE Trans. Acoustics, Speech, and Signal Processing*, **35,** no. 6 (June 1987), 724–735.
7. J. T-H. Lo and S. L. Marple, Jr., "Eigenstructure Methods for Array Sensor Localization," *ICASSP 1987*, Dallas, Texas, pp. 2260–2263.
8. A. Paulraj and T. Kailath, "Direction of Arrival Estimation by Eigenstructure Methods with Unknown Sensor Gain and Phase," *Proc. IEEE ICASSP'85*, Tampa, Florida, 1985, pp. 640–643.
9. Special issue on time delay estimation, *IEEE Trans. Acoustics, Speech, and Signal Processing*, **30** (June 1981).
10. S. Haykin, ed. *Array Signal Processing*, (Englewood Cliffs, N.J.: Prentice-Hall, 1985).
11. A. J. Barabell, J. Capon, D. F. Delong, J. R. Johnson, and K. D. Senne, "Performance Comparison of Superresolution Array Processing Algorithms," Project Report TST-72, Lincoln Laboratory, Massachusetts Institute of Technology, Lexington, Massachusetts, May 1984.
12. B. Porat and B. Friedlander, "Analysis of the Asymptotic Relative Efficiency of the MUSIC Algorithm," *IEEE Trans. Acoustics, Speech, and Signal Processing*, **36,** no. 4 (April 1988), pp. 532–544.
13. B. Friedlander, "A Sensitivity Analysis of the MUSIC Algorithm" *IEEE Trans. ASSP,* to appear 1990.
14. M. Kaveh and A. J. Barabell, "The Statistical Performance of the MUSIC and the Minimum-Norm Algorithms in Resolving Plane Waves in Noise," *IEEE Trans. Acoustics, Speech, and Signal Processing*, **34,** no. 2 (April 1986), 331–341.
15. B. Friedlander, and A. J. Weiss, "Eigenstructure Methods for Direction Finding with Sensor Gain and Phase Uncertainties," *Int. Conf. on Acoustics, Speech, and Signal Processing*, April 1988. Also to appear in *J. Circuits, Systems and Signal Processing,* 1990.
16. B. Friedlander and A. J. Weiss, "Direction Finding in the Presence of Mutual Coupling," 22nd Asilomar Conference on Signals, Systems and Computers, Pacific Grove, California, October 31–November 2, 1987. Also to appear in *IEEE Trans. on Antennas and Propagation,* 1990.
17. A. J. Weiss, and B. Friedlander, "Array Shape Calibration Using Sources in Unknown Locations: A Maximum Likelihood Approach," *Int. Conference on Acoustics, Speech, and Signal Processing*, April 1988. Also *IEEE Trans. Acoustics, Speech, and Signal Processing,* **37,** no. 12, 1958–1966, December 1989.
18. A. J. Weiss, A. S. Willsky, and B. C. Levy, "Eigenstructure Approach for Array Processing with Unknown Intensity Coefficients," *IEEE Trans. on Acoustics, Speech, and Signal Processing*, **36,** no. 10 (October 1988), 1613–1617.

APPENDIX A: PROOFS OF LEMMAS 6.3.1 – 6.3.3

Proof of Lemma 6.3.1

The proof of Lemma 6.3.1 can be obtained by direct multiplication.

Proof of Lemma 6.3.2

By definition an $M \times M$ symmetric circulant matrix $\mathbf{A}$, with i, jth element A_{ij}, has the following relations between its elements:

$$A_{ij} = A_{pq} \text{ if } (j - i)|M = (q - p)|M \tag{10.A.1}$$

and

$$A_{ij} = A_{ji} \tag{10.A.2}$$

where we used the notation:

$$(j - i)|M = \begin{cases} M + j - i & \text{when } i > j, \\ j - i & \text{when } i \leq j. \end{cases} \tag{10.A.3}$$

From these relations we see that a symmetric circulant matrix contains at most L distinct elements $a_1, a_2, \ldots, a_L$ where

$$L = \begin{cases} (M + 2)/2 & \text{when } M \text{ is even} \\ (M + 1)/2 & \text{when } M \text{ is odd} \end{cases} \tag{10.A.4}$$

For even M, the matrix $\mathbf{A}$ has the form

$$\mathbf{A} = \begin{bmatrix} a_1 & a_2 & \cdots & a_L & a_{L-1} & \cdots & a_2 \\ a_2 & a_1 & a_2 & & & & \\ \vdots & a_2 & \ddots & \ddots & & & \\ a_L & & \ddots & & & & \\ a_{L-1} & & & & & & \\ \vdots & & & & & & \\ a_2 & & & & & & \end{bmatrix}$$

while for odd M, $\mathbf{A}$ has the form

$$\mathbf{A} = \begin{bmatrix} a_1 & a_2 & \cdots & a_L & a_L & \cdots & a_2 \\ a_2 & a_1 & a_2 & & & & \\ \vdots & a_2 & \ddots & \ddots & & & \\ a_L & & \ddots & & & & \\ a_L & & & & & & \\ \vdots & & & & & & \\ a_2 & & & & & & \end{bmatrix}$$

It is easy to see that the matrix $\mathbf{A}$ may be represented by the sum

$$A = \sum_{k=-M}^{M} \text{diag}\,(A, k)$$

where diag (A, k) is an $M \times M$ zero matrix except for the kth diagonal which is equal to the corresponding diagonal of $\boldsymbol{A}$. Since the elements of each of the diagonals are equal, we have for even M:

$$\begin{aligned} \boldsymbol{A} = &\sum_{k=0}^{L-1} a_{k+1} \ \text{diag}(J_M, k) + \sum_{k=L}^{M-1} a_{2L-k-1} \ \text{diag}\ (J_M, k) \\ &+ \sum_{k=-1}^{-L+1} a_{-k+1} \ \text{diag}\ (J_M, k) + \sum_{k=-L}^{-M+1} a_{2L+k-1} \ \text{diag}\ (J_M, k), \end{aligned} \tag{10.A.5}$$

and for odd M:

$$\begin{aligned} \boldsymbol{A} = &\sum_{k=0}^{L-1} a_{k+1} \ \text{diag}(J_M, k) + \sum_{k=L}^{M-1} a_{2L-k} \ \text{diag}\ (J_M, k) \\ &\sum_{k=-1}^{-L+1} a_{-k+1} \ \text{diag}\ (J_M, k) + \sum_{k=-L}^{-M+1} a_{2L+k} \ \text{diag}\ (J_M, k); \end{aligned} \tag{10.A.6}$$

where J_M is an $M \times M$ matrix of ones. Now, note that

$$\text{diag}(J_M, k) \cdot \boldsymbol{x} = \begin{cases} [x_{k+1}, x_{k+2}, \cdots, x_M, 0, \cdots, 0]^T & k \geq 0 \\ [0, \cdots, 0, x_1, \cdots, x_{M+k}]^T & k \leq 0 \end{cases} \tag{10.A.7}$$

Using the notation $\mathbf{a} = [a_1, a_2, \cdots, a_L]^T$, we have

$$a_j \cdot \text{diag}\ (J_M, k) \cdot \mathbf{x} = \text{column}\ (\mathbf{x}_k^M, j) \cdot \mathbf{a} \tag{10.A.8}$$

where $\mathbf{x}_k^M$ is the right side of (10.A.7), and column $(\mathbf{x}, j)$ represents an $M \times L$ zero matrix except for the jth column which is equal to $\mathbf{x}$. Using (10.A.5) we obtain, for even M:

$$\begin{aligned} \boldsymbol{Ax} = &\{ \sum_{k=0}^{L-1} \text{column}\ (\mathbf{x}_k^M, k+1)\} \cdot \mathbf{a} + \{ \sum_{k=L}^{M-1} \text{column}\ (\mathbf{x}_k^M, 2L-k-1)\} \cdot \mathbf{a} \\ &+ \{ \sum_{k=-1}^{-L+1} \text{column}\ (\mathbf{x}_k^M, l-\text{k})\} \cdot \mathbf{a} + \{ \sum_{k=-L}^{-M+1} \text{column}\ (\mathbf{x}_k^M, 2L+k-1)\} \cdot \mathbf{a} \end{aligned} \tag{10.A.9}$$

and for odd M:

$$\begin{aligned} \mathbf{Ax} = &\left\{ \sum_{k=0}^{L-1} \text{column}\ (\mathbf{x}_k^M, k+1) \right\} \cdot \mathbf{a} + \left\{ \sum_{k=L}^{M-1} \text{column}\ (\mathbf{x}_k^M, 2L-k)\right\} \cdot \mathbf{a} \\ &+ \left\{ \sum_{k=-1}^{-L+1} \text{column}\ (\mathbf{x}_k^M, 1-k)\right\} \cdot \mathbf{a} + \left\{ \sum_{k=-L}^{-M+1} \text{column}\ (\mathbf{x}_k^M, 2L+k) \right\} \cdot \mathbf{a} \end{aligned} \tag{10.A.10}$$

It is now easy to verify that

$$W_1 = \sum_{k=0}^{L-1} \text{column}\,(\mathbf{x}_k^M, k+1)$$

$$W_2 = \sum_{k=-1}^{-L+1} \text{column}\,(\mathbf{x}_k^M, 1-k)$$

$$W_3 = \begin{cases} \displaystyle\sum_{k=L}^{M-1} \text{column}\,(\mathbf{x}_k^M, 2L-k-1) & \text{even } M \\ \displaystyle\sum_{k=L}^{M-1} \text{column}\,(\mathbf{x}_k^M, 2L-k) & \text{odd } M \end{cases}$$

$$W_4 = \begin{cases} \displaystyle\sum_{k=-L}^{-M+1} \text{column}\,(\mathbf{x}_k^M, 2L+k-1) & \text{even } M \\ \displaystyle\sum_{k=-L}^{-M+1} \text{column}\,(\mathbf{x}_k^M, 2L+k) & \text{odd } M \end{cases}$$

Hence,

$$\mathbf{Ax} = (W_1 + W_2 + W_3 + W_4)\mathbf{a} = Q_2(\mathbf{x})\mathbf{a}$$

Q.E.D.

Proof of Lemma 6.3.3

The proof of Lemma 6.3.3 goes along the same steps as the proof of Lemma 6.3.2.

APPENDIX B: THE CRAMÉR-RAO LOWER BOUND FOR GAUSSIAN SIGNALS

Consider a complex Gaussian vector $\mathbf{x}$ with zero mean,

$$E\{\mathbf{x}\} = 0, \tag{10.B.1}$$

and covariance

$$\boldsymbol{R} = E\{\mathbf{x}\mathbf{x}^\dagger\}. \tag{10.B.2}$$

The unknown parameters θ are imbedded in the covariance $\boldsymbol{R}$. The logarithm of the probability density function (PDF) for J_1 statistically independent observations can be written as

$$L(\theta) = -J_1 \ell n\{\det(\boldsymbol{R}\pi)\} - \sum_{j=1}^{J_1} \mathbf{x}(j)^\dagger \boldsymbol{R}^{-1}\mathbf{x}(j) \qquad (10.B.3)$$

$$= -J_1 \cdot \mathrm{tr}[\boldsymbol{R}^{-1}\hat{\boldsymbol{R}}] - J_1 \ell n\{\det(\boldsymbol{R}\pi)\},$$

where

$$\hat{\boldsymbol{R}} = \frac{1}{J_1} \sum_{j=1}^{J_1} \mathbf{x}(j)\mathbf{x}(j)^\dagger. \qquad (10.B.4)$$

The m, nth element of the Fisher information matrix (FIM) is given by

$$J_{mn} = -E\left\{\frac{\partial^2 L}{\partial\theta_m \partial\theta_n}\right\}. \qquad (10.B.5)$$

To evaluate (10.B.5) we use the following relations:

$$d(\boldsymbol{R}^{-1}) = -\boldsymbol{R}^{-1} \cdot d\boldsymbol{R} \cdot \boldsymbol{R}^{-1} \qquad (10.B.6)$$

$$d(\ell n\{\det(\boldsymbol{R})\}) = \mathrm{tr}[\boldsymbol{R}^{-1} \cdot d\boldsymbol{R}] \qquad (10.B.7)$$

$$E\{\hat{\boldsymbol{R}}\} = \boldsymbol{R}. \qquad (10.B.8)$$

Taking the first derivative of $L(\boldsymbol{\theta})$ we obtain

$$\partial L/\partial\theta_m = J_1 \cdot \mathrm{tr}[\boldsymbol{R}^{-1} \cdot \partial\boldsymbol{R}/\partial\theta_m \cdot \boldsymbol{R}^{-1} \cdot \hat{\boldsymbol{R}}] - J_1 \cdot \mathrm{tr}[\boldsymbol{R}^{-1} \cdot \partial\boldsymbol{R}/\partial\theta_m]$$

$$= J_1 \cdot \mathrm{tr}[\boldsymbol{R}^{-1} \cdot \partial\boldsymbol{R}/\partial\theta_m \cdot (\boldsymbol{R}^{-1}\hat{\boldsymbol{R}} - I)].$$

The second derivative is given by

$$\partial^2 L/\partial\theta_m\partial\theta_n = J_1 \cdot \mathrm{tr}[\partial(\boldsymbol{R}^{-1} \cdot \partial\boldsymbol{R}/\partial\theta_m)/\partial\theta_n \cdot (\boldsymbol{R}^{-1}\hat{\boldsymbol{R}} - I)$$

$$+ \boldsymbol{R}^{-1} \cdot \partial\boldsymbol{R}/\partial\theta_m \cdot (-\boldsymbol{R}^{-1} \cdot \partial\boldsymbol{R}/\partial\theta_n \cdot \boldsymbol{R}^{-1} \cdot \hat{\boldsymbol{R}})].$$

Taking the expectation of both sides we obtain

$$J_{mn} = -E\{\partial^2 L/\partial\theta_m\partial\theta_n\} = J_1 \cdot \mathrm{tr}[\boldsymbol{R}^{-1} \cdot \partial\boldsymbol{R}/\partial\theta_m \cdot \boldsymbol{R}^{-1} \cdot \partial\boldsymbol{R}/\partial\theta_n]. \qquad (10.B.9)$$

Thus, the number of observations enters the result only as a multiplicative constant. To simplify the exposition we assume $J_1 = 1$. The modification for other values of J_1 is straightforward.

The FIM for a General Passive Array

For the parameter estimation problem posed earlier in this work, the covariance matrix of the data vector $\mathbf{x}$ can be written as

$$\tilde{\boldsymbol{R}} = \boldsymbol{C}\boldsymbol{\Gamma}\tilde{\boldsymbol{A}}\boldsymbol{R}_s\tilde{\boldsymbol{A}}^\dagger\boldsymbol{\Gamma}^\dagger\boldsymbol{C}^\dagger + \sigma^2 \cdot I, \qquad (10.B.10)$$

where $\tilde{\boldsymbol{A}}$ is the direction matrix. To simplify the derivation we assume that σ^2 and $\boldsymbol{R}_s$ are known. Dividing both sides of (10.B.10) by σ^2, we obtain

$$\boldsymbol{R} = \tilde{\boldsymbol{R}}/\sigma^2 = \boldsymbol{C}\boldsymbol{\Gamma}\tilde{\boldsymbol{A}}\boldsymbol{P}\tilde{\boldsymbol{A}}^\dagger\boldsymbol{\Gamma}^\dagger\boldsymbol{C}^\dagger + \boldsymbol{I} \qquad (10.B.11)$$

where

$$P \triangleq R_s/\sigma^2. \qquad (10.B.12)$$

It is easy to verify that

$$J(\tilde{R}) = J(R), \qquad (10.B.13)$$

and therefore we use $\boldsymbol{R}$ instead of $\tilde{\boldsymbol{R}}$ to derive the FIM. Moreover, we use the following notation to simplify the formulas:

$$A \triangleq C\Gamma \tilde{A}, \qquad (10.B.14)$$

$$W \triangleq A^\dagger A, \qquad (10.B.15)$$

$$Q \triangleq (P^{-1} + W)^{-1}. \qquad (10.B.16)$$

We also use the relation

$$R^{-1} = I - AQA^\dagger, \qquad (10.B.17)$$

which implies that

$$PWQ = QWP = P - Q. \qquad (10.B.18)$$

Derivatives with Respect to DOA

In the ensuing development, we make use of the notational device

$$\dot{A}_{\gamma_j} \triangleq \partial A/\partial\gamma_j \qquad (10.B.19)$$

for the partial derivative of the matrix $\boldsymbol{A}$ with respect to the DOA γ_j of the jth sensor.

Using (10.B.19), we first write the partial derivative of the covariance matrix with respect to the jth DOA as

$$\partial R/\partial_{\gamma_j} = \dot{A}_{\gamma_j} PA^\dagger + \mathbf{AP}\dot{\mathbf{A}}^\dagger_{\gamma_j}. \qquad (10.B.20)$$

Substituting in (10.B.9) and noting that

$$\text{tr}[A^\dagger] = \text{conj}\{\text{tr}[A]\} \qquad (10.B.21)$$

we obtain

$$J_{\gamma i \gamma j} = 2\text{Re}\{\text{tr}[\dot{A}_{\gamma_i} PA^\dagger R^{-1} APA\dot{A}^\dagger_{\gamma_j} R^{-1}] + \text{tr}[\dot{A}_{\gamma i} PA^\dagger R^{-1} \dot{A}_{\gamma j} PA^\dagger R^{-1}]\}. \qquad (10.B.22)$$

Observe that

$$\dot{A}_{\gamma_i} = C\Gamma\partial\tilde{A}/\partial_{\gamma i} = C\Gamma\dot{\tilde{A}}\mathbf{e}_i\mathbf{e}_i^T, \qquad (10.B.23)$$

where the unit vector $\boldsymbol{e}_i$ is the ith column vector of the identity $M \times M$ identity matrix, and $\dot{\tilde{A}}$ is the matrix of derivatives given by

$$\dot{\tilde{A}} = [\dot{\mathbf{a}}(\gamma_1)\dot{\mathbf{a}}(\gamma_2)\cdots\dot{\mathbf{a}}(\gamma_N)]. \qquad (10.B.24)$$

We use the notation

$$\dot{A} \triangleq C\Gamma\dot{\tilde{A}} \tag{10.B.25}$$

to simplify our formulas.

Using (10.B.23) and (10.B.25), equation (10.B.22) becomes

$$\begin{aligned} J_{\gamma_i\gamma_j} &= 2\mathrm{Re}\{\mathrm{tr}\{\dot{A}\mathbf{e}_i\mathbf{e}_i^T PA^\dagger R^{-1}APe_j\mathbf{e}_j^T\dot{A}^\dagger R^{-1}\} \\ &\quad + \mathrm{tr}\{\dot{A}\mathbf{e}_i\mathbf{e}_i^T PA^\dagger R^{-1}\dot{A}\mathbf{e}_j\mathbf{e}_j^T PA^\dagger R^{-1}\}\} \\ &= 2\mathrm{Re}\{\mathbf{e}_i^T PA^\dagger R^{-1}APe_j\mathbf{e}_j^T\dot{A}^\dagger R^{-1}\dot{A}\mathbf{e}_i \\ &\quad + \mathbf{e}_i^T PA^\dagger R^{-1}\dot{A}\mathbf{e}_j\mathbf{e}_j^T PA^\dagger R^{-1}\dot{A}\mathbf{e}_i\}. \end{aligned} \tag{10.B.26}$$

Hence,

$$J_{\gamma\gamma} = 2\mathrm{Re}\{(PA^\dagger R^{-1}AP)\cdot(\dot{A}^\dagger R^{-1}\dot{A})^T + (PA^\dagger R^{-1}\dot{A})\cdot(PA^\dagger R^{-1}\dot{A})^T\} \tag{10.B.27}$$

where $J_{\gamma\gamma}$ is the submatrix of the FIM associated with the DOA derivatives and $\cdot$ denotes the *Hadamard product* of two matrices, defined by

$$(A\cdot B)_{ij} = A_{ij}B_{ij}. \tag{10.B.28}$$

Equation (10.B.27) may be further simplified using (10.B.16) as follows

$$J_{\gamma\gamma} = 2\mathrm{Re}\{(P-Q)\cdot(\dot{A}^\dagger R^{-1}\dot{A})^T + (QA^\dagger\dot{A})\cdot(QA^\dagger\dot{A})^T\}. \tag{10.B.29}$$

Derivatives with Respect to Sensor Phase

Repeating the same set of considerations leading to (10.B.22), we obtain

$$\begin{aligned} J_{\phi_i\phi_j} &= 2\mathrm{Re}\{\mathrm{tr}[\dot{A}_{\phi_i}PA^\dagger R^{-1}AP\dot{A}_{\phi_j}^\dagger R^{-1}] \\ &\quad + \mathrm{tr}[\dot{A}_{\phi_i}PA^\dagger R^{-1}\dot{A}_{\phi_j}PA^\dagger R^{-1}]\}, \end{aligned} \tag{10.B.30}$$

where

$$\dot{A}_{\phi_j} = \partial A/\partial\phi_j = C(\partial F/\partial\phi_j)G\tilde{A}. \tag{10.B.31}$$

Here F is a diagonal matrix containing the exponents of the sensors' phases while G is a diagonal matrix of the sensors' gains; thus

$$\Gamma = GF = FG. \tag{10.B.32}$$

It is useful to define $\dot{F}$ as the matrix of derivatives:

$$\dot{F} \triangleq \mathrm{diag}\{\dot{F}_{11}(\phi_1), \dot{F}_{22}(\phi_2), \ldots, \dot{F}_{MM}(\phi_M)\} = jF. \tag{10.B.33}$$

Thus (10.B.31) becomes

$$\dot{A}_{\phi_j} = C\mathbf{e}_j\mathbf{e}_j^T\dot{F}G\tilde{A} = jC\mathbf{e}_j\mathbf{e}_j^T\Gamma\tilde{A} = jC\mathbf{e}_j\mathbf{e}_j^T C^{-1}A. \tag{10.B.34}$$

Substituting (10.B.34) in (10.B.30) we obtain

$$J_{\phi_i\phi_j} = 2\mathrm{Re}\{\mathrm{tr}[C\mathbf{e}_i\mathbf{e}_i^T\Gamma\tilde{A}PA^\dagger R^{-1}AP\tilde{A}^\dagger\Gamma^\dagger\mathbf{e}_j\mathbf{e}_j^T C^\dagger R^{-1}]$$

$$+ \operatorname{tr}[\boldsymbol{C}\mathbf{e}_i\mathbf{e}_i^T\boldsymbol{\Gamma}\tilde{\boldsymbol{A}}\boldsymbol{P}\boldsymbol{A}^\dagger\boldsymbol{R}^{-1}\boldsymbol{C}\mathbf{e}_j\mathbf{e}_j^T\boldsymbol{\Gamma}\tilde{\boldsymbol{A}}\boldsymbol{P}\boldsymbol{A}^\dagger\boldsymbol{R}^{-1}]\}$$

$$\begin{aligned} J_{\phi\phi} &= 2\operatorname{Re}\{(\boldsymbol{\Gamma}\tilde{\boldsymbol{A}}\boldsymbol{P}\boldsymbol{A}^\dagger\boldsymbol{R}^{-1}\boldsymbol{A}\boldsymbol{P}\tilde{\boldsymbol{A}}^\dagger\boldsymbol{\Gamma}^\dagger)\cdot(\boldsymbol{C}^\dagger\boldsymbol{R}^{-1}\boldsymbol{C})^T \\ &\quad - (\boldsymbol{\Gamma}\tilde{\boldsymbol{A}}\boldsymbol{P}\boldsymbol{A}^\dagger\boldsymbol{R}^{-1}\boldsymbol{C})\cdot(\boldsymbol{\Gamma}\boldsymbol{A}\boldsymbol{P}\tilde{\boldsymbol{A}}^\dagger\boldsymbol{R}^{-1}\boldsymbol{C})^T\} \\ &= 2\operatorname{Re}\{(\boldsymbol{C}^{-1}\boldsymbol{P}\boldsymbol{A}^\dagger\boldsymbol{R}^{-1}\boldsymbol{A}\boldsymbol{P}\boldsymbol{A}^\dagger\boldsymbol{C}^{-1})\cdot(\boldsymbol{C}^\dagger\boldsymbol{R}^{-1}\boldsymbol{C})^T \\ &\quad - (\boldsymbol{C}^{-1}\boldsymbol{A}\boldsymbol{P}\boldsymbol{A}^\dagger\boldsymbol{R}^{-1}\boldsymbol{C})\cdot(\boldsymbol{C}^{-1}\boldsymbol{A}\boldsymbol{P}\boldsymbol{A}^\dagger\boldsymbol{R}^{-1}\boldsymbol{C})^T\} \\ &= 2\operatorname{Re}\{(\boldsymbol{C}^{-1}\boldsymbol{A}(\boldsymbol{P}-\boldsymbol{Q})\boldsymbol{A}^\dagger\boldsymbol{C}^{-\dagger})\cdot(\boldsymbol{C}^\dagger\boldsymbol{R}^{-1}\boldsymbol{C})^T \\ &\quad - (\boldsymbol{C}^{-1}\boldsymbol{A}\boldsymbol{Q}\boldsymbol{A}^\dagger\boldsymbol{C})\cdot(\boldsymbol{C}^{-1}\boldsymbol{A}\boldsymbol{Q}\boldsymbol{A}^\dagger\boldsymbol{C})^T\}. \end{aligned} \tag{10.B.35}$$

Derivatives with Respect to Gain

We first define

$$\dot{\boldsymbol{A}}_{\alpha_i} \triangleq \partial\boldsymbol{A}/\partial\alpha_i = \boldsymbol{C}(\partial\boldsymbol{G}/\partial\alpha_i)\boldsymbol{F}\tilde{\boldsymbol{A}} \tag{10.B.36}$$

and

$$\dot{\boldsymbol{G}} \triangleq \operatorname{diag}\{\dot{G}_{11}(\alpha_1),\dot{G}_{22}(\alpha_2),\ldots,\dot{G}_{MM}(\alpha_M)\} = \boldsymbol{I}_M. \tag{10.B.37}$$

Hence

$$\dot{\boldsymbol{A}}_{\alpha_i} = \boldsymbol{C}\mathbf{e}_i\mathbf{e}_i^T\dot{\boldsymbol{G}}\boldsymbol{F}\tilde{\boldsymbol{A}} = \boldsymbol{C}\mathbf{e}_i\mathbf{e}_i^T\boldsymbol{F}\tilde{\boldsymbol{A}} = \boldsymbol{C}\mathbf{e}_i\mathbf{e}_i^T\boldsymbol{G}^{-1}\boldsymbol{C}^{-1}\boldsymbol{A}. \tag{10.B.38}$$

Thus, repeating the considerations leading to (10.B.30) we obtain

$$\begin{aligned} J_{\alpha_i\alpha_j} &= 2\operatorname{Re}\{\operatorname{tr}[\dot{\boldsymbol{A}}_{\alpha_i}\boldsymbol{P}\boldsymbol{A}^\dagger\boldsymbol{R}^{-1}\boldsymbol{A}\boldsymbol{P}\dot{\boldsymbol{A}}_{\alpha_j}^\dagger\boldsymbol{R}^{-1}] \\ &\quad + \operatorname{tr}[\dot{\boldsymbol{A}}_{\alpha_i}\boldsymbol{P}\boldsymbol{A}^\dagger\boldsymbol{R}^{-1}\dot{\boldsymbol{A}}_{\alpha_j}\boldsymbol{P}\boldsymbol{A}^\dagger\boldsymbol{R}^{-1}]\}. \end{aligned} \tag{10.B.39}$$

Substituting (10.B.38) in (10.B.39) we obtain

$$\begin{aligned} J_{\alpha_i\alpha_j} &= 2\operatorname{Re}\{\operatorname{tr}[\boldsymbol{C}\mathbf{e}_i\mathbf{e}_i^T\boldsymbol{G}^{-1}\boldsymbol{C}^{-1}\boldsymbol{A}\boldsymbol{P}\boldsymbol{A}^\dagger\boldsymbol{R}^{-1}\boldsymbol{A}\boldsymbol{P}\boldsymbol{A}^\dagger\boldsymbol{C}^{-\dagger}\boldsymbol{G}^{-1}\mathbf{e}_j\mathbf{e}_j^T\boldsymbol{C}^\dagger\boldsymbol{R}^{-1}] \\ &\quad + \operatorname{tr}[\boldsymbol{C}\mathbf{e}_i\mathbf{e}_i^T\boldsymbol{G}^{-1}\boldsymbol{C}^{-1}\boldsymbol{A}\boldsymbol{P}\boldsymbol{A}^\dagger\boldsymbol{R}^{-1}\boldsymbol{C}\mathbf{e}_j\mathbf{e}_j^T\boldsymbol{G}^{-1}\boldsymbol{C}^{-1}\boldsymbol{A}\boldsymbol{P}\boldsymbol{A}^\dagger\boldsymbol{R}^{-1}]\} \end{aligned} \tag{10.B.40}$$

$$\begin{aligned} J_{\alpha\alpha} &= 2\operatorname{Re}\{(\boldsymbol{G}^{-1}\boldsymbol{C}^{-1}\boldsymbol{A}\boldsymbol{P}\boldsymbol{A}^\dagger\boldsymbol{R}^{-1}\boldsymbol{A}\boldsymbol{P}\boldsymbol{A}^\dagger\boldsymbol{C}^{-\dagger}\boldsymbol{G}^{-1})\cdot(\boldsymbol{C}^\dagger\boldsymbol{R}^{-1}\boldsymbol{C})^T \\ &\quad + (\boldsymbol{G}^{-1}\boldsymbol{C}^{-1}\boldsymbol{A}\boldsymbol{P}\boldsymbol{A}^\dagger\boldsymbol{R}^{-1}\boldsymbol{C})\cdot(\boldsymbol{G}^{-1}\boldsymbol{C}^{-1}\boldsymbol{A}\boldsymbol{P}\boldsymbol{A}^\dagger\boldsymbol{R}^{-1}\boldsymbol{C})^T\} \\ &= 2\operatorname{Re}\{(\boldsymbol{G}^{-1}\boldsymbol{C}^{-1}\boldsymbol{A}(\boldsymbol{P}-\boldsymbol{Q})(\boldsymbol{G}^{-1}\boldsymbol{C}^{-1}\boldsymbol{A})^\dagger\cdot(\boldsymbol{C}^\dagger\boldsymbol{R}^{-1}\boldsymbol{C})^T \\ &\quad + (\boldsymbol{G}^{-1}\boldsymbol{C}^{-1}\boldsymbol{A}\boldsymbol{Q}\boldsymbol{A}^\dagger\boldsymbol{C})\cdot(\boldsymbol{G}^{-1}\boldsymbol{C}^{-1}\boldsymbol{A}\boldsymbol{Q}\boldsymbol{A}^\dagger\boldsymbol{C})^T\}. \end{aligned} \tag{10.B.41}$$

DOA-Phase Cross-Terms

We first write the cross-term equivalent of (10.B.39),

$$\begin{aligned} J_{\gamma_i\phi_j} &= 2\operatorname{Re}\{\operatorname{tr}[\dot{\boldsymbol{A}}_{\gamma_i}\boldsymbol{P}\boldsymbol{A}^\dagger\boldsymbol{R}^{-1}\boldsymbol{A}\boldsymbol{P}\dot{\boldsymbol{A}}_{\phi_j}^\dagger\boldsymbol{R}^{-1} \\ &\quad + \operatorname{tr}[\dot{\boldsymbol{A}}_{\gamma_i}\boldsymbol{P}\boldsymbol{A}^\dagger\boldsymbol{R}^{-1}\dot{\boldsymbol{A}}_{\phi_j}\boldsymbol{P}\boldsymbol{A}^\dagger\boldsymbol{R}^{-1}] \end{aligned} \tag{10.B.42}$$

Substituting (10.B.34) and (10.B.23), (10.B.25) we obtain

$$J_{\gamma_i\phi_j} = 2\text{Re}\{\text{tr}[-j\dot{\boldsymbol{A}}\mathbf{e}_i\mathbf{e}_i^T\boldsymbol{P}\boldsymbol{A}^\dagger\boldsymbol{R}^{-1}\boldsymbol{A}\boldsymbol{P}\boldsymbol{A}^\dagger\boldsymbol{C}^{-\dagger}\mathbf{e}_j\mathbf{e}_j^T\boldsymbol{C}^\dagger\boldsymbol{R}^{-1}] + \text{tr}[j\dot{\boldsymbol{A}}\mathbf{e}_i\mathbf{e}_i^T\boldsymbol{P}\boldsymbol{A}^\dagger\boldsymbol{R}^{-1}\boldsymbol{C}\mathbf{e}_j\mathbf{e}_j^T\boldsymbol{C}^{+1}\boldsymbol{A}\boldsymbol{P}\boldsymbol{A}^\dagger\boldsymbol{R}^{-1}]\} \tag{10.B.43}$$

$$\begin{aligned} J_{\gamma\phi} &= 2\text{Re}\{-j\cdot(\boldsymbol{P}\boldsymbol{A}^\dagger\boldsymbol{R}^{-1}\boldsymbol{A}\boldsymbol{P}\boldsymbol{A}^\dagger\boldsymbol{C}^{-\dagger})\cdot(\boldsymbol{C}^\dagger\boldsymbol{R}^{-1}\dot{\boldsymbol{A}})^T \\ &\quad + j(\boldsymbol{P}\boldsymbol{A}^\dagger\boldsymbol{R}^{-1}\boldsymbol{C})\cdot(\boldsymbol{C}^{-1}\boldsymbol{A}\boldsymbol{P}\boldsymbol{A}^\dagger\boldsymbol{R}^{-1}\dot{\boldsymbol{A}})^T\} \\ &= 2\text{Re}\{-j[\boldsymbol{P}-\boldsymbol{Q})(\boldsymbol{C}^{-1}\boldsymbol{A})^\dagger]\cdot(\boldsymbol{C}^\dagger\boldsymbol{R}^{-1}\dot{\boldsymbol{A}})^T \\ &\quad + j(\boldsymbol{Q}\boldsymbol{A}^\dagger\boldsymbol{C})\cdot(\boldsymbol{C}^{-1}\boldsymbol{A}\boldsymbol{Q}\boldsymbol{A}^\dagger\dot{\boldsymbol{A}})^T\} \\ &= 2Im\{[(\boldsymbol{P}-\boldsymbol{Q})(\boldsymbol{C}^{-1}\boldsymbol{A})^\dagger]\cdot(\boldsymbol{C}^\dagger\boldsymbol{R}^{-1}\dot{\boldsymbol{A}})^T - (\boldsymbol{Q}\boldsymbol{A}^\dagger\boldsymbol{C})\cdot(\boldsymbol{C}^{-1}\boldsymbol{A}\boldsymbol{Q}\boldsymbol{A}^\dagger\dot{\boldsymbol{A}})^T\}. \end{aligned} \tag{10.B.44}$$

DOA-Gain Cross-Terms

The cross-term equivalent of (10.B.42) is given by

$$J_{\gamma_i\alpha_j} = 2\text{Re}\{\text{tr}[\dot{\boldsymbol{A}}_{\gamma_i}\boldsymbol{P}\boldsymbol{A}^\dagger\boldsymbol{R}^{-1}\boldsymbol{A}\boldsymbol{P}\dot{\boldsymbol{A}}_{\alpha_j}^\dagger\boldsymbol{R}^{-1}] + \text{tr}[\dot{\boldsymbol{A}}_{\gamma_i}\boldsymbol{P}\boldsymbol{A}^\dagger\boldsymbol{R}^{-1}\dot{\boldsymbol{A}}_{\alpha_j}\boldsymbol{P}\boldsymbol{A}^\dagger\boldsymbol{R}^{-1}]\}. \tag{10.B.45}$$

Substituting (10.B.23), (10.B.25), and (10.B.38), we obtain

$$J_{\gamma_i\alpha_j} = 2\text{Re}\{\text{tr}[\dot{\boldsymbol{A}}\mathbf{e}_i\mathbf{e}_i^T\boldsymbol{P}\boldsymbol{A}^\dagger\boldsymbol{R}^{-1}\boldsymbol{A}\boldsymbol{P}\boldsymbol{A}^\dagger\boldsymbol{C}^{-\dagger}\boldsymbol{G}^{-1}\mathbf{e}_j\mathbf{e}_j^T\boldsymbol{C}^\dagger\boldsymbol{R}^{-1}] + \text{tr}[\dot{\boldsymbol{A}}\mathbf{e}_i\mathbf{e}_i^T\boldsymbol{P}\boldsymbol{A}^H\boldsymbol{R}^{-1}\boldsymbol{C}\mathbf{e}_j\mathbf{e}_j^T\boldsymbol{G}^{-1}\boldsymbol{C}^{-1}\boldsymbol{A}\boldsymbol{P}\boldsymbol{A}^H\boldsymbol{R}^{-1}]\} \tag{10.B.46}$$

$$\begin{aligned} J_{\gamma\alpha} &= 2\text{Re}\{(\boldsymbol{P}\boldsymbol{A}^\dagger\boldsymbol{R}^{-1}\boldsymbol{A}\boldsymbol{P}\boldsymbol{A}^\dagger\boldsymbol{C}^{-\dagger}\boldsymbol{G}^{-1})\cdot(\boldsymbol{C}^\dagger\boldsymbol{R}^{-1}\dot{\boldsymbol{A}})^T \\ &\quad + (\boldsymbol{P}\boldsymbol{A}^\dagger\boldsymbol{R}^{-1}\boldsymbol{C})\cdot(\boldsymbol{G}^{-1}\boldsymbol{C}^{-1}\boldsymbol{A}\boldsymbol{P}\boldsymbol{A}^\dagger\boldsymbol{R}^{-1}\dot{\boldsymbol{A}})^T\} \\ &= 2\text{Re}\{[(\boldsymbol{P}-\boldsymbol{Q})(\boldsymbol{G}^{-1}\boldsymbol{C}^{-1}\boldsymbol{A})^\dagger]\cdot(\boldsymbol{C}^\dagger\boldsymbol{R}^{-1}\dot{\boldsymbol{A}})^T \\ &\quad + (\boldsymbol{Q}\boldsymbol{A}^\dagger\boldsymbol{C})\cdot(\boldsymbol{G}^{-1}\boldsymbol{C}^{-1}\boldsymbol{A}\boldsymbol{Q}\boldsymbol{A}^\dagger\dot{\boldsymbol{A}})^T\}. \end{aligned} \tag{10.B.47}$$

Gain-Phase Cross-Terms

The cross-term equivalent of (10.B.39) is given by

$$J_{\alpha_i\phi_j} = 2\text{Re}\{\text{tr}[\dot{\boldsymbol{A}}_{\alpha_i}\boldsymbol{P}\boldsymbol{A}^\dagger\boldsymbol{R}^{-1}\boldsymbol{A}\boldsymbol{P}\dot{\boldsymbol{A}}_{\phi_j}^\dagger\boldsymbol{R}^{-1}] + \text{tr}[\dot{\boldsymbol{A}}_{\alpha_i}\boldsymbol{P}\boldsymbol{A}^\dagger\boldsymbol{R}^{-1}\dot{\boldsymbol{A}}_{\phi_j}\boldsymbol{P}\boldsymbol{A}^\dagger\boldsymbol{R}^{-1}]\}. \tag{10.B.48}$$

Substituting (10.B.38) and (10.B.34) one obtains

$$J_{\alpha_i\phi_j} = 2\text{Re}\{-\text{tr}[\boldsymbol{C}\mathbf{e}_i\mathbf{e}_i^T\boldsymbol{G}^{-1}\boldsymbol{C}^{-1}\boldsymbol{A}\boldsymbol{P}\boldsymbol{A}^\dagger\boldsymbol{R}^{-1}\boldsymbol{A}\boldsymbol{P}\boldsymbol{A}^\dagger\boldsymbol{C}^{-\dagger}\mathbf{e}_j\mathbf{e}_j^T\boldsymbol{C}^\dagger\boldsymbol{R}^{-1}j] + \text{tr}[\boldsymbol{C}\mathbf{e}_i\mathbf{e}_i^T\boldsymbol{G}^{-1}\boldsymbol{C}^{-1}\boldsymbol{A}\boldsymbol{P}\boldsymbol{A}^\dagger\boldsymbol{R}^{-1}\boldsymbol{C}\mathbf{e}_j\mathbf{e}_j^T\boldsymbol{C}^{-1}\boldsymbol{A}\boldsymbol{P}\boldsymbol{A}^\dagger\boldsymbol{R}^{-1}j]\} \tag{10.B.49}$$

$$\begin{aligned} J_{\alpha\phi} &= 2\text{Re}\{-j(\boldsymbol{G}^{-1}\boldsymbol{C}^{-1}\boldsymbol{APA}^{\dagger}\boldsymbol{R}^{-1}\boldsymbol{APA}^{\dagger}\boldsymbol{C}^{-\dagger}) \cdot (\boldsymbol{C}^{\dagger}\boldsymbol{R}^{-1}\boldsymbol{C})^{T} \\ &\quad + j(\boldsymbol{G}^{-1}\boldsymbol{C}^{-1}\boldsymbol{APA}^{\dagger}\boldsymbol{R}^{-1}\boldsymbol{C}) \cdot (\boldsymbol{C}^{-1}\boldsymbol{APA}^{\dagger}\boldsymbol{R}^{-1}\boldsymbol{C})^{T}\} \\ &= 2Im\{[\boldsymbol{G}^{-1}\boldsymbol{C}^{-1}\boldsymbol{A}(\boldsymbol{P}-\boldsymbol{Q})(\boldsymbol{C}^{-1}\boldsymbol{A})^{\dagger}] \cdot (\boldsymbol{C}^{\dagger}\boldsymbol{R}^{-1}\boldsymbol{C})^{T} \\ &\quad - (\boldsymbol{G}^{-1}\boldsymbol{C}^{-1}\boldsymbol{AQA}^{\dagger}\boldsymbol{C}) \cdot (\boldsymbol{C}^{-1}\boldsymbol{AQA}^{\dagger}\boldsymbol{C})^{T}\}. \end{aligned} \qquad (10.B.50)$$

Derivatives with Respect to Mutual Coupling Coefficient

To simplify the analysis we concentrate here on a circulant matrix with only a single coupling coefficient given by

$$C_{12} = \mu \mathbf{e}^{j\zeta} \qquad (10.B.51)$$

where μ and ζ are real variables. Using the notation

$$\dot{\boldsymbol{A}}_{\mu} \triangleq (\partial \boldsymbol{C}/\partial\mu)\Gamma\tilde{\boldsymbol{A}} = \dot{\boldsymbol{C}}_{\mu}\boldsymbol{C}^{-1}\boldsymbol{A} \qquad (10.B.52)$$

$$\dot{\boldsymbol{A}}_{\zeta} \triangleq (\partial \boldsymbol{C}/\partial\zeta)\Gamma\tilde{\boldsymbol{A}} = \dot{\boldsymbol{C}}_{\zeta}\boldsymbol{C}^{-1}\boldsymbol{A}, \qquad (10.B.53)$$

we obtain

$$\begin{aligned} J_{\mu\mu} &= 2\text{Re}\{\text{tr}[\dot{\boldsymbol{A}}_{\mu}\boldsymbol{PA}^{\dagger}\boldsymbol{R}^{-1}\dot{\boldsymbol{A}}_{\mu}\boldsymbol{PA}^{\dagger}\boldsymbol{R}^{-1}] \\ &\quad + \text{tr}[\dot{\boldsymbol{A}}_{\mu}\boldsymbol{PA}^{\dagger}\boldsymbol{R}^{-1}\boldsymbol{AP}\dot{\boldsymbol{A}}_{\mu}^{\dagger}\boldsymbol{R}^{-1}]\} \end{aligned} \qquad (10.B.54)$$

$$\begin{aligned} J_{\zeta\zeta} &= 2\text{Re}\{\text{tr}[\dot{\boldsymbol{A}}_{\zeta}\boldsymbol{PA}^{\dagger}\boldsymbol{R}^{-1}\dot{\boldsymbol{A}}_{\zeta}\boldsymbol{PA}^{\dagger}\boldsymbol{R}^{-1}] \\ &\quad + \text{tr}[\dot{\boldsymbol{A}}_{\zeta}\boldsymbol{PA}^{\dagger}\boldsymbol{R}^{-1}\boldsymbol{AP}\dot{\boldsymbol{A}}_{\zeta}^{\dagger}\boldsymbol{R}^{-1}]\}. \end{aligned} \qquad (10.B.55)$$

$$\begin{aligned} J_{\mu\zeta} &= 2\text{Re}\{\text{tr}[\dot{\boldsymbol{A}}_{\mu}\boldsymbol{PA}^{\dagger}\boldsymbol{R}^{-1}\dot{\boldsymbol{A}}_{\zeta}\boldsymbol{PA}^{\dagger}\boldsymbol{R}^{-1}] \\ &\quad + \text{tr}[\dot{\boldsymbol{A}}_{\mu}\boldsymbol{PA}^{\dagger}\boldsymbol{R}^{-1}\boldsymbol{AP}\dot{\boldsymbol{A}}_{\zeta}^{\dagger}\boldsymbol{R}^{-1}]\}. \end{aligned} \qquad (10.B.56)$$

$$\begin{aligned} J_{\mu\gamma_j} &= 2\text{Re}\{\text{tr}[\dot{\boldsymbol{A}}_{\mu}\boldsymbol{PA}^{\dagger}\boldsymbol{R}^{-1}\dot{\boldsymbol{A}}_{\gamma_j}\boldsymbol{PA}^{\dagger}\boldsymbol{R}^{-1}] \\ &\quad + \text{tr}[\dot{\boldsymbol{A}}_{\mu}\boldsymbol{PA}^{\dagger}\boldsymbol{R}^{-1}\boldsymbol{AP}\dot{\boldsymbol{A}}_{\gamma_j}^{\dagger}\boldsymbol{R}^{-1}]\} \\ &= 2\text{Re}\{\text{tr}[\dot{\boldsymbol{A}}_{\mu}\boldsymbol{PA}^{\dagger}\boldsymbol{R}^{-1}\dot{\boldsymbol{A}}\mathbf{e}_j\mathbf{e}_j^{T}\boldsymbol{PA}^{\dagger}\boldsymbol{R}^{-1}] \\ &\quad + \text{tr}[\dot{\boldsymbol{A}}_{\mu}\boldsymbol{PA}^{\dagger}\boldsymbol{R}^{-1}\boldsymbol{AP}\mathbf{e}_j\mathbf{e}_j^{T}\dot{\boldsymbol{A}}^{\dagger}\boldsymbol{R}^{-1}]\} \\ &= 2\text{Re}\{\mathbf{e}_j^{T}\boldsymbol{PA}^{\dagger}\boldsymbol{R}^{-1}\dot{\boldsymbol{A}}_{\mu}\boldsymbol{PA}^{\dagger}\boldsymbol{R}^{-1}\dot{\boldsymbol{A}}\mathbf{e}_j + \mathbf{e}_j^{T}\dot{\boldsymbol{A}}^{\dagger}\boldsymbol{R}^{-1}\dot{\boldsymbol{A}}_{\mu}\boldsymbol{PA}^{\dagger}\boldsymbol{R}^{-1}\boldsymbol{AP}\mathbf{e}_j\}. \end{aligned} \qquad (10.B.57)$$

040 Introducing the notation diag($\boldsymbol{A}$) = $A_{11}, A_{22}, \ldots, A_{MM}$], we obtain,

$$\begin{aligned} J_{\mu\gamma} &= 2\text{Re}\{\text{diag}\{\boldsymbol{PA}^{\dagger}\boldsymbol{R}^{-1}\dot{\boldsymbol{A}}_{\mu}\boldsymbol{PA}^{\dagger}\boldsymbol{R}^{-1}\dot{A}\} \\ &\quad + \text{diag}\{\dot{\boldsymbol{A}}^{\dagger}\boldsymbol{R}^{-1}\dot{\boldsymbol{A}}_{\mu}\boldsymbol{PA}^{\dagger}\boldsymbol{R}^{-1}\boldsymbol{AP}\}\} \\ &= 2\text{Re}\{\text{diag}\{\boldsymbol{QA}^{\dagger}\dot{\boldsymbol{A}}_{\mu}\boldsymbol{QA}^{\dagger}\dot{\boldsymbol{A}}\} \\ &\quad + \text{diag}\{\dot{\boldsymbol{A}}^{\dagger}\boldsymbol{R}^{-1}\dot{\boldsymbol{A}}_{\mu}(\boldsymbol{P}-\boldsymbol{Q})\}\}. \end{aligned} \qquad (10.B.58)$$

$$\begin{aligned} J_{\zeta\gamma} &= 2\text{Re}\{\text{diag}\{\boldsymbol{QA}^{\dagger}\dot{\boldsymbol{A}}_{\zeta}\boldsymbol{QA}^{\dagger}\dot{\boldsymbol{A}}\} \\ &\quad + \text{diag}\{\dot{\boldsymbol{A}}^{\dagger}\,\boldsymbol{R}^{-1}\dot{\boldsymbol{A}}\zeta(\boldsymbol{P}-\boldsymbol{Q})\}\}. \end{aligned} \qquad (10.B.59)$$

$$\begin{aligned} J_{\mu\alpha_j} &= 2\mathrm{Re}\{\mathrm{tr}[\dot{A}_\mu PA^\dagger R^{-1}\dot{A}_{\alpha_j}PA^\dagger R^{-1}] \\ &\quad + \mathrm{tr}[\dot{A}_\mu PA^\dagger R^{-1}AP\dot{A}^\dagger_{\alpha_j}R^{-1}]\} \\ &= 2\mathrm{Re}\{\mathrm{tr}[\dot{A}_\mu PA^\dagger R^{-1}C\mathbf{e}_j\mathbf{e}_j^T G^{-1}C^{-1}APA^\dagger R^{-1}] \\ &\quad + \mathrm{tr}[\dot{A}_\mu PA^\dagger R^{-1}APA^\dagger C^{-\dagger}G^{-1}\mathbf{e}_j\mathbf{e}_j^T C^\dagger R^{-1}]\} \\ &= 2\mathrm{Re}\{\mathbf{e}_j^T G^{-1}C^{-1}APA^\dagger R^{-1}\dot{A}_\mu PA^\dagger R^{-1}C\mathbf{e}_j \\ &\quad + \mathbf{e}_j^T C^\dagger R^{-1}\dot{A}_\mu PA^\dagger R^{-1}APA^\dagger C^{-\dagger}G^{-1}\mathbf{e}_j\} \\ &= 2\mathrm{Re}\{\mathbf{e}_j^T G^{-1}C^{-1}AQA^\dagger\dot{A}_\mu QA^\dagger C\mathbf{e}_j \\ &\quad + \mathbf{e}_j^T C^\dagger R^{-1}\dot{A}_\mu(P-Q)A^\dagger C^{-\dagger}G^{-1}\mathbf{e}_j\} \end{aligned} \tag{10.B.60}$$

$$J_{\mu\alpha} = 2\mathrm{Re}\{\mathrm{diag}\{G^{-1}C^{-1}AQA^\dagger\dot{A}_\mu QA^\dagger C + C^\dagger R^{-1}\dot{A}_\mu(P-Q)A^\dagger C^{-\dagger}G^{-1}\}\} \tag{10.B.61}$$

$$J_{\zeta\alpha} = 2\mathrm{Re}\{\mathrm{diag}\{G^{-1}C^{-1}AQA^\dagger\dot{A}_\zeta QA^\dagger C + C^\dagger R^{-1}\dot{A}_\zeta(P-Q)A^\dagger C^{-\dagger}G^{-1}\}\} \tag{10.B.62}$$

$$\begin{aligned} J_{\mu\phi_j} &= 2\mathrm{Re}\{\mathrm{tr}[\dot{A}_\mu PA^\dagger R^{-1}\dot{A}_{\phi_j}PA^\dagger R^{-1}\} \\ &\quad + \mathrm{tr}[\dot{A}_\mu PA^\dagger R^{-1}AP\dot{A}^\dagger_{\phi_j}R^{-1}]\} \end{aligned} \tag{10.B.63}$$

$$\begin{aligned} J_{\mu\phi j} &= 2\mathrm{Re}\{\mathrm{tr}[j\dot{A}_\mu PA^\dagger R^{-1}C\mathbf{e}_j\mathbf{e}_j^T C^{-1}APA^\dagger R^{-1}\} \\ &\quad + \mathrm{tr}[-j\dot{A}_\mu PA^\dagger R^{-1}APA^\dagger C^{-\dagger}\mathbf{e}_j\mathbf{e}_j^T C^\dagger R^{-1}]\} \\ &= 2\mathrm{Re}\{j\mathbf{e}_j^T C^{-1}APA^\dagger R^{-1}\dot{A}_\mu PA^\dagger R^{-1}C\mathbf{e}_j\} \\ &\quad - j\mathbf{e}_j^T C^\dagger R^{-1}\dot{A}_\mu PA^\dagger R^{-1}APA^\dagger C^{-\dagger}\mathbf{e}_j\} \end{aligned} \tag{10.B.64}$$

$$\begin{aligned} J_{\mu\phi} &= 2Im\{\mathrm{diag}\{-C^{-1}APA^\dagger R^{-1}\dot{A}_\mu PA^\dagger R^{-1}C \\ &\quad + C^\dagger R^{-1}\dot{A}_\mu PA^\dagger R^{-1}APA^\dagger C^{-\dagger}\}\} \\ &= 2Im\{\mathrm{diag}\{-C^{-1}AQA^\dagger\dot{A}_\mu QA^\dagger C \\ &\quad + C^\dagger R^{-1}\dot{A}_\mu(P-Q)A^\dagger C^{-\dagger}\}\} \end{aligned} \tag{10.B.65}$$

$$\begin{aligned} J_{\zeta\phi} &= 2Im\{\mathrm{diag}\{-C^{-1}AQA^\dagger\dot{A}_\zeta QA^\dagger C \\ &\quad + C^\dagger R^{-1}\dot{A}_\zeta(P-Q)A^\dagger C^{-\dagger}\}\}. \end{aligned} \tag{10.B.66}$$

The FIM is given by

$$J = \begin{bmatrix} J_{\gamma\gamma} & J_{\gamma\alpha} & J_{\gamma\phi} & J_{\gamma\mu} & J_{\gamma\zeta} \\ J_{\alpha\gamma} & J_{\alpha\alpha} & J_{\alpha\phi} & J_{\alpha\mu} & J_{\alpha\zeta} \\ J_{\phi\gamma} & J_{\phi\alpha} & J_{\phi\phi} & J_{\phi\mu} & J_{\phi\zeta} \\ J_{\mu\gamma} & J_{\mu\alpha} & J_{\mu\phi} & J_{\mu\mu} & J_{\mu\zeta} \\ J_{\zeta\gamma} & J_{\zeta\alpha} & J_{\zeta\phi} & J_{\zeta\mu} & J_{\zeta\zeta} \end{bmatrix}. \tag{10.B.67}$$

Remark:

If one or more parameters are assumed known (e.g., gain-phase of a reference sensor), the corresponding columns and rows in J must be removed.

Index